Reelin Glycoprotein

Structure, Biology and Roles in Health and Disease

S. Hossein Fatemi

Reelin Glycoprotein

Structure, Biology and Roles in Health and Disease

Springer

Editor
S. Hossein Fatemi, M.D., Ph.D.
Professor of Psychiatry
Adjunct Professor of Pharmacology and Neuroscience
University of Minnesota
Medical School
Minneapolis, MN
USA
fatem002@umn.edu

ISBN 978-1-4899-8695-5 ISBN 978-0-387-76761-1 (eBook)
DOI 10.1007/978-0-387-76761-1

Softcover re-print of the Hardcover 1st edition 2008

Printed on acid-free paper

9 8 7 6 5 4 3 2 1

springer.com

To my father, S. Mehdi Fatemi, and to my family, S. Ali Fatemi, M.D., Naheed Fatemi, Parvin Fatemi, S. Mohammad Fatemi, Neelufaar Fatemi, Maryam Jalali-Mousavi, and last but not least, my mother, Fatemeh Parsa Moghaddam, whose love and support have enabled me to complete this work.

Foreword

By inviting me to contribute a short foreword to this superb collective work on the biology of Reelin, Dr Fatemi gives me a nice opportunity for some reminiscence and speculation. The first reeler mutation appeared in 1948 in Edinburgh, and reeler mice were for many years the only genetic model of cortical malformation. The model became popular in the mid sixties and seventies, thanks to the vision of Sidman and his colleagues, especially Caviness and Rakic. However, in the following years, reeler fell a bit out of fashion because the gene was not characterized. Only a few groups, such as those of Mikoshiba and myself, persisted in studying reeler mice. This changed dramatically when D'Arcangelo and Curran cloned the Reelin gene in 1995. The next few years saw rapid progress with the identification of the adaptor Dab1 by Howell and Cooper in 1997, and of reelin receptors by Herz in 1999. Since 1999, as is often the case after a golden period, progress has been slower. Still, we have witnessed very significant progress that is nicely covered in this book. We understand more about the biochemistry and structure of Reelin, and most probably a full structure of the native protein will be defined in the coming years. Reelin has been studied during brain development in several species, and this resulted in a somewhat clearer view of its importance during brain evolution. Similar comparative studies of Dab1 and receptors would be needed, however, and I doubt that they will be done soon. We know a lot about Reelin expression in the embryonic and adult brain, even though the basic mechanisms that control Reelin synthesis and secretion, particularly by Cajal-Retzius cells, need to be defined better. The proximal steps in the signal elicited by Reelin have basically been worked out, and interactions between proximal signaling by Reelin and other important pathways are being elucidated at increasing pace. In addition, several papers appear regularly that describe expression of Reelin where it is not really expected, generating new hypotheses on what are probably several different functions of a large protein. Whether these various functions use the canonical Reelin signaling pathway or additional pathways that remain to be identified is another theme of interest for the next few years. Last but not least, a large body of evidence has accumulated that hints at a role of Reelin in psychiatric disorders.

Notwithstanding this vast body of exciting data, however, we are still in the dark about some key issues, among which I would mention two that appear particularly important – at least to me. First and foremost, even though Reelin receptors and signaling components have been identified, we still do not know what Reelin does to immature neurons. The concept that Reelin provides a stop signal has been useful but is reaching its operational limit. Presumably, we all agree that Reelin somehow instructs neurons to arrest migration and take position in early architectonic patterns (cortical plate, Purkinje cell layer, etc. ...). However, how this happens remains unknown. Does Reelin modulate expression of adhesion molecules on the surface of neurons, radial glia, or both, as I always imagined? Alternatively, does Reelin signaling impact on the cytoskeleton, thereby "freezing" the architecture of end-migration neurons? Does Reelin signaling recruit or hijack other signaling pathways such as Notch, as recently proposed by the Rakic group? Second, what is the function of Reelin after maturation and in the mature brain. What does Reelin do when it is secreted by cortical interneurons? Related to this are clearly the questions about the actions of Reelin in learning and behavior, and in psychiatric diseases. To address this, we need to inactivate Reelin and/or its partners in neurons after normal maturation, using, for example, floxed alleles and Cre-ER technology. As far as I know, these tools are only being developed.

As this superb book outlines, the Reelin field is at a crossroad. Following a rapid initial phase, with cloning of Reelin, identification of Dab1 and receptors, each of which owed quite a lot to serendipity, progress has been much slower and difficult. Whether future breakthrough will also occur unexpectedly (maybe the next knockout...) or will result from more rational approaches, is anybody's guess. But one thing is more predictable, namely that this timely book will prove very useful and will end up on the bookshelf of all investigators with an interest in the Reelin puzzle. We should all be grateful to Dr Fatemi and his staff for editing it so carefully.

Andre M. Goffinet, MD, PhD
Brussels, Belgium

Preface

Reelin glycoprotein is a major secretory protein with important roles in embryogenesis and during adult life. Reelin gene mutations or deficiency of the protein product cause abnormal cortical development and Reelin signaling impairment in brain. Since the first discovery of the reelin mutant mouse in 1951 by Falconer, and later discovery of the gene for Reelin in 1995, there has been an explosion of new knowledge about this important molecule. As of this writing, a search of public library of medicine cites over 665 published papers on Reelin.

Thus, it became apparent that a book dealing with this topic and presenting contributions from an international panel of experts would be timely and necessary. In the following twenty eight chapters, various authors will present up-to-date discussions of the state of the knowledge on various aspects of Reelin such as reelin gene, its receptors, downstream effector molecules in Reelin signaling cascade, chemistry and structure of Reelin, comparative anatomy of reelin, presence of Reelin in various body tissues, Reelin mutations, and abnormalities of Reelin production in neuropsychiatric disorders and cancer.

It is hoped that this book serves as a foundation for analysis of this emerging novel protein for all interested neuroscientists and clinicians.

S. Hossein Fatemi, M.D., Ph.D.
Minneapolis, Minnesota

Acknowledgments

Many have helped to make the publication of this book possible. I am especially indebted to Ms. Teri Jane Reutiman, who faithfully reviewed all chapters for accuracy and worked as a liaison between the editor and the authors of the chapters, and Mr. Timothy D. Folsom for help with various aspects of editing this book. I am grateful to Ms. Laurie Iversen for clerical assistance. I am also grateful to the publishers and authors who have generously given approval for reproduction of tables and figures, as well as to Ms. Kathleen Lyons, Ms. Dana Andreachi, and Mr. Brian Halm at Springer Science + Business Media, and Ms. Padmasani Srimadhan at SPi Publisher Services for an excellent job in publishing this book.

S. Hossein Fatemi

Contents

Contributors

Hamid Mostafavi Abdolmaleky, MD
Senior Research Associate, Biomedical Engineering Department, Boston University, Boston; Laboratory of Nutrition and Metabolism at BIDMC, Department of Surgery, Harvard Medical School, Boston; Departments of Medicine (Genetics Program), Genetics & Genomics, and Pathology & Laboratory Medicine, Boston University School of Medicine, Boston, MA USA
and
Assistant Professor of Psychiatry, Department of Psychiatry, Tehran Psychiatric Institute and Mental Health Research Center, Iran University of Medical Sciences, Tehran, Iran

Nathaniel S. Allen, BS
Graduate Student, Division of Basic Sciences, Fred Hutchinson Cancer Research Center, Seattle, WA USA

Manuel Álvarez-Dolado, PhD
Ramón y Cajal Researcher, Team Leader of Cellular Regeneration Laboratory, Centro de Investigación Príncipe Felipe (CIPF), Valencia, Spain

Françoise Bleicher, PhD, MS
Professor of Biochemistry, University of Lyon 1, Odontoblast and Dental Tissue Regeneration Team, Faculty of Odontology, Lyon, France

Nelly Boehm, MD, PhD
Professor of Histology, Institut d'Histologie, Faculté de Médecine, Université Louis Pasteur, Strasbourg, France; _INSERM U666, Strasbourg, France; Hôpitaux Universitaires de Strasbourg, France

Hans H. Bock, MD
Research Fellow, Zentrum für Neurowissenschaften, Universität Freiburg, Freiburg, Germany

Arancha Botella-López, BS
Research Assistant, Instituto de Neurociencias de Alicante, Universidad Miguel Hernández-CSIC, Sant Joan d'Alacant, Spain, and Centro de Investigación Biomédica en Red sobre Enfermedades Neurodegenerativas (CIBERNED), Spain

C. Sue Carter, PhD
Professor of Psychiatry, Department of Psychiatry, Co-Director of The Brain Body Center, the Psychiatric Institute, College of Medicine, University of Illinois at Chicago, Chicago, IL USA

Ying Chen, MD
Research Specialist, the Psychiatric Institute, Department of Psychiatry, College of Medicine, University of Illinois at Chicago, Chicago, IL USA

Jonathan A. Cooper, PhD
Member, Division of Basic Sciences, Fred Hutchinson Cancer Research Center, Seattle, WA USA

Erminio Costa, MD
Distinguished Professor of Biochemistry and Psychiatry, Director of the Psychiatric Institute, Department of Psychiatry, College of Medicine, University of Illinois at Chicago, Chicago, IL USA

Marie-Lise Couble, MS
Research Assistant, University of Lyon 1, Odontoblast and Dental Tissue Regeneration Team, Faculty of Odontology, Lyon, France

Gabriella D'Arcangelo, PhD
Associate Professor, Department of Cell Biology and Neuroscience, Rutgers, the State University of New Jersey, Piscataway, NJ USA

Erbo Dong, PhD
Research Assistant Professor of Psychiatry, the Psychiatric Institute, Department of Psychiatry, College of Medicine, University of Illinois at Chicago, Chicago, IL USA

S. Hossein Fatemi, MD, PhD
Professor of Psychiatry, Adjunct Professor of Pharmacology and Neuroscience, Departments of Psychiatry, Pharmacology, and Neuroscience, University of Minnesota Medical School, Minneapolis, MN USA

Libing Feng, PhD
Postdoctoral Fellow, Division of Basic Sciences, Fred Hutchinson Cancer Research Center, Seattle, WA USA

Timothy D. Folsom, MS
Research Assistant, Department of Psychiatry, University of Minnesota Medical School, Minneapolis, MN USA

Eckart Förster, PhD
Professor of Anatomy, Institut für Anatomie I, Zelluläre Neurobiologie, Universität Hamburg, Hamburg, Germany

Michael Frotscher, MD
Professor of Anatomy, Institut für Anatomie und Zellbiologie, Abteilung für Neuroanatomie, Universität Freiburg, Freiburg, Germany

Michael Goggins, MD
Director of the Pancreatic Cancer Early Detection Laboratory, Associate Professor of Pathology, Medicine, and Oncology, Departments of Pathology, Medicine, and Oncology, The Sol Goldman Pancreatic Research Center, The Johns Hopkins Medical Institutions, Baltimore, MD USA

Dennis R. Grayson, PhD
Professor of Molecular Psychiatry, the Psychiatric Institute, Department of Psychiatry, College of Medicine, University of Illinois at Chicago, Chicago, IL USA

Renzo Guerrini, MD
Professor of Child Neurology and Psychiatry, Director - Child Neurology Unit, Pediatric Hospital A. Meyer-University of Firenze, Pediatric Neurology Unit and Laboratories, Children's Hospital A. Meyer-University of Florence, Florence, Italy

Alessandro Guidotti, MD
Professor of Psychiatry, Scientific Director, the Psychiatric Institute, Department of Psychiatry, College of Medicine, University of Illinois at Chicago, Chicago, IL USA

Mitsuharu Hattori, PhD
Associate Professor, Department of Biomedical Science, Graduate School of Pharmaceutical Sciences, Nagoya City University, Nagoya, Aichi, Japan

Joachim Herz, MD
Professor of Biophysics and Molecular Genetics, Thomas O. Hicks Family Distinguished Chair in Alzheimer's Disease Research, Department of Molecular Genetics, University of Texas Southwestern Medical Center, Dallas, TX USA
and
Zentrum für Neurowissenschaften, Universität Freiburg, Freiburg, Germany

Robert F. Hevner, MD, PhD
Associate Professor and Neuropathologist, Department of Pathology/ Neuropathology, University of Washington, Seattle, WA USA

Cheng-Chiu Huang, MS
Graduate Program in Developmental Biology, Baylor College of Medicine, Houston, TX USA; Graduate Student, PhD Program in Developmental Biology, Baylor College of Medicine, Houston, TX USA

Kunlin Jin, MD, PhD
Associate Research Professor, Buck Institute for Age Research, Novato, CA USA

Yves Jossin, PhD
Postdoctoral Fellow, Université Catholique de Louvain, Faculté de Médecine, Developmental Neurobiology Unit, Brussels, Belgium
and
Research Associate, Fred Hutchinson Cancer Research Center, Division of Basic Sciences, Seattle, WA USA

Carla Lintas, PhD
Research Scientist, Laboratory of Molecular Psychiatry and Neurogenetics, University "Campus Bio-Medico," and Department of Experimental Neurosciences, I.R.C.C.S. "Fondazione Santa Lucia," Rome, Italy

Henry Magloire, PhD, DDS
Professor of Basic Sciences in Oral Biology, University of Lyon 1, Odontoblast and Dental Tissue Regeneration Team, Faculty of Odontology, Lyon, France

Jean-Christophe Maurin, PhD, DDS
Assistant Professor of Operative Dentistry, Interfaces Biomatériaux/Tissus Hôtes, Faculty of Odontology of Reims, Reims, France

Gundela Meyer, MD, PhD
Professor of Anatomy, Departamento de Anatomía, Facultad de Medicina, Universidad de La Laguna, La Laguna, Spain

Jean-Marc Mienville, PhD
Professor, Laboratoire de Physiologie Cellulaire et Moléculaire, CNRS – UMR 6548, Université de Nice-Sophia Antipolis, Parc Valrose, Nice, France

Toshio Ohshima, MD, PhD
Professor and Head of Laboratory for Molecular Brain Science, Department of Life Science and Medical Bio-Science, Waseda University, Sinjuku, Tokyo, Japan

Eric C. Olson, PhD
Assistant Professor of Neuroscience and Physiology, Department of Neuroscience and Physiology, SUNY Upstate Medical University, Syracuse, NY USA

George D. Pappas, PhD
Professor of Psychiatry, Anatomy and Cell Biology, Department of Psychiatry, the Psychiatric Institute, College of Medicine, University of Illinois at Chicago, Chicago, IL USA

Elena Parrini, PhD
Neurogenetics Lab, Pediatric Neurology Unit and Laboratories, Children's Hospital A. Meyer-University of Florence, Florence, Italy

Emma Perez-Costas, PhD
Assistant Professor, Department of Psychiatry and Behavioral Neurobiology, University of Alabama at Birmingham, Birmingham, AL USA

Antonio Maria Persico, MD
Associate Professor of Physiology, Head of Laboratory of Molecular Psychiatry and Neurogenetics, University "Campus Bio-Medico," and Department of Experimental Neurosciences, I.R.C.C.S. "Fondazione Santa Lucia," Rome, Italy

Graziano Pinna, PhD
Research Assistant Professor of Psychiatry (Neuroendocrinology), the Psychiatric Institute, Department of Psychiatry, University of Illinois at Chicago, Chicago, IL USA

Shenfeng Qiu, PhD
Postdoctoral Fellow, Department of Molecular Physiology and Biophysics, Vanderbilt University Medical Center, Nashville, TN USA

Teri J. Reutiman, BA
Junior Scientist, Department of Psychiatry, University of Minnesota Medical School, Minneapolis, MN USA

Thomas Ringstedt, PhD
Assistant Professor of Developmental Neurobiology, Neonatal Unit, Karolinska Institutet, Astrid Lindgren Children's Hospital, Stockholm, Sweden

Rosalinda C. Roberts, PhD
Professor of Psychiatry, Kathy Ireland Endowed Chair for Psychiatric Research, Department of Psychiatry and Behavioral Neurobiology, University of Alabama at Birmingham, Birmingham, AL USA

Javier Sáez-Valero, PhD
Assistant Professor of Biochemistry, Instituto de Neurociencias de Alicante, Universidad Miguel Hernández-CSIC, Sant Joan d'Alacant, Spain and Centro de Investigación Biomédica en Red sobre Enfermedades Neurodegenerativas (CIBERNED), Spain

Brigitte Samama, PhD
Associate Professor of Histology, Institut d'Histologie, Faculté de Médecine, Université Louis Pasteur, Strasbourg, France; I_NSERM U666, Strasbourg, France; Hôpitaux Universitaires de Strasbourg, France

Cassandra L. Smith, PhD
Professor of Biomedical Engineering, Pharmacology and Experimental Therapeutics, School of Medicine, Director, Molecular Biotechnology Research Laboratory, Biomedical Engineering Department, Boston University, Boston, MA USA

Junichi Takagi, PhD
Professor, Institute for Protein Research, Laboratory of Protein Synthesis and Expression, Osaka University, Suita, Osaka, Japan

Sam Thiagalingam, PhD
Associate Professor of Medicine, Genetics/Genomics, Pathology and Laboratory Medicine, Departments of Medicine (Genetics Program), Genetics and Genomics, and Pathology and Laboratory Medicine, Boston University School of Medicine, Boston, MA USA

Patricia Tueting, PhD
Research Assistant Professor of Psychiatry (Neuroscience), the Psychiatric Institute, Department of Psychiatry, College of Medicine, University of Illinois at Chicago, Chicago, IL USA

Marin Veldic, MD
Research Assistant Professor of Psychiatry, the Psychiatric Institute, Department of Psychiatry, College of Medicine, University of Illinois at Chicago, Chicago, IL USA

Christopher A. Walsh, MD, PhD
Bullard Professor of Neurology at Harvard Medical School, Chief of the Division of Genetics at Boston Children's hospital, Director of the Harvard-MIT M.D.-Ph.D. Program, and Investigator of the Howard Hughes Medical Institute at Beth Israel Deaconess Medical Center, Howard Hughes Medical Institute, BIDMC, Division of Genetics, Boston Children's Hospital, Harvard Medical School, Boston, MA USA

Kimberly Walter, BS
Graduate Student in the Pathobiology and Disease Mechanisms Program at Johns Hopkins University, Department of Pathology, The Sol Goldman Pancreatic Research Center, The Johns Hopkins Medical Institutions, Baltimore, MD USA

Edwin John Weeber, PhD
Associate Professor, Department of Molecular Pharmacology and Physiology, University of South Florida, Tampa, FL USA

Shanting Zhao, MD
Senior Scientist, Institut für Anatomie und Zellbiologie, Abteilung für Neuroanatomie, Universität Freiburg, Freiburg, Germany

Jin-Rong Zhou, PhD
Assistant Professor of Surgery and Nutrition, Laboratory of Nutrition and Metabolism at BIDMC, Department of Surgery, Harvard Medical School, Boston, MA USA

Chapter 1
The *Reelin* Gene and Its Functions in Brain Development

Cheng-Chiu Huang and Gabriella D'Arcangelo

Contents

1 Introduction

Brain development and function requires the coordinated genesis, migration, and maturation of all of its cellular components. The product of the *Reelin* (*Reln*) gene has been identified as a major determinant of neuronal migration that also plays a significant role in cellular maturation and synaptic function. Thus, the *Reln* gene controls multiple aspects of brain development over the entire life span of a mammalian organism, from pre- to postnatal ages, and exerts distinct functions on migrating neuroblasts, radial progenitors, and postmigratory neurons. Some of the molecular mechanisms that mediate these functions have been elucidated by the analysis of mutant mice and biochemical interactions, but much remains to be discovered. In this chapter we will summarize the current state of our knowledge of *Reln* and its function in brain development.

C. -C. Huang
Graduate Program in Developmental Biology, Baylor College of Medicine, 1 Baylor Plaza, Houston, TX 77030
e-mail: ch147987@bcm.tmc.edu

G. D'Arcangelo
Department of Cell Biology and Neuroscience, Rutgers, The State University of New Jersey, 604 Allison Road B319, Piscataway, NJ 08854
e-mail: darcangelo@biology.rutgers.edu

S. H. Fatemi (ed.), *Reelin Glycoprotein: Structure, Biology and Roles in Health and Disease.*

2 The Reelin Gene and Its Product

The *Reln* gene was discovered based on the genetic analysis of *reeler* mutant mouse strains. These classical neurological mutants exhibit in homozygosity a distinct phenotype characterized by ataxia and the disruptions of all layered structures of the brain (reviewed by Lambert de Rouvroit and Goffinet, 1998). The *reeler* mutant trait is recessive and it maps to the distal region of mouse chromosome 5. Identification of *Reln* as the gene disrupted in two independent *reeler* mutant strains that completely lack expression (D'Arcangelo *et al.*, 1995) led to the full characterization of its genomic structure, transcript expression pattern, and amino acid composition of the predicted encoded product. The murine *Reln* gene is composed of 65 exons spanning a region of approximately 450 kb (Royaux *et al.*, 1997). The N-terminal exons are separated by large introns whereas the remaining exons are closer to each other. One microexon encoding just two amino acids near the C terminus is alternatively spliced to generate two *Reln* transcripts, but the significance of this alternative splicing event is not known. In addition, two major transcription initiation sites and two polyadenylation sites have also been identified (Royaux *et al.*, 1997). The only *Reln* mRNA detectable by Northern blot analysis is approximately 12 kb, is highly and predominately expressed in the brain, and is developmentally regulated (D'Arcangelo *et al.*, 1995). It first becomes detectable in the embryo, peaks between 1 and 2 postnatal weeks, and then declines at lower levels in the adult brain. The encoded Reelin protein is a large secreted protein that consists of an N-terminal region followed by eight unique repeats each containing an EGF-like motif (D'Arcangelo *et al.*, 1995). The N-terminal region contains the signal peptide and a small domain similar to F-spondin. Each Reelin repeat is composed of two related subrepeats, A and B, separated by an EGF-like motif. The C-terminal region contains a stretch of positively charged amino acids (D'Arcangelo *et al.*, 1997). Deletion of this region resulting from a retroviral insertion is responsible for the lack of secreted, functional Reelin and the appearance of the mutant phenotype in the naturally occurring *reeler* Orleans strain (de Bergeyck *et al.*, 1997; Takahara *et al.*, 1996). The full-length mouse protein is composed of 3461 amino acids, which is subjected to N- and O-glycosylation, resulting in a secreted protein of approximately 450 kDa (D'Arcangelo *et al.*, 1997). This protein is rapidly cleaved in the extracellular environment at two cleavage sites. The N-terminal site, located between repeats 2B and 3A, is cleaved by a metalloprotease, whereas the C-terminal site, located between repeats 6B and 7A, is cleaved by an unknown protease (Lambert de Rouvroit *et al.*, 1999; Jossin *et al.*, 2007). The activity of these proteases results in the generation of three major fragments, N terminal (N terminus to repeat 2; ~180 kDa), central (repeats 3 to 6; 190 kDa), and C terminal (repeats 7 and 8; 80 kDa) detectable with distinct specific antibodies (Jossin *et al.*, 2007). In addition, the full-length protein, as well as intermediate fragments of 370 kDa (N terminus to repeat 6) and 270 kDa (repeats 6 to 8) resulting from partial processing can also be detected in Reelin-containing culture medium or tissue lysates (Jossin *et al.*, 2007). Functional studies have demonstrated that the central (190 kDa) Reelin fragment is sufficient to activate a tyrosine phosphorylation-dependent signal transduction and to induce layer formation in cortical slice cultures, whereas the N-terminal and the C-terminal fragments appear to be inactive (Jossin *et al.*, 2004, 2007). The N-terminal region contains an epitope recognized by the functionally interfering antibody CR50 (D'Arcangelo *et al.*, 1997; Ogawa *et al.*, 1995). This region mediates the formation of stable homodimers composed of full-length Reelin proteins (Kubo *et al.*, 2002; Utsunomiya-Tate *et al.*, 2000). These homodimers appear to stimulate the tyrosine phosphorylation-dependent signaling pathway more efficiently than cleaved central fragments. On the other hand, the cleaved central fragment of Reelin can diffuse farther into the developing cortical plate than can full-length homodimers (Jossin *et al.*, 2007). Thus, proteolytic processing may modulate the strength and the range of Reelin activity *in vivo*.

Reln mRNA and protein are expressed at high levels in superficial layers of embryonic cortical structures (Alcantara *et al.*, 1998; D'Arcangelo *et al.*, 1995; Ogawa *et al.*, 1995; Schiffmann *et al.*, 1997). Cajal-Retzius cells in the marginal zone of the cortex, stratum lacunosum moleculare of the hippocampus, and outer marginal layer of the dentate gyrus are the major source of Reelin in the embryonic neocortex and early postnatal hippocampus. Young granule cells in the external granular layer of the embryonic and early postnatal cerebellum also synthesize high levels of Reelin. The pattern of *Reln* expression, however, changes dramatically during postnatal development. In the adult brain, this protein is no longer confined to superficial cortical layers, but it is expressed in all layers of the cortex and hippocampus by a subset of GABAergic interneurons (Alcantara *et al.*, 1998; Pesold *et al.*, 1998). In the adult cerebellum, Reelin continues to be expressed by granule cells, which have migrated inwardly to form the internal granule layer. This shift in expression pattern most likely reflects the different functions exerted by Reelin in the pre- and postnatal brain.

Reelin activities are mediated by a signaling pathway that has been partially elucidated through the genetic analysis of mutant mouse strains. Mutations or deletions in the *reelin* gene account for the majority of identified mouse strains that exhibit a *reeler* phenotype. However, mutant mice indistinguishable from *reeler* have also been identified or generated by disrupting genes that are essential to Reelin signal transduction (Fig. 1.1). In some cases, more than one gene has to be disrupted to overcome functional redundancy and reveal the phenotype. Double knockout mice lacking both the apolipoprotein E receptor 2 (ApoER2) and the very-low-density lipoprotein receptor (VLDLR) (Trommsdorff *et al.*, 1999), and double knockout mice lacking the two src-family kinases (SFKs) Fyn and Src (Kuo *et al.*, 2005) are similar to *reeler*, whereas single mutants only exhibit various degrees of cortical layering defects. On the other hand, disruption of a single gene encoding the nonredundant adapter protein Disabled-1 (Dab1) in the spontaneous mutants *scrambler* and *yotari* (Sheldon *et al.*, 1997; Ware *et al.*, 1997; Yoneshima *et al.*, 1997), the Dab1 null knockout (Howell *et al.*, 1997), or the knockin mutant Dab5F lacking SFK-dependent phosphorylation sites (Howell *et al.*, 2000), all result in a *reeler*-like phenotype. These genes thus define a signal transduction pathway that is crucial for Reelin activity on neuronal migration. Secreted full-length Reelin, or its active central

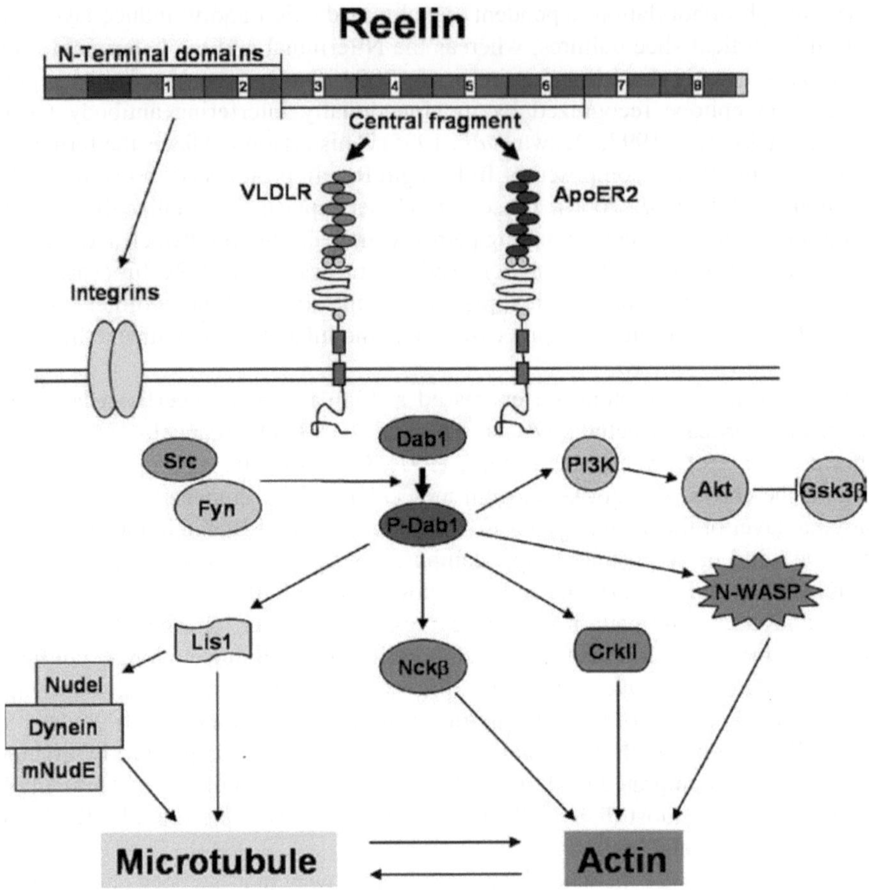

Fig. 1.1 The Reelin signaling pathway (*See Color Plates*)

fragment, bind to ApoER2 and VLDLR on the surface of target cells such as migrating neurons. These two Reelin receptors are members of the lipoprotein receptor superfamily and each is capable of binding Reelin with similar affinity (Benhayon *et al.*, 2003; D'Arcangelo *et al.*, 1999; Hiesberger *et al.*, 1999). They also bind lipoproteins and other extracellular ligands with lower affinity. Like all members of the lipoprotein receptor superfamily, ApoER2 and VLDLR internalize their ligand, including Reelin, using an internalization domain, the NPxY motif, present on their cytoplasmic tail. The Reelin receptors actively traffic between the plasma membrane and the endosomes. Their translocation to the plasma membrane is facilitated by the binding of Dab1 to the cytoplasmic tail of the receptors near their NPxY motif by virtue of its pleckstrin homology/phosphotyrosine binding domain (PH/PTB) (Morimura *et al.*, 2005; Trommsdorff *et al.*, 1998). Upon Reelin binding, the receptors cluster (Strasser *et al.*, 2004) causing the activation of Fyn and Src (Arnaud *et al.*, 2003b; Bock and Herz, 2003), which in turn phosphorylate Dab1 on specific tyrosine residues (Ballif *et al.*, 2004; Howell *et al.*, 1999; Keshvara *et al.*, 2001). This event results in the ubiquitination of Dab1 by the Cbl ubiquitin ligase and its degradation by the proteasome system (Arnaud *et al.*, 2003a; Suetsugu *et al.*, 2004). Thus, Reelin promotes the phosphorylation, as well as the degradation of Dab1. This observation explains why Dab1 protein accumulates in the brain of *reeler*, double Apoer2/Vldlr, and double Fyn/Src knockout mice (Kuo *et al.*, 2005; Sheldon *et al.*, 1997; Trommsdorff *et al.*, 1999). The short-lived phosphoDab1 is thought to function as a hub as it binds several intracellular signal transduction proteins including the PI3K regulatory subunit p85α (Bock *et al.*, 2003), the actin-binding N-WASP (Suetsugu *et al.*, 2004), Nckβ (Pramatarova *et al.*, 2003), Crk family proteins (Ballif *et al.*, 2004; Huang *et al.*, 2004), and the neuronal migration gene product Lis1 (Assadi *et al.*, 2003). All of these proteins potentially contribute to Reelin function in the control of neuronal migration by affecting cytoskeletal dynamics that determine cell motility and morphology. The binding of Dab1 to p85α correlates with Reelin-induction of PI3K activity, the downstream phosphorylation and activation of Akt, and the phosphorylation and inhibition of Gsk3β, which in turn results in a suppressed level of tau phosphorylation (Ballif *et al.*, 2003; Beffert *et al.*, 2002; Bock *et al.*, 2003; Ohkubo *et al.*, 2003). Consistent with a physiological role for Reelin in the regulation of this pathway, elevated levels of phosphorylated tau have been reported in *reeler*, double *Apoer2/Vldlr,* and *Dab1* mutant mice (Brich *et al.*, 2003; Hiesberger *et al.*, 1999). These findings could be important for the pathology of neurodegenerative disorders such as Alzheimer's disease, which are associated with accumulation of hyperphosphorylated tau. Binding of Dab1 to Crk family proteins such as CrkI, CrkII, and CrkL leads to the activation of a signaling pathway involving the GTP-exchange factor C3G and Rap1 (Ballif *et al.*, 2004). Activated Rap1 could be important for cytoskeletal changes required during neuronal migration. Similarly, Dab1 interactions with the actin-binding proteins N-WASP and Nckβ or with Lis1, a protein that associates with the microtubule dynamin–dynactin motor complex and with the Pafah1b enzymatic complex, could be important for neuronal migration. In the case of Lis1, there is also genetic evidence that this protein is important for Reelin-dependent cortical layer formation, and that it likely functions downstream of VLDLR through interactions mediated by the Pafah1b complex (Zhang *et al.*, 2007). In addition, β1 integrins have also been reported to bind Dab1 in a Reelin-dependent manner (Schmid *et al.*, 2005). Together with the finding that α3 integrins bind the N-terminal region of Reelin, these data suggest that α3β1 integrins may participate in Reelin functions. However, because genetic deletion of β1 integrin does not severely disrupt neuronal migration (Graus-Porta *et al.*, 2001), it is possible that these integrins may either stabilize newly formed cellular layers by promoting cell–cell adhesion (Schmid *et al.*, 2004) or participate in other postnatal functions of Reelin such as synaptogenesis or synaptic activity (Dong *et al.*, 2003; Rodriguez *et al.*, 2000). Alternatively, the binding of Dab1 to α3β1 integrins may promote their degradation in response to Reelin, thus allowing the detachment of neurons from radial fibers and the formation of cellular layers (Sanada *et al.*, 2004).

3 Functions of Reelin in Brain Development

The best-characterized function of Reelin is the control of radial neuronal migration and the formation of cellular layers during prenatal brain development. Layer formation is a distinct feature of all cortical structures including the cerebral cortex, the hippocampus, and the cerebellum. In the neocortex, principal neurons are born from the asymmetric division of progenitor cells near the ventricular zone, the radial glia (Malatesta *et al.*, 2000; Noctor *et al.*, 2001). These young neurons then migrate radially toward the pial surface, and stop just underneath the marginal zone, a relatively cell-free superficial layer, to form tight cellular layers with other neurons born approximately at the same time. The first cohort of radially migrating neurons splits an earlier transient structure called the preplate, consisting of Cajal-Retzius cells and subplate cells. Each new cohort of migrating neurons bypasses older layers en route toward the marginal zone generating the typical six cellular layers of the mammalian cortex in an inside-out fashion (Angevine and Sidman, 1961). Inhibitory neurons originating from extracortical regions such as the ganglionic eminences enter the developing cortical plate by tangential migration and distribute in specific cellular layers according to their specific subtype, thus contributing to the establishment of the cortical circuitry (Anderson *et al.*, 1997; Lavdas *et al.*, 1999). Similar mechanisms operate in other regions of the brain, such as the hippocampus or the cerebellum. In the hippocampus, principal (pyramidal) neurons are born near the ventricle and migrate radially to form a single, multicellular but compact layer at a considerable distance from the pial surface. In the cerebellum, principal neurons (the Purkinje cells) are born near the ventricle and migrate radially along so-called Bergmann fibers to form a single cell layer underneath a superficial layer composed initially by tangentially migrating granule cells, the external granular layer. Later in development, these cells actually migrate inwardly to form an internal granular layer underneath the Purkinje cell layer. The observation that layer formation in all cortical structures is disrupted in *reeler* and *reeler*-like mutants demonstrated that Reelin and its signaling pathway are crucial for this function (Caviness and Sidman, 1973; Caviness, 1973; Goffinet, 1983) (Fig. 1.2). However, the exact mechanisms governing layer formation are still not entirely understood. A current model suggests that a soluble form of Reelin (such as its central fragment) is produced near the pial surface and diffuses down into the developing cortical plate, hippocampus, or cerebellum in a smooth or step-gradient manner (D'Arcangelo, 2005; Jossin *et al.*, 2007). There, at low levels, it promotes the extension of a leading edge and the radial migration of Reelin target cells such as cortical and hippocampal principal neurons or cerebellar Purkinje cells. Once these cells reach the top of the cortex, they encounter either high levels or a matrix-immobilized form of Reelin (such as the full-length homodimers), which prompt them to stop migration, detach from radial fibers, and associate into tight layers. Indeed, Reelin has been shown to cause migration arrest and detachment from radial glia *in vitro* using cortical imprint assays, and *in vivo* when injected focally into the neocortex using immobilized beads (Dulabon *et al.*, 2000). Thus, Reelin could either increase or arrest motility, depending on the concentration or other

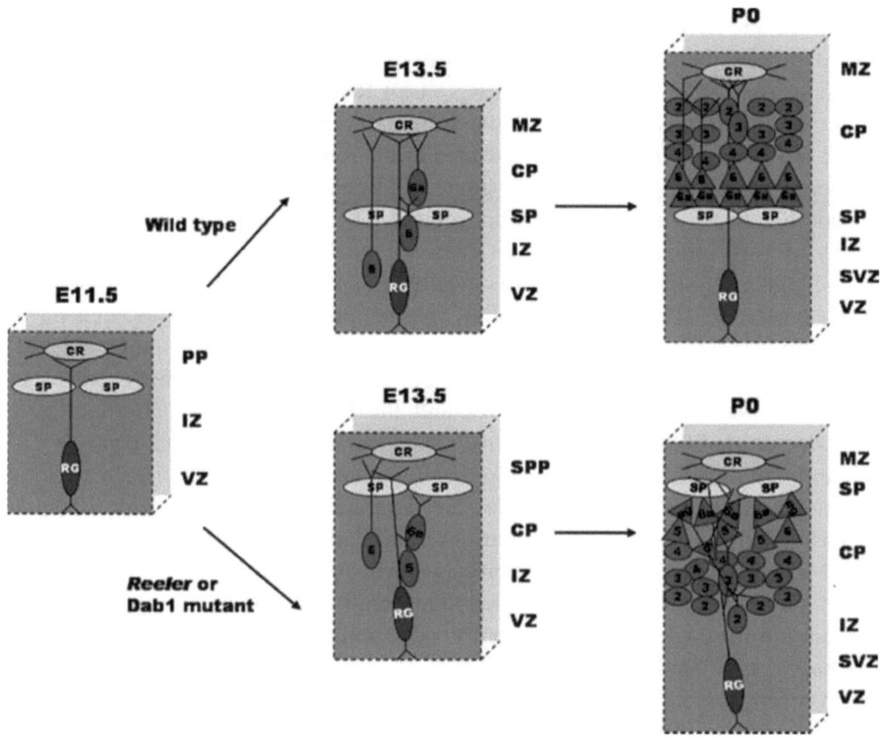

Fig. 1.2 Cortical development in normal and *reeler* and Dab1 mutant mice. In the embryonic cortex of normal mice, the preplate (PP) is split by the arrival of early radially migrating neurons, whereas in the *reeler* cortex, this does not happen and cells form a superplate structure (SPP). Cellular layers in the cortical plate (CP) are also disrupted in *reeler*. Other abbreviations: MZ, marginal zone; IZ, intermediate zone; VZ, ventricular zone; SVZ, subventricular zone; RG, radial glia; CR, Cajal-Retzius cells (*See Color Plates*)

environmental factor controling its diffusion. The different signal could be mediated by the differential recruitment of Dab1-interacting proteins that interact with the actin and microtubule cytoskeleton.

In recent years, it has become clear that Reelin not only controls neuronal migration during embryogenesis but also promotes neuronal maturation and function at postnatal ages. Heterozygous *reeler* mice express half the *Reln* mRNA levels of wild-type mice and are phenotypically normal, that is, they do not present any cortical layer defect nor are they ataxic like the homozygous mutants. However, they display a variety of behavioral and cognitive defects reminiscent of those found in human psychosis (reviewed by Tueting *et al.*, 2006). Anatomically, they exhibit a stunted growth of dendritic processes in hippocampal neurons (Niu *et al.*, 2004) and a reduction in the density of synaptic contacts in the frontal cortex (Liu *et al.*, 2001). Because cellular layers are intact in heterozygous *reeler* mice, these developmental defects reflect a direct function of the *Reln* gene product on neuronal maturation and synaptic formation. This postnatal function of Reelin appears to be mediated by the same signaling pathway that mediates its function in neuronal migration, that is, involves the ApoER2/VLDLR receptors and Dab1 (Niu *et al.*, 2004). The importance of Reelin for the normal development of synaptic connectivity and function was initially demonstrated in the retina (Rice *et al.*, 2001), but it is now appreciated in the brain as well (reviewed by Herz and Chen, 2006). Addition of recombinant Reelin promotes hippocampal LTP and this function requires the activity of both lipoprotein receptors (Weeber *et al.*, 2002). A splicing variant of Apoer2 containing exon 19, which is capable of interacting with the postsynaptic density protein 95 (PSD95) and the JNK interacting protein (JIP), has been shown to be important for Reelin-induced LTP and the formation of spatial memory (Beffert *et al.*, 2005). The role of Reelin in synaptic function is mediated in part through interactions between ApoER2 and the NMDA receptor (Beffert *et al.*, 2005; Hoe *et al.*, 2006). These proteins form a synaptic complex that controls Ca^{2+} entry through the NMDA receptor and thus regulate synaptic plasticity. In addition, Reelin signaling is also important for the regulation of NMDA receptor subunit composition during hippocampal neuronal maturation (Sinagra *et al.*, 2005), and the NMDA receptor-mediated activity in cortical neurons (Chen *et al.*, 2005). Recent physiological studies revealed that Reelin also enhances glutamatergic transmission through AMPA receptors. The enhancement of AMPA receptor responses is mediated by increased surface expression and increased amplitude of AMPA receptor-mediated excitatory postsynaptic currents (Qiu and Weeber, 2007; Qiu *et al.*, 2006b). These results demonstrate that Reelin functions in the developing postnatal, as well as in the adult hippocampus by affecting synaptic strength and plasticity.

4 Reelin in Human Diseases

Reelin is highly conserved among vertebrate species, especially in mammals, suggesting a conserved function related to the development of layered structures that are particularly prominent in mammalian species. Reelin homologues have been identified in humans, rat, chicken, turtle, and *Xenopus*, but not in invertebrate species such as *Drosophila*. The amino acid sequence of human Reelin is 94.8% identical to that of the murine homologue indicating striking functional conservation (DeSilva *et al.*, 1997). Homozygous mutations in the *RELN* gene in humans result in a phenotype strikingly similar to that of *reeler* mice, featuring severe ataxia, cognitive dysfunction, cerebellar hypoplasia, and cortical neuronal migration defects leading to a reduced number of cortical gyri (lissencephaly) (Hong *et al.*, 2000). This severe phenotype reflects the essential function of Reelin in neuronal migration during prenatal brain development. However, even reduced levels of *RELN* expression may be deleterious for human brain development. For example, *RELN* expression is downregulated in inhibitory cortical neurons of patients with schizophrenia and psychotic bipolar disorder (Guidotti *et al.*, 2000; Fatemi *et al.*,

2000; Impagnatiello *et al.*, 1998). This decrease is due to epigenetic mechanisms involving the increased expression of DNA methyltransferase (DNMT1) (Grayson *et al.*, 2005; Veldic *et al.*, 2004). Reduced Reelin expression has also been reported in other cognitive disorders, such as autism (Fatemi *et al.*, 2001, 2005), nonpsychotic bipolar disorder, major depression (Fatemi *et al.*, 2000), and Alzheimer's disease (Chin *et al.*, 2007), suggesting that a dysfunction in Reelin-regulated neuronal maturation and synaptic activity in the postnatal brain may contribute to these disorders. These subtle defects are recapitulated in the heterozygous *reeler* mice, which exhibit defects in learning and memory (Larson *et al.*, 2003; Qiu *et al.*, 2006a), as well as behavioral performance (Krueger *et al.*, 2006; Ognibene *et al.*, 2007; Tueting *et al.*, 1999). Thus, heterozygous *reeler* mice may serve as models to investigate the molecular and physiological basis of cognitive dysfunction linked to Reelin deficiency. These animal models may now be just as valuable as homozygous *reeler* mice have been during the past few decades in aiding our understanding of the molecular mechanisms of cortical layer formation.

Acknowledgment We thank M. Cecilia Ljungberg for critical reading of the manuscript.

References

Alcantara, S., Ruiz, M., D'Arcangelo, G., Ezan, F., de Lecea, L., Curran, T., Sotelo, C., and Soriano, E. (1998). Regional and cellular patterns of *reelin* mRNA expression in the forebrain of the developing and adult mouse. *J. Neurosci.* 18:7779–7799.

Anderson, S. A., Eisenstat, D. D., Shi, L., and Rubenstein, J. L. R. (1997). Interneuron migration from basal forebrain to neocortex: dependence on Dlx genes. *Science* 278:474–476.

Angevine, J. G., and Sidman, R. L. (1961). Autoradiographic study of cell migration during histogenesis of cerebral cortex in the mouse. *Nature* 192:766–768.

Arnaud, L., Ballif, B. A., and Cooper, J. A. (2003a). Regulation of protein tyrosine kinase signaling by substrate degradation during brain development. *Mol. Cell. Biol.* 23:9293–9302.

Arnaud, L., Ballif, B. A., Forster, E., and Cooper, J. A. (2003b). Fyn tyrosine kinase is a critical regulator of disabled-1 during brain development. *Curr. Biol.* 13:9–17.

Assadi, A. H., Zhang, G., Beffert, U., McNeil, R. S., Renfro, A. L., Niu, S., Quattrocchi, C. C., Antalffy, B. A., Sheldon, M., Armstrong, D. D., Wynshaw-Boris, A., Herz, J., D'Arcangelo, G., and Clark, G. D. (2003). Interaction of reelin signaling and Lis1 in brain development. *Nat. Genet.* 35:270–276.

Ballif, B. A., Arnaud, L., and Cooper, J. A. (2003). Tyrosine phosphorylation of disabled-1 is essential for reelin-stimulated activation of Akt and Src family kinases. *Brain Res. Mol. Brain Res.* 117:152–159.

Ballif, B. A., Arnaud, L., Arthur, W. T., Guris, D., Imamoto, A., and Cooper, J. A. (2004). Activation of a Dab1/CrkL/C3G/Rap1 pathway in reelin-stimulated neurons. *Curr. Biol.* 14:606–610.

Beffert, U., Morfini, G., Bock, H. H., Reyna, H., Brady, S. T., and Herz, J. (2002). Reelin-mediated signaling locally regulates protein kinase B/Akt and glycogen synthase kinase 3beta. *J. Biol. Chem.* 277:49958–49964.

Beffert, U., Weeber, E. J., Durudas, A., Qiu, S., Masiulis, I., Sweatt, J. D., Li, W.-P., Adelmann, G., Frotscher, M., Hammer, R. E., and Herz, J. (2005). Modulation of synaptic plasticity and memory by reelin involves differential splicing of the lipoprotein receptor ApoER2. *Neuron* 47:567–579.

Benhayon, D., Magdaleno, S., and Curran, T. (2003). Binding of purified reelin to ApoER2 and VLDLR mediates tyrosine phosphorylation of disabled-1. *Mol. Brain Res.* 112:33–45.

Bock, H. H., and Herz, J. (2003). Reelin activates SRC family tyrosine kinases in neurons. *Curr. Biol.* 13:18–26.

Bock, H. H., Jossin, Y., Liu, P., Forster, E., May, P., Goffinet, A. M., and Herz, J. (2003). PI3-kinase interacts with the adaptor protein Dab1 in response to reelin signaling and is required for normal cortical lamination. *J. Biol. Chem.* 278:38772–38779.

Brich, J., Shie, F. S., Howell, B. W., Li, R., Tus, K., Wakeland, E. K., Jin, L. W., Mumby, M., Churchill, G., Herz, J., and Cooper, J. A. (2003). Genetic modulation of tau phosphorylation in the mouse. *J. Neurosci.* 23:187–192.

Caviness, V. S. J. (1973). Time of neuron origin in the hippocampus and dentate gyrus of normal and reeler mutant mice: an autoradiographic analysis. *J. Comp. Neurol.* 151113–120.

Caviness, V. S. J., and Sidman, R. L. (1973). Time of origin of corresponding cell classes in the cerebral cortex of normal and reeler mutant mice: an autoradiographic analysis. *J. Comp. Neurol.* 148:141–152.

Chen, Y., Beffert, U., Ertunc, M., Tang, T. S., Kavalali, E. T., Bezprozvanny, I., and Herz, J. (2005). Reelin modulates NMDA receptor activity in cortical neurons. *J. Neurosci.* 25:8209–8216.

Chin, J., Massaro, C. M., Palop, J. J., Thwin, M. T., Yu, G. Q., Bien-Ly, N., Bender, A., and Mucke, L. (2007). Reelin depletion in the entorhinal cortex of human amyloid precursor protein transgenic mice and humans with Alzheimer's disease. *J. Neurosci.* 27:2727–2733.

D'Arcangelo, G. (2005). Reelin mouse mutants as models of cortical development disorders. *Epilepsy Behav.* 8:81–90.

D'Arcangelo, G., Miao, G. G., Chen, S. C., Soares, H. D., Morgan, J. I., and Curran, T. (1995). A protein related to extracellular matrix proteins deleted in the mouse mutant reeler. *Nature* 374:719–723.

D'Arcangelo, G., Nakajima, K., Miyata, T., Ogawa, M., Mikoshiba, K., and Curran, T. (1997). Reelin is a secreted glycoprotein recognized by the CR-50 monoclonal antibody. *J. Neurosci.* 17:23–31.

D'Arcangelo, G., Homayouni, R., Keshvara, L., Rice, D. S., Sheldon, M., and Curran, T. (1999). Reelin is a ligand for lipoprotein receptors. *Neuron* 24:471–479.

de Bergeyck, V., Nakajima, K., Lambert de Rouvroit, C., Naerhuyzen, B., Goffinet, A. M., Miyata, T., Ogawa, M., and Mikoshiba, K. (1997). A truncated reelin protein is produced but not secreted in the "Orleans" reeler mutation ($Reln^{rl\text{-}Orl}$). *Mol. Brain Res.* 50:85–90.

DeSilva, U., D'Arcangelo, G., Braden, V. V., Chen, J., Miao, G., Curran, T., and Green, E. D. (1997). The human reelin gene: isolation, sequencing, and mapping on chromosome 7. *Genome Res.* 7:157–164.

Dong, E., Caruncho, H., Liu, W. S., Smalheiser, N. R., Grayson, D. R., Costa, E., and Guidotti, A. (2003). A reelin–integrin receptor interaction regulates Arc mRNA translation in synaptoneurosomes. *Proc. Natl. Acad. Sci. USA* 100:5479–5484.

Dulabon, L., Olson, E. C., Taglienti, M. G., Eisenhuth, S., McGrath, B., Walsh, C. A., Kreidberg, J. A., and Anton, E. S. (2000). Reelin binds alpha3beta1 integrin and inhibits neuronal migration. *Neuron* 27:33–44.

Fatemi, S. H., Earle, J. A., and McMenomy, T. (2000). Reduction in reelin immunoreactivity in hippocampus of subjects with schizophrenia, bipolar disorder and major depression. *Mol. Psychiatry* 571:654–663.

Fatemi, S. H., Stary, J. M., Halt, A. R., and Realmuto, G. R. (2001). Dysregulation of reelin and Bcl-2 proteins in autistic cerebellum. *J. Autism Dev. Disord.* 31:529–535.

Fatemi, S. H., Snow, A. V., Stary, J. M., Araghi-Niknam, M., Reutiman, T. J., Lee, S., Brooks, A. I., and Pearce, D. A. (2005). Reelin signaling is impaired in autism. *Biol. Psychiatry* 57:777–787.

Goffinet, A. M. (1983). The embryonic development of the cerebellum in normal and reeler mutant mice. *Anat. Embryol. (Berl.)* 168:73–86.

Graus-Porta, D., Blaess, S., Senften, M., Littlewood-Evans, A., Damsky, C., Huang, Z., Orban, P., Klein, R., Schittny, J. C., and Muller, U. (2001). Beta1-class integrins regulate the development of laminae and folia in the cerebral and cerebellar cortex. *Neuron* 31:367–379.

Grayson, D. R., Jia, X., Chen, Y., Sharma, R. P., Mitchell, C. P., Guidotti, A., and Costa, E. (2005). Reelin promoter hypermethylation in schizophrenia. *Proc. Natl. Acad. Sci. USA* 102:9341–9346.

Guidotti, A., Auta, J., Davis, J. M., DiGiorgi Gerevini, V., Dwivedi, Y., Grayson, D. R., Impagnatiello, F., Pandey, G., Pesold, C., Sharma, R., Uzunov, D., and Costa, E. (2000). Decrease in reelin and glutamic acid decarboxylase67 (GAD67) expression in schizophrenia and bipolar disorder: a postmortem brain study. *Arch. Gen. Psychiatry* 57:1061–1069.

Herz, J., and Chen, Y. (2006). Reelin, lipoprotein receptors and synaptic plasticity. *Nature Rev.* 7:850–859.

Hiesberger, T., Trommsdorff, M., Howell, B. W., Goffinet, A. M., Mumby, M. C., Cooper, J. A., and Herz, J. (1999). Direct binding of reelin to VLDL receptor and ApoE receptor 2 induces tyrosine phosphorylation of disabled-1 and modulates tau phosphorylation. *Neuron* 24:481–489.

Hoe, H. S., Tran, T. S., Matsuoka, Y., Howell, B. W., and Rebeck, G. W. (2006). DAB1 and reelin effects on amyloid precursor protein and ApoE receptor 2 trafficking and processing. *J. Biol. Chem.* 281:35176–35185.

Hong, S. E., Shugart, Y. Y., Huang, D. T., Al Shahwan, S., Grant, P. E., Hourihane, J. O. B., Martin, N. D. T., and Walsh, C. A. (2000). Autosomal recessive lissencephaly with cerebellar hypoplasia is associated with human RELN mutations. *Nature Genet.* 26:93–96.

Howell, B. W., Hawkes, R., Soriano, P., and Cooper, J. A. (1997). Neuronal position in the developing brain is regulated by mouse disabled-1. *Nature* 389:733–736.

Howell, B. W., Herrick, T. M., and Cooper, J. A. (1999). Reelin-induced tyrosine phosphorylation of disabled 1 during neuronal positioning. *Genes Dev.* 13:643–648.

Howell, B. W., Herrick, T. M., Hildebrand, J. D., Zhang, Y., and Cooper, J. A. (2000). Dab1 tyrosine phosphorylation sites relay positional signals during mouse brain development. *Curr. Biol.* 10:877–885.

Huang, Y., Magdaleno, S., Hopkins, R., Slaughter, C., Curran, T., and Keshvara, L. (2004). Tyrosine phosphorylated disabled 1 recruits Crk family adapter proteins. *Biochem. Biophys. Res. Commun.* 318:204–212.

Impagnatiello, F., Guidotti, A. R., Pesold, C., Dwivedi, Y., Caruncho, H., Pisu, M. G., Uzunov, D. P., Smalheiser, N. R., Davis, J. M., Pandey, G. N., Pappas, G. D., Tueting, P., Sharma, R. P., and Costa, E. (1998). A decrease of reelin expression as a putative vulnerability factor in schizophrenia. *Proc. Natl. Acad. Sci. USA* 95:15718–15723.

Jossin, Y., Ignatova, N., Hiesberger, T., Herz, J., Lambert de Rouvroit, C., and Goffinet, A. M. (2004). The central fragment of reelin, generated by proteolytic processing in vivo, is critical to its function during cortical plate development. *J. Neurosci.* 24:514–521.

Jossin, Y., Gui, L., and Goffinet, A. M. (2007). Processing of reelin by embryonic neurons is important for function in tissue, but not in dissociated cultured neurons. *J. Neurosci.* 27:4243–4352.

Keshvara, L., Benhayon, D., Magdaleno, S., and Curran, T. (2001). Identification of reelin-induced sites of tyrosyl phosphorylation on disabled 1. *J. Biol. Chem.* 276:16008–16014.

Krueger, D. D., Howell, J. L., Hebert, B. F., Olausson, P., Taylor, J. R., and Nairn, A. C. (2006). Assessment of cognitive function in the heterozygous reeler mouse. *Psychopharmacology* 189:95–104.

Kubo, K., Mikoshiba, K., and Nakajima, K. (2002). Secreted reelin molecules form homodimers. *Neurosci. Res.* 43:381–388.

Kuo, G., Arnaud, L., Kronstad-O'Brien, P., and Cooper, J. A. (2005). Absence of Fyn and Src causes a reeler-like phenotype. *J. Neurosci.* 25:8578–8586.

Lambert de Rouvroit, C., and Goffinet, A. M. (1998). The reeler mouse as a model of brain development. *Adv. Anat. Embryol. Cell Biol.* 150:1–108.

Lambert de Rouvroit, C., de Bergeyck, V., Cortvrindt, C., Bar, I., Eeckhout, Y., and Goffinet, A. M. (1999). Reelin, the extracellular matrix protein deficient in reeler mutant mice, is processed by a metalloproteinase. *Exp. Neurol.* 156:214–217.

Larson, J., Hoffman, J. S., Guidotti, A., and Costa, E. (2003). Olfactory discrimination learning deficit in heterozygous reeler mice. *Brain Res.* 971:40–46.

Lavdas, A. A., Grigoriou, M., Pachnis, V., and Parnavelas, J. G. (1999). The medial ganglionic eminence gives rise to a population of early neurons in the developing cerebral cortex. *J. Neurosci.* 99:7881–7888.

Liu, W. S., Pesold, C., Rodriguez, M. A., Carboni, G., Auta, J., Lacor, P., Larson, J., Condie, B. G., Guidotti, A., and Costa, E. (2001). Down-regulation of dendritic spine and glutamic acid decarboxylase 67 expressions in the reelin haploinsufficient heterozygous reeler mouse. *Proc. Natl. Acad. Sci. USA* 98:3477–3482.

Malatesta, P., Hartfuss, E., and Gotz, M. (2000). Isolation of radial glial cells by fluorescent-activated cell sorting reveals a neuronal lineage. *Development* 127:5253–5263.

Morimura, T., Hattori, M., Ogawa, M., and Mikoshiba, K. (2005). Disabled1 regulates the intracellular trafficking of reelin receptors. *J. Biol. Chem.* 280:16901–16908.

Niu, S., Renfro, A., Quattrocchi, C. C., Sheldon, M., and D'Arcangelo, G. (2004). Reelin promotes hippocampal dendrite development through the VLDLR/ApoER2-Dab1 pathway. *Neuron* 41:71–84.

Noctor, S. C., Flint, A. C., Weissman, T. A., Dammerman, R. S., and Kriegstein, A. R. (2001). Neurons derived from radial glial cells establish radial units in neocortex. *Nature* 409:714–720.

Ogawa, M., Miyata, T., Nakajima, K., Yagyu, K., Seike, M., Ikenaka, K., Yamamoto, H., and Mikoshiba, K. (1995). The reeler gene-associated antigen on Cajal-Retzius neurons is a crucial molecule for laminar organization of cortical neurons. *Neuron* 14:899–912.

Ognibene, E., Adriani, W., Granstrem, O., Pieretti, S., and Laviola, G. (2007). Impulsivity-anxiety-related behavior and profiles of morphine-induced analgesia in heterozygous reeler mice. *Brain Res.* 1131:173–180.

Ohkubo, N., Lee, Y. D., Morishima, A., Terashima, T., Kikkawa, S., Tohyama, M., Sakanaka, M., Tanaka, J., Maeda, N., Vitek, M. P., and Mitsuda, N. (2003). Apolipoprotein E and reelin ligands modulate tau phosphorylation through an apolipoprotein E receptor/disabled-1/glycogen synthase kinase-3beta cascade. *FASEB J.* 17:295–297.

Pesold, C., Impagnatiello, F., Pisu, M. G., Uzunov, D. P., Costa, E., Guidotti, A., and Caruncho, H. J. (1998). Reelin is preferentially expressed in neurons synthesizing g-aminobutyric acid in cortex and hippocampus of adult rats. *Proc. Natl. Acad. Sci. USA* 95:3221–3226.

Pramatarova, A., Ochalski, P. G., Chen, K., Gropman, A., Myers, S., Min, K. T., and Howell, B. W. (2003). Nck beta interacts with tyrosine-phosphorylated disabled 1 and redistributes in reelin-stimulated neurons. *Mol. Cell Biol.* 23:7210–7221.

Qiu, S., and Weeber, E. J. (2007). Reelin signaling facilitates maturation of CA1 glutamatergic synapses. *J. Neurophys.* 97:2312–2321.

Qiu, S., Korwek, K. M., Pratt-Davis, A. R., Peters, M., Bergman, M. Y., and Weeber, E. J. (2006a). Cognitive disruption and altered hippocampus synaptic function in reelin haploinsufficient mice. *Neurobiol. Learning Memory* 85:228–242.

Qiu, S., Zhao, L. F., Korwek, K. M., and Weeber, E. J. (2006b). Differential reelin-induced enhancement of NMDA and AMPA receptor activity in the adult hippocampus. *J. Neurosci.* 26:12943–12955.

Rice, D. S., Nusinowitz, S., Azimi, A. M., Martinez, A., Soriano, E., and Curran, T. (2001). The reelin pathway modulates the structure and function of retinal synaptic circuitry. *Neuron* 31:929–941.

Rodriguez, M. A., Pesold, C., Liu, W. S., Khrino, V., Guidotti, A., Pappas, G. D., and Costa, E. (2000). Colocalization of integrin receptors and reelin in dendritic spine postsynaptic densities of adult nonhuman primate cortex. *Proc. Natl. Acad. Sci. USA* 97:3550–3555.

Royaux, I., Lambert de Rouvroit, C., D'Arcangelo, G., Demirov, D., and Goffinet, A. M. (1997). Genomic organization of the mouse *reelin* gene. *Genomics* 46:240–250.

Sanada, K., Gupta, A., and Tsai, L. H. (2004). Disabled-1-regulated adhesion of migrating neurons to radial glial fiber contributes to neuronal positioning during early corticogenesis. *Neuron* 42:197–211.

Schiffmann, S. N., Bernier, B., and Goffinet, A. M. (1997). Reelin mRNA expression during mouse brain development. *Eur. J. Neurosci.* 9:1055–1071.

Schmid, R. S., Shelton, S., Stanco, A., Yokota, Y., Kreidberg, J. A., and Anton, E. S. (2004). alpha3beta1 integrin modulates neuronal migration and placement during early stages of cerebral cortical development. *Development* 131:6023–6031.

Schmid, R. S., Jo, R., Shelton, S., Kreidberg, J. A., and Anton, E. S. (2005). Reelin, integrin and Dab1 interactions during embryonic cerebral cortical development. *Cerebral Cortex* 15:1632–1636.

Sheldon, M., Rice, D. S., D'Arcangelo, G., Yoneshima, H., Nakajima, K., Mikoshiba, K., Howell, B. W., Cooper, J. A., Goldowitz, D., and Curran, T. (1997). *Scrambler* and *yotari* disrupt the *disabled* gene and produce a *reeler*-like phenotype in mice. *Nature* 389:730–733.

Sinagra, M., Verrier, D., Frankova, D., Korwek, K. M., Blahos, J., Weeber, E. J., Manzoni, O. J., and Chavis, P. (2005). Reelin, very-low-density lipoprotein receptor, and apolipoprotein E receptor 2 control somatic NMDA receptor composition during hippocampal maturation in vitro. *J. Neurosci.* 25:6127–6136.

Strasser, V., Fasching, D., Hauser, C., Mayer, H., Bock, H. H., Hiesberger, T., Herz, J., Weeber, E. J., Sweatt, J. D., Pramatarova, A., Howell, B., Schneider, W. J., and Nimpf, J. (2004). Receptor clustering is involved in reelin signaling. *Mol. Cell Biol.* 24:1378–1386.

Suetsugu, S., Tezuka, T., Morimura, T., Hattori, M., Mikoshiba, K., Yamamoto, T., and Takenawa, T. (2004). Regulation of actin cytoskeleton by mDab1 through N-WASP and ubiquitination of mDab1. *Biochem. J.* 384:1–8.

Takahara, T., Ohsumi, T., Kuromitsu, J., Shibata, K., Sasaki, N., Okazaki, Y., Shibata, H., Sato, S., Yoshiki, A., Kusakabe, M., Muramatsu, M., Ueki, M., Okuda, K., and Hayashizaki, Y. (1996). Dysfunction of the Orleans *reeler* gene arising from exon skipping due to transposition of a full-length copy of an active L1 sequence into the skipped exon. *Hum. Mol. Genet.* 5:989–993.

Trommsdorff, M., Borg, J. P., Margolis, B., and Herz, J. (1998). Interaction of cytosolic adaptor proteins with neuronal apolipoprotein E receptors and the amyloid precursor protein. *J. Biol. Chem.* 273:33556–33560.

Trommsdorff, M., Gotthardt, M., Hiesberger, T., Shelton, J., Stockinger, W., Nimpf, J., Hammer, R. E., Richardson, J. A., and Herz, J. (1999). Reeler/disabled-like disruption of neuronal migration in knockout mice lacking the VLDL receptor and ApoE receptor 2. *Cell* 97:689–701.

Tueting, P., Costa, E., Dwivedi, Y., Guidotti, A., Impagnatiello, F., Manev, R., and Pesold, C. (1999). The phenotypic characteristics of heterozygous reeler mouse. *Neuroreport* 10:1329–1334.

Tueting, P., Doueiri, M. S., Guidotti, A., Davis, J. M., and Costa, E. (2006). Reelin down-regulation in mice and psychosis endophenotypes. *Neurosci. Biobehav. Rev.* 30:1065–1077.

Utsunomiya-Tate, N., Kubo, K., Tate, S., Kainosho, M., Katayama, E., Nakajima, K., and Mikoshiba, K. (2000). Reelin molecules assemble together to form a large protein complex, which is inhibited by the function-blocking CR-50 antibody. *Proc. Natl. Acad. Sci. USA* 97:9729–9734.

Veldic, M., Caruncho, H. J., Liu, W. S., Davis, J., Satta, R., Grayson, D.R., Guidotti, A., and Costa, E. (2004). DNA-methyltransferase 1 mRNA is selectively overexpressed in telencephalic GABAergic interneurons of schizophrenia brains. *Proc. Natl. Acad. Sci. USA* 101:348–353.

Ware, M., Fox, J. W., Gonzales, J. L., Davis, N. M., Lambert de Rouvroit, C., Russo, C. J., Chua, S. C. J., Goffinet, A. M., and Walsh, C. A. (1997). Aberrant splicing of a mouse *disabled* homolog, *mdab1*, in the *scrambler* mouse. *Neuron* 19:239–249.

Weeber, E. J., Beffert, U., Jones, C., Christian, J. M., Forster, E., Sweatt, J. D., and Herz, J. (2002). Reelin and ApoE receptors cooperate to enhance hippocampal synaptic plasticity and learning. *J. Biol. Chem.* 277:39944–39952.

Yoneshima, H., Nagata, E., Matsumoto, M., Yamada, M., Nakajima, K., Miyata, T., Ogawa, M., and Mikoshiba, K. (1997). A novel neurological mutation of mouse, *yotari*, which exhibits *reeler*-like phenotype but expresses *reelin*. *Neurosci. Res.* 29:217–223.

Zhang, G., Assadi, A. H., McNeil, R. S., Beffert, U., Wynshaw-Boris, A., Herz, J., Clark, G. D., and D'Arcangelo, G. (2007). The PAFAH1B complex interacts with the reelin receptor VLDLR. *PlosOne* 2:e252.

Chapter 2
Apolipoprotein E Receptor 2 and Very-Low-Density Lipoprotein Receptor: An Overview

Hans H. Bock and Joachim Herz

Contents

1 The LDL Receptor Gene Family

1.1 Functions of Lipoprotein Receptors in Neurobiology

It is now well established that members of the low-density lipoprotein (LDL) receptor gene family are crucial regulators of different aspects of neuronal development, synaptic plasticity, maintenance of neuronal homeostasis, and neurodegeneration. This was highlighted in particular by the discovery that the lipoprotein receptors apolipoprotein E receptor-2 (ApoER2) and very-low-density lipoprotein receptor (VLDLR) function as receptors for the neuronal signaling protein Reelin. In this

H. H. Bock
Zentrum für Neurowissenschaften, Universität Freiburg, Albertstrasse 23, 79104 Freiburg, Germany
e-mail: hans.bock@zfn.uni-freiburg.de

J. Herz
Department of Molecular Genetics, University of Texas Southwestern Medical Center, 5323 Harry Hines Boulevard, Dallas, TX 75390 and Zentrum für Neurowissenschaften, Universität Freiburg, 79104 Freiburg, Germany

S. H. Fatemi (ed.), *Reelin Glycoprotein: Structure, Biology and Roles in Health and Disease.*

chapter, we will briefly introduce the family of LDL receptor-related proteins and review the functions of its members ApoER2 and VLDLR as Reelin receptors and their role in the developing and adult brain.

1.2 Structural Organization and Common Features of LDL Receptor Gene Family Members

The LDL receptor gene family consists of seven core family members in vertebrates that share common structural features. They are type I transmembrane receptors with an extracellular domain containing complement-type repeats and epidermal growth factor (EGF) homology domains in a modular arrangement followed by the single transmembrane-spanning sequence and a comparatively short cytoplasmic domain that interacts with adapter and scaffolding proteins through various sequence motifs (Fig. 2.1A; reviewed by Herz and Bock, 2002; Nykjaer and Willnow, 2002). The intracellular domains of all core members contain at least one Asn-Pro-X-Tyr (NPXY, where X designates any amino acid) tetra-amino acid motif, which is crucial for their involvement in receptor-mediated endocytosis and cellular signal transduction (reviewed by May and Herz, 2003; Schneider and Nimpf, 2003; Stolt and Bock, 2006). Another commonality of all LDL receptor-related proteins (LRPs) is their binding to the receptor-associated protein (RAP), an endoplasmic reticulum-associated chaperone that acts as a universal antagonist of all LRP ligands (Bu and Schwartz, 1998; Herz, 2006). Other ligands that are bound by members of the receptor family include lipoproteins, protease–protease inhibitor complexes, matrix metalloproteinases, plasma proteins, growth factors, and cytokines (reviewed by Willnow *et al.*, 1999; May *et al.*, 2005).

1.2.1 Intracellular Adapter Proteins Bind to the NPXY Motif in the Intracellular Domains of LRPs

The presence of at least one NPXY motif, which was discovered as a signal for receptor-mediated endocytosis (reviewed by Bonifacino and Traub, 2003), in the intracellular domains of all LDL receptor gene family members suggested that they might function as mere cargo receptors. This view was challenged by observations that a range of modular cytosolic adapter proteins bind to the NPXY motifs of different receptors with their phosphotyrosine binding (PTB) / protein interaction domains (Gotthardt *et al.*, 2000), including the neuronally enriched proteins FE65 and Disabled-1 (Dab1). These studies (Trommsdorff *et al.*, 1998; Howell *et al.*, 1999b) set the stage for the seminal discovery that two family members, ApoER2 and VLDLR, directly act as signal transduction receptors for the neuronal signaling molecule Reelin (D'Arcangelo *et al.*, 1999; Hiesberger *et al.*, 1999; Trommsdorff *et al.*, 1999).

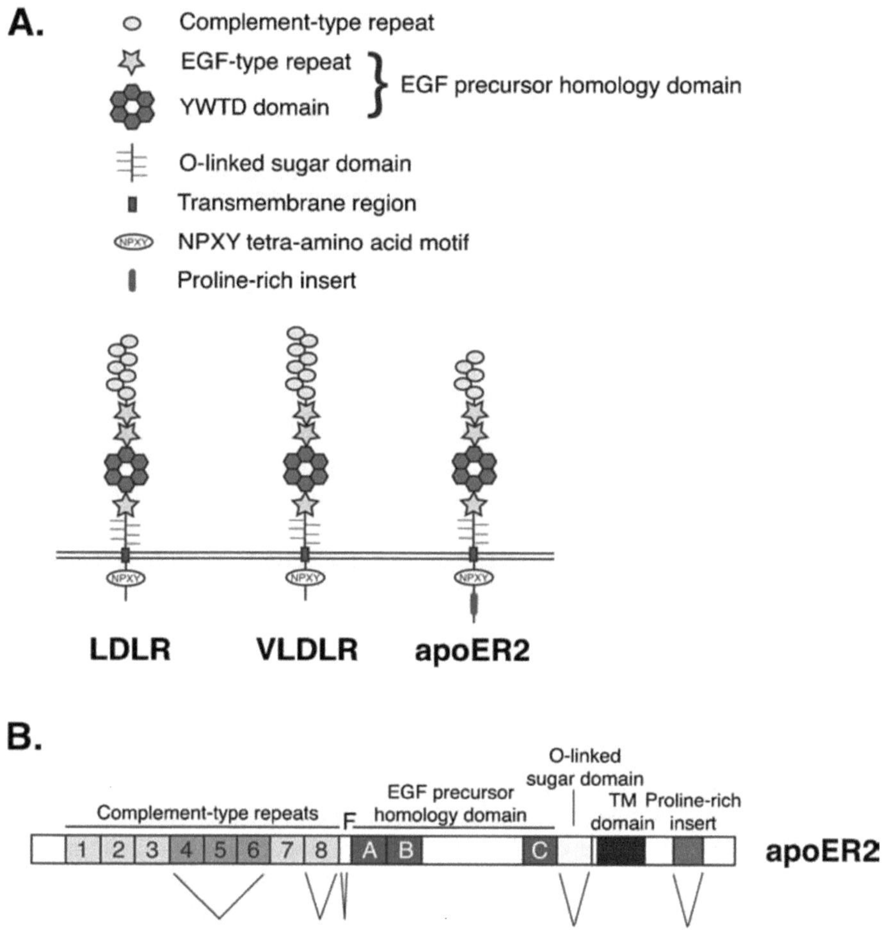

Fig. 2.1 Structure of ApoER2 and VLDLR. (**A**) ApoER2 and VLDLR, two close relatives of the LDL receptor. ApoER2 and VLDLR are two members of the LDL receptor gene family that closely resemble the namesake of the family, the LDL receptor. They are modular type I transmembrane proteins that consist of one ligand-binding domain comprising several complement-type (a.k.a. ligand-type or type A) repeats, an epidermal growth factor (EGF) precursor homology domain, and an O-linked sugar domain followed by a single transmembrane stretch and a short cytoplasmic tail. The EGF precursor homology domain consists of three EGF-like (type B) repeats flanking a domain containing modules of consensus YWTD tetrapeptides (ß-propeller). The NPXY motif in the cytoplasmic tail participates in endocytosis and signal transduction. Additional, larger members of the LDL receptor gene family that share this modular structure are the LDL receptor-related proteins LRP1, LRP1b, megalin (a.k.a. gp330), and MEGF7 (not depicted here). The more distantly related receptors LRP5, LRP6, and LR11/SorLA display different domain structures. **(B)** ApoER2 undergoes extensive tissue-specific splicing. Spliced exons are indicated below the bar representing the mRNA encoding murine ApoER2. The exon encoding the complement-type repeats 4–6 is constitutively spliced out in mouse. Other alternatively spliced exons encode the ligand-type repeat 8, a small insertion including the furin cleavage site (F) just upstream of the first EGF-like repeat (A), the O-linked sugar domain, and the proline-rich cytoplasmic insert. The type I and type II variants of VLDLR differ from each other by the presence or absence of the O-linked sugar domain, which is always present in the LDLR -TM domain, transmembrane domain

1.2.2 LRPs Are Receptors for Apolipoprotein E

All members of the receptor family also interact with the cholesterol transport protein apolipoprotein E (ApoE); therefore, they are commonly referred to as apolipoprotein E (ApoE) receptors (Herz and Beffert, 2000). ApoE was discovered as a polymorphic lipid transport protein that exists in three major isoforms in humans, named ApoE2, ApoE3, and ApoE4 (Mahley *et al.*, 1984; Mahley and Rall, 2000; Strittmatter and Bova Hill, 2002). Besides its functions in lipoprotein metabolism and cardiovascular physiology, ApoE plays important roles in neurobiology (Mahley *et al.*, 2006). Although neurons are capable of synthesizing ApoE, astrocytes are the main cell type secreting ApoE-containing high-density lipoprotein (HDL)-like lipid particles in the brain (Boyles *et al.*, 1985; Pitas *et al.*, 1987), which participate in lipoprotein receptor-mediated lipid redistribution during neuronal maintenance, plasticity, and repair (Ignatius *et al.*, 1987; Boyles *et al.*, 1989; Mauch *et al.*, 2001). The seminal finding that the ε4 allele of *APOE* is a major risk factor for Alzheimer's disease (reviewed by Roses, 1996) together with the unanticipated discovery that ApoE receptors serve as pivotal mediators of neuronal signaling cascades (reviewed by Herz and Beffert, 2000) fueled an enormous interest in isoform-specific effects of ApoE on neuronal homeostasis beyond mere lipid redistribution.

1.2.3 Proteolytic Processing of LRPs by Gamma Secretase

One of the neuropathological hallmarks of Alzheimer's disease is the accumulation of amyloid plaques in the cerebrovasculature and brain parenchyma (Goedert and Spillantini, 2006). A major component of these plaques is the 42-amino-acid amyloid-beta (A-beta) peptide, which is derived from the type I transmembrane amyloid precursor protein (APP) by a sequence of proteolytic processing steps (Sinha and Lieberburg, 1999). In the last step, which is mediated by a complex of integral membrane proteins called gamma secretase (Wolfe, 2006), the APP intracellular domain (AICD) is released into the cytosol after regulated intramembrane proteolysis of the membrane-attached carboxyl-terminal stub. Clearance of amyloid-beta from the extracellular space and amyloid plaque formation might both be influenced by its isoform-specific interaction with ApoE (Mahley and Rall, 2000). Remarkably, several members of the LDL receptor gene family, including the Reelin receptors ApoER2 and VLDLR, are cleaved by gamma secretase as well (May *et al.*, 2002, 2003; Zou *et al.*, 2004; Hoe and Rebeck, 2005). This, together with the observation that several of the intracellular adapter proteins that bind to LRPs also interact with the cytoplasmic domain of APP, adds to a growing body of evidence for a critical role of ApoE receptors in the pathogenesis of Alzheimer's disease that involves different but converging cellular mechanisms (Fig. 2.3; discussed by Andersen and Willnow, 2006; Herz and Chen, 2006).

2 APOE Receptors as Mediators of Reelin Signaling in the Brain

In this section, we will review the molecular and cellular mechanisms of how ApoER2 and VLDLR might regulate neuronal positioning during neurodevelopment and discuss evidence for a role of these receptors in the control of synaptic transmission in the adult brain.

2.1 *Structural Characteristics of ApoER2 and VLDLR*

ApoER2 and VLDLR are of a comparable size and display the same overall structure as the LDL receptor (Fig. 2.1A). Both receptors display a high degree of conservation across species and are highly expressed in the brain (Schneider *et al.*, 1997). In contrast to the LDL receptor, the VLDLR exists in two major splice forms designated type I and type II that differentially express the O-linked sugar domain just outside the plasma membrane, and its ligand binding domain contains eight, rather than seven, complement-type repeats (Fig. 2.1A). The type I VLDLR containing the O-linked sugar domain is the preferentially expressed form in the brain (reviewed by Takahashi *et al.*, 2004). ApoER2, on the other hand, occurs in various species- and tissue-specific splice variants (Fig. 2.1B) (Schneider and Nimpf, 2003). These include constitutive deletion of the complement-type repeats 4–6 in the murine *apoer2* gene by alternative splicing, the variable expression of a furin cleavage site that allows for the secretion of the ligand-binding domain, splicing of the O-linked sugar domain, and a differentially expressed unique proline-rich insert in the cytoplasmic tail that specifically interacts with JNK-interacting proteins (JIP) and the postsynaptic density protein PSD-95 (Brandes *et al.*, 1997, 2001; Clatworthy *et al.*, 1999; Koch *et al.*, 2002). Of note, tissue-specific glycosylation due to the presence or absence of the O-linked sugar domain turned out to be an essential regulator of ApoER2 cleavage by gamma secretase (May *et al.*, 2003).

2.2 *Reelin Signaling During Neurodevelopment*

2.2.1 Phenotype of ApoER2- and Vldl-Receptor-Deficient Mice

Due to the structural similarities of ApoER2 and VLDL receptor to the LDL receptor, their closest relative which is defective in familial hypercholesterolemia, it was initially assumed that both receptors might function primarily as uptake receptors for cholesterol in the brain and other tissues (see, e.g., Frykman *et al.*, 1995; Yamamoto and Bujo, 1996). A major breakthrough in the understanding of VLDLR and ApoER2 function in the brain came from a study by

Trommsdorff and colleagues. They showed that simultaneous inactivation of both receptors by gene targeting in mice resulted in a reeler-like phenotype, which is characterized by severe neuronal positioning defects in laminated structures of the brain (for details on the phenotype of reeler mice, see chapters by Mienville, Huang and D'Arcangelo, Förster *et al.*, Hevner), suggesting a direct involvement of ApoE receptors in neuronal signaling (Trommsdorff *et al.*, 1999). Subsequent studies demonstrated that Reelin directly binds to VLDLR and ApoER2 (D'Arcangelo *et al.*, 1999; Hiesberger *et al.*, 1999), with the latter exhibiting a slighty higher apparent affinity for Reelin (Benhayon *et al.*, 2003). Analysis of single receptor mutant mice revealed milder neuroanatomical phenotypes, with neuronal positioning defects in the neocortex and hippocampus in the absence of ApoER2, and Vldlr deficiency resulting in mild cerebellar abnormalities (Trommsdorff *et al.*, 1999; Benhayon *et al.*, 2003; Beffert *et al.*, 2006b). These observations might be explained by a combination of mere expression differences of both receptors in diverse regions of the developing brain and different binding affinities for Reelin. However, assuming that both receptors are functionally redundant with regard to the transmission of the Reelin signal, this leaves open the question why two structurally closely related receptors with different but overlapping expression patterns for the same ligand should have evolved (Cooper and Howell, 1999). Possible alternative explanations for the different neurodevelopmental phenotypes of single receptor mutants include region- and/or receptor-specific modulators of the Reelin response, e.g., co-receptors, co-ligands, or intracellular adapter proteins. Of note, ApoE receptor-independent effects of Reelin on neuronal migration of hindbrain efferent nuclei and gonadotropin-releasing hormone-secreting hypothalamic neurons have been described (Cariboni *et al.*, 2005; Rossel *et al.*, 2005).

2.2.2 Nonneuronal Manifestations of ApoER2 and Vldlr Deficiency

Besides their neurodevelopmental abnormalities, ApoER2- and Vldlr-deficient mice display additional defects according to their expression pattern in peripheral tissues. Loss of Vldlr in mice, which is expressed in the heart, muscle, adipose tissue, endothelium, and on macrophages, leads to reduced body weight (Frykman *et al.*, 1995), resistance to diet-induced obesity (Goudriaan *et al.*, 2001), and altered triglyceride metabolism in the concurrent absence of the LDL receptor (Tacken *et al.*, 2000). Moreover, subretinal neovascularization is common in Vldlr-deficient mice (Heckenlively *et al.*, 2003). An autosomal-recessive human genetic defect of the *VLDLR* gene was recently identified in the Hutterite population (Boycott *et al.*, 2005). Patients homozygous for a large deletion including the *VLDLR* locus present with inferior cerebellar hypoplasia, nonprogressive ataxia, mild cerebral gyral simplification, and mental retardation, reminiscent of, but less severe than, the lissencephaly syndrome caused by a null mutation of the human *RELN* gene (Hong *et al.*, 2000). In addition, a polymorphic triplet repeat in the 5′ untranslated region of *VLDLR* might predispose to cognitive impairment and Alzheimer's disease

(Okuizumi *et al.*, 1995; Helbecque and Amouyel, 2000; Helbecque *et al.*, 2001). Finally, homozygous deletion of *VLDLR* by genomic loss or promoter hypermethylation was observed in gastric cancer cell lines (Takada *et al.*, 2006). This suggests a possible role as a tumor suppressor gene and adds to the finding that RELN is frequently silenced in pancreatic cancers (Sato *et al.*, 2006; see chapter by Walter and Goggins).

ApoER2 is mainly expressed in the brain and in reproductive organs. ApoER2-deficient male mice are infertile as a result of abnormal sperm morphology and motility, which could be linked to the reduced expression of a selenoperoxidase important for spermatogenesis (Andersen *et al.*, 2003). A significant association between a maternal *APOER2* polymorphism and fetal growth restriction was recently reported among African-American women (Wang *et al.*, 2006). As ApoER2 is highly expressed in the placenta, this might reflect a requirement for ApoER2 in the regulation of the microenvironment for fetal growth. Other than for VLDLR, a human ApoER2 brain malformation syndrome has not been described. However, a possible association of *APOER2* with Alzheimer's disease may exist (Ma *et al.*, 2002).

2.3 The Reelin–ApoER2/VLDLR–Dab1 Signaling Cascade

2.3.1 Evidence for a Linear Signaling Pathway Involving Reelin, ApoE Receptors, and Dab1

Biochemical and genetic evidence resulting from independent studies of several groups led to the delination of a signaling cascade (Fig. 2.2) where the extracellular protein Reelin binds to ApoER2 and VLDLR, thereby inducing the tyrosine phosphorylation and activation of the intracellular adapter protein Dab1 (see chapters by Olson and Walsh and Cooper *et al.*). This was based on the reeler-like phenotype of Dab1-mutant mice (Howell *et al.*, 1997b; Sheldon *et al.*, 1997; Ware *et al.*, 1997), the identification of Dab1 as an intracellular binding partner of ApoER2 and VLDLR (Trommsdorff *et al.*, 1999), the observation that Dab1 protein levels are upregulated in brains lacking Reelin or Vldlr and ApoER2 (Rice *et al.*, 1998; Trommsdorff *et al.*, 1998), the development of a neuronal Reelin signaling assay (Howell *et al.*, 1999a), and the demonstration that mice expressing a nonphosphorylatable form of Dab1 recapitulate the reeler phenotype (Howell *et al.*, 2000). Expression of these components in a heterologous cell line proved sufficient to reconstitute the signaling pathway (Mayer *et al.*, 2006).

2.3.2 Modulation of Cytoskeletal Components in Response to Reelin

The identification of an ApoE receptor-dependent Reelin signaling cascade involving tyrosine phosphorylation of Dab1 raised the urgent question how this might translate

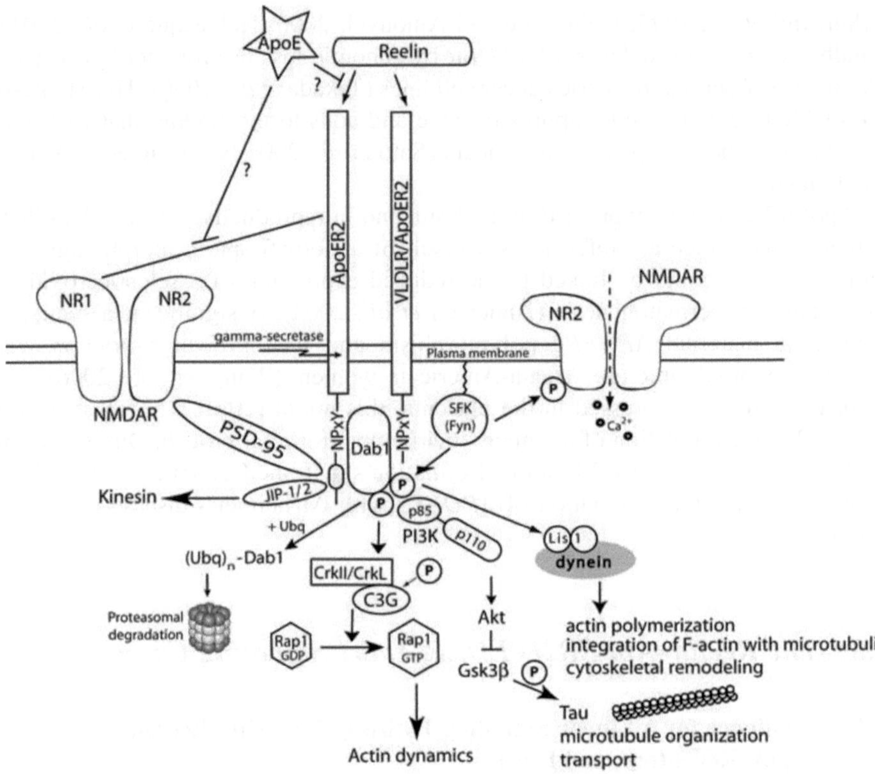

Fig. 2.2 Molecular mechanisms of Reelin signaling. The diagram summarizes mechanisms of ApoE receptor-mediated Reelin signaling that are important during neurodevelopment and for synaptic transmission in the adult brain (for details, see Sections 2.3 and 2.4). The cytoplasmic tails of ApoER2 and VLDLR serve as scaffolds for different signaling complexes, many of which target the cytoskeleton. Gamma secretase-mediated release of the receptor tail might therefore profoundly influence these signaling cascades. The adapter protein Dab1, which interacts with the NPXY motif of different transmembrane proteins, including the Reelin receptors, plays a crucial role as a signaling relay whose binding preferences are regulated by its Reelin-dependent phosphorylation state. Clustering of different transmembrane receptors by Reelin or other ligands like F-spondin represents another possibility of fine-tuning the signaling responses mediated by ApoER2 and VLDLR

into cytoskeletal rearrangement required for neuronal positioning (Feng and Walsh, 2001) in response to Reelin. An important clue came from the observation that the microtubule-associated protein tau is hyperphosphorylated in mice with genetic defects in components of the Reelin signaling cascade (Hiesberger *et al.*, 1999). This led to the identification of a continuous Reelin-dependent pathway from its receptors to the cytoskeleton, involving activation of class I phosphatidylinositol-3 kinase (PI3K), a central regulator of diverse cellular functions, and Akt/PKB, which results in the downregulation of the tau kinase glycogen synthase kinase-3β

(GSK3β) (Beffert *et al.*, 2002; Ballif *et al.*, 2003; Bock *et al.*, 2003). The modulation of tau phosphorylation by Reelin adds to the growing evidence linking ApoE receptors and Alzheimer's disease (see Section 1.2.3 and Fig. 2.3), where neurofibrillary tangles that result from the aggregation of abnormally phosphorylated tau are a neuropathological feature (Goedert and Spillantini, 2006). Of note, Reelin did not regulate the activity of Cdk5 (Beffert *et al.*, 2002), another major tau kinase and regulator of neuronal positioning (reviewed by Dhavan and Tsai, 2001; see chapter by Ohshima). Further *in vivo* evidence for an independent but synergistic effect of Reelin and Cdk5 signaling on neuronal migration came from the phenotypic analysis of compound mutant mice lacking the Cdk5 activator p35 and Dab1 (Ohshima *et al.*, 2001) or p35 and either Vldlr or ApoER2 (Beffert *et al.*, 2004). On the other hand, Cdk5 signaling might modulate the Reelin response by phosphorylating Dab1 independently of Reelin signaling (Ohshima *et al.*, 2007), and both signaling cascades converge on the microtubule-associated Ndel1/Lis1/dynein complex (Niethammer *et al.*, 2000; Sasaki *et al.*, 2000; Assadi *et al.*, 2003) that regulates nucleokinesis, a critical cellular event in neuronal migration (Tsai and Gleeson,

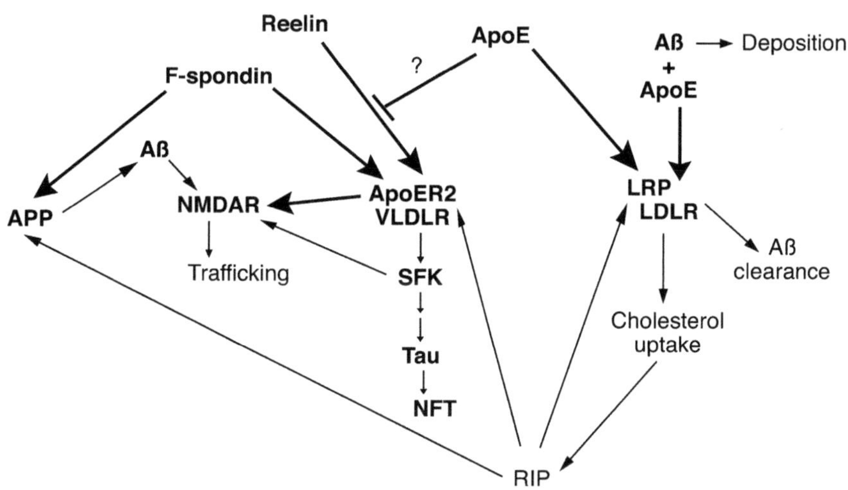

Fig. 2.3 Possible interactions of ApoE and amyloid-beta at the synapse. This diagram highlights possible interactions between ApoE, ApoE receptors, amyloid-beta, and NMDA receptors at the level of the synapse. Isoform-specific interactions of ApoE with its receptors might modulate Reelin signaling and endocytic uptake of ApoE-containing lipid particles. This would influence NMDAR-mediated neurotransmission, as well as the formation of neurofibrillary tangles and amyloid-beta deposition. ApoE-mediated uptake of extracellular cholesterol contributes to cholesterol homeostasis, which in turn has a marked impact on the regulated intramembrane proteolysis (RIP) of APP, ApoER2, and LRP. Proteolytic cleavage of APP yields amyloid-beta, the main component of amyloid plaques. Soluble amyloid-beta regulates the trafficking of NMDA receptors, which interact with Reelin receptors both intra- and extracellularly. The accumulation of subtle dysregulations of these complex events during the lifetime of an individual might contribute to neurodegenerative processes

2005). Another cytoskeletal element, the actin network, is also targeted by Reelin signaling. This is accomplished by the interaction of tyrosine-phosphorylated Dab1 with SH2/SH3 domain-containing scaffold proteins of the Nck and Crk families (Pramatarova *et al.*, 2003; Ballif *et al.*, 2004; Chen *et al.*, 2004; Huang *et al.*, 2004), which involves the Reelin-dependent phosphorylation at the positions Y220 and 232 (Howell *et al.*, 2000; Keshvara *et al.*, 2001) in a cluster of tyrosine residues proximal of the PTB domain of Dab1 (see chapter by Cooper *et al.*). Binding of phospho-Dab1 to CrkL promotes the phosphorylation of the guanine nucleotide exchange factor C3G (see Fig. 2.2), which results in the activation of the small GTPase Rap1 (Ballif *et al.*, 2004) and subsequent modulation of the actin cytoskeleton (reviewed by Bos, 2005). The interaction of a splice variant of ApoER2 with the molecular motor kinesin is mediated by the adapter proteins c-jun NH2-terminal kinase (JNK)-interacting proteins (JIPs) JIP1 and JIP2 (Verhey *et al.*, 2001), which bind to the spliced proline-rich insert of the intracellular domain (Gotthardt *et al.*, 2000; Stockinger *et al.*, 2000). Recruitment of a JIP1-JNK3 module to the ApoER2 insert plays a role in the control of neuronal survival in the normal aging brain (Beffert *et al.*, 2006a).

2.3.3 Molecular Mechanism of ApoE Receptor-Mediated Dab1 Activation by Reelin

Since LRPs do not display an intrinsic kinase domain within their intracellular tails, it remained elusive how Reelin activates Dab1 *in vivo*, which had been shown before to be a substrate of nonreceptor tyrosine kinases of the Src family *in vitro* (Howell *et al.*, 1997a). A series of genetic and biochemical studies from several groups (Arnaud *et al.*, 2003b; Bock and Herz, 2003; Jossin *et al.*, 2003) established that the Src family kinase (SFK) Fyn is the physiological Dab1 kinase (Fig. 2.2). Whereas Fyn knockout mice (Grant *et al.*, 1992; Yagi *et al.*, 1993), which express increased neuronal Dab1 levels, display only minor cortical lamination defects (Yuasa *et al.*, 2004), simultaneous inactivation of Fyn and Src results in a reeler-like phenotype (Kuo *et al.*, 2005), providing evidence for functional redundancy among SFKs during brain development. The activation of SFKs requires clustering of ApoER2 and VLDLR at the plasma membrane by oligomeric Reelin or receptor-activating antibodies, which brings the receptor-bound Dab1 into close proximity of SFK-enriched membrane subdomains (Jossin *et al.*, 2004; Strasser *et al.*, 2004). The resulting reciprocal activation of SFKs and Dab1 (Bock and Herz, 2003) is limited by the targeting of tyrosine-phosphorylated Dab1 for polyubiquitination and proteasomal degradation (Arnaud *et al.*, 2003a; Bock *et al.*, 2004). This highly efficient feedback regulation explains the observed accumulation of Dab1 protein in mice deficient for components of the Reelin signaling cascade.

Structural analysis of the Dab1 PTB domain bound to NPXY-containing peptides corresponding to parts of the ApoER2 (Stolt *et al.*, 2003) or APP (Yun *et al.*, 2003) intracellular domains and complexed with phosphoinositides provided a mechanistic insight into the simultaneous and noncooperative interaction of the PTB domain with both ligands (Stolt *et al.*, 2004). Further investigations using lentiviral transduction of primary neurons with Dab1 mutant constructs (Stolt *et al.*, 2005) demonstrated the importance of phosphoinoside binding for the membrane targeting of Dab1 in neurons (Bock *et al.*, 2003) and showed that both receptor binding and phosphoinositide binding of the Dab1 PTB domain are important for Reelin signal transduction (Stolt *et al.*, 2005). Along with two subsequent studies (Huang *et al.*, 2005; Xu *et al.*, 2005), these data suggest that phosphoinositides target Dab1 to the plasma membrane, where its binding to the ApoER2 and VLDL receptor tails facilitates the Reelin-induced oligomerization and interaction with membrane-associated Src family kinases (reviewed by Stolt and Bock, 2006). Mice carrying a site-directed disruption of the Dab1 interaction motif in the ApoER2 intracellular domain display a neurodevelopmental phenotype that is virtually indistinguishable from reeler when bred on a *vldlr* knockout background, providing compelling *in vivo* evidence for the importance of the ApoER2–Dab1 interaction (Beffert *et al.*, 2006b).

According to the model that emerges from these studies, tyrosine phosphorylation of Dab1, a key event in Reelin-mediated signal transduction, does not require the presence of an additional co-receptor providing tyrosine kinase activity. Interestingly, VLDLR- or ApoER2-dimerizing antibodies stimulated Dab1 tyrosine phosphorylation but failed to rescue the reeler phenotype in an *in vitro* embryonic slice culture assay of cortical plate development (Jossin *et al.*, 2003). This suggests a functional role for other Reelin-binding, Dab1-interacting receptors like α3β1 integrin (Dulabon *et al.*, 2000; Calderwood *et al.*, 2003) in a developmental, cell-type or tissue-specific context (Forster *et al.*, 2002; Sanada *et al.*, 2004).

2.3.4 Trafficking of Reelin Receptors

In addition to its interaction with ApoE receptors, Dab1 also binds to the NPXY motif in the intracellular tail of APP and APP-like proteins (Trommsdorff *et al.*, 1998; Homayouni *et al.*, 1999; Howell *et al.*, 1999b). Further evidence for a functional relationship between APP and the Reelin–ApoE receptor–Dab1 pathway came from genetic studies. A gene-mapping approach for quantitative trait loci (QTLs) that influence the variation of tau hyperphosphorylation in Dab1 knockout mice on different strain backgrounds identified a highly significant QTL on mouse chromosome 16 centered over the *App* gene (Brich *et al.*, 2003). In support of this, a genetic interaction between *appl* (amyloid precursor protein-like), the *Drosophila* homologue of *APP*, and Disabled has been described in two studies (Merdes *et al.*, 2004; Pramatarova *et al.*, 2006). Interestingly, Reelin and F-spondin, an extracellular matrix protein that binds to APP (Ho and Sudhof, 2004) and ApoER2 (Hoe *et al.*, 2005), alter the trafficking and proteolytic processing of ApoER2 and APP (Morimura *et al.*, 2005; Hoe *et al.*, 2006b). Of note, other ApoE receptors, such as LRP1, LRP1b, and the distantly related receptor SorLA, regulate the endocytosis and intracellular trafficking of APP in a partly opposing manner as well (Andersen *et al.*, 2005; Andersen and Willnow, 2006; Cam and Bu, 2006). An additional level of complexity in the interactions between ApoE, its receptors, and APP comes from the putative transcriptional activity of the intracellular transmembrane receptor tails released by gamma secretase activity (Cao and Sudhof, 2001; Kinoshita *et al.*, 2003).

2.4 Control of Synaptic Functions by ApoER2 and VLDLR in the Adult Brain

Most of the studies on the functions of ApoER2 and VLDLR as Reelin receptors have focused on the role of this signaling cascade in the developing brain. However, the molecules that are essential to this pathway continue to be highly expressed in the adult brain, suggesting additional functions beyond neurodevelopment. In this section, we will describe how Reelin and its ApoE receptors contribute to glutamatergic signaling at the synapse.

2.4.1 ApoER2 and VLDLR Modulate Neurotransmission

The emerging link between ApoE, cholesterol, and brain physiology (Mahley, 1988; Dietschy and Turley, 2004) sparked increasing interest in the possible regulation of neurotransmission by ApoE receptors, especially by LRP1, a multifunctional LDL receptor gene family member that is highly expressed in the brain (Bacskai *et al.*, 2000; Zhuo *et al.*, 2000; May *et al.*, 2004). Defects in long-term potentiation (LTP), considered as an electrophysiological correlate of learning and memory, were recorded in hippocampal slices of mice lacking either Vldlr or ApoER2, with ApoER2 knockouts showing a more pronounced decay (Weeber *et al.*, 2002). These defects were attributable to a defect in Reelin signaling, because perfusion of hippocampal slices with Reelin augments baseline LTP in wild-type, but not in ApoER2- or Vldlr-deficient mice. Both mutants also displayed behavioral deficiencies in fear-conditioned memory tests (Weeber *et al.*, 2002). Mice that express a mutated form of ApoER2 with a defective intracellular Dab1 binding site show a severe LTP deficit and lack a Reelin-induced LTP enhancement, indicating a critical role for the Dab1 adapter protein (Beffert *et al.*, 2006b). However, Reelin prevented the rapid degradation of LTP seen in mice expressing Dab1 binding-mutant ApoER2, which suggests the involvement of additional signaling mechanisms in the Reelin-mediated, ApoER2-dependent increase of synaptic plasticity (Beffert *et al.*, 2006b).

2.4.2 Molecular Mechanisms of ApoE Receptor Signaling at the Synapse

The generation of knockin mice that constitutively either express or lack the proline-rich intracellular sequence encoded by the alternatively spliced exon 19 (plus-insert and minus-insert mice) provided important insights into the molecular mechanisms that are involved in the Reelin-dependent modulation of LTP and behavior

(Beffert *et al.*, 2005). The ApoER2 insert mediates the interaction of ApoER2 with the modular adapter proteins JIP1 and JIP2 (Gotthardt *et al.*, 2000; Stockinger *et al.*, 2000), which can aggregate components of a mitogen-activated protein (MAP) kinase module on the intracellular receptor tail (Yasuda *et al.*, 1999) and can also bind to the NPXY motif of APP with their carboxyl-terminal PTB domain (Taru *et al.*, 2002). Furthermore, it mediates the splice form-specific interaction of ApoER2 with postsynaptic density-95 (PSD95) (Gotthardt *et al.*, 2000; Beffert *et al.*, 2005), a multidomain synaptic scaffolding protein that also interacts with another LDL receptor gene family member, LRP1 (May *et al.*, 2004). Whereas baseline LTP was not dependent on the constitutive absence or presence of the alternatively spliced exon 19, the Reelin-mediated increase in LTP was absent in the minus-insert mice (Beffert *et al.*, 2005). Interestingly, the alternatively spliced ApoER2 tail insert is dispensable during neurodevelopment, as indicated by the absence of neuronal positioning defects in the minus-insert knockin mice; accordingly, the activation of Dab1 by Reelin was unaltered in these mice (Beffert *et al.*, 2005). The intracellular proline-rich sequence of ApoER2 seems to be required for the Reelin-induced and Dab1-dependent activation and recruitmemt of the SFK Fyn to the *N*-methyl-D-aspartate (NMDA) receptor complex, whose activity is regulated by the tyrosine phosphorylation of its subunits NR2A and NR2B (Beffert *et al.*, 2005; Chen *et al.*, 2005; Qiu *et al.*, 2006). In addition to the PSD95-dependent recruitment of the insert-containing receptor variant to the postsynaptic density (Beffert *et al.*, 2005), coupling of ApoER2 to the NMDA receptor complex can also occur in the absence of the insert, by interactions of the extracellular domains of ApoER2 and the NR1 subunit of NMDAR (Hoe *et al.*, 2006a). The regulation of exon 19 alternative splicing by physical activity (Beffert *et al.*, 2005), the possible interference of secreted soluble ApoER2 (Koch *et al.*, 2002) with NMDAR, the coupling of the insert-containing form of ApoER2 with the kinesin motor via JIP1/2, the isoform-specific modulation of ApoER2–ligand interaction by ApoE, and the association of additional LDL receptor gene family members, including LRP1 (May *et al.*, 2004), with postsynaptic NMDA receptors illustrate the high level of complexity of ApoE receptor signaling at the synapse (Fig. 2.3). Potential implications of this emerging field for neurological and psychiatric human disorders, including neurodegeneration, schizophrenia, autism, and temporal lobe epilepsy, are discussed in detail in several other chapters of this volume (see chapters by Lintas and Persico, Fatemi *et al.*, Costa *et al.*, Qiu and Weeber, and Abdolmaleky *et al.*).

3 Perspective

Although significant progress in the understanding of the role of ApoER2 and VLDLR as Reelin receptors during neurodevelopment and in the regulation of synaptic plasticity has been achieved within the past few years, and ascribing novel and unanticipitated functions to LDL receptor gene family members in signal transduction, important questions remain unsolved. For example, it is not clear why both

ApoER2 and VLDLR are required for Reelin-induced enhancement of LTP, whereas both receptors can compensate at least partially for each other during neuronal positioning in the developing brain. Another unresolved issue is the involvement of co-factors, which could represent co-receptors, co-ligands, or additional intracellular adapter molecules, in the modulation of the Reelin response. As mentioned above, induction of Dab1 tyrosine phosphorylation, albeit absolutely essential for the regulation of neuronal positioning in the neocortex (Howell *et al.*, 2000), is not sufficient to correct the reeler phenotype in embryonic neocortical slices (Jossin *et al.*, 2004). How do the Reelin–Dab1 and the Cdk5 signaling pathways interact? Interestingly, genetic ablation of either of the Cdk5 activators p35 or p39 inhibits the effect of Reelin on LTP, although no discernible effect of Cdk5 inhibition on Reelin signaling was observed (Beffert *et al.*, 2004).

Finally, the biochemical and functional interaction of ApoER2, VLDLR, and LRP1 with the NMDA receptor at the synapse points toward the crucial question of how the different isoforms of ApoE, its ApoE receptors, and modulating ligands such as Reelin or amyloid-beta, which regulates NMDA receptor trafficking (Snyder *et al.*, 2005), cooperate to control synaptic transmission at the molecular and cellular level (Fig 2.3).

References

Andersen, O. M., and Willnow, T. E. (2006). Lipoprotein receptors in Alzheimer's disease. *Trends Neurosci.* 29:687–694.

Andersen, O. M., Yeung, C. H., Vorum, H., Wellner, M., Andreassen, T. K., Erdmann, B., Mueller, E. C., Herz, J., Otto, A., Cooper, T. G., and Willnow, T. E. (2003). Essential role of the apolipoprotein E receptor-2 in sperm development. *J. Biol. Chem.* 278:23989–23995.

Andersen, O. M., Reiche, J., Schmidt, V., Gotthardt, M., Spoelgen, R., Behlke, J., von Arnim, C. A., Breiderhoff, T., Jansen, P., Wu, X., Bales, K. R., Cappai, R., Masters, C. L., Gliemann, J., Mufson, E. J., Hyman, B. T., Paul, S. M., Nykjaer, A., and Willnow, T. E. (2005). Neuronal sorting protein-related receptor sorLA/LR11 regulates processing of the amyloid precursor protein. *Proc. Natl. Acad. Sci. USA* 102:13461–13466.

Arnaud, L., Ballif, B. A., and Cooper, J. A. (2003a). Regulation of protein tyrosine kinase signaling by substrate degradation during brain development. *Mol. Cell. Biol* 23:9293–9302.

Arnaud, L., Ballif, B. A., Forster, E., and Cooper, J. A. (2003b). Fyn tyrosine kinase is a critical regulator of disabled-1 during brain development. *Curr. Biol.* 13:9–17.

Assadi, A. H., Zhang, G., Beffert, U., McNeil, R. S., Renfro, A. L., Niu, S., Quattrocchi, C. C., Antalffy, B. A., Sheldon, M., Armstrong, D. D., Wynshaw-Boris, A., Herz, J., D'Arcangelo, G., and Clark, G. D. (2003). Interaction of reelin signaling and Lis1 in brain development. *Nature Genet.* 35:270–276.

Bacskai, B. J., Xia, M. Q., Strickland, D. K., Rebeck, G. W., and Hyman, B. T. (2000). The endocytic receptor protein LRP also mediates neuronal calcium signaling via N-methyl-D-aspartate receptors. *Proc. Natl. Acad. Sci. USA* 97:11551–11556.

Ballif, B. A., Arnaud, L., and Cooper, J. A. (2003). Tyrosine phosphorylation of disabled-1 is essential for reelin-stimulated activation of Akt and Src family kinases. *Brain Res. Mol. Brain Res.* 117:152–159.

Ballif, B. A., Arnaud, L., Arthur, W. T., Guris, D., Imamoto, A., and Cooper, J. A. (2004). Activation of a Dab1/CrkL/C3G/Rap1 pathway in reelin-stimulated neurons. *Curr. Biol.* 14:606–610.

Beffert, U., Morfini, G., Bock, H. H., Reyna, H., Brady, S. T., and Herz, J. (2002). Reelin-mediated signaling locally regulates protein kinase B/Akt and glycogen synthase kinase 3beta. *J. Biol. Chem.* 277:49958–49964.

Beffert, U., Weeber, E. J., Morfini, G., Ko, J., Brady, S. T., Tsai, L. H., Sweatt, J. D., and Herz, J. (2004). Reelin and cyclin-dependent kinase 5-dependent signals cooperate in regulating neuronal migration and synaptic transmission. *J. Neurosci.* 24:1897–1906.

Beffert, U., Weeber, E. J., Durudas, A., Qiu, S., Masiulis, I., Sweatt, J. D., Li, W. P., Adelmann, G., Frotscher, M., Hammer, R. E., and Herz, J. (2005). Modulation of synaptic plasticity and memory by reelin involves differential splicing of the lipoprotein receptor apoer2. *Neuron* 47:567–579.

Beffert, U., Nematollah Farsian, F., Masiulis, I., Hammer, R. E., Yoon, S. O., Giehl, K. M., and Herz, J. (2006a). ApoE receptor 2 controls neuronal survival in the adult brain. *Curr. Biol.* 16:2446–2452.

Beffert, U., Durudas, A., Weeber, E. J., Stolt, P. C., Giehl, K. M., Sweatt, J. D., Hammer, R. E., and Herz, J. (2006b). Functional dissection of reelin signaling by site-directed disruption of disabled-1 adaptor binding to apolipoprotein E receptor 2: Distinct roles in development and synaptic plasticity. *J. Neurosci.* 26:2041–2052.

Benhayon, D., Magdaleno, S., and Curran, T. (2003). Binding of purified reelin to ApoER2 and VLDLR mediates tyrosine phosphorylation of disabled-1. *Brain Res. Mol. Brain Res.* 112:33–45.

Bock, H. H., and Herz, J. (2003). Reelin activates Src family tyrosine kinases in neurons. *Curr. Biol.* 13:18–26.

Bock, H. H., Jossin, Y., Liu, P., Forster, E., May, P., Goffinet, A. M., and Herz, J. (2003). Phosphatidylinositol 3-kinase interacts with the adaptor protein Dab1 in response to reelin signaling and is required for normal cortical lamination. *J. Biol. Chem.* 278:38772–38779.

Bock, H. H., Jossin, Y., May, P., Bergner, O., and Herz, J. (2004). Apolipoprotein E receptors are required for reelin-induced proteasomal degradation of the neuronal adaptor protein disabled-1. *J. Biol. Chem.* 279:33471–33479.

Bonifacino, J. S., and Traub, L. M. (2003). Signals for sorting of transmembrane proteins to endosomes and lysosomes. *Annu. Rev. Biochem.* 72:395–447.

Bos, J. L. (2005). Linking Rap to cell adhesion. *Curr. Opin. Cell Biol.* 17:123–128.

Boycott, K. M., Flavelle, S., Bureau, A., Glass, H. C., Fujiwara, T. M., Wirrell, E., Davey, K., Chudley, A. E., Scott, J. N., McLeod, D. R., and Parboosingh, J. S. (2005). Homozygous deletion of the very low density lipoprotein receptor gene causes autosomal recessive cerebellar hypoplasia with cerebral gyral simplification. *Am. J. Hum. Genet.* 77:477–483.

Boyles, J. K., Pitas, R. E., Wilson, E., Mahley, R. W., and Taylor, J. M. (1985). Apolipoprotein E associated with astrocytic glia of the central nervous system and with nonmyelinating glia of the peripheral nervous system. *J. Clin. Invest.* 76:1501–1513.

Boyles, J. K., Zoellner, C. D., Anderson, L. J., Kosik, L. M., Pitas, R. E., Weisgraber, K. H., Hui, D. Y., Mahley, R. W., Gebicke-Haerter, P. J., Ignatius, M. J., and Shooter, E. M. (1989). A role for apolipoprotein E, apolipoprotein A-I, and low density lipoprotein receptors in cholesterol transport during regeneration and remyelination of the rat sciatic nerve. *J. Clin. Invest.* 83:1015–1031.

Brandes, C., Novak, S., Stockinger, W., Herz, J., Schneider, W. J., and Nimpf, J. (1997). Avian and murine LR8B and human apolipoprotein E receptor 2: differentially spliced products from corresponding genes. *Genomics* 42:185–191.

Brandes, C., Kahr, L., Stockinger, W., Hiesberger, T., Schneider, W. J., and Nimpf, J. (2001). Alternative splicing in the ligand binding domain of mouse ApoE receptor-2 produces receptor variants binding reelin but not alpha 2-macroglobulin. *J. Biol. Chem.* 276:22160–22169.

Brich, J., Shie, F. S., Howell, B. W., Li, R., Tus, K., Wakeland, E. K., Jin, L. W., Mumby, M., Churchill, G., Herz, J., and Cooper, J. A. (2003). Genetic modulation of tau phosphorylation in the mouse. *J. Neurosci.* 23:187–192.

Bu, G., and Schwartz, A. L. (1998). RAP, a novel type of ER chaperone. *Trends Cell Biol.* 8:272–276.

Calderwood, D. A., Fujioka, Y., de Pereda, J. M., Garcia-Alvarez, B., Nakamoto, T., Margolis, B., McGlade, C. J., Liddington, R. C., and Ginsberg, M. H. (2003). Integrin beta cytoplasmic domain interactions with phosphotyrosine-binding domains: A structural prototype for diversity in integrin signaling. *Proc. Natl. Acad. Sci. USA*100:2272–2277.

Cam, J. A., and Bu, G. (2006). Modulation of beta-amyloid precursor protein trafficking and processing by the low density lipoprotein receptor family. *Mol. Neurodegener*. 1:8.

Cao, X., and Sudhof, T. C. (2001). A transcriptionally [correction of transcriptively] active complex of APP with Fe65 and histone acetyltransferase Tip60. *Science* 293:115–120.

Cariboni, A., Rakic, S., Liapi, A., Maggi, R., Goffinet, A., and Parnavelas, J. G. (2005). Reelin provides an inhibitory signal in the migration of gonadotropin-releasing hormone neurons. *Development* 132:4709–4718.

Chen, K., Ochalski, P. G., Tran, T. S., Sahir, N., Schubert, M., Pramatarova, A., and Howell, B. W. (2004). Interaction between Dab1 and CrkII is promoted by reelin signaling. *J. Cell Sci.* 117:4527–4536.

Chen, Y., Beffert, U., Ertunc, M., Tang, T. S., Kavalali, E. T., Bezprozvanny, I., and Herz, J. (2005). Reelin modulates NMDA receptor activity in cortical neurons. *J. Neurosci.* 25:8209–8216.

Clatworthy, A. E., Stockinger, W., Christie, R. H., Schneider, W. J., Nimpf, J., Hyman, B. T., and Rebeck, G. W. (1999). Expression and alternate splicing of apolipoprotein E receptor 2 in brain. *Neuroscience* 90:903–911.

Cooper, J. A., and Howell, B. W. (1999). Lipoprotein receptors: signaling functions in the brain? *Cell* 97:671–674.

D'Arcangelo, G., Homayouni, R., Keshvara, L., Rice, D. S., Sheldon, M., and Curran, T. (1999). Reelin is a ligand for lipoprotein receptors. *Neuron* 24:471–479.

Dhavan, R., and Tsai, L. H. (2001). A decade of CDK5. *Nature Rev. Mol. Cell Biol.* 2:749–759.

Dietschy, J. M., and Turley, S. D. (2004). Thematic review series: Brain lipids. Cholesterol metabolism in the central nervous system during early development and in the mature animal. *J. Lipid Res.* 45:1375–1397.

Dulabon, L., Olson, E. C., Taglienti, M. G., Eisenhuth, S., McGrath, B., Walsh, C. A., Kreidberg, J. A., and Anton, E. S. (2000). Reelin binds alpha3beta1 integrin and inhibits neuronal migration. *Neuron* 27:33–44.

Feng, Y., and Walsh, C. A. (2001). Protein–protein interactions, cytoskeletal regulation and neuronal migration. *Nature Rev. Neurosci.* 2:408–416.

Forster, E., Tielsch, A., Saum, B., Weiss, K. H., Johanssen, C., Graus-Porta, D., Muller, U., and Frotscher, M. (2002). Reelin, disabled 1, and beta 1 integrins are required for the formation of the radial glial scaffold in the hippocampus. *Proc. Natl. Acad. Sci. USA*99:13178–13183.

Frykman, P. K., Brown, M. S., Yamamoto, T., Goldstein, J. L., and Herz, J. (1995). Normal plasma lipoproteins and fertility in gene-targeted mice homozygous for a disruption in the gene encoding very low density lipoprotein receptor. *Proc. Natl. Acad. Sci. USA* 92:8453–8457.

Goedert, M., and Spillantini, M. G. (2006). A century of Alzheimer's disease. *Science* 314:777–781.

Gotthardt, M., Trommsdorff, M., Nevitt, M. F., Shelton, J., Richardson, J. A., Stockinger, W., Nimpf, J., and Herz, J. (2000). Interactions of the low density lipoprotein receptor gene family with cytosolic adaptor and scaffold proteins suggest diverse biological functions in cellular communication and signal transduction. *J. Biol. Chem.* 275:25616–25624.

Goudriaan, J. R., Tacken, P. J., Dahlmans, V. E., Gijbels, M. J., van Dijk, K. W., Havekes, L. M., and Jong, M. C. (2001). Protection from obesity in mice lacking the VLDL receptor. *Arterioscler. Thromb. Vasc. Biol.* 21:1488–1493.

Grant, S. G., O'Dell, T. J., Karl, K. A., Stein, P. L., Soriano, P., and Kandel, E. R. (1992). Impaired long-term potentiation, spatial learning, and hippocampal development in fyn mutant mice. *Science* 258:1903–1910.

Heckenlively, J. R., Hawes, N. L., Friedlander, M., Nusinowitz, S., Hurd, R., Davisson, M., and Chang, B. (2003). Mouse model of subretinal neovascularization with choroidal anastomosis. *Retina* 23:518–522.

Helbecque, N., and Amouyel, P. (2000). Very low density lipoprotein receptor in Alzheimer disease. *Microsc. Res. Tech.* 50:273–277.
Helbecque, N., Berr, C., Cottel, D., Fromentin-David, I., Sazdovitch, V., Ricolfi, F., Ducimetiere, P., Di Menza, C., and Amouyel, P. (2001). VLDL receptor polymorphism, cognitive impairment, and dementia. *Neurology* 56:1183–1188.
Herz, J. (2006). The switch on the RAPper's necklace. *Mol. Cell* 23:451–455.
Herz, J., and Beffert, U. (2000). Apolipoprotein E receptors: linking brain development and Alzheimer's disease. *Nature Rev. Neurosci.* 1:51–58.
Herz, J., and Bock, H. H. (2002). Lipoprotein receptors in the nervous system. *Annu. Rev. Biochem.* 71:405–434.
Herz, J., and Chen, Y. (2006). Reelin, lipoprotein receptors and synaptic plasticity. *Nature Rev. Neurosci.* 7:850–859.
Hiesberger, T., Trommsdorff, M., Howell, B. W., Goffinet, A., Mumby, M. C., Cooper, J. A., and Herz, J. (1999). Direct binding of reelin to VLDL receptor and ApoE receptor 2 induces tyrosine phosphorylation of disabled-1 and modulates tau phosphorylation. *Neuron* 24: 481–489.
Ho, A., and Sudhof, T. C. (2004). Binding of F-spondin to amyloid-beta precursor protein: a candidate amyloid-beta precursor protein ligand that modulates amyloid-beta precursor protein cleavage. *Proc. Natl. Acad. Sci. USA* 101:2548–2553.
Hoe, H. S., and Rebeck, G. W. (2005). Regulation of ApoE receptor proteolysis by ligand binding. *Brain Res. Mol. Brain Res.* 137:31–39.
Hoe, H. S., Wessner, D., Beffert, U., Becker, A. G., Matsuoka, Y., and Rebeck, G. W. (2005). F-spondin interaction with the apolipoprotein E receptor ApoEr2 affects processing of amyloid precursor protein. *Mol. Cell Biol.* 25:9259–9268.
Hoe, H. S., Pocivavsek, A., Chakraborty, G., Fu, Z., Vicini, S., Ehlers, M. D., and Rebeck, G. W. (2006a). Apolipoprotein E receptor 2 interactions with the N-methyl-D-aspartate receptor. *J. Biol. Chem.* 281:3425–3431.
Hoe, H. S., Tran, T. S., Matsuoka, Y., Howell, B. W., and Rebeck, G. W. (2006b). DAB1 and reelin effects on amyloid precursor protein and ApoE receptor 2 trafficking and processing. *J. Biol. Chem.* 281:35176–35185.
Homayouni, R., Rice, D. S., Sheldon, M., and Curran, T. (1999). Disabled-1 binds to the cytoplasmic domain of amyloid precursor-like protein 1. *J. Neurosci.* 19:7507–7515.
Hong, S. E., Shugart, Y. Y., Huang, D. T., Shahwan, S. A., Grant, P. E., Hourihane, J. O., Martin, N. D., and Walsh, C. A. (2000). Autosomal recessive lissencephaly with cerebellar hypoplasia is associated with human RELN mutations. *Nature Genet.* 26:93–96.
Howell, B. W., Gertler, F. B., and Cooper, J. A. (1997a). Mouse disabled (mDab1): a Src binding protein implicated in neuronal development. *EMBO J.* 16:121–132.
Howell, B. W., Hawkes, R., Soriano, P., and Cooper, J. A. (1997b). Neuronal position in the developing brain is regulated by mouse disabled-1. *Nature* 389:733–737.
Howell, B. W., Herrick, T. M., and Cooper, J. A. (1999a). Reelin-induced tryosine phosphorylation of disabled 1 during neuronal positioning. *Genes Dev.* 13:643–648.
Howell, B. W., Lanier, L. M., Frank, R., Gertler, F. B., and Cooper, J. A. (1999b). The disabled 1 phosphotyrosine-binding domain binds to the internalization signals of transmembrane glycoproteins and to phospholipids. *Mol. Cell Biol.* 19:5179–5188.
Howell, B. W., Herrick, T. M., Hildebrand, J. D., Zhang, Y., and Cooper, J. A. (2000). Dab1 tyrosine phosphorylation sites relay positional signals during mouse brain development. *Curr. Biol.* 10:877–885.
Huang, Y., Magdaleno, S., Hopkins, R., Slaughter, C., Curran, T., and Keshvara, L. (2004). Tyrosine phosphorylated disabled 1 recruits Crk family adapter proteins. *Biochem. Biophys. Res. Commun.* 318:204–212.
Huang, Y., Shah, V., Liu, T., and Keshvara, L. (2005). Signaling through disabled 1 requires phosphoinositide binding. *Biochem. Biophys. Res. Commun.* 331:1460–1468.
Ignatius, M. J., Shooter, E. M., Pitas, R. E., and Mahley, R. W. (1987). Lipoprotein uptake by neuronal growth cones in vitro. *Science* 236:959–962.

Jossin, Y., Ogawa, M., Metin, C., Tissir, F., and Goffinet, A. M. (2003). Inhibition of SRC family kinases and non-classical protein kinases C induce a reeler-like malformation of cortical plate development. *J. Neurosci.* 23:9953–9959.

Jossin, Y., Ignatova, N., Hiesberger, T., Herz, J., Lambert de Rouvroit, C., and Goffinet, A. M. (2004). The central fragment of reelin, generated by proteolytic processing in vivo, is critical to its function during cortical plate development. *J. Neurosci.* 24:514–521.

Keshvara, L., Benhayon, D., Magdaleno, S., and Curran, T. (2001). Identification of reelin-induced sites of tyrosyl phosphorylation on disabled 1. *J. Biol. Chem.* 276:16008–16014.

Kinoshita, A., Shah, T., Tangredi, M. M., Strickland, D. K., and Hyman, B. T. (2003). The intracellular domain of the low density lipoprotein receptor-related protein modulates transactivation mediated by amyloid precursor protein and Fe65. *J. Biol. Chem.* 278:41182–41188.

Koch, S., Strasser, V., Hauser, C., Fasching, D., Brandes, C., Bajari, T. M., Schneider, W. J., and Nimpf, J. (2002). A secreted soluble form of ApoE receptor 2 acts as a dominant-negative receptor and inhibits reelin signaling. **EMBO** J. 21:5996–6004.

Kuo, G., Arnaud, L., Kronstad-O'Brien, P., and Cooper, J. A. (2005). Absence of Fyn and Src causes a reeler-like phenotype. *J. Neurosci.* 25:8578–8586.

Ma, S. L., Ng, H. K., Baum, L., Pang, J. C., Chiu, H. F., Woo, J., Tang, N. L., and Lam, L. C. (2002). Low-density lipoprotein receptor-related protein 8 (apolipoprotein E receptor 2) gene polymorphisms in Alzheimer's disease. *Neurosci. Lett.* 332:216–218.

Mahley, R. W. (1988). Apolipoprotein E: cholesterol transport protein with expanding role in cell biology. *Science* 240:622–630.

Mahley, R. W., and Rall, S. C., Jr. (2000). Apolipoprotein E: far more than a lipid transport protein. *Annu. Rev. Genomics Hum. Genet.* 1:507–537.

Mahley, R. W., Innerarity, T. L., Rall, S. C., Jr., and Weisgraber, K. H. (1984). Plasma lipoproteins: apolipoprotein structure and function. *J. Lipid Res.* 25:1277–1294.

Mahley, R. W., Weisgraber, K. H., and Huang, Y. (2006). Apolipoprotein E4: a causative factor and therapeutic target in neuropathology, including Alzheimer's disease. *Proc. Natl. Acad. Sci. USA* 103:5644–5651.

Mauch, D. H., Nagler, K., Schumacher, S., Goritz, C., Muller, E. C., Otto, A., and Pfrieger, F. W. (2001). CNS synaptogenesis promoted by glia-derived cholesterol. *Science* 294:1354–1357.

May, P., and Herz, J. (2003). LDL receptor-related proteins in neurodevelopment. *Traffic* 4:291–301.

May, P., Reddy, Y. K., and Herz, J. (2002). Proteolytic processing of low density lipoprotein receptor-related protein mediates regulated release of its intracellular domain. *J. Biol. Chem.* 277:18736–18743.

May, P., Bock, H. H., Nimpf, J., and Herz, J. (2003). Differential glycosylation regulates processing of lipoprotein receptors by gamma-secretase. *J. Biol. Chem.* 278:37386–37392.

May, P., Rohlmann, A., Bock, H. H., Zurhove, K., Marth, J. D., Schomburg, E. D., Noebels, J. L., Beffert, U., Sweatt, J. D., Weeber, E. J., and Herz, J. (2004). Neuronal LRP1 functionally associates with postsynaptic proteins and is required for normal motor function in mice. *Mol. Cell Biol.* 24:8872–8883.

May, P., Herz, J., and Bock, H. H. (2005). Molecular mechanisms of lipoprotein receptor signalling. *Cell. Mol. Life Sci.* 62:2325–2338.

Mayer, H., Duit, S., Hauser, C., Schneider, W. J., and Nimpf, J. (2006). Reconstitution of the reelin signaling pathway in fibroblasts demonstrates that Dab1 phosphorylation is independent of receptor localization in lipid rafts. *Mol. Cell Biol.* 26:19–27.

Merdes, G., Soba, P., Loewer, A., Bilic, M. V., Beyreuther, K., and Paro, R. (2004). Interference of human and Drosophila APP and APP-like proteins with PNS development in Drosophila. *EMBO J.* 23:4082–4095.

Morimura, T., Hattori, M., Ogawa, M., and Mikoshiba, K. (2005). Disabled1 regulates the intracellular trafficking of reelin receptors. *J. Biol. Chem.* 280:16901–16908.

Niethammer, M., Smith, D. S., Ayala, R., Peng, J., Ko, J., Lee, M. S., Morabito, M., and Tsai, L. H. (2000). NUDEL is a novel Cdk5 substrate that associates with LIS1 and cytoplasmic dynein. *Neuron* 28:697–711.

Nykjaer, A., and Willnow, T. E. (2002). The low-density lipoprotein receptor gene family: a cellular Swiss army knife? *Trends Cell Biol.* 12:273–280.

Ohshima, T., Ogawa, M., Veeranna, Hirasawa, M., Longenecker, G., Ishiguro, K., Pant, H. C., Brady, R. O., Kulkarni, A. B., and Mikoshiba, K. (2001). Synergistic contributions of cyclin-dependent kinase 5/p35 and reelin/Dab1 to the positioning of cortical neurons in the developing mouse brain. *Proc. Natl. Acad. Sci. USA* 98:2764–2769.

Ohshima, T., Suzuki, H., Morimura, T., Ogawa, M., and Mikoshiba, K. (2007). Modulation of reelin signaling by cyclin-dependent kinase 5. *Brain Res.* 1140:84–95.

Okuizumi, K., Onodera, O., Namba, Y., Ikeda, K., Yamamoto, T., Seki, K., Ueki, A., Nanko, S., Tanaka, H., Takahashi, H., Oyanagi, K., Mizusawa, H., Kanazawa, I., and Tsuji, S. (1995). Genetic association of the very low density lipoprotein (VLDL) receptor gene with sporadic Alzheimer's disease. *Nature Genet.* 11:207–209.

Pitas, R. E., Boyles, J. K., Lee, S. H., Hui, D., and Weisgraber, K. H. (1987). Lipoproteins and their receptors in the central nervous system. Characterization of the lipoproteins in cerebrospinal fluid and identification of apolipoprotein B,E(LDL) receptors in the brain. *J. Biol. Chem.* 262:14352–14360.

Pramatarova, A., Ochalski, P. G., Chen, K., Gropman, A., Myers, S., Min, K. T., and Howell, B. W. (2003). Nck beta interacts with tyrosine-phosphorylated disabled 1 and redistributes in reelin-stimulated neurons. *Mol. Cell Biol.* 23:7210–7221.

Pramatarova, A., Ochalski, P. G., Lee, C. H., and Howell, B. W. (2006). Mouse disabled 1 regulates the nuclear position of neurons in a Drosophila eye model. *Mol. Cell Biol.* 26:1510–1517.

Qiu, S., Zhao, L. F., Korwek, K. M., and Weeber, E. J. (2006). Differential reelin-induced enhancement of NMDA and AMPA receptor activity in the adult hippocampus. *J. Neurosci.* 26:12943–12955.

Rice, D. S., Sheldon, M., D'Arcangelo, G., Nakajima, K., Goldowitz, D., and Curran, T. (1998). Disabled-1 acts downstream of reelin in a signaling pathway that controls laminar organization in the mammalian brain. *Development* 125:3719–3729.

Roses, A. D. (1996). Apolipoprotein E alleles as risk factors in Alzheimer's disease. *Annu. Rev. Med.* 47:387–400.

Rossel, M., Loulier, K., Feuillet, C., Alonso, S., and Carroll, P. (2005). Reelin signaling is necessary for a specific step in the migration of hindbrain efferent neurons. *Development* 132:1175–1185.

Sanada, K., Gupta, A., and Tsai, L. H. (2004). Disabled-1-regulated adhesion of migrating neurons to radial glial fiber contributes to neuronal positioning during early corticogenesis. *Neuron* 42:197–211.

Sasaki, S., Shionoya, A., Ishida, M., Gambello, M.J., Yingling, J., Wynshaw-Boris, A., and Hirotsune, S. (2000). A LIS1/NUDEL/cytoplasmic dynein heavy chain complex in the developing and adult nervous system. *Neuron* 28:681–696.

Sato, N., Fukushima, N., Chang, R., Matsubayashi, H., and Goggins, M. (2006). Differential and epigenetic gene expression profiling identifies frequent disruption of the RELN pathway in pancreatic cancers. *Gastroenterology* 130:548–565.

Schneider, W. J., and Nimpf, J. (2003). LDL receptor relatives at the crossroad of endocytosis and signaling. *Cell. Mol. Life Sci.* 60:892–903.

Schneider, W. J., Nimpf, J., and Bujo, H. (1997). Novel members of the low density lipoprotein receptor superfamily and their potential roles in lipid metabolism. *Curr. Opin. Lipidol.* 8:315–319.

Sheldon, M., Rice, D. S., D'Arcangelo, G., Yoneshima, H., Nakajima, K., Mikoshiba, K., Howell, B. W., Cooper, J. A., Goldowitz, D., and Curran, T. (1997). Scrambler and yotari disrupt the disabled gene and produce a reeler-like phenotype in mice. *Nature* 389:730–733.

Sinha, S., and Lieberburg, I. (1999). Cellular mechanisms of beta-amyloid production and secretion. *Proc. Natl. Acad. Sci. USA* 96:11049–11053.

Snyder, E. M., Nong, Y., Almeida, C. G., Paul, S., Moran, T., Choi, E. Y., Nairn, A. C., Salter, M. W., Lombroso, P. J., Gouras, G. K., and Greengard, P. (2005). Regulation of NMDA receptor trafficking by amyloid-beta. *Nature Neurosci.* 8:1051–1058.

Stockinger, W., Brandes, C., Fasching, D., Hermann, M., Gotthardt, M., Herz, J., Schneider, W. J., and Nimpf, J. (2000). The reelin receptor ApoER2 recruits JNK-interacting proteins-1 and -2. *J. Biol. Chem.* 275:25625–25632.

Stolt, P. C., and Bock, H. H. (2006). Modulation of lipoprotein receptor functions by intracellular adaptor proteins. *Cell Signal* 18:1560–1571.

Stolt, P. C., Jeon, H., Song, H. K., Herz, J., Eck, M. J., and Blacklow, S. C. (2003). Origins of peptide selectivity and phosphoinositide binding revealed by structures of disabled-1 PTB domain complexes. *Structure (Cambridge)* 11:569–579.

Stolt, P. C., Vardar, D., and Blacklow, S. C. (2004). The dual-function disabled-1 PTB domain exhibits site independence in binding phosphoinositide and peptide ligands. *Biochemistry* 43:10979–10987.

Stolt, P. C., Chen, Y., Liu, P., Bock, H. H., Blacklow, S. C., and Herz, J. (2005). Phosphoinositide binding by the disabled-1 PTB domain is necessary for membrane localization and reelin signal transduction. *J. Biol. Chem.* 280:9671–9677.

Strasser, V., Fasching, D., Hauser, C., Mayer, H., Bock, H. H., Hiesberger, T., Herz, J., Weeber, E. J., Sweatt, J. D., Pramatarova, A., Howell, B., Schneider, W. J., and Nimpf, J. (2004). Receptor clustering is involved in reelin signaling. *Mol. Cell Biol.* 24:1378–1386.

Strittmatter, W. J., and Bova Hill, C. (2002). Molecular biology of apolipoprotein E. *Curr. Opin. Lipidol.* 13:119–123.

Tacken, P. J., Teusink, B., Jong, M. C., Harats, D., Havekes, L. M., van Dijk, K. W., and Hofker, M. H. (2000). LDL receptor deficiency unmasks altered VLDL triglyceride metabolism in VLDL receptor transgenic and knockout mice. *J. Lipid Res.* 41:2055–2062.

Takada, H., Imoto, I., Tsuda, H., Nakanishi, Y., Sakakura, C., Mitsufuji, S., Hirohashi, S., and Inazawa, J. (2006). Genomic loss and epigenetic silencing of very-low-density lipoprotein receptor involved in gastric carcinogenesis. *Oncogene* 25:6554–6562.

Takahashi, S., Sakai, J., Fujino, T., Hattori, H., Zenimaru, Y., Suzuki, J., Miyamori, I., and Yamamoto, T. T. (2004). The very low-density lipoprotein (VLDL) receptor: characterization and functions as a peripheral lipoprotein receptor. *J. Atheroscler. Thromb.* 11:200–208.

Taru, H., Iijima, K., Hase, M., Kirino, Y., Yagi, Y., and Suzuki, T. (2002). Interaction of Alzheimer's beta-amyloid precursor family proteins with scaffold proteins of the JNK signaling cascade. *J. Biol. Chem.* 277:20070–20078.

Trommsdorff, M., Borg, J. P., Margolis, B., and Herz, J. (1998). Interaction of cytosolic adaptor proteins with neuronal apolipoprotein E receptors and the amyloid precursor protein. *J. Biol. Chem.* 273:33556–33560.

Trommsdorff, M., Gotthardt, M., Hiesberger, T., Shelton, J., Stockinger, W., Nimpf, J., Hammer, R. E., Richardson, J. A., and Herz, J. (1999). Reeler/disabled-like disruption of neuronal migration in knockout mice lacking the VLDL receptor and ApoE receptor 2. *Cell* 97:689–701.

Tsai, L. H., and Gleeson, J. G. (2005). Nucleokinesis in neuronal migration. *Neuron* 46:383–388.

Verhey, K. J., Meyer, D., Deehan, R., Blenis, J., Schnapp, B. J., Rapoport, T. A., and Margolis, B. (2001). Cargo of kinesin identified as JIP scaffolding proteins and associated signaling molecules. *J. Cell Biol.* 152:959–970.

Wang, L., Wang, X., Laird, N., Zuckerman, B., Stubblefield, P., and Xu, X. (2006). Polymorphism in maternal LRP8 gene is associated with fetal growth. *Am. J. Hum. Genet.* 78:770–777.

Ware, M. L., Fox, J. W., Gonzalez, J. L., Davis, N. M., Lambert de Rouvroit, C., Russo, C. J., Chua, S. C., Jr., Goffinet, A. M., and Walsh, C. A. (1997). Aberrant splicing of a mouse disabled homolog, mdab1, in the scrambler mouse. *Neuron* 19:239–249.

Weeber, E. J., Beffert, U., Jones, C., Christian, J. M., Forster, E., Sweatt, J. D., and Herz, J. (2002). Reelin and ApoE receptors cooperate to enhance hippocampal synaptic plasticity and learning. *J. Biol. Chem.* 277:39944–39952.

Willnow, T. E., Nykjaer, A., and Herz, J. (1999). Lipoprotein receptors: new roles for ancient proteins. *Nature Cell Biol.* 1:E157–162.

Wolfe, M. S. (2006). The gamma-secretase complex: membrane-embedded proteolytic ensemble. *Biochemistry* 45:7931–7939.

Xu, M., Arnaud, L., and Cooper, J. A. (2005). Both the phosphoinositide and receptor binding activities of Dab1 are required for reelin-stimulated Dab1 tyrosine phosphorylation. *Brain Res. Mol. Brain Res.* 139:300–305.

Yagi, T., Aizawa, S., Tokunaga, T., Shigetani, Y., Takeda, N., and Ikawa, Y. (1993). A role for Fyn tyrosine kinase in the suckling behaviour of neonatal mice. *Nature* 366:742–745.

Yamamoto, T., and Bujo, H. (1996). Close encounters with apolipoprotein E receptors. *Curr. Opin. Lipidol.* 7:298–302.

Yasuda, J., Whitmarsh, A. J., Cavanagh, J., Sharma, M., and Davis, R. J. (1999). The JIP group of mitogen-activated protein kinase scaffold proteins. *Mol. Cell Biol.* 19:7245–7254.

Yuasa, S., Hattori, K., and Yagi, T. (2004). Defective neocortical development in Fyn-tyrosine-kinase-deficient mice. *Neuroreport* 15:819–822.

Yun, M., Keshvara, L., Park, C. G., Zhang, Y. M., Dickerson, J. B., Zheng, J., Rock, C. O., Curran, T., and Park, H. W. (2003). Crystal structures of the Dab homology domains of mouse disabled 1 and 2. *J. Biol. Chem.* 278:36572–36581.

Zhuo, M., Holtzman, D. M., Li, Y., Osaka, H., DeMaro, J., Jacquin, M., and Bu, G. (2000). Role of tissue plasminogen activator receptor LRP in hippocampal long-term potentiation. *J. Neurosci.* 20:542–549.

Zou, Z., Chung, B., Nguyen, T., Mentone, S., Thomson, B., and Biemesderfer, D. (2004). Linking receptor-mediated endocytosis and cell signaling: evidence for regulated intramembrane proteolysis of megalin in proximal tubule. *J. Biol. Chem.* 279:34302–34310.

Chapter 3
Chemistry of Reelin

Yves Jossin

Contents

1 Introduction

Reelin is a large extracellular protein involved in several aspects of brain development, such as cell positioning, dendrite growth, synaptic plasticity, and memory, and may be implicated as a susceptibility factor in psychoses (Caviness and Rakic, 1978; Impagnatiello *et al.*, 1998; Liu *et al.*, 2001; Weeber *et al.*, 2002; Jossin, 2004; Beffert *et al.*, 2005; Fatemi, 2005). This wide array of functions indicates that Reelin is able to trigger different intracellular signaling pathways depending on the maturation state or the type of target cell that may express different receptors or intracellular signaling modules. In this chapter, I will review the current state of knowledge on the best established and some other putative partners of the Reelin pathway (Fig. 3.1).

2 Reelin

Reelin is a bulky glycoprotein, the structure of which starts with a large N-terminal segment that contains an F-spondin homology domain (163 residues) and a unique region (309 residues). The main body of Reelin consists of 8 repeats of 350–390

Y. Jossin
Université Catholique de Louvain, Faculté de Médecine, Developmental Neurobiology Unit, Avenue E. Mounier, 73 Box DENE7382, B1200 Brussels, Belgium
e-mail: Yves.Jossin@uclouvain.be
and
Fred Hutchinson Cancer Research Center, Division of Basic Sciences, 1100 Fairview Avenue North, PO Box 19024, Seattle, WA 98109
e-mail: yjossin@fhcrc.org

S. H. Fatemi (ed.), *Reelin Glycoprotein: Structure, Biology and Roles in Health and Disease.*

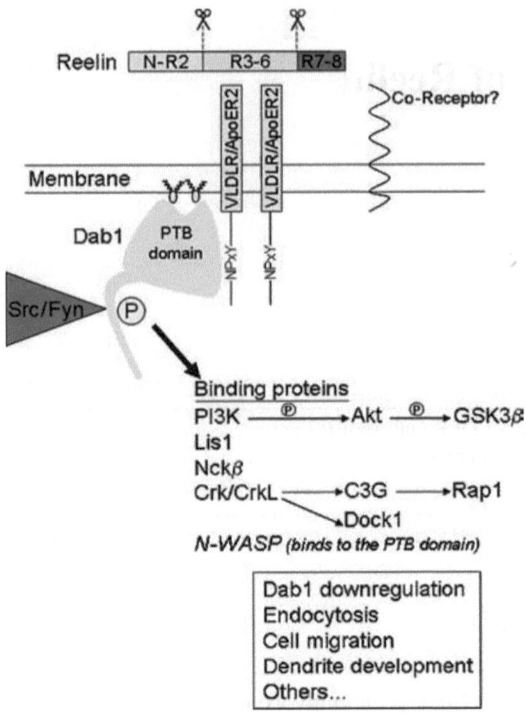

Fig. 3.1 Summary of the Reelin signaling pathway. See text for details (*See Color Plates*)

amino acids. Each repeat contains two related subdomains (A and B) flanking a pattern of conserved cysteine residues that form EGF-like motifs. The protein ends with a basic stretch of 33 amino acids (D'Arcangelo *et al.*, 1995) (Fig. 3.2). It is secreted by several different classes of cells, particularly Cajal-Retzius cells (CR) (Schiffmann *et al.*, 1997).

Reelin is processed in the extracellular environment at two sites by unidentified, and presumably redundant, metalloproteinases, resulting in the production of five processed fragments (Lambert de Rouvroit *et al.*, 1999) (Fig. 3.2). Because of the difficulty in estimating high-molecular-weight proteins in gels, those shown in Fig. 3.2 may slightly differ from previously published sizes. We estimated the sizes in Western blot using C-terminal antibodies that revealed three bands (approx. 450, 370, and 180 kDa), N-terminal antibodies that revealed three fragments (approx. 450, 270, and 80 kDa), and new unpublished antibodies (Jossin *et al.*, unpublished data) directed against epitopes in the central fragment that revealed four fragments (approx. 450, 370, 270, and 190 kDa). Processing does not decrease activity, as recombinant proteins made of the predicted central fragments (repeats 3–6) bind to lipoprotein receptors, trigger Disabled-1 (Dab1, see below) phosphorylation, and mimic functions of Reelin during cortical plate development (Jossin *et al.*, 2004). Electron micrographs of this active R3–6 fragment exhibit a rodlike shape (Nogi *et al.*, 2006). Rather

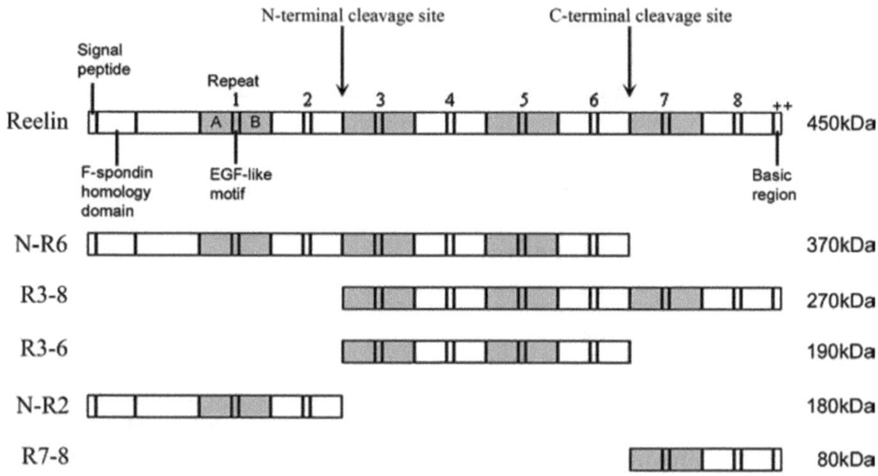

Fig. 3.2 Reelin is cleaved at two sites (N-t and C-t), thus generating five fragments named according to their composition. The corresponding size observed in SDS-PAGE gels is shown on the right-hand side. (Adapted from Jossin, 2004)

surprisingly, although a protein corresponding to the N-terminal fragment of Reelin (N-R2) does not bind to the lipoprotein receptors VLDLR and ApoER2, and a Reelin form amputated of the N-R2 fragment is biologically active (Jossin *et al.*, 2004), the CR50 antibody which is directed against an N-terminal epitope is able to block the action of Reelin (Miyata *et al.*, 1997). This antibody interferes with the aggregation of Reelin, which may affect its function (Utsunomiya-Tate *et al.*, 2000). Clearly, there are still many questions about the role of the different parts of the protein that remain to be studied further.

3 Reelin Receptors: VLDLR and ApoER2

Reelin binds directly to the ectodomains of the lipoprotein receptor Very-Low-Density Lipoprotein Receptor (VLDLR) and Apolipoprotein E Receptor Type 2 (ApoER2) (D'Arcangelo *et al.*, 1999; Hiesberger *et al.*, 1999), and the repeats 3 to 6 are necessary and sufficient for this binding (Jossin *et al.*, 2004). Both receptors are expressed by the neurons, in the cortex and elsewhere, that are Reelin target cells, and mice with double inactivation of the *Vldlr* and *Apoer2* genes have a reelerlike phenotype, whereas single receptor gene mutations generate subtle phenotypes (Trommsdorff *et al.*, 1999).

Mice lacking an alternatively spliced exon in the intracellular domain of ApoER2 have normal cortical development but perform poorly in learning and memory tasks. Reelin enhances LTP through ApoER2 in complex with NMDA receptors in postsynaptic densities of excitatory synapses by inducing the tyrosine phosphorylation of NMDA receptor subunits, which is dependent on the presence of this alternatively spliced exon (Beffert *et al.*, 2005).

4 Other Candidate Receptors

The involvement of a Reelin coreceptor in complex or in collaboration with VLDLR and ApoER2 has been proposed but never convincingly demonstrated. Reelin associates with two or more lipoprotein receptor molecules simultaneously to achieve high-affinity interaction (Andersen *et al.*, 2003). However, the addition of cross-linking antibodies directed toward the extracellular domain of VLDLR and ApoER2 induces the phosphorylation of Dab1 but is not sufficient to correct the mutant phenotype, whereas a similar induction of Dab1 phosphorylation by recombinant Reelin is able to rescue the phenotype, thus indicating the possible requirement of a coreceptor for full signaling, but not for Dab1 phosphorylation (Jossin *et al.*, 2004). Results using a divalent RAP-Fc fusion protein support this view (Strasser *et al.*, 2004). However, these negative results could have other explanations and the requirement for a coreceptor still remains unclear.

The proposal that the protocadherin cadherin-related neuronal receptor-1 (CNR1) is a Reelin receptor (Senzaki *et al.*, 1999) has been disproved (Jossin *et al.*, 2004). The Reelin-alkaline phosphatase fusion protein used in the binding study of Senzaki *et al.* (1999) was extracted from cell lysates and not secreted, suggesting that it may not be properly folded and may bind nonspecifically to CNR1. Our results with identical CNR1 ectodomain fusion proteins, but a secreted and biologically active Reelin, demonstrated no detectable interaction between CNR1 and Reelin, whereas binding of Reelin to VLDLR and ApoER2 was consistently observed in the same conditions (Jossin *et al.*, 2004).

α3β1 integrins can bind to the N-terminal region of Reelin, and this has been proposed to inhibit neuronal migration by stimulating the detachment of neurons from radial glia (Dulabon *et al.*, 2000; Schmid *et al.*, 2005). The relatively high proportion of the N-terminal Reelin fragment and the reduction in Dab1 protein levels in cerebral cortices of α3 integrin-deficient mice (Dulabon *et al.*, 2000), and the increased level of α3 integrins in Dab1 mutant cortices (Sanada *et al.*, 2004) suggest an interaction with the Reelin pathway. However, the phenotypes of mice with brain-specific inactivation of β1 integrins (Graus-Porta *et al.*, 2001) or the α3 subunit (Anton *et al.*, 1999) are not reeler-like. The N-terminal fragment of Reelin which binds β1 integrin is not necessary for the action of Reelin during cortical plate development (Jossin *et al.*, 2004). Still, integrins may be involved in other functions of Reelin particularly in synaptic plasticity (Dong *et al.*, 2003) or in normal positioning of dentate granule cells (Forster *et al.*, 2002).

5 Disabled-1

Mutations in the gene encoding the intracellular adapter protein Dab1, either induced (Howell *et al.*, 1997) or spontaneous (in *scrambler* and *yotari* mutant mice), generate a reeler-like phenotype (Sheldon *et al.*, 1997). There is no additional cortical defect in mice lacking both Reelin and Dab1, suggesting that the two proteins function in a linear pathway (Howell *et al.*, 1999a). Similar to VLDLR and ApoER2, Dab1 is mostly expressed by neurons of the cortex and elsewhere that are targets of Reelin action (Walsh and Goffinet, 2000). Additionally, Dab1 is also expressed together with ApoER2 and VLDLR in the ventricular and subventricular zones by radial glial precursors and newborn neurons (Luque *et al.*, 2003).

Dab1 possess a PI/PTB domain (Protein Interaction/Phospho-Tyrosine Binding domain) that docks to the NPxY (Asp-Pro-any amino acid-Tyr) sequence in the intracellular tail of VLDLR and ApoER2, as well as other NPxY-containing transmembrane proteins, such as LDLR, LRP, megalin and amyloid precursor proteins (APP, APLP1, and APLP2) (Trommsdorff *et al.*, 1998; Howell *et al.*, 1999b; Trommsdorff *et al.*, 1999), and integrins (Schmid *et al.*, 2005). However, unlike other PI/PTB domains (i.e., Shc, IRS1), the Dab1 PTB motif binds preferentially to unphosphorylated NPxY sites (Howell *et al.*, 1999a).

Some phosphatidylinositols, especially phosphatidylinositol-4,5-bisphosphate ($PI4,5P_2$), can bind the PTB domain of Dab1 without affecting its interaction with lipoprotein receptors (Howell *et al.*, 1999b; Stolt *et al.*, 2003). This binding is required for membrane localization and basal tyrosine phosphorylation of Dab1 independently of VLDLR and ApoER2 and is necessary for effective transduction of the Reelin signal (Huang *et al.*, 2005; Stolt *et al.*, 2005; Xu *et al.*, 2005). These observations suggest that phospholipids recruit Dab1 to membranes but do not play a direct role in relaying the Reelin signal, whereas direct Dab1–receptor interaction is responsible for the signal but not for membrane recruitment.

The NPxY motif is involved in internalization of lipoprotein receptors, but whether receptor internalization plays a role in Reelin signaling remains unclear. One group, however, showed that Dab1 is recruited to the plasma membrane upon Reelin addition, and then the complex is internalized (Morimura *et al.*, 2005). Additionally, when the phosphorylation of Dab1 is inhibited, Reelin remains associated with Dab1 near the plasma membrane for a prolonged period (Morimura *et al.*, 2005).

Src, Fyn, and Abl are able to phosphorylate Dab1 *in vitro,* but only Fyn, and to a lesser extent Src, are involved in Dab1 phosphorylation *in vivo* (Howell *et al.*, 1997; Arnaud *et al.*, 2003b; Bock and Herz, 2003; Jossin *et al.*, 2003). The combined deficit of Src and Fyn generates a reeler-like phenotype, providing definitive proof that these two kinases play a crucial role (Kuo *et al.*, 2005). Dab1 tyrosine residues 198, 220, and 232 are phosphorylated *in vivo* in response to the binding of Reelin to VLDLR and ApoER2 (Keshvara *et al.*, 2001) and are indispensable to Reelin signaling (Howell *et al.*, 2000). The replacement of the *Dab1* gene by a partial cDNA encoding its PTB domain and the stretch containing five tyrosine residues is able to rescue most (but not all) features of the Dab1 mutant phenotype (Herrick and Cooper, 2002), suggesting unidentified roles of the C-terminal part of Reelin. Interestingly, the C-terminal segment contains consensus S/T phosphorylation sites, some of which can be phosphorylated by cyclin-dependent kinase 5 (Cdk5) independently of Reelin (Keshvara *et al.*, 2002). Following tyrosine phosphorylation, Dab1 is polyubiquitinated and degraded by the proteasome and this may be important to ensure a transient response to Reelin (Arnaud *et al.*, 2003a; Bock *et al.*, 2004).

6 Downstream of *Dab1*

Mutations in genes coding for the proteins discussed above (single mutation for *Reln* and *Dab1* or double inactivation for *Vldlr* and *Apoer2* or *Src* and *Fyn*) induce a similar phenotype, leaving little doubt about their involvement in the same signaling pathway. Here, I will describe putative downstream partners, identified either as phospho-Dab1 interacting proteins or as enzymes activated in cultured neurons after Reelin stimulation. All of these proteins are known to be able to modulate the cytoskeletal dynamics and thereby may affect neuronal migration or cellular adhesion.

Reelin stimulation of cultured neurons triggers the interaction of Y-phosphorylated Dab1 with the regulatory p85α subunit of phosphatidylinositol 3-kinase (PI3K). This results in increased protein kinase B (PKB) phosphorylation at S473 (known to activate the enzyme) and glycogen synthase kinase-3 beta (GSK3β at S9 (known to inhibit the enzyme) (Beffert *et al.*, 2002; Bock *et al.*, 2003). GSK3β acts on several targets that include tau and microtubule-associated protein 1b (Map1b). Increased phosphorylation of tau is present in mice deficient in Reelin, Dab1, or both VLDLR and ApoER2, and this may reflect the inhibition of GSK3β (Hiesberger *et al.*, 1999). However, strain background has a major influence on this phenomenon, suggesting the presence of modifier genes (Brich *et al.*, 2003) and *tau* knockout animals display normal brain organization and neuronal migration (Harada *et al.*, 1994). Despite reported inhibition of GSK3β by Reelin, a recent report suggested that Reelin induces Map1b phosphorylation through activation of GSK3 and Cdk5 (Gonzalez-Billault *et al.*, 2005). Although *map1b*$^{-/-}$ and *tau*$^{-/-}$*map1b*$^{-/-}$ mutant mice have reeler-like defective layering of CA1 hippocampal pyramidal cells, they lack neocortical anomalies (Takei *et al.*, 2000). However, *map2*$^{-/-}$*map1b*$^{-/-}$ mice exhibit retarded neuronal migration but display normal inside-outside cortical layering (Teng *et al.*, 2001). The suggestion that Reelin could activate Cdk5 is not supported by several reports. First, the activity of Cdk5 is not affected by Reelin in primary neuronal culture and is unchanged in mice with mutations of Reelin, Dab1, or VLDLR and ApoER2 (Beffert *et al.*, 2004). Second, the phosphorylation of Dab1 by Cdk5 is independent of Reelin (Keshvara *et al.*, 2002). And finally, mice lacking both Cdk5/p35 and Reelin/Dab1 exhibit some similar phenotypes, but most defects are additive, suggesting that both pathways work in parallel rather than sequentially (Ohshima *et al.*, 2001).

Lissencephaly-1 (Lis1), the gene mutated in the Miller-Dieker syndrome and type 1 lissencephaly, plays a key role in neuronal migration. Interactions between

Lis1 and Reelin signaling were investigated by studies of compound mutant mice with disruptions in the Reelin pathway and heterozygous Lis1 mutations. These animals had a higher incidence of hydrocephalus and enhanced cortical and hippocampal layering defects. Dab1 and Lis1 bound in a Reelin-induced, phosphorylation-dependent manner, clearly points to convergence of genetic and biochemical interactions (Assadi *et al.*, 2003).

Nck (noncatalytic region of tyrosine kinase) α and Nckβ adapter proteins are regulators of the actin cytoskeleton implicated in cell movement and axon guidance. Nckβ (but not Nckα) interacts, through its SH2 domain, with Dab1 when Dab1 is phosphorylated on the Reelin-regulated sites Y220 or Y232. Nckβ redistributes from the soma to the leading edge of neuronal processes in about 5% of Reelin-treated primary forebrain neurons, and coexpression of Nckβ with Y-phosphorylated Dab1-RFP (Dab1 fused with a red fluorescent protein) in fibroblasts disrupts the actin cytoskeleton (Pramatarova *et al.*, 2003).

The search for proteins that bind to tyrosine-phosphorylated Dab1 in embryonic brain extracts led to the identification of CT10 regulator of kinase (Crk) family proteins (CrkL, CrkI, and CrkII) (Ballif *et al.*, 2004; Chen *et al.*, 2004; Huang *et al.*, 2004). CrkL and CrkII binding to Dab1 involve two critical tyrosine phosphorylation sites, Y220 and Y232. The Crks and CrkL proteins are SH2 and SH3 containing adapter molecules that have been shown to regulate cell migration. Chen *et al.* (2004) showed that CrkII mediates the interaction of dedicator of cytokinesis 1 (Dock1), an exchange factor (GEF) for Ras-related C3 botulinum toxin substrate 1 (Rac1), with phosphorylated Dab1 and proposed a model in which tyrosine phosphorylated Dab1 engages the conserved CrkII-Dock1-Rac signaling cassette, but, when bound to Dab1, this complex would not support neuronal migration. On the other hand, another group showed that Reelin stimulates the tyrosine phosphorylation of C3G, a GEF for Ras-related protein 1 (Rap1), on a site known to be required for activation of Rap1 by C3G, and that this phosphorylation is downstream to Dab1 phosphorylation (Ballif *et al.*, 2004). Moreover, Rap1 GTP levels, but not Rac1 GTP levels, in cultured neurons were increased after 15 minutes of stimulation by Reelin, indicating that these two signaling enzymes may be activated in a different way either temporally or spatially in the cell (Ballif *et al.*, 2004). Finally, a third group showed that CrkI and CrkII, but not CrkL, induced Dab1 phosphorylation by Src family kinases at Y220, but not Y198 and Y232, when expressed with Dab1 in HEK293T cells (Huang *et al.*, 2004).

Dab1 co-immunoprecipitates with N-WASP (neuronal Wiskott-Aldrich syndrome protein) in E14 mouse brain lysates. *In vitro*, an overexpression of the PTB domain of Dab1 in nonneuronal Cos-7 cells causes formation of microspikes and this effect was interpreted to be due to an interaction between Dab1 with an NRFY (Asn-Arg-Phe-Tyr) sequence in N-WASP (Suetsugu *et al.*, 2004).

Despite the identification of several candidate partners of Reelin signaling, stronger evidence is clearly necessary to implicate unequivocally any of them in Reelin's functions. The understanding of the mechanism of action of Reelin on target cells remains a formidable experimental challenge.

Acknowledgments The work of Dr. Jossin is supported by the Fonds National de la Recherche Scientifique.

References

Andersen, O. M., Benhayon, D., Curran, T., and Willnow, T. E. (2003). Differential binding of ligands to the apolipoprotein E receptor 2. *Biochemistry* 42:9355–9364.

Anton, E. S., Kreidberg, J. A., and Rakic, P. (1999). Distinct functions of alpha3 and alpha(v) integrin receptors in neuronal migration and laminar organization of the cerebral cortex. *Neuron* 22:277–289.

Arnaud, L., Ballif, B. A., and Cooper, J. A. (2003a). Regulation of protein tyrosine kinase signaling by substrate degradation during brain development. *Mol. Cell. Biol.* 23:9293–9302.

Arnaud, L., Ballif, B. A., Forster, E., and Cooper, J. A. (2003b). Fyn tyrosine kinase is a critical regulator of disabled-1 during brain development. *Curr. Biol.* 13:9–17.

Assadi, A. H., Zhang, G., Beffert, U., McNeil, R. S., Renfro, A. L., Niu, S., Quattrocchi, C. C., Antalffy, B. A., Sheldon, M., Armstrong, D. D., Wynshaw-Boris, A., Herz, J., D'Arcangelo, G., and Clark, G. D. (2003). Interaction of reelin signaling and Lis1 in brain development. *Nat. Genet.* 35:270–276.

Ballif, B. A., Arnaud, L., Arthur, W. T., Guris, D., Imamoto, A., and Cooper, J. A. (2004). Activation of a Dab1/CrkL/C3G/Rap1 pathway in Reelin-stimulated neurons. *Curr. Biol.* 14:606–610.

Beffert, U., Morfini, G., Bock, H. H., Reyna, H., Brady, S. T., and Herz, J. (2002). Reelin-mediated signaling locally regulates protein kinase B/Akt and glycogen synthase kinase 3beta. *J. Biol. Chem.* 277:49958–49964.

Beffert, U., Weeber, E. J., Morfini, G., Ko, J., Brady, S. T., Tsai, L. H., Sweatt, J. D., and Herz, J. (2004). Reelin and cyclin-dependent kinase 5-dependent signals cooperate in regulating neuronal migration and synaptic transmission. *J. Neurosci.* 24:1897–1906.

Beffert, U., Weeber, E. J., Durudas, A., Qiu, S., Masiulis, I., Sweatt, J. D., Li, W. P., Adelmann, G., Frotscher, M., Hammer, R. E., and Herz, J. (2005). Modulation of synaptic plasticity and memory by reelin involves differential splicing of the lipoprotein receptor apoer2. *Neuron* 47:567–579.

Bock, H. H., and Herz, J. (2003). Reelin activates SRC family tyrosine kinases in neurons. *Curr. Biol.* 13:18–26.

Bock, H. H., Jossin, Y., Liu, P., Forster, E., May, P., Goffinet, A. M., and Herz, J. (2003). Phosphatidylinositol 3-kinase interacts with the adaptor protein Dab1 in response to Reelin signaling and is required for normal cortical lamination. *J. Biol. Chem.* 278:38772–38779.

Bock, H. H., Jossin, Y., May, P., Bergner, O., and Herz, J. (2004). Apolipoprotein E receptors are required for reelin-induced proteasomal degradation of the neuronal adaptor protein Disabled-1. *J. Biol. Chem.* 279:33471–33479.

Brich, J., Shie, F. S., Howell, B. W., Li, R., Tus, K., Wakeland, E. K., Jin, L. W., Mumby, M., Churchill, G., Herz, J., and Cooper, J. A. (2003). Genetic modulation of tau phosphorylation in the mouse. *J. Neurosci.* 23:187–192.

Caviness, V. S., Jr., and Rakic, P. (1978). Mechanisms of cortical development: a view from mutations in mice. *Annu. Rev. Neurosci.* 1:297–326.

Chen, K., Ochalski, P. G., Tran, T. S., Sahir, N., Schubert, M., Pramatarova, A., and Howell, B. W. (2004). Interaction between Dab1 and CrkII is promoted by Reelin signaling. *J. Cell. Sci.* 117:4527–4536.

D'Arcangelo, G., Miao, G. G., Chen, S. C., Soares, H. D., Morgan, J. I., and Curran, T. (1995). A protein related to extracellular matrix proteins deleted in the mouse mutant reeler. *Nature* 374:719–723.

D'Arcangelo, G., Homayouni, R., Keshvara, L., Rice, D. S., Sheldon, M., and Curran, T. (1999). Reelin is a ligand for lipoprotein receptors. *Neuron* 24:471–479.

Dong, E., Caruncho, H., Liu, W. S., Smalheiser, N. R., Grayson, D. R., Costa, E., and Guidotti, A. (2003). A reelin–integrin receptor interaction regulates Arc mRNA translation in synaptoneurosomes. *Proc. Natl. Acad. Sci. USA* 100:5479–5484.

Dulabon, L., Olson, E. C., Taglienti, M. G., Eisenhuth, S., McGrath, B., Walsh, C. A., Kreidberg, J. A., and Anton, E. S. (2000). Reelin binds alpha3beta1 integrin and inhibits neuronal migration. *Neuron* 27:33–44.

Fatemi, S. H. (2005). Reelin glycoprotein: structure, biology and roles in health and disease. *Mol. Psychiatry* 10:251–257.

Forster, E., Tielsch, A., Saum, B., Weiss, K. H., Johanssen, C., Graus-Porta, D., Muller, U., and Frotscher, M. (2002). Reelin, disabled 1, and beta 1 integrins are required for the formation of the radial glial scaffold in the hippocampus. *Proc. Natl. Acad. Sci. USA* 99:13178--3183.

Gonzalez-Billault, C., Del Rio, J. A., Urena, J. M., Jimenez-Mateos, E. M., Barallobre, M. J., Pascual. M., Pujadas, L., Simo, S., Torre, A. L., Gavin, R., Wandosell, F., Soriano, E., and Avila, J. (2005). A role of MAP1B in reelin-dependent neuronal migration. *Cereb. Cortex* 15:1134–1145.

Graus-Porta, D., Blaess, S., Senften, M., Littlewood-Evans, A., Damsky, C., Huang, Z., Orban, P., Klein, R., Schittny, J. C., and Muller, U. (2001). Beta1-class integrins regulate the development of laminae and folia in the cerebral and cerebellar cortex. *Neuron* 31:367–379.

Harada, A., Oguchi, K., Okabe, S., Kuno, J., Terada, S., Ohshima, T., Sato-Yoshitake, R., Takei, Y., Noda, T., and Hirokawa, N. (1994). Altered microtubule organization in small-calibre axons of mice lacking tau protein. *Nature* 369:488–491.

Herrick, T. M., and Cooper, J. A. (2002). A hypomorphic allele of dab1 reveals regional differences in reelin-Dab1 signaling during brain development. *Development* 129:787–796.

Hiesberger, T., Trommsdorff, M., Howell, B. W., Goffinet, A., Mumby, M. C., Cooper, J. A., and Herz, J. (1999). Direct binding of reelin to VLDL receptor and ApoE receptor 2 induces tyrosine phosphorylation of disabled-1 and modulates tau phosphorylation. *Neuron* 24:481–489.

Howell, B. W., Hawkes, R., Soriano, P., and Cooper, J. A. (1997). Neuronal position in the developing brain is regulated by mouse disabled-1. *Nature* 389:733–737.

Howell, B. W., Herrick, T. M., and Cooper, J. A. (1999a). Reelin-induced tryosine phosphorylation of disabled 1 during neuronal positioning. *Genes Dev.* 13:643–648.

Howell, B. W., Lanier, L. M., Frank, R., Gertler, F. B., and Cooper, J. A. (1999b). The disabled 1 phosphotyrosine-binding domain binds to the internalization signals of transmembrane glycoproteins and to phospholipids. *Mol. Cell. Biol.* 19:5179–5188.

Howell, B. W., Herrick, T. M., Hildebrand, J. D., Zhang, Y., and Cooper, J. A. (2000). Dab1 tyrosine phosphorylation sites relay positional signals during mouse brain development. *Curr. Biol.* 10:877–885.

Huang, Y., Magdaleno, S., Hopkins, R., Slaughter, C., Curran, T., and Keshvara, L. (2004). Tyrosine phosphorylated disabled 1 recruits Crk family adapter proteins. *Biochem. Biophys. Res. Commun.* 318:204–212.

Huang, Y., Shah, V., Liu, T., and Keshvara, L. (2005). Signaling through disabled 1 requires phosphoinositide binding. *Biochem. Biophys. Res. Commun.* 331:1460–1468.

Impagnatiello, F., Guidotti, A. R., Pesold, C., Dwivedi, Y., Caruncho, H., Pisu, M. G., Uzunov, D. P., Smalheiser, N. R., Davis, J. M., Pandey, G. N., Pappas, G. D., Tueting, P., Sharma, R. P., and Costa, E. (1998). A decrease of reelin expression as a putative vulnerability factor in schizophrenia. *Proc. Natl. Acad. Sci. USA* 95:15718–15723.

Jossin, Y. (2004). Neuronal migration and the role of reelin during early development of the cerebral cortex. *Mol. Neurobiol.* 30:225–251.

Jossin, Y., Ogawa, M., Metin, C., Tissir, F., and Goffinet, A. M. (2003). Inhibition of SRC family kinases and non-classical protein kinases C induce a reeler-like malformation of cortical plate development. *J. Neurosci.* 23:9953–9959.

Jossin, Y., Ignatova, N., Hiesberger, T., Herz, J., Lambert de Rouvroit, C., and Goffinet, A. M. (2004). The central fragment of reelin, generated by proteolytic processing in vivo, is critical to its function during cortical plate development. *J. Neurosci.* 24:514–521.

Keshvara, L., Benhayon, D., Magdaleno, S., and Curran, T. (2001). Identification of reelin-induced sites of tyrosyl phosphorylation on disabled 1. *J. Biol. Chem.* 276:16008–16014.

Keshvara, L., Magdaleno, S., Benhayon, D., and Curran, T. (2002). Cyclin-dependent kinase 5 phosphorylates disabled 1 independently of reelin signaling. *J. Neurosci.* 22:4869–4877.

Kuo, G., Arnaud, L., Kronstad-O'Brien, P., and Cooper, J. A. (2005). Absence of Fyn and Src causes a reeler-like phenotype. *J. Neurosci.* 25:8578–8586.

Lambert de Rouvroit, C., de Bergeyck, V., Cortvrindt, C., Bar, I., Eeckhout, Y., and Goffinet, A. M. (1999). Reelin, the extracellular matrix protein deficient in reeler mutant mice, is processed by a metalloproteinase. *Exp. Neurol.* 156:214–217.

Liu, W. S., Pesold, C., Rodriguez, M. A., Carboni, G., Auta, J., Lacor, P., Larson, J., Condie, B. G., Guidotti, A., and Costa, E. (2001). Down-regulation of dendritic spine and glutamic acid decarboxylase 67 expressions in the reelin haploinsufficient heterozygous reeler mouse. *Proc. Natl. Acad. Sci. US.* 98:3477–3482.

Luque, J. M., Morante-Oria, J., and Fairen, A. (2003). Localization of ApoER2, VLDLR and Dab1 in radial glia: groundwork for a new model of reelin action during cortical development. *Brain Res. Dev. Brain Res.* 140:195–203.

Miyata, T., Nakajima, K., Mikoshiba, K., and Ogawa, M. (1997). Regulation of Purkinje cell alignment by reelin as revealed with CR-50 antibody. *J. Neurosci.* 17:3599–3609.

Morimura, T., Hattori, M., Ogawa, M., and Mikoshiba, K. (2005). Disabled1 regulates the intracellular trafficking of reelin receptors. *J. Biol. Chem.* 280:16901–16908.

Nogi, T., Yasui, N., Hattori, M., Iwasaki, K., and Takagi, J. (2006). Structure of a signaling-competent reelin fragment revealed by X-ray crystallography and electron tomography. *EMBO J.* 25:3675–3683.

Ohshima, T., Ogawa, M., Veeranna, Hirasawa, M., Longenecker, G., Ishiguro, K., Pant, H. C., Brady, R. O., Kulkarni, A. B., and Mikoshiba, K. (2001). Synergistic contributions of cyclin-dependent kinase 5/p35 and reelin/Dab1 to the positioning of cortical neurons in the developing mouse brain. *Proc. Natl. Acad. Sci. USA* 98:2764–2769.

Pramatarova, A., Ochalski, P. G., Chen, K., Gropman, A., Myers, S., Min, K. T., and Howell, B. W. (2003). Nck beta interacts with tyrosine-phosphorylated disabled 1 and redistributes in reelin-stimulated neurons. *Mol. Cell. Biol.* 23:7210–7221.

Sanada, K., Gupta, A., and Tsai, L. H. (2004). Disabled-1-regulated adhesion of migrating neurons to radial glial fiber contributes to neuronal positioning during early corticogenesis. *Neuron* 42:197–211.

Schiffmann, S. N., Bernier, B., and Goffinet, A. M. (1997). Reelin mRNA expression during mouse brain development. *Eur. J. Neurosci.* 9:1055–1071.

Schmid, R. S., Jo, R., Shelton, S., Kreidberg, J. A., and Anton, E. S. (2005). Reelin, integrin and DAB1 interactions during embryonic cerebral cortical development. *Cereb. Cortex* 15:1632–1636.

Senzaki, K., Ogawa, M., and Yagi, T. (1999). Proteins of the CNR family are multiple receptors for reelin. *Cell* 99:635–647.

Sheldon, M., Rice, D. S., D'Arcangelo, G., Yoneshima, H., Nakajima, K., Mikoshiba, K., Howell, B. W., Cooper, J. A., Goldowitz, D., and Curran, T. (1997). Scrambler and yotari disrupt the disabled gene and produce a reeler-like phenotype in mice. *Nature* 389:730–733.

Stolt, P. C., Jeon, H., Song, H. K., Herz, J., Eck, M. J., and Blacklow, S. C. (2003). Origins of peptide selectivity and phosphoinositide binding revealed by structures of disabled-1 PTB domain complexes. *Structure* 11:569–579.

Stolt, P. C., Chen, Y., Liu, P., Bock, H. H., Blacklow, S. C., and Herz, J. (2005). Phosphoinositide binding by the disabled-1 PTB domain is necessary for membrane localization and reelin signal transduction. *J. Biol. Chem.* 280:9671–9677.

Strasser, V., Fasching, D., Hauser, C., Mayer, H., Bock, H. H., Hiesberger, T., Herz, J., Weeber, E. J., Sweatt, J. D., Pramatarova, A., Howell, B., Schneider, W. J., and Nimpf, J. (2004). Receptor clustering is involved in reelin signaling. *Mol. Cell. Biol.* 24:1378–1386.

Suetsugu, S., Tezuka, T., Morimura, T., Hattori, M., Mikoshiba, K., Yamamoto, T., and Takenawa, T. (2004). Regulation of actin cytoskeleton by mDab1 through N-WASP and ubiquitination of mDab1. *Biochem. J.* 384:1–8.

Takei, Y., Teng, J., Harada, A., and Hirokawa, N. (2000). Defects in axonal elongation and neuronal migration in mice with disrupted tau and map1b genes. *J. Cell. Biol.* 150:989–1000.

Teng, J., Takei, Y., Harada, A., Nakata, T., Chen, J., and Hirokawa, N. (2001). Synergistic effects of MAP2 and MAP1B knockout in neuronal migration, dendritic outgrowth, and microtubule organization. *J. Cell Biol.* 155:65–76.

Trommsdorff, M., Borg, J. P., Margolis, B., and Herz, J. (1998). Interaction of cytosolic adaptor proteins with neuronal apolipoprotein E receptors and the amyloid precursor protein. *J. Biol. Chem.* 273:33556–33560.

Trommsdorff, M., Gotthardt, M., Hiesberger, T., Shelton, J., Stockinger, W., Nimpf, J., Hammer, R. E., Richardson, J. A., and Herz, J. (1999). Reeler/disabled-like disruption of neuronal migration in knockout mice lacking the VLDL receptor and ApoE receptor 2. *Cell* 97:689–701.

Utsunomiya-Tate, N., Kubo, K., Tate, S., Kainosho, M., Katayama, E., Nakajima, K., and Mikoshiba, K. (2000). Reelin molecules assemble together to form a large protein complex, which is inhibited by the function-blocking CR-50 antibody. *Proc. Natl. Acad. Sci. USA* 97:9729–9734.

Walsh, C. A., and Goffinet, A. M. (2000). Potential mechanisms of mutations that affect neuronal migration in man and mouse. *Curr. Opin. Genet. Dev.* 10:270–274.

Weeber, E. J., Beffert, U., Jones, C., Christian, J. M., Forster, E., Sweatt, J. D., and Herz, J. (2002). Reelin and ApoE receptors cooperate to enhance hippocampal synaptic plasticity and learning. *J. Biol. Chem.* 277:39944–39952.

Xu, M., Arnaud, L., and Cooper, J. A. (2005). Both the phosphoinositide and receptor binding activities of Dab1 are required for reelin-stimulated Dab1 tyrosine phosphorylation. *Brain Res. Mol. Brain Res.* 139:300–305.

Chapter 4
The C-Terminal Region of Reelin: Structure and Function

Mitsuharu Hattori

Contents

1 The Structure of the C-Terminal Region (CTR) of Reelin

1.1 Definition of CTR

Reelin is a very large secreted glycoprotein of about 3460 amino acids in mammalian species. Reelin is divided into three subdomains from its primary sequence (D'Arcangelo *et al.*, 1995): the N-terminal F-spondin-like domain, the eight tandem of Reelin repeat, and the short and highly basic C-terminal region (CTR). The N-terminal F-spondin-like domain is proposed to be necessary and sufficient for multimerization of Reelin (Utsunomiya-Tate *et al.*, 2000). The function of each Reelin repeat is not fully understood, but the region between the fifth and sixth repeat is sufficient for binding to the Reelin receptors, very-low-density lipoprotein receptor (VLDLR) and apolipoprotein receptor 2 (ApoER2) (Yasui *et al.*, 2007). The CTR, which is about 30 amino acids long, comprises less than 1% of the whole Reelin protein.

M. Hattori
Department of Biomedical Science, Graduate School of Pharmaceutical Sciences, Nagoya City University, 3-1, Tanabe-dori, Mizuho-ku, Nagoya, Aichi 467-8603, Japan
e-mail: mhattori@phar.nagoya-cu.ac.jp

S. H. Fatemi (ed.), *Reelin Glycoprotein: Structure, Biology and Roles in Health and Disease.*

The exact definition of CTR has not been determined and is closely related with the definition of Reelin repeat. In the original work that reported the first identification of mouse Reelin (D'Arcangelo *et al.*, 1995), the eighth Reelin repeat was proposed to end at Val^{3427}, making the last 34 amino acid residues CTR. The reason for choosing this boundary was not given. Subsequently, Ichihara *et al.* (2001) found some new repetitive units at its N-terminal region by computerized search and realigned the Reelin repeat. In their alignment, the eighth Reelin repeat ended at Val^{3429} (i.e., CTR = 32 amino acids, Ichihara *et al.*, 2001). Deciding the boundary between the eighth Reelin repeat from the alignment of the amino acid sequence has limitations, and it should be determined by functional examinations. Hereafter in this chapter, we adopt the definition by Ichihara *et al.* (2001) solely to avoid confusion, and it should be noted that the exact definition of CTR is an open question. It is also worth mentioning that the results discussed in this chapter are unlikely to be affected by moving the boundary by a few amino acids.

1.2 Characteristics of CTR

CTR is highly positively charged: among 32 amino acid residues of CTR, 12 (38%) are basic (Fig. 4.1), while none are acidic. It does not appear to have any motif that can be a clue to its function, and no posttranslational modification has been reported in this region.

Most of CTR (from Thr^{3431} to the C-terminal end) is encoded in a single exon (exon 65; Royaux *et al.*, 1997). Exon 64 is a microexon of 6 nucleotides, coding Val^{3429} and Ser^{3430} (Royaux *et al.*, 1997) (Fig. 4.1). Exon 64 may be skipped by alternative splicing (Lambert de Rouvroit *et al.*, 1999; Royaux *et al.*, 1997), but its biological significance remains unexplored. Importantly, alternative polyadenylation generates two types of exon 63 (Lambert de Rouvroit *et al.*, 1999). The "larger" exon 63 (exon 63a) contains termination codons and thus it eventually gives rise to a truncated Reelin protein without the C-terminal 33 amino acids (Lambert de Rouvroit *et al.*, 1999). We will discuss this short-form, CTR-less Reelin protein later.

The primary sequence of CTR is completely (100%) conserved in all the mammalian species whose CTR sequences can be found in the available databases, namely, human, chimpanzee, macaque, dog, mouse, rat, cow, and gray short-tailed opossum (Nakano *et al.*, 2007). It is also perfectly conserved in chicken and turtle. The sequence of crocodile Reelin found in the database seems to lack the microexon-derived two residues but, apart from that, it is completely conserved with that of mammals. As a matter of course, DNA sequence encoding CTR is very well (>90%) conserved among these organisms (Nakano *et al.*, 2007). From the evolutionary point of view, these observations strongly suggest that CTR has an essential physiological function(s) in the land vertebrates. Reelin gene is also found in genome databases of fishes and sea urchins. The sequences of CTR in these species are quite variable and, in some cases, appear to be missing. At present, it remains

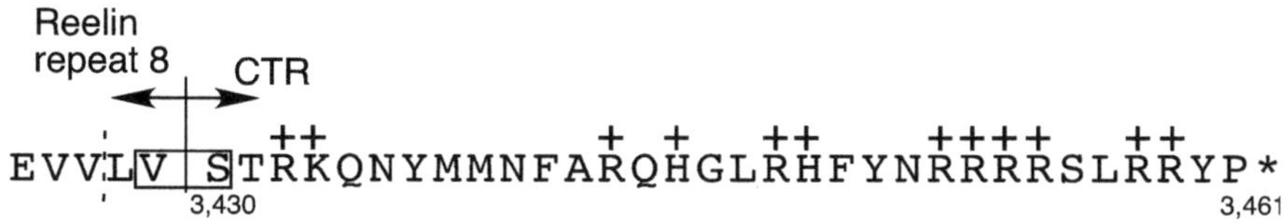

Fig. 4.1 The primary sequence of Reelin CTR. The boundary between the eighth Reelin repeat and CTR proposed by Ichihara *et al.* (2001) and by D'Arcangelo *et al.* (1995) are shown by the vertical and the dashed lines, respectively. Basic residues are indicated by "+" symbols. Two residues encoded by the microexon (exon 64) are boxed. The residue numbers counting from the first methionine are indicated below the sequence

obscure whether Reelin is actually transcribed in these species, except in zebrafish (Costagli *et al.*, 2002). Function of Reelin in non-mammalian species has not been elucidated.

2 Is CTR Required for Secretion of Reelin?

2.1 Observations from Orleans Reeler Mouse and Earlier Mutational Study

In one of the murine Reelin mutations called Orleans, a frame shift results in the production of aberrant Reelin protein that lacks the C-terminal half of the eighth Reelin repeat as well as CTR, and has 70 unrelated amino acid residues instead (ReelinOrl) (de Bergeyck *et al.*, 1997). This protein is not secreted (de Bergeyck *et al.*, 1997) and accumulates in the endoplasmic reticulum (ER) (Derer *et al.*, 2001). It was also reported that an artificial Reelin mutant protein in which a stop codon was inserted in the middle of the eighth Reelin repeat (lacking the C-terminal 133 residues) was not secreted from transfected COS-7 cells (D'Arcangelo *et al.*, 1997). These results suggested that secretion of Reelin was regulated between ER and Golgi apparatus and that this regulation required the C-terminal half of the eighth Reelin repeat and/or the CTR. As the C-terminal half of the eighth Reelin repeat did not seem to have any particular motif and no other role was assigned to CTR at that time, it was not unreasonable to assume that CTR was essential for Reelin secretion.

In 1999, Lambert de Rouvroit *et al.* reported that an alternative polyadenylation can give rise to a truncated Reelin protein without the C-terminal 33 amino acids, and this short-form Reelin is secreted when overexpressed in COS-7 cells (Lambert de Rouvroit *et al.*, 1999). From this result, it was proposed that Reelin secretion required residues from 3328 to 3428 (Lambert de Rouvroit *et al.*, 1999). However, the amount of secreted CTR-less Reelin was apparently much lower than that of the wild-type Reelin (Lambert de Rouvroit *et al.*, 1999), and the expression level of the respective protein was not provided. Therefore, it was somewhat ambiguous whether CTR is involved in secretion or if it affects expression or stability of Reelin protein.

2.2 Detailed Mutational Study

The regulatory mechanism of Reelin secretion is important not only for understanding its role in the developing brain, but also for elucidation of its significance in adult brain function, especially in synaptic plasticity (Beffert *et al.*, 2005), for which the action must be spatiotemporally regulated. Therefore, we set out to analyze the role of CTR in Reelin secretion by making series of mutants and expressing them in various cell lines (Nakano *et al.*, 2007). All of the results discussed below are essentially independent of the cell type used.

A mutant Reelin protein that lacked only CTR (ReelinΔC) was secreted, although its efficiency was lower than that of wild-type Reelin, suggesting that CTR is not essential for, but may play a role in, secretion of Reelin. On the other hand, a mutant that lacked the C-terminal 20 residues (including 10 of 12 basic residues) of the CTR was secreted as efficiently as wild-type Reelin. Moreover, replacing CTR with FLAG epitope (DYKDDDDK, ReelinΔC-FLAG), eight arginine (ReelinΔC-Arg8), or glutamate residues had little effect on the secretion efficiency. A mutant in which CTR was replaced with Venus (a variant of yellow fluorescent protein with 239 amino acids (Nagai *et al.*, 2002)) was secreted quite efficiently, but the mutants, in which CTR is replaced with eight alanine or histidine, residues were not secreted. These results indicated three important points. First, CTR is not absolutely indispensable for secretion. In particular, basic residues in CTR are totally dispensable. Second, many, but not all, amino acid sequences are able to substitute CTR in facilitating secretion. We presume that a hydrophilic structure is necessary in this region for efficient secretion of Reelin protein. Third, CTR is likely to have another important role, considering that it is highly conserved among species. It should be mentioned that these results did not seem to be cell-type specific. However, we do not exclude the possibility that Reelin secretion *in vivo*, such as by Cajal-Retzius cells, is in fact regulated in a CTR-dependent manner.

We also obtained a clue as to why ReelinOrl is not secreted (Nakano *et al.*, 2007). First, a truncated Reelin mutant that terminated just after the seventh Reelin repeat was efficiently secreted, while none of the truncated mutants that terminated in the middle of Reelin repeat were secreted. Thus, it is suggested that abrupt termination in the middle of the eighth Reelin repeat, not the absence of a certain sequence, is the main cause of secretion failure of ReelinOrl. Second, immunocytochemical analysis of overexpressed cells revealed that wild-type Reelin (and all of the mutants that were efficiently secreted) was present mainly in the ER and presumably in secretory vesicles, while ReelinOrl mutant tended to accumulate Reelin around the nucleus. This was consistent with a previous report that ReelinOrl accumulates in the rough ER in Cajal-Retzius cells (Derer *et al.*, 2001). Therefore, it is strongly suggested that ReelinOrl is unable to go beyond the ER in its secretory pathway. On the other hand, ReelinΔC accumulated in ER and dilated it. It was thus indicated that ReelinOrl and ReelinΔC have distinct intracellular fates in the secretory pathway. These observations further support the idea that lack of CTR is not the direct cause of the secretion failure of ReelinOrl.

3 Functions of CTR

3.1 Role of CTR in Activation of Downstream Signaling

The fact that most of the CTR is unnecessary for secretion prompted us to investigate the other physiological function of CTR. We thought that CTR may be involved in activation of downstream signaling and thus stimulated the primary cortical neurons from embryonic mice with conditioned media containing wild-type or mutant Reelin protein (Nakano *et al.*, 2007). To our surprise, Reelin mutants that lack CTR were generally much less potent than wild-type Reelin in inducing Dab1 phosphorylation. Importantly, ReelinΔC-Arg8 induced Dab1 phosphorylation more strongly than other mutants that lacked CTR, but not as strongly as wild-type Reelin. Moreover, Reelin mutants with the insertion of an unrelated sequence (FLAG epitope or Venus) between the eighth Reelin repeat and CTR were no more potent than other mutants without CTR. These results highlight three important points. First, CTR is necessary for efficient activation of downstream signaling. Second, the highly basic nature of CTR contributes to activation of downstream signaling. Third, CTR must be located just after the eighth Reelin repeat to have full activity. In other words, CTR functions in concert with the eighth Reelin repeat for activation of downstream signaling.

Our results seem to contradict the previous reports (Jossin *et al.*, 2004, 2007), because they argue that the region between the third and sixth Reelin repeats is as potent as full-length Reelin in Dab1 phosphorylation. The reason for the discrepancy is unknown at present, but there are some possibilities. First, concentrations of Reelin and its mutants/fragments may be different between the two groups. Both groups adjusted the concentration of the samples in their own assays, but it is difficult to compare them between the two distant groups. Second, small differences in the methodologies used to make the recombinant samples (e.g., culturing conditions, transfection reagents, serum, and so on) may affect their signaling capacity, for example, by affecting posttranslational modifications. We recently performed fairly quantitative analysis by collaboration with Junichi Takagi's group and found that the full-length Reelin is approximately 100 times as potent as the recombinant protein consisting of the third and sixth Reelin repeats (Nogi *et al.*, 2006). This estimation is, however, based on the assumption that all the recombinant proteins in the solutions are potent, but they might be partially misfolded or lack modifications necessary for full potency. Third, both groups use "full-length Reelin" that contained various amounts of proteolytic fragments. The fragments, particularly the

N- and C-terminal fragments whose functions are not well defined, may affect the assay. Fourth, as preparation methods for primary cortical neurons are not completely the same between two groups, expression levels of the receptors, Dab1, kinases, and other factors may be different. Finally, and perhaps most importantly, Dab1 phosphorylation and activation of Src family protein kinase form a positive feedback loop (Bock and Herz, 2003). Therefore, a small fluctuation could easily result in the all-or-none responses. Elucidation of all the molecular mechanisms will solve the discrepancy, and phenotypic analysis of the ReelinΔC-FLAG knockin mice (see Section 4) will reveal the physiological importance of CTR *in vivo*.

3.2 How Is CTR Involved in Activation of Downstream Signaling?

Why does deletion of CTR or insertion of unrelated residues just prior to CTR reduce the Dab1-phosphorylating ability? We first checked whether deletion of CTR affects the dimer formation or oligomerization of Reelin, as oligomerization of Reelin is a prerequisite for Dab1 phosphorylation. However, the results obtained indicated that CTR is not involved in oligomerization of Reelin (Nakano *et al.*, 2007).

We next investigated whether deletion of CTR affects interaction between Reelin and its receptors. For this purpose, we first employed the recombinant, soluble ligand-binding domain of ApoER2 fused to human growth hormone (GH-ApoER2LBD) to pull down Reelin protein. Conditioned media containing either wild-type Reelin or ReelinΔC-FLAG were mixed with GH-ApoER2LBD, anti-GH, and Protein-G Sepharose, and the precipitates were analyzed by Western blotting. No difference in binding was observed between wild-type Reelin and ReelinΔC-FLAG, indicating that CTR is not directly involved in the interaction between Reelin and the extracellular domain of ApoER2.

Negatively charged proteoglycans and extracellular matrices are known to play important roles in signaling machinery of many secreted molecules including Wnt (Lin, 2004) and Semaphorin 5A (Kantor *et al.*, 2004). As CTR of Reelin is positively charged, we investigated whether it is involved in the interaction between Reelin and its receptors on the cell surface. We found that wild-type Reelin bound much more strongly (or more stably) to the receptor-bearing cell membrane than did Reelin mutants that lacked CTR (Nakano *et al.*, 2007). Among the mutants, ReelinΔC-Arg8 bound to the receptor-bearing cell membrane slightly more strongly than other mutants, indicating that Reelin–receptor interaction on the cell membrane is partly, but not solely, mediated by the positive charges of the CTR.

We also investigated binding of Reelin to COS-7 cells expressing ApoER2 by immunostaining without permeabilization (in order to detect cell-surface, but not internalized, Reelin) (Nakano *et al.*, 2007). Wild-type Reelin bound strongly to the surface of ApoER2-expressing COS-7 cells, while most mutants lacking CTR did so only weakly. Consistent with the pull-down experiments using isolated cell membrane, ReelinΔC-Arg8 bound to ApoER2-expressing cells more strongly than mutants lacking the CTR but more weakly than wild-type Reelin. The same phenomena were observed when primary cortical neurons that endogenously express Reelin receptors were used. These results demonstrated that CTR is necessary for a stable association between Reelin and its receptors on the plasma membranes of live cells and positive charges of CTR are partly involved in it.

4 Concluding Remarks

The CTR of Reelin is not necessary for its secretion in any of the cell lines tested. In addition, at least in cerebellar granule cells in culture, Reelin secretion does not appear to be regulated (Lacor *et al.*, 2000). Whether secretion of Reelin is regulated or not in other systems remains unknown, however, and requires further investigation.

Our results from *in vitro* assays indicated that CTR is important for efficient induction of Dab1 phosphorylation. This effect of CTR is likely due to its interaction with a co-receptor molecule on the plasma membrane, but the identity of this co-receptor is unknown presently. Jossin *et al.* (2007) proposed that it is the central fragment of Reelin that carries the signaling capacity and that the C-terminal part may contribute to Reelin's binding to extracellular matrix (Jossin *et al.*, 2007). This is a quite reasonable model, and we also agree that the CTR may impact the diffusion of Reelin by binding to certain molecules on the cell membrane (Nakano *et al.*, 2007). More detailed examinations, ideally using an *in vivo* system, are necessary to understand the role of CTR. In this regard, we recently established ReelinΔC-FLAG knockin mice in which the genome sequence coding CTR (exon 65) is replaced with that coding FLAG epitope. By analyzing them, we will be able to answer the questions such as: (1) Is CTR dispensable for secretion *in vivo*? (2) Is CTR involved in Reelin's action during correct development of cerebral cortex and cerebellum? (3) Is CTR important for localization and/or diffusion of Reelin? It is also important to understand the molecular mechanism by which CTR augments the activation of downstream signaling. Clarification of all of these issues will help in understanding the physiological and pathological roles of Reelin.

References

Beffert, U., Weeber, E. J., Durudas, A., Qiu, S., Masiulis, I., Sweatt, J. D., Li, W. P., Adelmann, G., Frotscher, M., Hammer, R. E., and Herz, J. (2005). Modulation of synaptic plasticity and memory by reelin involves differential splicing of the lipoprotein receptor ApoER2. *Neuron* 47:567–579.

Bock, H. H., and Herz, J. (2003). Reelin activates Src family tyrosine kinases in neurons. *Curr. Biol.* 13:18–26.

Costagli, A., Kapsimali, M., Wilson, S. W., and Mione, M. (2002). Conserved and divergent patterns of reelin expression in the zebrafish central nervous system. *J. Comp. Neurol.* 450:73–93.

D'Arcangelo, G., Miao, G. G., Chen, S. C., Soares, H. D., Morgan, J. I., and Curran, T. (1995). A protein related to extracellular matrix proteins deleted in the mouse mutant reeler. *Nature* 374:719–723.

D'Arcangelo, G., Nakajima, K., Miyata, T., Ogawa, M., Mikoshiba, K., and Curran, T. (1997). Reelin is a secreted glycoprotein recognized by the CR-50 monoclonal antibody. *J. Neurosci.* 17: 23–31.

de Bergeyck, V., Nakajima, K., Lambert de Rouvroit, C., Naerhuyzen, B., Goffinet, A. M., Miyata, T., Ogawa, M., and Mikoshiba, K. (1997). A truncated reelin protein is produced but not secreted in the 'Orleans' reeler mutation (Reln[rl-Orl]). *Brain Res. Mol. Brain Res.* 50:85–90.

Derer, P., Derer, M., and Goffinet, A. (2001). Axonal secretion of reelin by Cajal-Retzius cells: evidence from comparison of normal and Reln(Orl) mutant mice. *J. Comp. Neurol.* 440:136–143.

Ichihara, H., Jingami, H., and Toh, H. (2001). Three novel repetitive units of reelin. *Brain Res. Mol. Brain Res.* 97:190–193.

Jossin, Y., Ignatova, N., Hiesberger, T., Herz, J., Lambert de Rouvroit, C., and Goffinet, A. M. (2004). The central fragment of reelin, generated by proteolytic processing in vivo, is critical to its function during cortical plate development. *J. Neurosci.* 24:514–521.

Jossin, Y., Gui, L., and Goffinet, A. M. (2007). Processing of reelin by embryonic neurons is important for function in tissue but not in dissociated cultured neurons. *J. Neurosci.* 27:4243–4252.

Kantor, D. B., Chivatakarn, O., Peer, K. L., Oster, S. F., Inatani, M., Hansen, M. J., Flanagan, J. G., Yamaguchi, Y., Sretavan, D. W., Giger, R. J., and Kolodkin, A. L. (2004). Semaphorin 5A is a bifunctional axon guidance cue regulated by heparan and chondroitin sulfate proteoglycans. *Neuron* 44:961–975.

Lacor, P. N., Grayson, D. R., Auta, J., Sugaya, I., Costa, E., and Guidotti, A. (2000). Reelin secretion from glutamatergic neurons in culture is independent from neurotransmitter regulation. *Proc. Natl. Acad. Sci. USA* 97:3556–3561.

Lambert de Rouvroit, C., Bernier, B., Royaux, I., de Bergeyck, V., and Goffinet, A. M. (1999). Evolutionarily conserved, alternative splicing of reelin during brain development. *Exp. Neurol.* 156:229–238.

Lin, X. (2004). Functions of heparan sulfate proteoglycans in cell signaling during development. *Development* 131:6009–6021.

Nagai, T., Ibata, K., Park, E. S., Kubota, M., Mikoshiba, K., and Miyawaki, A. (2002). A variant of yellow fluorescent protein with fast and efficient maturation for cell-biological applications. *Nature Biotechnol.* 20:87–90.

Nakano, Y., Kohno, T., Hibi, T., Kohno, S., Baba, A., Mikoshiba, K., Nakajima, K., and Hattori, M. (2007). The extremely conserved C-terminal region of reelin is not necessary for secretion but is required for efficient activation of downstream signaling. *J. Biol. Chem.* 282:20544–20552.

Nogi, T., Yasui, N., Hattori, M., Iwasaki, K., and Takagi, J. (2006). Structure of a signaling-competent reelin fragment revealed by X-ray crystallography and electron tomography. *EMBO J.* 25:3675–3683.

Royaux, I., Lambert de Rouvroit, C., D'Arcangelo, G., Demirov, D., and Goffinet, A. M. (1997). Genomic organization of the mouse reelin gene. *Genomics* 46:240–250.

Utsunomiya-Tate, N., Kubo, K., Tate, S., Kainosho, M., Katayama, E., Nakajima, K., and Mikoshiba, K. (2000). Reelin molecules assemble together to form a large protein complex, which is inhibited by the function-blocking CR-50 antibody. *Proc. Natl. Acad. Sci. USA* 97:9729–9734.

Yasui, N., Nogi, T., Kitao, T., Nakano, Y., Hattori, M., and Takagi, J. (2007). Structure of a receptor-binding fragment of reelin and mutational analysis reveal a recognition mechanism similar to endocytic receptors. *Proc. Natl. Acad. Sci. USA* 104:9988–9993.

Chapter 5
Crystal Structure of Reelin Repeats

Junichi Takagi

Contents

1 Reelin Primary Structure

1.1 Domain Architecture of Reelin

Reelin is a large glycoprotein with a repetitive modular structure (D'Arcangelo and Curran, 1999). It consists of a signal sequence, an F-spondin-like region, another region containing at least eight "reelin repeats," and a C-terminal basic peptide of ~30 residues (Fig. 5.1A). Each complete reelin repeat contains a central EGF module flanked by two subrepeats of 150–190 amino acids (Fig. 5.1A, inset). The central EGF module is relatively short in length (~30 residues), but nevertheless has consensus signatures, including the spacings between cysteines (Campbell and Bork, 1993). Although the homology between subrepeat A and subrepeat B was first

J. Takagi
Laboratory of Protein Synthesis and Expression, Institute for Protein Research, Osaka University, 3-2 Yamadaoka, Suita, Osaka 565-0871, Japan
e-mail: takagi@protein.osaka-u.ac.jp

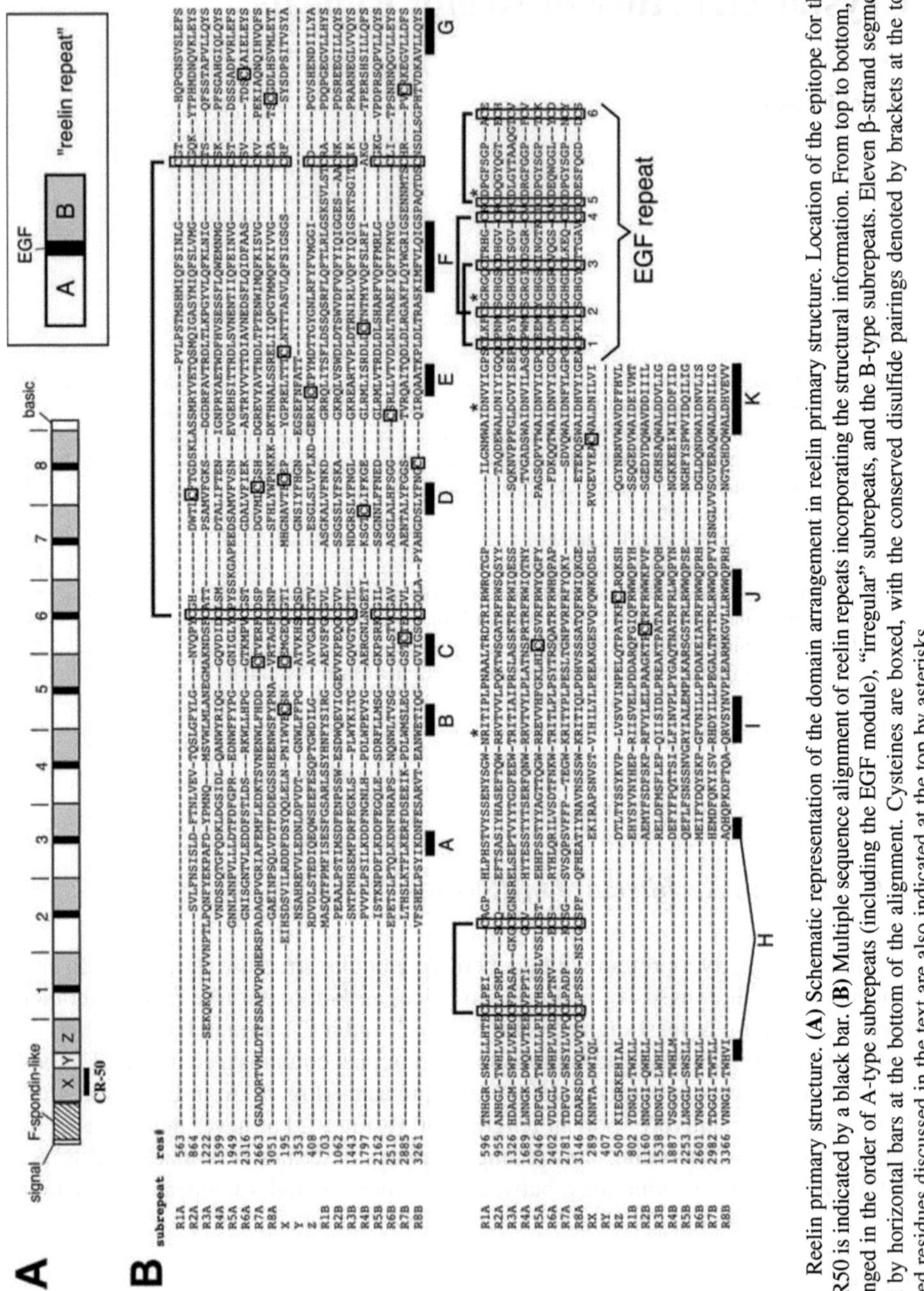

Fig. 5.1 Reelin primary structure. **(A)** Schematic representation of the domain arrangement in reelin primary structure. Location of the epitope for the antibody CR50 is indicated by a black bar. **(B)** Multiple sequence alignment of reelin repeats incorporating the structural information. From top to bottom, repeats are arranged in the order of A-type subrepeats (including the EGF module), "irregular" subrepeats, and the B-type subrepeats. Eleven β-strand segments are denoted by horizontal bars at the bottom of the alignment. Cysteines are boxed, with the conserved disulfide pairings denoted by brackets at the top. The conserved residues discussed in the text are also indicated at the top by asterisks

noted during the cloning of the reelin gene (D'Arcangelo *et al.*, 1995), these repeats have failed to show any sequence homology to known protein domains.

1.2 Definition of a "Reelin Repeat"

Sequence alignment (Fig. 5.1B) clearly shows that the A and B subrepeats are related, with the critical difference being the insertion of a disulfide-bonded loop in the A subrepeat. Because of the regularity of the "A-EGF-B" concatenation, it became common to regard this as a unit. However, reexamination of the sequence identified additional reelin repeat-like sequences before the first repeat (Ichihara *et al.*, 2001). This segment contains three stretches of sequences (denoted repeats X, Y, and Z) related to the reelin subrepeat, although without the intervening EGF module. Among these, repeats X and Z belong to the "B" type subrepeat, while repeat Y contains only a portion (~50 amino acids) of a subrepeat (Fig. 5.1B). The presence of such "incomplete" or "irregular" repeat segment argues against the idea that the "A-EGF-B" repeating unit corresponds to an inseparable single folding unit. In fact, it seemed entirely possible that the whole reelin molecule was merely a concatenation of independent A, B, and EGF modules. To determine the molecular architecture of reelin protein, we clearly needed to resolve the three-dimensional structure of the reelin repeat or subrepeats.

2 Crystal Structure of a Single Reelin Repeat

2.1 Overall Structure

Thus far, structures of three reelin repeats have been determined using X-ray crystallography. First, the crystal structure of reelin repeat 3 (R3) was solved at a resolution of 2.05 Å (Nogi *et al.*, 2006). The crystals had one R3 molecule per asymmetric unit, and 303 out of 387 residues of the R3 construct were visible in the electron density map derived from X-ray diffraction. The electron density corresponding to the N-terminal segment (1222–1293) was poor, indicating that this segment may be mobile with regard to the rest of the domain. The most striking feature of the structure is that the three subdomains (i.e., subrepeat A, EGF, and subrepeat B) are arranged in a horseshoe-like manner, making intimate contact with one another (Fig. 5.2A). Subrepeats A and B lie distal to each other at the ends of the central EGF module, although an abrupt bend at the subdomain junction places two subrepeats in a close proximity, creating a direct A–B contact. Structures for R5 and R6 have also been determined (to be published elsewhere), revealing identical overall subdomain arrangements (Fig. 5.2A). In particular, the inter-subrepeat contact within R6 was further stabilized by a disulfide bridge whereby Cys2393 and Cys2559 directly connected subrepeats A and B at the "bottom," thus encircling the

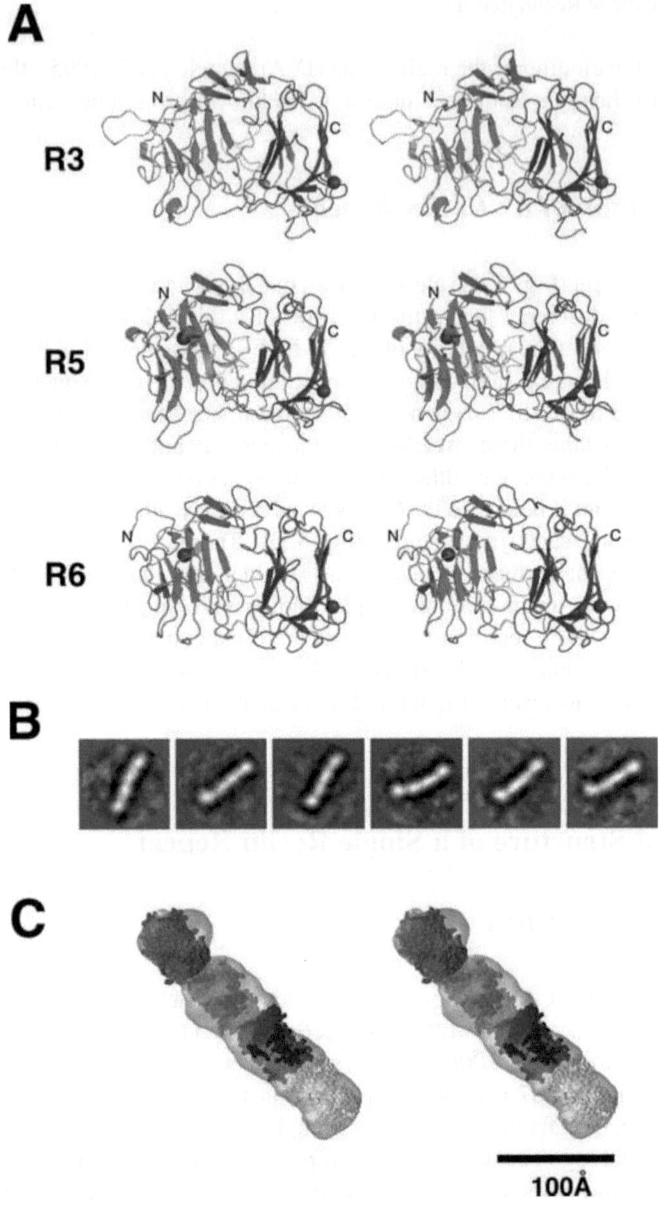

Fig. 5.2 Reelin repeat structure. (**A**) Crystal structures of single reelin repeat domains. Each panel shows a stereo presentation of R3 (top), R5 (middle), and R6 (bottom) structures. Subdomains are differently colored; subrepeat A (cyan), EGF (green), subrepeat B (magenta), and N- and C-termini are labeled. Bound calcium ions and disulfide bridges are shown as red spheres and yellow stick models, respectively. In R3, segments missing in the crystal structure are modeled and shown in gray. (**B**) Two-dimensional averages from representative particle classes obtained from the untilted electron micrographs of the R3–6 fragment. The width of each panel corresponds to 376 Å. (**C**) Three-dimensional volume map of an R3–6 fragment derived from single-particle tomography (gray) in a stereo representation. Four complete space-filling models for reelin repeats (R3, red; R4, green; R5, blue; and R6, yellow) are fitted into the envelope (*See Color Plates*)

subdomain loop. It is evident from these structures that a reelin repeat, represented by A-EGF-B set, does indeed constitute an independent structural unit, making a single globular domain with dimensions of 65×45×35 Å, as opposed to an elongated domain with a linear subdomain concatenation. An important question then arises as to whether the individual submodule can exist on its own. Expression experiments using numerous truncation fragments favor the notion that, at least in R3 through R6, the three submodules are inseparable and form part of a single structural domain.

2.2 *Subdomain Structure*

2.2.1 Subrepeats A and B

Both subrepeats are composed of 11 β strands that form two antiparallel five-stranded β sheets in a jelly-roll fold. The two β sheets are approximately parallel to each other and curved, producing concave and convex sides. The first and the last strands are next to each other in the outer (i.e., convex) sheet, bringing the N- and C-termini close together. The presence of an "Asp-box" motif, a short structural element that has a unique β-hairpin configuration in many different structural contexts, was previously noted in a reelin repeat (Copley *et al.*, 2001). The motif present in the loop connecting the G- and the H-strands (G-H loop) does indeed assume its backbone configuration identical to the other Asp-box, creating a curled protrusion that serves as an upper ridge of the concave surface. A single structural Ca^{2+} is found at the middle of the convex surface, such that it "sews together" the discontinuous sheet between the A-B and C-D loops. It is important to note that the presence of this calcium site could not be predicted by sequence analysis, and its identification was made possible by direct structural determination. Examination of the sequence alignment revealed that the Asp residue in the last (K) strand, which provides the most important bidentate Ca^{2+} ligand, was conserved in all reelin subrepeats including repeats X and Z (Fig. 5.1B). Therefore, a reelin monomer would contain a total of 18 calcium ions. The calcium is buried in the protein's interior and appears to play an important structural role in stabilizing the domain. Indeed, the expression level of R3 diminishes when this calcium site is disrupted by mutation. It was also reported that the full-length reelin becomes more susceptible to protease digestion in the presence of EDTA (Lugli *et al.*, 2003). It is not known, however, whether there is any physiological function for these calcium ions other than domain stabilization.

As predicted from the sequence alignment, the only structural difference between subrepeats A and B was the insertion of a disulfide-bonded loop. The loops make long excursions such that they "sprout out" from the middle of strand H. Furthermore, this segment plays a major role in the interaction with subrepeat B (see below). Other than this loop insertion, subrepeats A and B are very similar and can be superimposed onto each other with root mean square (RMS) deviations of 1.57–1.99 Å for matched alpha carbon atoms.

Three-dimensional comparison revealed that the reelin subrepeat shows an unexpected similarity to the carbohydrate-binding domains of numerous enzymes and nonenzymes. The top hit in a DALI search is a noncatalytic domain of xylanase from *Clostridium thermocellum* (PDB ID; 1dyo, Z-score = 12.5) (Charnock *et al.*, 2000), followed by numerous jelly-roll modules all classified as "galactose-binding domain-like" in the SCOP database. When superimposed, the xylanase structure shows remarkable similarity to subrepeat B with RMS deviations of 2.6–3.0 Å for ~120 residues. Moreover, xylanase has one Ca^{2+} site that is structurally equivalent to that of the reelin repeats. Nevertheless, it shows only ~10% identity with reelin repeats at the amino acid level, making it difficult to evaluate any evolutionary relationship. As reelin is not known to bind any carbohydrates, the physiological significance of its resemblance to carbohydrate-binding domains remains unclear. It is important to note, however, that the concave side of these β jelly rolls is, in many cases, implicated in molecular recognition (Weis and Drickamer, 1996).

2.2.2 Central EGF Module

The EGF or EGF-like modules, which are extremely abundant and widespread among extracellular proteins in multicellular organisms, are frequently found in receptors and matrix proteins (Bork *et al.*, 1996). Their three-dimensional structure, which has been determined for many different proteins, generally exhibits a small ellipsoid with its N- and C-termini located at the opposite ends of its major axis (Campbell and Bork, 1993). They appear in widely different structural contexts and in varying numbers. In contrast, the EGF modules in reelin are unique in that they are invariably positioned between subrepeats A and B. The EGF modules within the reelin repeat assume a typical EGF-like fold with the disulfide-bonding pattern of 1–3, 2–4, 5–6. The uniqueness of the EGF modules in reelin compared with regular EGF modules stems from their involvement in compact subdomain packing. For example, an Asp residue immediately following the fifth Cys is completely conserved among all reelin EGF modules, although this position is often occupied by hydrophobic residues in typical EGF modules. This residue forms a bidentate hydrogen bond to an invariable Arg residue present at strand I of subrepeat A, forcing the ~90° bend at the subrepeat A–EGF interface. Another unique feature conserved across the reelin EGF is the presence of a Gly residue in the +2 position from the second Cys, which is obligated to accommodate contact with the Asp-box turn in subrepeat B.

2.2.3 Inter-subrepeat Interaction

Within a single reelin repeat structure, the concave side of subrepeat B approaches from the "side" of subrepeat A and grabs the disulfide-bonded loop, causing the axes of the β-sheets to become significantly twisted toward each other. In effect, the A-specific loop functions as a joint that enables a horseshoe-like arrangement. Because of inter-subrepeat interaction, the concave side of subrepeat B is occluded within the domain, while that of subrepeat A is freely accessible to the solvent. This interface, however, does not seem to be particularly stable since it is discontinuous and mainly hydrophilic in nature. It may therefore be possible that the repeat relaxes to assume an "open" conformation under certain conditions. Nevertheless, the compact subdomain packing is maintained in solution when multiple repeats are concatenated (see below).

3 Structure of a Four-Repeat Fragment

3.1 Two-Dimensional Electron Microscopy

Structures of single reelin repeats alone provide little information regarding the higher-order structure of a full-length reelin promoter, i.e., how each domain is organized in the context of the larger protein. It would be highly desirable to resolve the structure of larger reelin fragments or even the entire (i.e., full-length) molecule. However, protein production and crystallization on such a scale remains extremely difficult, if not impossible. Electron microscopy (EM) offers an excellent alternative to X-ray crystallography, as it requires relatively small amounts of protein and can be applied to proteins that assume multiple conformations. Negative staining EM of a recombinant fragment containing reelin repeats 3–6 (R3–6) revealed a very homogeneous molecular shape, i.e., an elongated rod with an average length of ~25 nm (Fig. 5.2B). Although a partial segment, this fragment covers about 40% of the total molecule and most importantly was capable of binding to the receptor and transducing the signal, as confirmed by both receptor binding and Dab1 phosphorylation assays (Jossin *et al.*, 2004; Nogi *et al.*, 2006). In these two-dimensional averaged images, particles showed largely straight, sometimes bent rod-like shapes, with four densities separated by segmentations. The most important feature of the EM images is the size of the fragment. The length of the rod was roughly the same as, or even shorter than, the longest dimension of the crystal structure of the single repeat multiplied by 4, strongly arguing against the possibility that the reelin repeat assumes a more extended conformation in the native protein. Three segmentations divide the rod into four parts, which most likely correspond to each reelin repeat.

3.2 Three-Dimensional Structure as Revealed by Electron Tomography

The conventional two-dimensionally averaged images can derive an overall molecular shape that is "projected" onto a plane. In order to examine the spatial arrangements of domains within a molecule, however, 3D information must be obtained. In fact, a methodology called "single-particle tomography" can extract 3D information from the same specimen prepared for conventional EM (Iwasaki *et al.*, 2005). It collects the image data at different tilt angles and reconstructs a 3D volume map for individual particles. From a pool of the representative tomograms, an averaged 3D volume map was constructed (Fig. 5.2C). It showed a flattened rod having dimensions of 240×50×30 Å. Segmentation is no longer visible, probably due to the slight axial offset that occurred among different particles during the averaging process. Having the 3D structures for a single reelin repeat domain in hand, one can now interpret these EM-derived structures at atomic resolution.

Because the first and last strands in a subrepeat are adjacent, the N- and C-termini (of a subrepeat) are in close proximity. As a result, the N- and C-termini of the entire repeat are located at the "upper" corners of the domain, roughly aligned with the axis of the EGF module (Fig. 5.2A). As the "linker" segments connecting the two consecutive repeats are usually only several residues long, there would not be much available space at the repeat–repeat junction. The EM images proved to be in complete agreement with these predictions; they do not show the "beads-on-a-string" conformation; instead, each repeat is stacked close together.

Atomic coordinates for R3, R5, R6, and modeled R4 can be very successfully fitted in the density (Fig. 5.2C). Four repeats are arbitrarily placed in the main body of the density such that they are related by a translational movement, with their longest dimension parallel to the rod axis. This arrangement of the domains is plausible, since both the 2D and 3D EM images did not show any signs of twists or meandering. It is, therefore, very likely that reelin repeats are related by translation without significant rotation, as shown in the model in Fig. 5.2C.

4 Structural Model for the Full-Length Reelin Molecule

4.1 Arrangement of Reelin Repeats

Using the experimentally derived structural information, the 3D architecture of reelin fragment R3–6 was deduced. However, this still leaves unresolved the rest of the molecule. We are in a good position to begin building a realistic structural model of the full-length reelin protein. The sequence alignment shown in Fig. 5.1B was modified from the alignment made by Ichihara *et al.* (2001) by incorporating structural information from the crystal structures. This "structural alignment" takes into account the alignment of the secondary structural elements rather than just maximizing the residue-wise matches. By using this alignment, we can define the (sub)repeat boundary and predict the 3D structure of the unknown parts more accurately.

It is obvious from the alignment that repeats 3 and 7 have unusually long N-terminal extensions (Fig. 5.1B). The polypeptide lengths between the last strand of subrepeat B and the first strand of the following subrepeat A are 7–9 residues for most repeats, except for repeat 3 (23 residues) and repeat 7 (34 residues). This indicates that there are long linker regions present before repeat 3 and after repeat 6. This configuration is in perfect agreement with the fact that the full-length reelin protein is cleaved at sites roughly located between repeats 2 and 3, as well as between repeats 6 and 7 by metalloproteinases *in vivo* (Jossin *et al.*, 2004; Lugli *et al.*, 2003). The EM images of the R3–6 fragment indicate that this segment behaves as a relatively rigid rod, with little interdomain flexibility. Therefore, it is natural to speculate that the reelin repeat region can be separated into three rods, R1–2, R3–6, and R7–8, joined by flexible and protease-susceptible linkers (Fig. 5.3A).

Another valuable piece of information extracted from the sequence alignment is the ~50-residue truncation of repeat 1A. This is not caused by a misalignment, since the segment immediately preceding the R1A (i.e., repeat Z) aligns fully to the "B-type" subrepeats, and, hence, cannot be offset. The missing part would contain five β strands participating in both the inner and outer sheets, making it difficult to imagine subrepeat 1A folding correctly without this piece. Curiously, the irregular "repeat Y" corresponds exactly to this segment (Fig. 5.1B). In fact, only "Y" and "1A" are imperfect among the 19 subrepeat segments in reelin, and they are complementary to each other. This points to the possibility that repeat Y is part of repeat 1 in the 3D structure, although they are separated by repeat Z in the primary structure (Fig. 5.3A).

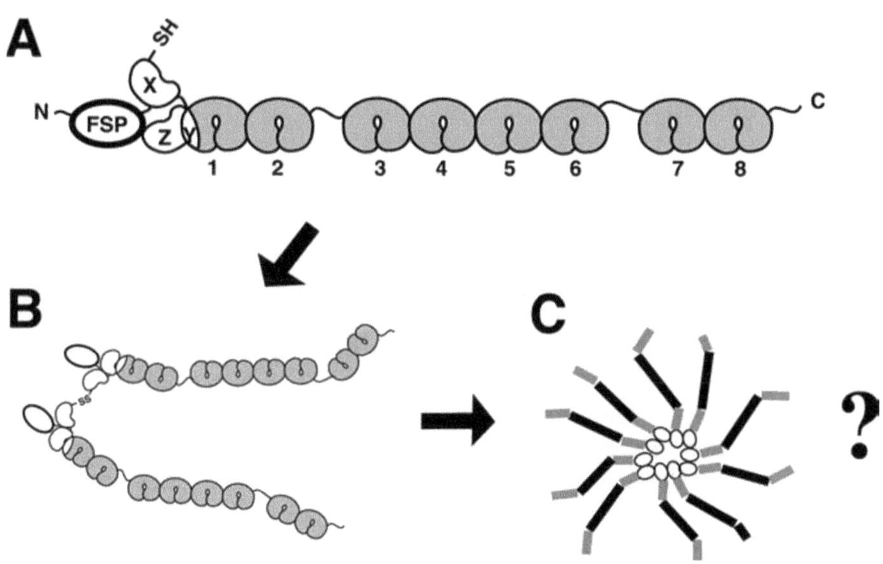

Fig. 5.3 Hypothetical model of native reelin protein. Predicted domain arrangement in a reelin protomer (**A**) suggests the presence of a segmented rod region (gray) and a dimerizing region at the N-terminus (white). The monomer may undergo dimerization using the free Cys residue in repeat X (**B**), and finally into higher-order multimerization (**C**) with radially presented rod regions containing receptor-binding sites (black)

4.2 A Model for the Reelin Monomer and Its Higher-Order Assembly

If we accept the above hypothesis that repeat Y constitutes the N-terminal portion of repeat 1, the primary sequence connectivity would place not only repeat Z but also repeat X adjacent to the A-side of repeat 1. What would then be the roles for these "lone" subrepeats?

Reelin is known to exist as a disulfide-bonded homodimer, and the Cys residue responsible for dimerization is thought to be located in the first 368-residue portion (Kubo *et al.*, 2002). This region is comprised of an F-spondin domain and the repeat X, which contain 4 and 7 cysteines, respectively. Sequence consideration and model building predict that Cys256 in repeat X is likely to be missing a bonding partner. It is therefore likely that the repeat X portion serves as a dimerizing point in the native protein (Fig. 5.3B).

In addition to covalent homodimerization, secreted reelin is also known to exist as noncovalently associated multimers (Utsunomiya-Tate *et al.*, 2000). In this regard, two additional B-type subrepeats with unoccupied concave sides (i.e., repeats X and Z) may provide a platform for multimerization. It is noteworthy that repeat X contains the epitope for the function-blocking antibody CR50, which has been shown to inhibit higher-order multimerization of reelin (D'Arcangelo *et al.*, 1997; Utsunomiya-Tate *et al.*, 2000). A multimerizing point located at the beginning of the linear reelin repeat segment, away from the receptor binding site (i.e., R3–6), could facilitate the assembly of a radially arranged reelin multimer (Fig. 5.3C), which would be ideal for transducing signals to neurons via receptor clustering (Strasser *et al.*, 2004).

5 Concluding Remarks

Despite the major advances in our understanding of the genetic, physiological, and cell biological aspects of reelin made in recent years, structural and biochemical analyses have proven slow to develop. Some of the reasons for this difficulty include the unusually large size of the protein, the lack of a panel of anti-reelin monoclonal antibodies encompassing a wide range of epitope regions, the presence of multiple protease-processed products, and the instability of the full-length protein after isolation. The structural determination of representative reelin repeat fragments has laid the foundation for discussion of the true molecular architecture of this protein. A single reelin repeat domain consists of a central EGF module flanked by two related subrepeats that assume an 11-strand β jelly-roll fold. Moreover, direct contact between the subrepeats ensures that this unit maintains a compact overall structure. Repeats 3 to 6 are linearly arranged and stacked together with no twists or meandering, making a stiff rod. Combining these structural data with a detailed sequence analysis, a picture of the full-length reelin structure has emerged in which the long eight-repeat region, subdivided into three segments

(R1–2, R3–6, and R7–8), are assembled into a disulfide-bonded homodimer and additional noncovalent multimers (Fig. 5.3). The validity of this model should be tested by further ultrastructural and biochemical analyses.

References

Bork, P., Downing, A. K., Kieffer, B., and Campbell, I. D. (1996). Structure and distribution of modules in extracellular proteins. *Q. Rev. Biophys.* 29:119–167.

Campbell, I. D., and Bork, P. (1993). Epidermal growth factor-like modules. *Curr. Opin. Struct. Biol.* 3:385–392.

Charnock, S. J., Bolam, D. N., Turkenburg, J. P., Gilbert, H. J., Ferreira, L. M., Davies, G. J., and Fontes, C. M. (2000). The X6 "thermostabilizing" domains of xylanases are carbohydrate-binding modules: structure and biochemistry of the Clostridium thermocellum X6b domain. *Biochemistry* 39:5013–5021.

Copley, R. R., Russell, R. B., and Ponting, C. P. (2001). Sialidase-like Asp-boxes: sequence-similar structures within different protein folds. *Protein Sci.* 10:285–292.

D'Arcangelo, G., and Curran, T. (1999). Reelin. In: Kreis, T., and Vale, R. (eds.), *Guidebook to the Extracellular Matrix, Anchor, and Adhesion Receptors*. Oxford University Press, Oxford, pp. 465–467.

D'Arcangelo, G., Miao, G. G., Cheng, S. C., Soares, H. D., Morgen, J. I., and Curren, T. (1995). A protein related to extracellular matrix proteins mutant reeler. *Nature* 374: 719–723.

D'Arcangelo, G., Nakajima, K., Miyata, T., Ogawa, M., Mikoshiba, K., and Curran, T. (1997). Reelin is a secreted glycoprotein recognized by the CR-50 monoclonal antibody. *J. Neurosci.* 17:23–31.

Ichihara, H., Jingami, H., and Toh, H. (2001). Three novel repetitive units of reelin. *Brain Res. Mol. Brain Res.* 97:190–193.

Iwasaki, K., Mitsuoka, K., Fujiyoshi, Y., Fujisawa, Y., Kikuchi, M., Sekiguchi, K., and Yamada, T. (2005). Electron tomography reveals diverse conformations of integrin alphaIIbbeta3 in the active state. *J. Struct. Biol.* 150:259–267.

Jossin, Y., Ignatova, N., Hiesberger, T., Herz, J., Lambert de Rouvroit, C., and Goffinet, A. M. (2004). The central fragment of reelin, generated by proteolytic processing in vivo, is critical to its function during cortical plate development. *J. Neurosci.* 24:514–521.

Kubo, K., Mikoshiba, K., and Nakajima, K. (2002). Secreted reelin molecules form homodimers. *Neurosci. Res.* 43:381–388.

Lugli, G., Krueger, J. M., Davis, J. M., Persico, A. M., Keller, F., and Smalheiser, N. R. (2003). Methodological factors influencing measurement and processing of plasma reelin in humans. *BMC Biochem.* 4:9.

Nogi, T., Yasui, N., Hattori, M., Iwasaki, K., and Takagi, J. (2006). Structure of a signaling-competent reelin fragment revealed by X-ray crystallography and electron tomography. *EMBO J.* 25:3675–3683.

Strasser, V., Fasching, D., Hauser, C., Mayer, H., Bock, H. H., Hiesberger, T., Herz, J., Weeber, E. J., Sweatt, J. D., Pramatarova, A., Howell, B., Schneider, W. J., and Nimpf, J. (2004). Receptor clustering is involved in reelin signaling. *Mol. Cell Biol.* 24:1378–1386.

Utsunomiya-Tate, N., Kubo, K., Tate, S., Kainosho, M., Katayama, E., Nakajima, K., and Mikoshiba, K. (2000). Reelin molecules assemble together to form a large protein complex, which is inhibited by the function-blocking CR-50 antibody. *Proc. Natl. Acad. Sci. USA* 97:9729–9734.

Weis, W. I., and Drickamer, K. (1996). Structural basis of lectin-carbohydrate recognition. *Annu. Rev. Biochem.* 65:441–473.

Chapter 6
Comparative Anatomy and Evolutionary Roles of Reelin

Gundela Meyer

Contents

1 The *Reelin* Gene Is Evolutionarily Preserved

The *reelin* gene maps to mouse chromosome 5 and human chromosome 7q22 (DeSilva *et al.*, 1997; Royaux *et al.*, 1997). The mouse *reelin* gene has a large size, about 450 kb, principally due to the presence of some very large introns. It is composed of 65 exons, 51 of which encode the eight reelin repeats. At the 3´-terminal portion of the gene, alternative splicing involves the inclusion of a hexanucleotide AGTAAG encoding amino acids Val-Ser, which create a potential phosphorylation site. This sequence is flanked by two introns and considered a *bona fide* exon (exon 64) (Royaux *et al.*, 1997). The hexanucleotide sequence is evolutionarily conserved, because it is observed in the same relative location in the turtle and lizard cDNA, while the similar sequence AATAAG is present in chick (Lambert de Rouvroit *et al.*, 1999). An alternative, polyadenylated product corresponds to the alternative exon 63a, expressed in the embryonic mouse brain, that codes for a truncated protein lacking the C-terminal region. This alternative mRNA represents between 10 and 25% of total reelin message

G. Meyer
Departamento de Anatomía, Facultad de Medicina, Universidad de La Laguna, 38071 La Laguna, Spain
e-mail: gmeyer@ull.es

in the embryonic mouse brain and is most abundant in Cajal-Retzius neurons of the cerebral cortex and hippocampus and in granule cells of the cerebellum; highly similar sequences are also found in human and rat. While reelin mRNA containing the microexon 64 is the major form in the brain of mouse, rat, man, turtle, and lizard, reelin transcripts in liver and kidney lack the hexanucleotide (Lambert de Rouvroit *et al.*, 1999).

The human *RELN* 5'-untranslated region (UTR) has been related to genetic susceptibility to autism (Persico *et al.*, 2001); the proximal *RELN* promoter is CG-rich, and it has been proposed that hypermethylation of the promoter is associated with decreased expression of reelin in psychiatric patients (M. L. Chen *et al.*, 2002; Y. Chen et al., 2002).

The reelin protein is a large (3461 amino acids long), secreted glycoprotein (D'Arcangelo *et al.*, 1995, 1997). N-terminal antibodies reveal two fragments of about 320 and 180 kDa, whereas C-terminal antibodies show fragments of about 240 and 100 kDa. The antibody most widely used in comparative neuroanatomical studies, which reacts with reelin in almost all vertebrates, is the N-terminal monoclonal antibody 142 (de Bergeyck *et al.*, 1998), which has also been used for the figures of this review. In most brain extracts and body fluids, full-length reelin is rarely detected. The central portion of reelin (Jossin *et al.*, 2004) is involved in binding of VLDLR (very-low-density lipoprotein receptor) and ApoER2 (apolipoprotein E receptor type 2) and signal activation, eliciting phosphorylation of the cytoplasmic adapter protein disabled 1 (Dab1), the target of the Reelin signal (Rice *et al.*, 1998; Hiesberger *et al.*, 1999; Howell *et al.*, 1999a; Trommsdorff *et al.*, 1999; Rice and Curran, 2001). Comparison of the *Dab1* gene of zebrafish, mouse, and human also shows an overall conservation of the genomic organization, although this very complex gene has shorter introns and some variations in the exonic sequences in the zebrafish compared to the mammalian *Dab1* gene (Bar *et al.*, 2003; Costagli *et al.*, 2006). The lipoprotein receptors are highly conserved in vertebrates and invertebrates (Willnow *et al.*, 1999), and the *disabled* gene is present in *Drosophila melanogaster* (Gertler *et al.*, 1989). By contrast, *reelin*-related cDNA sequences have been described in turtles, lizards, and chicks, but were not detected in *Caenorhabditis elegans* or *Drosophila* (Bar *et al.*, 2000), and may be absent in invertebrate genomes.

2 Conserved Reelin Expression Pattern in the Central Nervous System

2.1 Reelin in the Adult CNS

Reelin is present in the nervous system of all vertebrates, from amphioxus (C. G. Pérez-García and G. Meyer, unpublished), lamprey, and fish to amphibians, reptiles, birds, and mammals. Reelin is widely expressed throughout the central nervous system (CNS), and the overall expression pattern is surprisingly conserved.

The highest divergence from the common expression pattern is observed in the dorsal telencephalon of the zebrafish, which probably corresponds to the pallium of other vertebrates (Nieuwenhuys and Meek, 1990) and will be discussed separately. In the adult lamprey (*Petromyzon marinus* L.), the most ancient representative of living vertebrates, reelin-positive neurons are present in the olfactory bulb, in pallial and subpallial regions of the telencephalon, in some hypothalamic nuclei and habenula, in nerve brain stem motor nuclei and neurons of the reticular formation, as well as in the rostral spinal cord (Pérez-Costas *et al.*, 2004). In the adult zebrafish (*Danio rerio*), reelin is strongly expressed in various areas of the dorsal and ventral telencephalon, but not in the olfactory bulb. Reelin mRNA is also present in several nuclei of the dorsal thalamus and in most hypothalamic regions, in the pretectum, optic tectum, tegmentum, and throughout the meso-rhomboencephalic reticular formation. In the spinal cord, reelin expression is confined to a subpopulation of interneurons. Strong *reelin* mRNA signal is in granule cells of the cerebellum and, outside the CNS, in the retina (Costagli *et al.*, 2002).

Remarkably, many of the expression sites of reelin are common to nonmammalian and mammalian vertebrates. The presence of reelin in the adult mammalian brain has been examined in mouse and rat (Alcantara *et al.*, 1998; Ramos-Moreno *et al.*, 2006). Reelin transcripts outside the cerebral cortex are prominent in the olfactory bulb, mainly in mitral cells but also in some periglomerular neurons. In general, expression seems to decline during the postnatal period and is rather weak in adult medial septum/diagonal band, amygdaloid area and hypothalamus, as well as in the pretectum (Alcantara *et al.*, 1998).

In general, the expression of reelin is highly conserved in laminated structures [e.g., olfactory bulb (Fig. 6.1C), retina (Fig. 6.1A), cerebral cortex (Fig. 6.3B), cerebellum, and optic tectum/superior colliculus], where the expression of reelin is usually complementary to that of the reelin receptors and the effector protein Dab1 (Rice and Curran, 2001). Lamination of many of these centers, in particular the cerebral cortex and cerebellum, is disturbed in *reelin*, *Dab1*, and double receptor-deficient mice, indicating the important developmental role of the reelin signaling pathway (Lambert de Rouvroit and Goffinet, 1998; Rice *et al.*, 1999; Trommsdorff *et al.*, 1999). However, reelin expression is also conserved in nonlaminated structures, such as in the hypothalamus, where we were able to detect the protein in widely disparate species, such as adult lizards (Fig. 6.1B), adult mice, and young adult cats.

A wider distribution of reelin in cortical and subcortical structures, including long-projecting fiber tracts, has been reported in the ferret, along with the suggestion that reelin may be anterogradely transported by axons and secreted at their terminal arborizations (Martinez-Cerdeño *et al.*, 2003). A similarly widespread, almost generalized presence of reelin has been described in the macaque brain (Martinez-Cerdeño *et al.*, 2002). In contrast to previous studies that emphasized the presence of reelin in interneurons and the negativity of most long-projecting neurons such as cortical pyramidal cells and cerebellar Purkinje cells, Lambert de Rouvroit and Goffinet (1998) reported reelin immunoreactivity in the great majority of brain neurons and in their axonal projections. Further *in situ* hybridization studies are needed to confirm such abundancy of reelin in the adult mammalian brain.

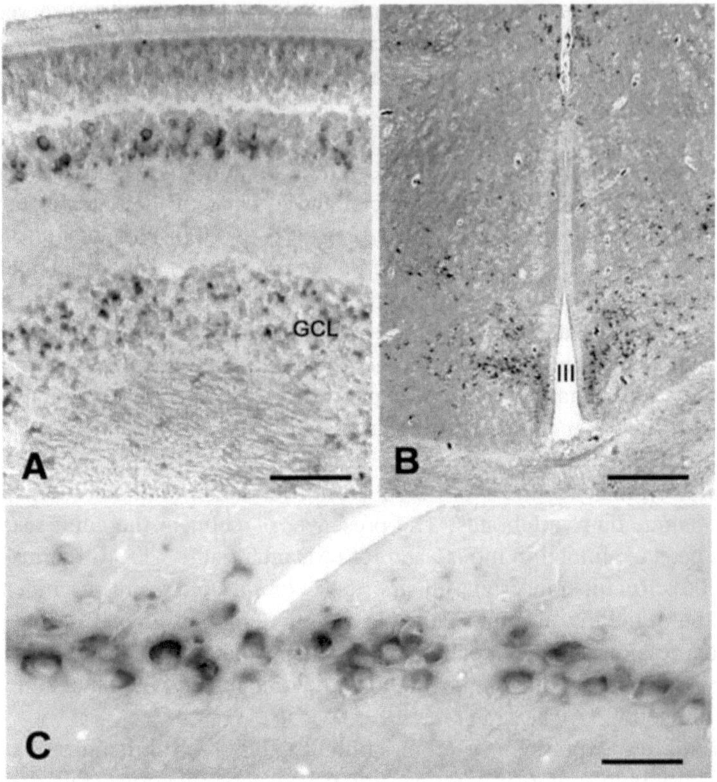

Fig. 6.1 Conserved expression pattern of reelin in the nervous system of amniotes. (**A**) Reelin-immunoreactive (ir) cells in the retina of the lizard *Lacerta galloti*. GCL, ganglion cell layer. (**B**) Reelin-ir neurons in the lizard hypothalamus. III, third ventricle. (**C**) Mitral cells in the olfactory bulb of the cat. Bars: 30 μm (**A**), 50 μm (**B**), 20 μm (**C**)

2.2 Reelin in the Developing CNS

As mentioned above, reelin is widely expressed throughout the CNS. Comparative studies in a variety of developing vertebrate species show that the reelin signal is usually higher during development than in the adult, and that more brain centers are reelin-positive. Among the centers that display *reelin* mRNA or protein expression during development of the crocodile are the septum, dorsal ventricular ridge, the subventricular telencephalic zone, lateral geniculate nucleus, cochleovestibular, and sensory trigeminal nuclei (Tissir *et al.*, 2003). Similar distributions are noted in the embryonic mouse (Schiffmann *et al.*, 1997; Alcantara *et al.*, 1998), where reelin is also expressed in the preoptic area, striatum, zona incerta, habenula, lateral geniculate nucleus, and superior colliculus. In the zebrafish, reelin expression is particularly dynamic during the first 24–72 hours of development, with early and intense expression in the telencephalon, ventral areas of the diencephalon, mesencephalon, hindbrain, and spinal cord. Later on, the pattern of reelin becomes restricted to specific CNS regions and cell populations where expression persists unchanged to 1–3 months of age (Costagli *et al.*, 2002). Similarly, in the larval stages of the sea lamprey, reelin is expressed in the olfactory bulb, pallium, habenula, hypothalamus, and optic tectum (Pérez-Costas *et al.*, 2002). On the whole, the early and widespread expression of reelin is in keeping with important roles of the reelin--Dab1 signaling pathway in the migration and final positioning of newly born neurons and in the formation of laminated structures (Tissir *et al.*, 2002; Tissir and Goffinet, 2003).

In addition to abundant reelin expression in many brain structures, reelin is highly and specifically expressed in the Cajal-Retzius cells of the developing cerebral cortex (D'Arcangelo *et al.*, 1995; Ogawa *et al.*, 1995; Meyer *et al.*, 1999). Since these cells play a key role in the development and evolution of the cortex, they will be discussed in detail in Section 3.3.

3 The Evolution of Reelin Expression from the Submammalian Pallium to the Mammalian Neocortex

3.1 Reelin Expression in the Adult Nonmammalian Pallium

In the course of evolution, the cortical mantle of the telencephalon undergoes significant changes in architectonic organization, and reelin expression reflects these morphological and functional differences. The most divergent arrangement of pallial neurons is found in teleosts. The teleostean pallium develops through a unique process of eversion, during which the cerebral hemispheres evert and the dorsal and internal regions bend outwards. In the zebrafish pallium, no laminar arrangement can be recognized, and many neurons are reelin-positive (Pérez-García *et al.*, 2001; Costagli *et al.*, 2002). Fig. 6.2A shows reelin-immunoreactive neurons in the dorsolateral pallium of an adult zebrafish, where groups of reelin-positive neurons lie side by side with groups of reelin-negative neurons. Since most neurons in the dorsal telencephalon are small and their cytoplasm is sparse, the reelin signal is usually not very pronounced. The most intense reelin immunoreactivity is present along the everted ventricular layer (Pérez-García *et al.*, 1991). Costagli *et al.* (2002) described *reelin* mRNA in medial, lateral, central, and dorsal regions of the dorsal telencephalon, whereas in the ventral telencephalon only a few cells showed weak positivity.

The amphibian pallium forms through evagination of the cerebral hemispheres from the prosencephalic vesicle, which is the developmental mechanism common to all vertebrates except teleosts, and comprises several subregions. The dorsal and lateral pallium consists of several rows of neurons close to the ventricle, whereas the wide molecular layer is almost cell free. In adult *Hyla meridionalis* (Mediterranean Treefrog), reelin-expressing neurons are located at the periphery of the periventricular cell layer, whereas the molecular layer is basically reelin-negative (Fig. 6.2B).

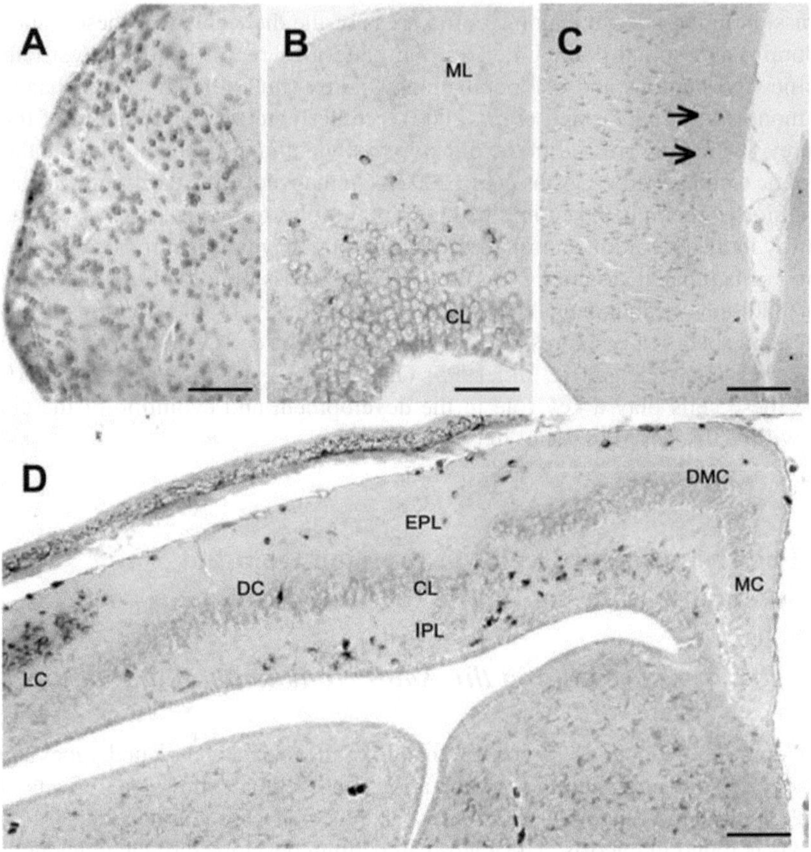

Fig. 6.2 Variable expression patterns of reelin in the adult nonmammalian pallium. (**A**) Reelin in the dorsal telencephalon of the zebrafish *Danio rerio*. Many diffusely distributed neurons are reelin-positive. (**B**) In the dorsal pallium of the Mediterranean Treefrog (*Hyla meridionalis*), few neurons outside the main cell layer (CL) show faint reelin-immunoreactivity. The molecular layer (ML) is almost cell-free. (**C**) The medial pallium of the pigeon *Columba livia* displays very few small reelin-ir neurons in the cell-free outer layer. (**D**) The pallium of the lizard *Lacerta galloti* is subdivided into a medial cortex (MC), dorsomedial cortex (DMC), dorsal cortex (DC), and lateral cortex (LC). The main cell layer is reelin-negative and sandwiched between the external plexiform and internal plexiform layers (EPL, IPL), both populated by reelin-positive neurons. In the lateral cortex, neurons in the main cell layer express reelin. Bars: 20 µm (**A**), 25 µm (**B**, **C**), 50 µm (**D**)

Although the amphibian pallium has an extremely rudimentary architecture, it already displays a basic feature of the vertebrate cortex: the principal, long-projecting neurons are usually reelin-negative, whereas reelin is confined to a subpopulation of interneurons. In the pigeon pallium (*Columba livia*) (Fig. 6.2C), reelin-positive neurons are rare, with only a few in the molecular layer of the medial cortex.

The adult lizard displays a clearly laminated cortex composed of an external plexiform or molecular layer, an intermediate cellular layer formed by densely packed cell somata, and an internal plexiform layer (Fig. 6.2D). The plexiform layers contain the dendrites of the bitufted pyramidal cells, and a number of interneurons which are usually reelin-positive. Architectonic differences allow a subdivision into four zones, namely, the medial, dorsomedial, dorsal, and lateral cortex (Ulinski, 1990), but the three-layered organization is a common feature. Examination of three squamate reptiles (*Lacerta galloti, Tarentola delalandii,* and *Chalcides ocellatus*) showed a very similar architecture and distribution of reelin-positive neurons. The compact cell layer is reelin-negative but Dab1-positive, and sandwiched between the two plexiform layers containing small numbers of reelin-positive neurons (Pérez-García *et al.*, 2001). The exception is the lateral cortex where even in adult animals the main cell layer is reelin-positive. We will discuss this peculiar feature of the lateral cortex in Section 4, in the context of reelin expression in the entorhinal cortex and hippocampus.

In the turtle *Clemmys caspica*, lamination is less pronounced than in the lizard, and the principal neurons form a loosely organized cell layer close to the ventricle. Very few reelin-positive cells lie in the wide external plexiform layer, while the internal plexiform layer is devoid of reelin (Pérez-García *et al.*, 2001).

In sum, the examples presented here —teleosts, amphibians, birds, and reptiles— demonstrate that reelin expression in the adult pallium is extremely variable and ranges from abundant expression in the everted pallium of the zebrafish to an extremely sparse presence of the protein in the amphibian, bird, and turtle pallium. The fact that in the evaginated amniote pallium the main projection neurons are usually reelin-negative seems to be the most conserved trait.

3.2 Reelin Expression in the Adult Mammalian Cortex

In most mammalian species examined, the presence of reelin in the cerebral neocortex is confined to GABAergic interneurons (Alcantara *et al.*, 1998; Pesold *et al.*, 1998; Fatemi *et al.*, 2000; Martinez-Cerdeño and Clascá, 2002), while the main projection neurons, the pyramidal cells, express Dab1 (Rice *et al.*, 1998; Rice and Curran, 2001). The highest density of reelin-positive interneurons is consistently observed in the molecular layer (layer I), but they are also common in layers II–VI (Fig. 6.3B). In the molecular layer, reelin-expressing neurons are usually small and rounded, although sometimes they are larger and horizontally oriented, resembling the Cajal-Retzius cells of the fetal stage (Fig. 6.3A; see Section 3.3). Cortical layer I has a variable width in the different mammalian species, and also the reelin-positive cell populations in this layer are to some extent variable. For instance, in the wide odontocete (*Tursiops truncatus* and *Ziphius cavirostris*) molecular layer, reelin-positive neurons display a large variety of shapes and sizes and are particularly numerous in the depths of the numerous sulci and cortical folds. The opposite pattern is found in the hedgehog cortex, also characterized by a wide molecular layer, but which contains few small reelin-positive neurons located close to the pial surface (Pérez-García *et al.*, 2001). In human, small reelin-positive neurons are very

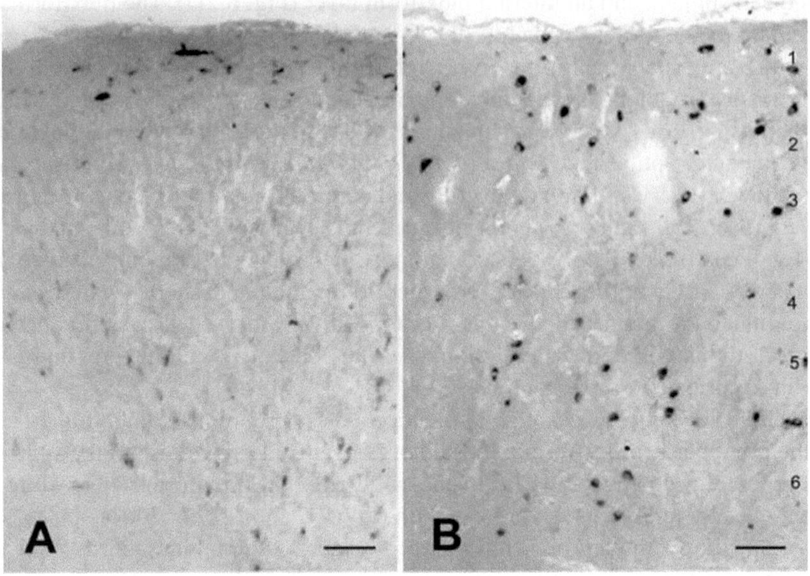

Fig. 6.3 Cajal-Retzius cells and interneurons in the cerebral cortex of the cat. (**A**) Reelin-immunoreactive (ir) neurons in a late prenatal stage. Interneurons are still faintly stained, and a Cajal-Retzius cell is still prominent in the marginal zone (layer 1). (**B**) At 1 postnatal month, interneurons in all layers are strongly reelin-ir, and Cajal-Retzius cells have disappeared. Bars: 30 µm

numerous in the postnatal layer I, but decrease in number during the first years of life. The adult human molecular layer (Fatemi *et al.*, 2000) is populated by few rounded interneurons, similar to those present in the adult rodent and carnivore cortex (Fig. 6.4E).

The reelin-positive interneurons of cortical layers II–VI were examined in detail in mouse (Alcantara *et al.*, 1998; Fatemi *et al.*, 1999), rat (Pesold *et al.*, 1998), and monkey (Rodriguez *et al.*, 2000, 2002). In mice, they appear at birth, first in layers V and VI of the neocortex and later on also in layers II–IV. After a peak during the first postnatal week, *reelin* mRNA expression gradually decreases, although reelin-positive interneurons persist throughout adult life. In the developing cat cortex, reelin-positive interneurons are faintly stained before birth, and acquire intense expression only postnatally (Fig. 6.3A,B). Coexpression experiments showed that reelin is expressed in subsets of GABAergic interneurons, some of which express also the interneuron markers calbindin, calretinin, somatostatin, and NPY, indicating that *reelin* transcripts are expressed in a heterogeneous population of nonpyramidal neurons (Alcantara *et al.*, 1998; Pesold *et al.*, 1999). Similarly, in the adult rat, reelin-positive neurons are scattered throughout all cortical layers, and a majority are GABAergic (Pesold *et al.*, 1998). Combined reelin immunolabeling and electron microscopy revealed that reelin is also present in the extracellular space surrounding dendritic

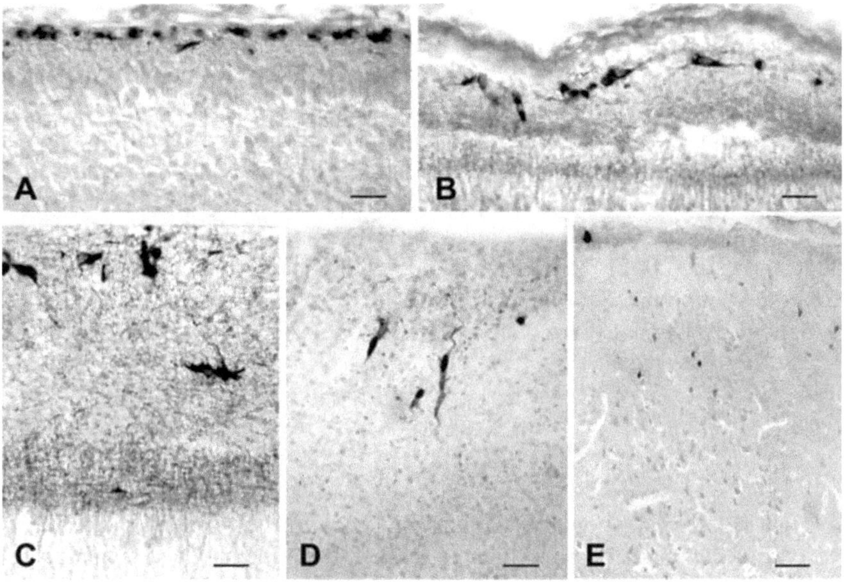

Fig. 6.4 The unique differentiation of human Cajal-Retzius cells. Human fetuses at (**A**) 9 gestational weeks (GW), (**B**) 14 GW, (**C**) 20 GW, (**D**) 25 GW, stained with anti-reelin antibody 142; in **D** doubly stained for p73 (in the nucleus) and reelin (in the cytoplasm). Cajal-Retzius are initially horizontally oriented and attached to the pial surface. They change morphology and descend to deeper levels of the marginal zone, extending a reelin-ir axonal plexus (PL in **C**). They degenerate and die around 25–30 GW. In the adult (**E**), they have disappeared, and layer 1 contains few small reelin-ir interneurons. Bars: 30 μm

shafts and dendritic spines in layers I and II of wild-type mice cortex in a discontinuous pattern; in the hippocampus, the neuropil is heavily and uniformly reelin-immunoreactive (Pappas *et al.*, 2001). In the patas monkey, *reelin* mRNA is expressed in all cortical areas and layers, with the highest numbers in layer II. GABAergic interneurons have been classified according to their ability to express reelin, which is secreted into the extracellular matrix where it may interact with dendritic shafts, dendritic spines, and spine postsynaptic densities (Rodriguez *et al.*, 2002).

There are significant discrepancies regarding reelin expression in adult human cortex. Whereas some authors observe protein/mRNA only in nonpyramidal neurons (Fatemi et al., 2000; Pérez-García *et al.*, 2001; Martinez-Cerdeño and Clascá, 2002; Eastwood and Harrison, 2006), others describe the presence of reelin also in pyramidal cells, as well as in the neuropil, dendritic spines, small axon terminals, and postsynaptic densities (Deguchi *et al.*, 2003; Roberts *et al.*, 2005). The latter distribution is in line with a study of the adult macaque cortex (Martinez-Cerdeño *et al.*, 2002) which reports that the majority of cortical neurons express reelin, suggesting that reelin can influence most brain circuits of the adult primate brain. It would be important to solve these discrepancies, because recent studies have involved the reelin–dab1 signaling pathway in cortical and hippocampal synaptic plasticity (Weeber *et al.*, 2002). Reelin enhances long-term potentiation (LTP), and this function is abolished if either VLDLR or ApoER2 is absent. The impairment of ApoE receptor-dependent neuromodulation may contribute to cognitive impairment and synaptic loss in Alzheimer's disease. Furthermore, synaptic strength and activity-dependent synaptic plasticity depend to a large extent on NMDA receptors on the postsynaptic side of excitatory synapses. Reelin potentiates calcium influx through NMDA receptors and may thus modulate learning and memory (Beffert *et al.*, 2005, 2006; Chen *et al.*, 2005). The activities of the reelin pathway in cognitive functions are of utmost importance for understanding the possible relationship between reelin and Alzheimer's disease (Saez-Valero *et al.*, 2003; Botella-Lopez *et al.*, 2006).

3.3 *Evolutionary Aspects of the Cajal-Retzius Cells*

The Cajal-Retzius cells populate the marginal zone of the developing cerebral cortex and hippocampus and are intimately related to the reelin–Dab1 signaling pathway. First described by Retzius and Cajal more than one century ago, they have attracted the interest of many researchers, sometimes with controversial opinions (reviewed by Meyer *et al.*, 1999; Meyer, 2001). Today, there is an overall consensus that Cajal-Retzius cells are a heterogeneous group of neurons that appear from the earliest stages of cortical development, derive from several birth places, most notably the cortical hem, and invade the neocortical marginal zone by subpial tangential migration (Meyer and Goffinet, 1998; Meyer and Wahle, 1999; Meyer *et al.*, 2002; Bielle *et al.*, 2005; Yoshida *et al.*, 2006; Meyer, 2007).

The most interesting feature of Cajal-Retzius cells is that they secrete high levels of reelin, which binds to the reelin receptors expressed in the upper part of the developing cortical plate (Pérez-García *et al.*, 2004). In turn, Dab1 is present in radially migrating neurons and in the radial glia cells in the ventricular zone (Luque *et al.*, 2003; Meyer *et al.*, 2003); it docks to an NPxY sequence in the intracellular domain of VLDLR and ApoER2 (Howell *et al.*, 1999b) and becomes phosphorylated on key Tyr residues when reelin binds to its receptors (Keshvara *et al.*, 2001). Cajal-Retzius cells are crucial for cortical lamination and cell positioning, and degenerate and die once they have completed their developmental function (Derer and Derer, 1990; Meyer *et al.*, 2002).

Comparative studies of the developing pallium of chick (Bernier *et al.*, 2000), turtle (Bernier *et al.*, 1999), lizard (Goffinet *et al.*, 1999), and crocodile (Tissir *et al.*, 2003) have shown that reelin-positive neurons are a consistent feature of the outer, cell-sparse layer of the developing cortex of all amniotes and may have evolved from an ancestral cell type present in stem amniotes (Bar *et al.*, 2000). In reptiles and rodents, Cajal-Retzius cells have a horizontal shape and lie in a subpial position. In rats and particularly in humans, Cajal-Retzius cells undergo complex morphological changes during development, and settle at progressively deeper levels of the marginal zone (Fig. 6.4A–D) (Meyer and González-Hernández, 1993; Meyer *et al.*,

1998). In turtles and apparently all mammals, Cajal-Retzius cells coexpress transcription factors such as p73 (Fig. 6.4D, 6.5) (Yang *et al.*, 2000; Meyer *et al.*, 2002; Tissir *et al.*, 2003) and Tbr1 (Hevner *et al.*, 2003). Furthermore, the intensity of the reelin signal is much higher in mammals than in nonmammalian species, suggesting that the reelin pathway played a driving role in cortical evolution in mammalian and squamate lineages (Bar *et al.*, 2000). In keeping with this hypothesis is the observation that in the human cortex Cajal-Retzius cells are more numerous than in other mammalian species and display highly differentiated morphological features (Meyer and González-Hernández, 1993; Meyer and Goffinet, 1998; Meyer *et al.*, 2002). Most importantly, they extend a huge axonal fiber plexus at the boundary between the cortical plate and the marginal zone (Meyer and González-Hernández, 1993; Cabrera-Socorro *et al.*, 2007). In the rat, the axons of Cajal-Retzius cells show secretory activity (Derer *et al.*, 2001). Likewise, the human Cajal-Retzius plexus is reelin-immunoreactive (Fig. 6.4C), suggesting that it may also secrete reelin and lead to a further diffusion and amplification of the reelin signal (Cabrera-Socorro *et al.*, 2007).

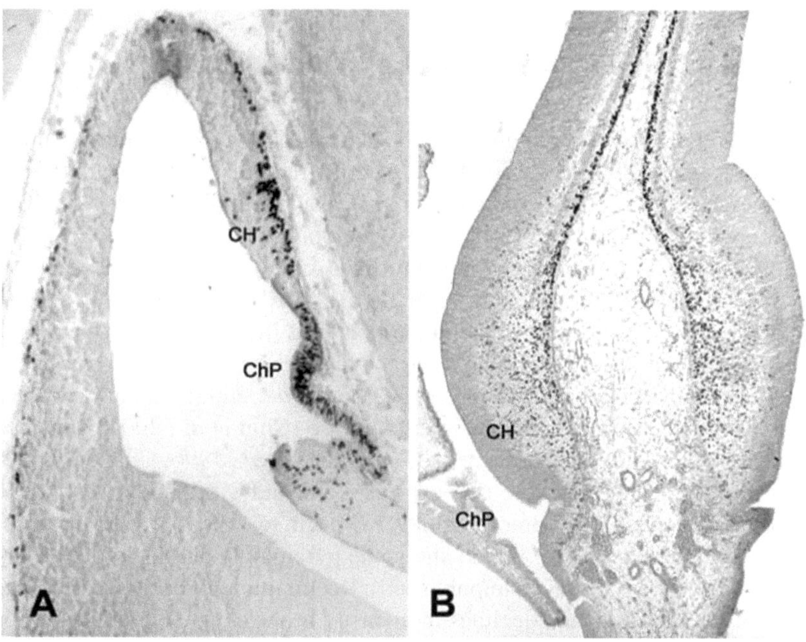

Fig. 6.5 The evolution of the cortical hem, the main source of Cajal-Retzius cells. (**A**) The cortical hem in a mouse at embryonic day 12, and (**B**) in a human embryo at 9 GW.The transcription factor p73 is highly expressed in Cajal-Retzius cells and may determine their survival and death. Note the size increase of the human hem, and the large numbers of human Cajal-Retzius cells, which lead to a high reelin signal in the human cortex. Ch, choroid plexus; CH, cortical hem. Bars: 50 μm (**A**), 100 μm (**B**)

The unique differentiation of human Cajal-Retzius cells may be related to the presence of p73 in their nucleus. Cajal-Retzius cells express Delta Np73, a truncated p73 isoform with anti-apoptotic activities (Pozniak *et al.*, 2000; Yang *et al.*, 2002; Meyer *et al.*, 2004), which may allow Cajal-Retzius cells to survive during the protracted period of human cortical migration and to acquire their complex morphology. The majority of p73-expressing Cajal-Retzius cells have their origin in the cortical hem (Fig. 6.5A,B), a signaling center at the interface of the choroid plexus and the hippocampus (Meyer *et al.*, 2002; Yoshida *et al.*, 2006), although there are additional sources at the boundaries of the telencephalon (Bielle *et al.*, 2005). Interestingly enough, the cortical hem undergoes a significant increase in size during evolution (Fig. 6.5), in parallel with an increase in number of p73/reelin-expressing Cajal-Retzius cells in the cortex (Cabrera-Socorro *et al.*, 2007). Altogether, these findings support the view that the increasing intensity of the reelin signal in Cajal-Retzius cells of the marginal zone during phylogenesis may correlate with the increasing complexity of cortical architecture and cognitive functions in the course of evolution. The evolutionary significance of Cajal-Retzius cells is also reflected by the novel "human accelerated regions" RNA gene (HAR1F) that is expressed specifically in human Cajal-Retzius neurons (Pollard *et al.*, 2006).

4 Reelin in Hippocampus and Entorhinal Cortex

The expression pattern of reelin in the hippocampal formation and entorhinal cortex is particularly interesting, because both are core structures of the explicit memory system, which is crucial for learning and cognitive abilities (Qiu and Weeber, 2007). Reelin is, in fact, highly expressed in the developing and adult hippocampus and entorhinal cortex. In the adult human entorhinal cortex, reelin is present in large polygonal cells that form clusters in layer II (Fig. 6.6C) (layer Pre-α of Braak and Braak, 1992). These neurons are important insofar as they give rise to the perforant path connecting the entorhinal cortex with the hippocampus (Steward and Scoville, 1976; Witter and Groenewegen, 1984; van Groen *et al.*, 2003), and are particularly vulnerable for degeneration in Alzheimer's disease (Chin *et al.*, 2007; Van Hoesen and Hyman, 1990; Jellinger *et al.*, 1991; Gomez-Isla *et al.*, 1996; Chin *et al.*, 2007). Neurons in adult human layer Pre-α coexpress reelin and Dab1 (Meyer *et al.*, 2003). In human fetuses, neurons of entorhinal layer II begin to express low levels of reelin already around 20 gestational weeks, and some pyramidal neurons in deeper layers also become positive (G. Meyer, unpublished). Reelin immunoreactivity is high after birth and in young adults and declines in intensity in the course of adult life (Pérez-García *et al.*, 2001). Layer II cells lose reelin and/or die in Alzheimer's disease (Chin *et al.*, 2007), suggesting a possible correlation between reelin expression and Alzheimer-related degenerative processes. This hypothesis is supported by the finding that the numbers of reelin-expressing neurons in layer II are reduced in human amyloid precursor protein (hAPP) transgenic mice, whereas reelin levels in entorhinal interneurons are not affected (Chin *et al.*, 2007).

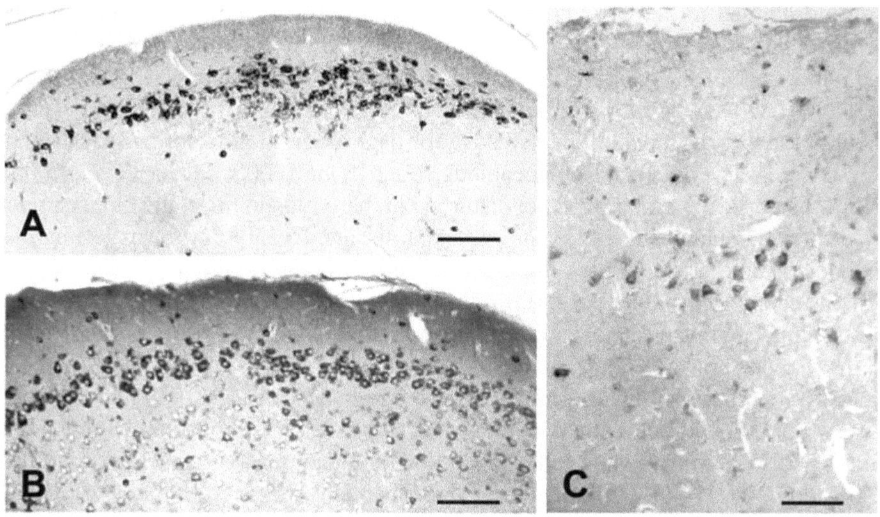

Fig. 6.6 The evolution of reelin expression in the entorhinal cortex. (**A**) The lateral cortex of the lizard *Lacerta galloti*, where the principal cells are intensely reelin-immunoreactive. (**B**) Reelin in layer 2 of the cat entorhinal cortex. There are also reelin-positive pyramidal neurons in deeper layers. (**C**) Adult human (49 years) entorhinal cortex. Reelin is expressed in clusters of large modified pyramidal neurons in layer 2; the intensity of expression declines during adult life and is moderate at 49 years. Bars: 50 μm (**A**), 60 μm (**B**), 30 μm (**C**)

The positivity of entorhinal layer II neurons is remarkably conserved in phylogenesis. All mammalian species examined so far display a usually strong reelin signal in this specific cell type, i.e., mouse (Alcantara *et al.*, 1998), rat (Drakew *et al.*, 1998; Pesold *et al.*, 1998; Ramos-Moreno *et al.*, 2006), ferret (Martinez-Cerdeño *et al.*, 2003), cat (Fig. 6.6B), and gerbil (Pérez-García *et al.*, 2001). Even in the lizard *Lacerta galloti*, the lateral cortex is strongly reelin-immunoreactive and closely resembles the mammalian entorhinal cortex (Fig. 6.6A). Also, the outer fiber layer of the pallium connecting the lateral with the medial cortex, the homologue of the hippocampus, shows faint reelin-positivity (Cabrera-Socorro *et al.*, 2007), and may be the lacertilian equivalent of the perforant path. In mammals, reelin may be transported by axons in the perforant path and released in the target layers in the dentate gyrus and Ammon's horn, where it may contribute to the synaptic plasticity of the hippocampal circuits.

In the hippocampal subregions, some layers display high levels of reelin even in adulthood, especially the stratum lacunosum-moleculare of Ammon's horn, and the molecular layer of the dentate gyrus, which are populated by numerous small interneurons (Fatemi *et al.*, 2000; Pappas *et al.*, 2001; Abraham and Meyer, 2003; Abraham *et al.*, 2004; Ramos-Moreno *et al.*, 2006). This enrichment of reelin is again in accord with its proposed activity in modulating synaptic plasticity underlying long-term hippocampus-dependent learning.

As discussed in this chapter, reelin has multiple functions in the brain, and the adult activities are not necessarily the same as during development. In keeping with the importance of the reelin–Dab1 signaling pathway for neuronal positioning, reelin in Cajal-Retzius cells of the hippocampal fissure is crucial for the correct development of the hippocampus. Mice defective for the transcription factor p73 lack Cajal-Retzius cells in cortex and hippocampus (Yang *et al.*, 2000; Meyer *et al.*, 2002). While neocortical lamination is not disturbed in these mutant mice, the hippocampus is severely disorganized, and the hippocampal fissure does not form, suggesting that hippocampal architectonic development depends strongly on the presence of Cajal-Retzius cells (Meyer *et al.*, 2004). Furthermore, early generated Cajal-Retzius cells of the hippocampus provide a template for entorhinal fibers to find their target layers in the hippocampus (Ceranik *et al.*, 1999). In the *reeler* mouse, the granule cells fail to form a compact cell layer (Lambert de Rouvroit and Goffinet, 1998), perhaps due to a twofold function of reelin which may act as a differentiation factor for radial glia cells and also as a positional cue for radial fiber orientation and granule cell migration (Zhao *et al.*, 2004). In the human hippocampus, there are several sources for Cajal-Retzius cells, which are particularly numerous in the head of the hippocampus, which is highly differentiated in man, in accordance with the cognitive functions of the human hippocampus (Abraham *et al.*, 2004). The hippocampus is thus another example for the variety of reelin-related mechanisms involved in the development and evolution of the telencephalon, and demonstrates the importance of this protein in the formation and maintenance of the CNS.

References

Abraham, H., and Meyer, G. (2003). Reelin-expressing neurons in the postnatal and adult human hippocampal formation. *Hippocampus* 13:715–727.

Abraham, H., Perez-Garcia, C. G., and Meyer, G. (2004). p73 and reelin in Cajal-Retzius cells of the developing human hippocampal formation. *Cerebral Cortex* 14:484–495.

Alcantara, S., Ruiz, M., D'Arcangelo, G., Ezan, F., de Lecea, L., Curran, T., Sotelo, C., and Soriano E. (1998). Regional and cellular patterns of reelin mRNA expression in the forebrain of the developing and adult mouse. *J. Neurosci.* 18:7779–7799.

Bar, I., Lambert de Rouvroit, C., and Goffinet, A. M. (2000). The evolution of cortical development. An hypothesis based on the role of the reelin signaling pathway. *Trends Neurosci.* 23:633–638.

Bar, I., Tissir, F., Lambert de Rouvroit, C., De Backer, O., and Goffinet, A. M. (2003). The gene encoding disabled-1 (DAB1), the intracellular adaptor of the reelin pathway, reveals unusual complexity in human and mouse. *J. Biol. Chem.* 278:5802–5812.

Beffert, U., Weeber, E. J., Durudas, A., Qiu, S., Masiulis, I., Sweatt, J. D., Li, W. P., Adelmann, G., Frotscher, M., Hammer, R. E., and Herz, J. (2005). Modulation of synaptic plasticity and memory by reelin involves differential splicing of the lipoprotein receptor Apoer2. *Neuron* 47:567–579.

Beffert, U., Durudas, A., Weeber, E. J., Stolt, P. C., Giehl, K. M., Sweatt, J. D., Hammer, R. E., and Herz, J. (2006). Functional dissection of reelin signaling by site-directed disruption of disabled-1 adaptor binding to apolipoprotein E receptor 2: distinct roles in development and synaptic plasticity. *J. Neurosci.* 26:2041–2052.

Bernier, B., Bar, I., Pieau, C., Lambert de Rouvroit, C., and Goffinet, A. M. (1999). Reelin mRNA expression during embryonic brain development in the turtle Emys orbicularis. *J. Comp. Neurol.* 413:463–479.

Bernier, B., Bar, I., D'Arcangelo, G., Curran, T., and Goffinet, A. M. (2000). Reelin mRNA expression during embryonic brain development in the chick. *J. Comp. Neurol.* 422:448–463.

Bielle, F., Griveau, A., Narboux-Neme, N., Vigneau, S., Sigrist, M., Arber, S., Wassef, M., and Pierani, A. (2005). Multiple origins of Cajal-Retzius cells at the borders of the developing pallium. *Nature Neurosci.* 8:1002–1012.

Botella-Lopez, A., Burgaya, F., Gavin, R., Garcia-Ayllon, M. S., Gomez-Tortosa, E., Pena-Casanova, J., Urena, J. M., Del Rio, J. A., Blesa, R., Soriano, E., and Saez-Valero, J. (2006). Reelin expression and glycosylation patterns are altered in Alzheimer's disease. *Proc. Natl. Acad. Sci. USA* 103:5573–5578.

Braak, H., and Braak, E. (1992). The human entorhinal cortex: normal morphology and lamina-specific pathology in various diseases. *Neurosci. Res.* 15:6–31.

Cabrera-Socorro, A., Hernandez-Acosta, N. C., Gonzalez-Gomez, M., and Meyer, G. (2007). Comparative aspects of p73 and reelin expression in Cajal-Retzius cells and the cortical hem in lizard, mouse and human. *Brain Res.* 1132:59–70.

Ceranik, K., Deng, J., Heimrich, B., Lubke, J., Zhao, S., Forster, E., and Frotscher, M. (1999). Hippocampal Cajal-Retzius cells project to the entorhinal cortex: retrograde tracing and intracellular labelling studies. *Eur. J. Neurosci.* 11:4278–4290.

Chen, M. L., Chen, S. Y., Huang, C. H., and Chen, C. H. (2002). Identification of a single nucleotide polymorphism at the 5 promoter region of human reelin gene and association study with schizophrenia. *Mol. Psychiatry* 7:447–448.

Chen, Y., Sharma, R. P., Costa, R. H., Costa, E., and Grayson, D. R. (2002). On the epigenetic regulation of the human reelin promoter. *Nucleic Acids Res.* 30:2930–2939.

Chen, Y., Beffert, U., Ertunc, M., Tang, T. S., Kavalali, E. T., Bezprozvanny, I., and Herz, J. (2005). Reelin modulates NMDA receptor activity in cortical neurons. *J. Neurosci.* 25:8209–8216.

Chin, J., Massaro, C. M., Palop, J. J., Thwin, M. T., Yu, G. Q., Bien-Ly, N., Bender, A., and Mucke, L. (2007). Reelin depletion in the entorhinal cortex of human amyloid precursor protein transgenic mice and humans with Alzheimer's disease. *J. Neurosci.* 27:2727–2733.

Costagli, A., Kapsimali, M., Wilson, S. W., and Mione, M. (2002). Conserved and divergent patterns of reelin expression in the zebrafish central nervous system. *J. Comp. Neurol.* 450:73–93.

Costagli, A., Felice, B., Guffanti, A., Wilson, S. W., and Mione, M. (2006). Identification of alternatively spliced dab1 isoforms in zebrafish. *Dev.Genes Evol.* 216:291–299.

D'Arcangelo G., Miao, G. G., Chen, S. C., Soares, H. D., Morgan, J. I., and Curran, T. (1995). A protein related to extracellular matrix proteins deleted in the mouse mutant reeler. *Nature* 374:719–723.

D'Arcangelo, G., Nakajima, K., Miyata, T., Ogawa, M., Mikoshiba, K., and Curran, T. (1997). Reelin is a secreted glycoprotein recognized by the CR-50 monoclonal antibody. *J. Neurosci.* 17:23–31.

de Bergeyck, V., Naerhuyzen, B., Goffinet, A. M., and Lambert de Rouvroit, C. (1998). A panel of monoclonal antibodies against reelin, the extracellular matrix protein defective in reeler mutant mice. *J. Neurosci. Methods* 82:17–24.

Deguchi, K., Inoue, K., Avila, W. E., Lopez-Terrada, D., Antalffy, B. A., Quattrocchi, C. C., Sheldon, M., Mikoshiba, K., D'Arcangelo, G., and Armstrong, D. L. (2003). Reelin and disabled-1 expression in developing and mature human cortical neurons. *J. Neuropathol. Exp. Neurol.* 62:676–684.

Derer, P., and Derer, M. (1990). Cajal-Retzius cell ontogenesis and death in mouse brain visualized with horseradish peroxidase and electron microscopy. *Neuroscience* 36:839–856.

Derer, P., Derer, M., and Goffinet, A. (2001). Axonal secretion of reelin by Cajal-Retzius cells: evidence from comparison of normal and Reln(Orl) mutant mice. *J. Comp. Neurol.* 440:136–143.

DeSilva, U., D'Arcangelo, G., Braden, V. V., Chen, J., Miao, G. G., Curran, T., and Green, E. D. (1997). The human *reelin* gene: isolation, sequencing, and mapping on chromosome 7. *Genome Res*. 199:157–164.

Drakew, A., Frotscher, M., Deller, T., Ogawa, M., and Heimrich, B. (1998). Developmental distribution of a *reeler* gene-related antigen in the rat hippocampal formation visualized by CR-50 immunocytochemistry. *Neuroscience* 82:1079–1086.

Eastwood, S. L., and Harrison, P. J. (2006). Cellular basis of reduced cortical reelin expression in schizophrenia. *Am. J. Psychiatry* 163:540–542.

Fatemi, S. H., Emamian, E. S., Kist, D., Sidwell, R. W., Nakajima, K., Akhter, P., Shier, A., Sheikh, S., and Bailey, K. (1999). Defective corticogenesis and reduction in reelin immunoreactivity in cortex and hippocampus of prenatally infected neonatal mice. *Mol. Psychiatry* 4:145–154.

Fatemi, S. H., Earle, J. A., and McMenomy, T. (2000). Reduction in reelin immunoreactivity in hippocampus of subjects with schizophrenia, bipolar disorder and major depression. *Mol. Psychiatry* 5:654–663.

Gertler, F. B., Bennett, R. L., Clark, M. J., and Hoffmann, F. M. (1989). Drosophila abl tyrosine kinase in embryonic CNS axons: a role in axonogenesis is revealed through dosage-sensitive interactions with disabled. *Cell* 58:103–113.

Goffinet, A. M., Bar, I., Bernier, B., Trujillo, C., Raynaud, A., and Meyer, G. (1999). Reelin expression during embryonic brain development in lacertilian lizards. *J. Comp. Neurol.* 414:533–550.

Gomez-Isla, T., Price, J. L., McKeel, D. W., Jr., Morris, J. C., Growdon, J. H., and Hyman, B. T. (1996). Profound loss of layer II entorhinal cortex neurons occurs in very mild Alzheimer's disease. *J. Neurosci.* 16:4491–4500.

Hevner, R. F., Neogi, T., Englund, C., Daza, R. A., and Fink, A. (2003). Cajal-Retzius cells in the mouse: transcription factors, neurotransmitters, and birthdays suggest a pallial origin. *Brain Res. Dev. Brain Res.* 141:39–53.

Hiesberger, T., Trommsdorff, M., Howell, B. W., Goffinet, A., Mumby, M. C., Cooper, J. A., and Herz, J. (1999). Direct binding of reelin to VLDL receptor and ApoE receptor 2 induces tyrosine phosphorylation of disabled-1 and modulates tau phosphorylation. *Neuron* 24:481–489.

Howell, B. W., Herrick, T. M., and Cooper, J. A. (1999a). Reelin-induced tryosine phosphorylation of disabled 1 during neuronal positioning. *Genes Dev.* 13:643–648.

Howell, B. W., Lanier, L. M., Frank, R., Gertler, F. B., and Cooper, J. A. (1999b). The disabled 1 phosphotyrosine-binding domain binds to the internalization signals of transmembrane glycoproteins and to phospholipids. *Mol. Cell Biol.*19:5179–5188.

Jellinger, K., Braak, H., Braak, E., and Fischer, P. (1991). Alzheimer lesions in the entorhinal region and isocortex in Parkinson's and Alzheimer's diseases. *Ann. N.Y. Acad. Sci.* 640:203–209.

Jossin, Y., Ignatova, N., Hiesberger, T., Herz, J., Lambert de Rouvroit, C., and Goffinet, A. M. (2004). The central fragment of reelin, generated by proteolytic processing in vivo, is critical to its function during cortical plate development. *J. Neurosci.* 24:514–521.

Keshvara, L., Benhayon, D., Magdaleno, S., and Curran, T. (2001). Identification of reelin-induced sites of tyrosyl phosphorylation on disabled 1. *J. Biol. Chem.* 276:16008–16014.

Lambert de Rouvroit, C., and Goffinet, A. M. (1998). The reeler mouse as a model of brain development. *Adv. Anat. Embryol.Cell Biol.* 50:1–106.

Lambert de Rouvroit, C., Bernier, B., Royaux, I., de Bergeyck, V., and Goffinet, A. M. (1999). Evolutionarily conserved, alternative splicing of reelin during brain development. *Exp. Neurol.* 156:229–238.

Luque, J. M., Morante-Oria, J., and Fairen, A. (2003). Localization of ApoER2, VLDLR and Dab1 in radial glia: groundwork for a new model of reelin action during cortical development. *Brain Res. Dev. Brain Res.* 140:195–203.

Martinez-Cerdeno, V., and Clasca, F. (2002). Reelin immunoreactivity in the adult neocortex: a comparative study in rodents, carnivores, and non-human primates. *Brain Res. Bull.* 57:485–488.

Martinez-Cerdeno, V., Galazo, M. J., Cavada, C., and Clasca, F. (2002). Reelin immunoreactivity in the adult primate brain: intracellular localization in projecting and local circuit neurons of the cerebral cortex, hippocampus and subcortical regions. *Cerebral Cortex* 12:1298–1311.

Martinez-Cerdeno, V., Galazo, M. J., and Clasca, F. (2003). Reelin-immunoreactive neurons, axons, and neuropil in the adult ferret brain: evidence for axonal secretion of reelin in long axonal pathways. *J. Comp. Neurol.* 463:92–116.

Meyer, G. (2001). Human neocortical development: the importance of embryonic and early fetal events. *Neuroscientist* 7:303–314.

Meyer, G. (2007). Genetic control of neuronal migrations in human cortical development. *Adv. Anat. Embryol. Cell Biol.*189:1–111.

Meyer, G., and Goffinet, A. M. (1998). Prenatal development of reelin-immunoreactive neurons in the human neocortex. *J. Comp. Neurol.* 397:29–40.

Meyer, G., and González-Hernández, T. (1993). Developmental changes in layer I of the human neocortex during prenatal life: a DiI-tracing and AChE and NADPH-d histochemistry study. *J. Comp. Neurol.* 338:317–336.

Meyer, G., and Wahle, P. (1999). The paleocortical ventricle is the origin of reelin-expressing neurons in the marginal zone of the foetal human neocortex. *Eur. J. Neurosci.* 11:3937–3944.

Meyer, G., Soria, J. M., Martinez-Galan, J. R., Martin-Clemente, B., and Fairen, A. (1998). Different origins and developmental histories of transient neurons in the marginal zone of the fetal and neonatal rat cortex. *J. Comp. Neurol.* 397:493–518.

Meyer, G., Goffinet, A. M., and Fairen, A. (1999). What is a Cajal-Retzius cell? A reassessment of a classical cell type based on recent observations in the developing neocortex. *Cerebral Cortex* 9:765–775.

Meyer, G., Pérez-García, C. G., Abraham, H., and Caput, D. (2002). Expression of p73 and reelin in the developing human cortex. *J.Neurosci.* 22:4973–4986.

Meyer, G., de Rouvroit, C. L., Goffinet, A. M., and Wahle, P. (2003). Disabled-1 mRNA and protein expression in developing human cortex. *Eur. J. Neurosci.* 17:517–525.

Meyer, G., Cabrera-Socorro, A., Pérez-García, C. G., Martínez-Millán, L., Walker, N., and Caput, D. (2004). Developmental roles of p73 in Cajal-Retzius cells and cortical patterning. *J. Neurosci.* 24:9878–9887.

Nieuwenhuys, R., and Meek, J. (1990). The telencephalon of actinopterygian fishes. In Jones, G. G., and Peters, A. (Eds.), *Cerebral Cortex*, Vol. 8A. Plenum Press, New York, pp.31–73.

Ogawa, M., Miyata, T., Nakajima, K., Yagyu, K., Seike, M., Ikenaka, K., Yamamoto, H., and Mikoshiba, K. (1995). The reeler gene-associated antigen on Cajal-Retzius neurons is a crucial molecule for laminar organization of cortical neurons. *Neuron* 14:899–912.

Pappas, G. D., Kriho, V., and Pesold, C. (2001). Reelin in the extracellular matrix and dendritic spines of the cortex and hippocampus: a comparison between wild type and heterozygous reeler mice by immunoelectron microscopy. *J. Neurocytol.* 30:413–425.

Pérez-Costas, E., Melendez-Ferro, M., Santos, Y., Anadon, R., Rodicio, M. C., and Caruncho, H. J. (2002). Reelin immunoreactivity in the larval sea lamprey brain. *J. Chem. Neuroanat.* 23:211–221.

Pérez-Costas, E., Meléndez-Ferro, M., Pérez-García, C. G., Caruncho, H. J., and Rodicio, M. C. (2004). Reelin immunoreactivity in the adult sea lamprey brain. *J. Chem. Neuroanat.* 27:7–21.

Pérez-García, C. G., González-Delgado, F. J., Suárez-Solá, M. L., Castro-Fuentes, R., Martín-Trujillo, J. M., Ferres-Torres, R., and Meyer, G. (2001). Reelin-immunoreactive neurons in the adult vertebrate pallium. *J. Chem. Neuroanat.* 21:41–51.

Pérez-García, C. G., Tissir, F., Goffinet, A. M., and Meyer, G. (2004). Reelin receptors in developing laminated brain structures of mouse and human. *Eur. J. Neurosci.* 20:2827–2832.

Persico, A. M., D'Agruma, L., Maiorano, N., Totaro, A., Militerni, R., Bravaccio, C., Wassink, T. H., Schneider, C., Melmed, R., Trillo, S., Montecchi, F., Palermo, M., Pascucci, T., Puglisi-Allegra, S., Reichelt, K. L., Conciatori, M., Marino, R., Quattrocchi, C. C., Baldi, A., Zelante, L., Gasparini, P., and Keller, F. (2001). Reelin gene alleles and haplotypes as a factor predisposing to autistic disorder. *Mol. Psychiatry* 6:150–159.

Pesold, C., Impagnatiello, F., Pisu, M. G., Uzunov, D. P., Costa, E., Guidotti, A., and Caruncho, H. J. (1998). Reelin is preferentially expressed in neurons synthesizing gamma-aminobutyric acid in cortex and hippocampus of adult rats. *Proc. Natl. Acad. Sci. USA* 95:3221–3226.

Pesold, C., Liu, W. S., Guidotti, A., Costa, E., and Caruncho, H. J. (1999). Cortical bitufted, horizontal, and Martinotti cells preferentially express and secrete reelin into perineuronal nets, nonsynaptically modulating gene expression. *Proc. Natl. Acad. Sci. USA* 96:3217–3222.

Pollard, K. S., Salama, S. R., Lambert, N., Lambot, M. A., Coppens, S., Pedersen, J. S., Katzman, S., King, B., Onodera, C., Siepel, A., Kern, A. D., Dehay, C., Igel, H., Ares, M., Jr., Vanderhaeghen, P., and Haussler, D. (2006). An RNA gene expressed during cortical development evolved rapidly in humans. *Nature* 443:167–172.

Pozniak, C. D., Radinovic, S., Yang, A., McKeon, F., Kaplan, D. R., and Miller, F. D. (2000). An anti-apoptotic role for the p53 family member, p73, during developmental neuron death. *Science* 289:304–306.

Qiu, S., and Weeber, E. J. (2007). Reelin signaling facilitates maturation of CA1 glutamatergic synapses. *J. Neurophysiol.* 97:2312–2321.

Ramos-Moreno, T., Galazo, M. J., Porrero, C., Martinez-Cerdeno, V., and Clasca, F. (2006). Extracellular matrix molecules and synaptic plasticity: immunomapping of intracellular and secreted reelin in the adult rat brain. *Eur. J. Neurosci.* 23:401–422.

Rice, D. S., and Curran, T. (2001). Role of the reelin signaling pathway in central nervous system development. *Annu. Rev. Neurosci.* 24:1005–1039.

Rice, D. S., Sheldon, M., D'Arcangelo, G., Nakajima, K., Goldowitz, D., and Curran, T. (1998). Disabled-1 acts downstream of reelin in a signaling pathway that controls laminar organization in the mammalian brain. *Development* 125:3719–3729.

Rice, D. S., Nusinowitz, S., Azimi, A. M., Martinez, A., Soriano, E., and Curran, T. (2001). The reelin pathway modulates the structure and function of retinal synaptic circuitry. *Neuron* 31:929–941.

Roberts, R. C., Xu, L., Roche, J. K., and Kirkpatrick, B. (2005). Ultrastructural localization of reelin in the cortex in post-mortem human brain. *J. Comp. Neurol.* 482:294–308.

Rodriguez, M. A., Pesold, C., Liu, W. S., Kriho, V., Guidotti, A., Pappas, G. D., and Costa, E. (2000). Colocalization of integrin receptors and reelin in dendritic spine postsynaptic densities of adult nonhuman primate cortex. *Proc. Natl. Acad. Sci. USA* 97:3550–3555.

Rodríguez, M. A., Caruncho, H. J., Costa, E., Pesold, C., Liu, W. S., and Guidotti, A. (2002). In Patas monkey, glutamic acid decarboxylase-67 and reelin mRNA coexpression varies in a manner dependent on layers and cortical areas. *J. Comp. Neurol.* 451:279–288.

Royaux, I., Lambert de Rouvroit, C., D'Arcangelo, G., Demirov, D., and Goffinet, A. M. (1997). Genomic organization of the mouse reelin gene. *Genomics* 46:240–250.

Saez-Valero, J., Costell, M., Sjogren, M., Andreasen, N., Blennow, K., and Luque, J. M. (2003). Altered levels of cerebrospinal fluid reelin in frontotemporal dementia and Alzheimer's disease. *J. Neurosci. Res.* 72:132–136.

Schiffmann, S. N., Bernier, B., and Goffinet, A. M. (1997). Reelin mRNA expression during mouse brain development. *Eur. J. Neurosci.* 9:1055–1071.

Steward, O., and Scoville, S. A. (1976). Cells of origin of entorhinal cortical afferents to the hippocampus and fascia dentata of the rat. *J. Comp. Neurol.* 169:347–370.

Tissir, F., and Goffinet, A. M. (2003). Reelin in brain development. *Nature Rev. Neurosci.* 4:496–505.

Tissir, F., Lambert de Rouvroit, C., and Goffinet, A. M. (2002). The role of reelin in the development and evolution of the cerebral cortex. *Braz. J. Med. Biol. Res.* 35:1473–1484.

Tissir, F., Lambert de Rouvroit, C., Sire, J. Y., Meyer, G., and Goffinet, A. M. (2003). Reelin expression during embryonic brain development in Crocodylus niloticus. *J. Comp. Neurol.* 457:250–262.

Trommsdorff, M., Gotthardt, M., Hiesberger, T., Shelton, J., Stockinger, W., Nimpf, J., Hammer, R. E., Richardson, J. A., and Herz, J. (1999). Reeler/disabled-like disruption of neuronal migration in knockout mice lacking the VLDL receptor and ApoE receptor 2. *Cell* 97:689–701.

Ulinski, P. S. (1990). The cerebral cortex of reptiles. In: Jones, E. G., and Peters, A. (Eds.), *Cerebral Cortex,* Vol. 8A. Plenum Press, New York, pp. 139–215.

van Groen, T., Miettinen, P., and Kadish, I. (2003). The entorhinal cortex of the mouse: organization of the projection to the hippocampal formation. *Hippocampus* 13:133–149.

Van Hoesen, G. W., and Hyman, B. T. (1990). Hippocampal formation: anatomy and the patterns of pathology in Alzheimer's disease. *Prog. Brain Res.* 83:445–457.

Weeber, E. J., Beffert, U., Jones, C., Christian, J. M., Forster, E., Sweatt, J. D., and Herz, J. (2002). Reelin and ApoE receptors cooperate to enhance hippocampal synaptic plasticity and learning. *J. Biol. Chem.* 277:39944–39952.

Willnow, T. E., Nykjaer, A., and Herz, J. (1999). Lipoprotein receptors: new roles for ancient proteins. *Nature Cell Biol.* 1:157–162.

Witter, M. P., and Groenewegen, H. J. (1984). Laminar origin and septotemporal distribution of entorhinal and perirhinal projections to the hippocampus in the cat. *J. Comp. Neurol.* 224:371–385.

Yang, A., Walker, N., Bronson, R., Kaghad, M., Oosterwegel, M., Bonnin, J., Vagner, C., Bonnet, H., Dikkes, P., Sharpe, A., McKeon, F., and Caput, D. (2000). p73-deficient mice have neurological, pheromonal and inflammatory defects but lack spontaneous tumours. *Nature* 404:99–103.

Yoshida, M., Assimacopoulos, S., Jones, K. R., and Grove, E. A. (2006). Massive loss of Cajal-Retzius cells does not disrupt neocortical layer order. *Development* 133:537–545.

Zhao, S., Chai, X., Forster, E., and Frotscher, M. (2004). Reelin is a positional signal for the lamination of dentate granule cells. *Development* 131:5117–5125.

Chapter 7
Reelin/Dab1 Signaling in the Developing Cerebral Cortex

Eric C. Olson and Christopher A. Walsh

Contents

Abstract Mice lacking the cytoplasmic adapter protein Dab1 (Disabled homolog-1) display histological defects in the central nervous system (CNS) that are essentially indistinguishable from those observed in the *reeler* mouse. Dab1 is expressed in virtually all Reelin-responsive cells and is rapidly phosphorylated in response to

E. C. Olson
Department of Neuroscience and Physiology, SUNY Upstate Medical University,
750 E. Adams Street, Syracuse, NY 13035
e-mail: olsone@upstate.edu

C. A. Walsh
Howard Hughes Medical Institute, BIDMC, Division of Genetics, Boston Children's Hospital,
Harvard Medical School, NRB 266, 77 Avenue Louis Pasteur, Boston, MA 02115
e-mail: cwalsh@bidmc.harvard.edu

S. H. Fatemi (ed.), *Reelin Glycoprotein: Structure, Biology and Roles in Health and Disease.*

Reelin application. The finding of a near identity in phenotype, coupled with a direct biochemical response to Reelin, has raised great interest in understanding Dab1 function, both as an exemplar of an adapter protein with a profound phenotypic contribution, and as a means of decoding mechanisms of Reelin signaling. What has emerged from these studies is a surprisingly complex picture of Dab1 at the genomic, mRNA, protein, and functional levels. This chapter will summarize some of the key features of Dab1, and its role as a transducer of the Reelin signal in the developing cerebral cortex.

1 Historical Context

The *reeler* mouse is one of the most fascinating and comprehensively studied neurological mutants, with a literature extending back over 50 years (Falconer, 1951). *Reeler* mice, which lack Reelin, show disordered cellular positioning in the major laminated structures of the brain, including the cerebral cortex, hippocampus, and cerebellum. In the *reeler* cortex, migrating neurons fail to split the preplate (Sheppard and Pearlman, 1997), and subsequently generated neurons "pile up" behind previously generated neurons, forming an approximate inversion of the normal cellular layering of the six-layered cortex (Caviness and Sidman, 1973). It was therefore of considerable interest when a new spontaneous mutant mouse, termed *scrambler*, was identified in 1996, which was phenotypically indistinguishable from *reeler*, but for which a recessive mutation mapped to chromosome 4 and not to chromosome 5, the location of the *reeler* mutation (Sweet *et al.*, 1996; Goldowitz *et al.*, 1997; Gonzalez *et al.*, 1997). The characterization and mapping of the *scrambler* locus followed soon after the identification and cloning of *Reln*, the novel gene mutated in the *reeler* mouse that encodes the secreted protein Reelin (D'Arcangelo *et al.*, 1995). The close phenotypic match between *scrambler* and *reeler* mice raised hope that the chromosome 4 defect would involve a gene encoding the Reelin receptor, and an effort in several labs was initiated to identify the genetic disruption in the *scrambler* mouse. About this time, a gene then called mDab1 (Howell *et al.*, 1997a), now called Dab1, was "knocked out" on chromosome 4 and the resultant mouse displayed a *reeler*-like phenotype (Howell *et al.*, 1997b). This observation facilitated the rapid identification of mutations in Dab1 that were within the mapped intervals in the *scrambler* and related *yotari* mice (Sheldon *et al.*, 1997; Ware *et al.*, 1997).

2 Mutations in *Scrambler* and *Yotari* Mice

The mutations identified in *scrambler* and *yotari* mice are both associated with aberrant *dab1* mRNA message and little or no Dab1 protein. In wild-type animals, *dab1* message is expressed as a primary transcript of 5.5 kb (Howell *et al.*, 1997a). In *scrambler*, very small amounts of this 5.5-kb message are detected, along with the presence of a larger, aberrant transcript of ~7 kb. The *scm* mutation causes the splicing of an intracisternal-A particle (IAP) retrotransposon element into the Dab1 message. This leads to the insertion of an ~1.5-kb noncoding insert into the Dab1 mRNA and a near absence of Dab1 protein (Sheldon *et al.*, 1997; Ware *et al.*, 1997). To date, the exact mutation in *scrambler* that promotes the splicing of the IAP into the Dab1 message has not been identified. In contrast, the *yotari* mouse mutation is caused by a 962-base-pair insertion of an L1 retrotransposon fragment into chromosome 4 that disrupts three exons of the *Dab1* gene (Sheldon *et al.*, 1997). In *yotari* mice, there is no detectable Dab1 protein. To date, there are no identified human mutations within the *DAB1* locus, although three distinct *RELN* mutations have been identified in humans (Chang *et al.*, 2006), along with one mutation in the Reelin receptor *VLDLR* (Boycott *et al.*, 2005).

The identified gene *dab1* encodes a cytoplasmic adapter protein that lacks enzymatic activity *per se* and appears instead to act as a scaffold for the assembly of a multiprotein signaling complex (Howell *et al.*, 1997a). Dab1 contains an ~150-amino-acid PTB/PID (phosphotyrosine binding/phosphotyrosine interacting domain) (Howell *et al.*, 1997a) that preferentially binds NPXY motifs in target proteins. Near this PTB/PID domain on the C-terminal side are a series of five tyrosine residues that are potential target sites for the Src family of tyrosine kinases (Howell *et al.*, 1997a). The Dab1 PTB/PID preferentially binds F/YXNPXY amino acid sequences found in the cytoplasmic domains of a number of transmembrane proteins (Trommsdorff *et al.*, 1998; Howell *et al.*, 1999b), including the Reelin receptors.

3 Dab1 and the Reelin Receptor Complex

Confirmation of a biochemical relationship between Reelin and Dab1 *in vivo* is found in the expression pattern of Dab1 protein in *reeler* mice. The absolute level of Dab1 protein is 5- to 10-fold higher in the embryonic *reeler* neocortex compared to wild-type littermate controls (Rice *et al.*, 1998). Elevated expression of Dab1 protein is now thought to be a diagnostic feature of Reelin signaling abnormalities. In addition, Dab1 protein displays reduced levels of tyrosine phosphorylation in the *reeler* neocortex, indicating that one consequence of Reelin deficiency is reduction in tyrosine phosphorylation of Dab1. Surprisingly, Dab1 mRNA levels are similar in mutant and control brains, suggesting that that Dab1 misregulation is posttranscriptional (Rice *et al.*, 1998). In contrast, when the reciprocal experiment is performed—namely, examination of Reelin expression in *scrambler* mice—approximately normal Reelin expression is observed (Gonzalez *et al.*, 1997), with the interpretation being that Dab1 is not required for the expression or secretion of Reelin. The expression studies of Dab1, particularly those performed in *reeler* mice, position Dab1 on the receptive side of the Reelin signaling pathway.

Dramatic evidence for the identity of the Reelin receptors followed from the *reeler*-like phenotype of mice doubly deficient in VLDLR (very-low-density lipoprotein receptor) and ApoER2 (ApoE receptor 2) (Trommsdorff *et al.*, 1999). Mice singly

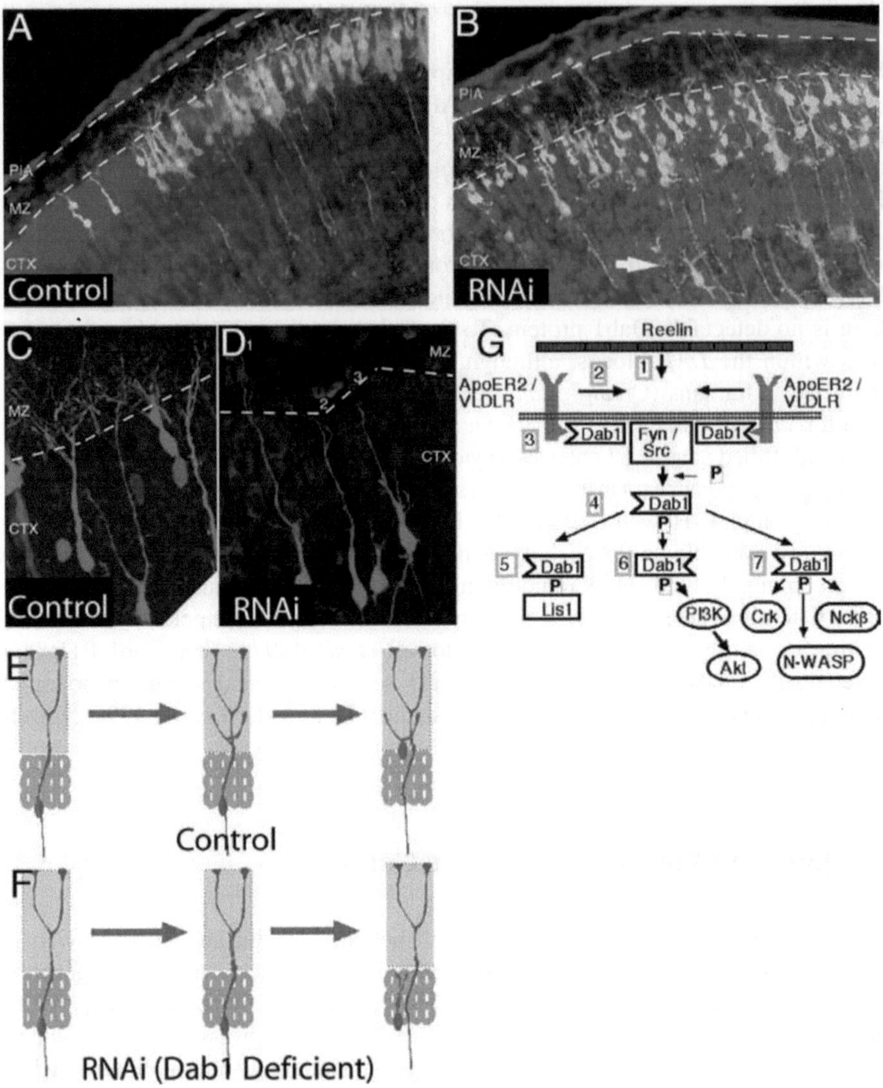

Fig. 7.1 Reelin Dab1 signaling in upper layer cortical neurons. (**A, B**) Low-magnification images of layer 2/3 cortical neurons on postnatal day 2 (P2), 7 days after *in utero* electroporation on E16 with either RNAi that suppresses Dab1 (RNAi) or control RNAi vector (Control). (**A**) Control electroporated neurons show precise lamination and exuberant dendritic growth in the MZ (dashed lines) on P2, whereas (**B**) Dab1-suppressed cells (RNAi) show disrupted lamination with occasional ectopic deep cells (arrow) and sparse dendrites in the MZ. (**C, D**) Higher-magnification images revealing extensive dendrites in (**C**) control cells and stunted dendrites in (**D**) RNAi-treated cells that either do not penetrate the MZ (cells 2 and 3) or stunted dendrites that do not show extensive secondary and tertiary branching in the MZ (cell 1). Scale bars: 50 μm (**A, B**); 20 μm (**D**). (**E, F**) Model of cell positioning and dendritogenesis in the developing cortex. (**E**) A control neuron (dark green) migrating on a radial glial process (red) extends a branched leading process into the MZ and then translocates through the upper ~50 μm of the CP, arresting migration at the first branch point of the leading process. (**F**) Dab1-deficient cells extend a leading process into the MZ but it remains simplified and the neuron does not translocate efficiently. (**G**) Dab1 interactions (after D'Arcangelo, 2006). Reelin secreted by CR cells (1) binds Reelin receptors (ApoER2 and VLDLR) in the migrating neuron causing (2) the clustering of Reelin receptors and Dab1.

deficient for either receptor show much more subtle phenotypes and are overtly normal (not ataxic), whereas mice deficient for both receptors phenocopy the *reeler* mouse (Trommsdorff *et al.*, 1999). Both of these receptors are members of the LDLR (low-density-lipoprotein receptor) family and contain a single Dab1 binding site (NPXY) on their cytoplasmic tails. Upon binding of Reelin to VLDLR and ApoER2 (Hiesberger *et al.*, 1999; D'Arcangelo *et al.*, 1999), a rapid phosphorylation of Dab1 on tyrosine residues is induced. Neither ApoER2 nor VLDLR possesses kinase domains or displays kinase activity, however, leading to the suggestion that the Reelin-induced phosphorylation of Dab1 is achieved through Src family kinases (SFKs). Two SFKs, Fyn and Src, bind Dab1 (Howell *et al.*, 1997a), phosphorylate Dab1, and are activated by Reelin binding to its receptors (Arnaud *et al.*, 2003b; Bock and Herz, 2003; Jossin *et al.*, 2003). Conclusive evidence for the involvement of SFKs in Reelin signaling is found in mice deficient in both Fyn and Src. These "double knockout" animals recapitulate major aspects of the *reeler* phenotype including cortical layer inversion and cerebellar abnormalities (Kuo *et al.*, 2005). These genetic and biochemical findings establish Dab1 as a central link in an entirely novel receptor complex that includes Reelin, VLDLR or ApoER2, and the tyrosine kinases Fyn or Src (Fig. 7.1G).

4 Dab1 Isoforms and the Importance of p80

In both mouse and human, the genomic organization of *dab1* is complex with multiple isoforms generated from alternative poly-adenylation and splicing events (Bar *et al.*, 2003). In the developing mouse nervous system, Dab1 appears to exist in a number of distinct isoforms represented in cDNAs encoding 555-, 271-, and 217-amino-acid proteins that all share the N-terminal PTB/PID domain (Howell *et al.*, 1997a). Immunoblotting of whole brain lysates during the neurodevelopmental period reveals proteins with weights of p120, p80, and p60 (Howell *et al.*, 1997a,b); however, the correspondence between the cDNAs and these different protein forms of Dab1 is only known for the p80 form, which is encoded by the Dab1 555 cDNA. The p80 form appears to be the only isoform expressed in brain that is required for normal histological development. The essential nature of p80 was confirmed in a "knockin" experiment; the Dab1 555 cDNA was inserted into the mouse Dab1 locus, knocking out the endogenous splice forms of Dab1 and creating an obligate p80-expressing mouse (Howell *et al.*, 2000). This obligate p80-expressing mouse was histologically indistinguishable from normal mice that express multiple splice

Fig. 7.1 (continued) (3) The cytoplasmic clustering of Dab1 activates two SFKs (Fyn and Src) leading to (4) tyrosine phosphorylation of Dab1. (5) Phospho-Dab1 binds Lis1, a cytoplasmic dynein interacting protein encoded by *Lis1*, the gene underlying Miller-Dieker lissencephaly. (6) Phospho-Dab1 also activates PI3 kinase and Akt kinase and (7) binds adapter proteins Crk, Nckβ as well as N-WASP. Reelin signaling may regulate multiple cellular events including glial adhesion, somal positioning, and dendritogenesis. Panels **A–F** modified from Olson *et al.* (2006), copyright 2006 by the Society for Neuroscience (*See Color Plates*)

forms of the Dab1 message. Given the evidence pointing to the importance of the p80 isoform of Dab1 in development, it was surprising to find that a similarly constructed "knockin" mouse that was an obligate expresser of a p45 allele of Dab1, that lacks 284 amino acids at the C-terminus, showed normal cortical development (Herrick and Cooper, 2002). Although animals with a single copy of the p45 allele over a null allele (Dab1$^{p45/-}$) showed disruptions in the upper cortical layers, the observation that Dab1$^{p45/p45}$ homozygotes appear normal focuses efforts at understanding Dab1's role in cortical development at the N-terminal 271 amino acids.

5 PTB/PI Domains

The Dab1 PTB/PID binds F/YXNPXY domains in the cytoplasmic tails of the Reelin receptors ApoER2 and VLDLR. The PTB/PID is the major recognized feature within the N-terminal 271 amino acids and extends from amino acids 23 to 174 (Howell *et al.*, 1997a). A new appreciation of the PTB/PID domain has emerged since the prototype PTB was identified in 1994 in the adapter Shc (Src homology 2 domain-containing protein; Blaikie *et al.*, 1994; Kavanaugh and Williams, 1994). Originally, the PTB/PID was so named because Shc preferentially bound NPXY motifs when the tyrosine residue on the ligand was phosphorylated (NPXpY) (Kavanaugh *et al.*, 1995; Songyang *et al.*, 1995). For example, Shc binds the epidermal growth factor receptor (EGFR) when the NPXY sequence in the receptor's cytoplasmic tail is tyrosine phosphorylated (O'Bryan *et al.*, 1998). One might assume that this phosphorylation-dependent binding of the PTB/PID would be a fundamental property of the domain, but subsequently identified PTB/PID proteins, including Dab1, appear to preferentially bind the dephosphorylated amino acid sequence NPXY sequences (Howell *et al.*, 1999b). In fact, the dephosphorylated tyrosine-preferring PTB/PIDs appear to be the most common, comprising approximately 75% of the characterized PTB/PIDs (Uhlik *et al.*, 2005). This emerging picture of the PTB/PID has complicated the terminology (Margolis, 1999) and has led to further studies of PTB/PID–ligand interactions.

6 Structure of the Dab1 PTB/PI Domain

The Dab1 PTB/PID has been crystallized in bound states with peptides corresponding to the cytoplasmic domains of ApoER2 (Stolt *et al.*, 2003) and another Dab1 binding partner, the amyloid precursor protein (APP) (Yun *et al.*, 2003). The Dab1 PTB/PID, like each of the 10 different solved PTB domains, adopts a structure similar to that of the pleckstrin homology (PH) superfold. Whereas the Shc PTB/PID has three critical basic amino acids in the binding pocket that interact with the phosphorylated tyrosine residue in NPXpY, Dab1 lacks two of these critical basic amino acids. Instead, binding of nonphosphorylated tyrosine is favored by van der Waals attraction between the PTB/PID and the NPXY target. The distinct characteristics of the Dab1

PTB/PID constitute a third PTB/PID family separate from the Shc-like and IRS-like PTB/PID domains (Uhlik *et al.*, 2005) and have led to the term Dab homology (DH) domain to describe this particular class of PTB/PID (Yun *et al.*, 2003).

6.1 Phospholipid Binding by the Dab1 PTB/PID

Another shared feature among many PTB/PID domain family members is the phospholipid binding subdomain within the PTB/PID. This phospholipid binding domain is a highly basic sequence that in Dab1 "crowns" the PTB/PID and binds the phosphatidylinositol (PtdIns) phosphate head groups of plasma membrane lipids. This binding localizes Dab1 to the plasma membrane and is apparently independent of NPXY peptide binding, as mutations have been identified that can separately and selectively abolish phospholipid or NPXY binding (Xu *et al.*, 2005; Huang *et al.*, 2005; Stolt *et al.*, 2004). The crystal structure of Dab1 PTB/PID–ApoER2–PtdIns-4,5-P2 suggests that binding of PtdIns-4,5-P2 confines the peptide binding pocket to a vertical orientation with respect to the plasma membrane, possibly leading to more favorable interactions between the peptide binding pocket and the NPXY motifs in the cytoplasmic tails of the receptors (Stolt *et al.*, 2003; Uhlik *et al.*, 2005).

7 Dab1 Phosphorylation by Src Family Kinases

As mentioned above, a defining feature of the Reelin–Dab1 interaction is the rapid, Reelin-dependent tyrosine phosphorylation of Dab1. Although it was apparent early on that Dab1 is hypophosphorylated in the *reeler* mouse (Rice *et al.*, 1998), it was not initially clear if Dab1 phosphorylation was directly dependent on Reelin. The development of an *in vitro* assay using cultured embryonic neurons enabled the demonstration that Dab1 is rapidly phosphorylated in response to recombinant Reelin application (Howell *et al.*, 1999a). The essential nature of tyrosine phosphorylation of Dab1 was later demonstrated in an elegant "knockin" experiment, where a p80 allele of Dab1 was constructed that substituted phenylalanine (F) for tyrosine (Y) at five candidate phosphorylation sites on Dab1 (Howell *et al.*, 2000). This so-called "5F" allele was then "knocked-in" to the Dab1 locus, thereby rendering a Dab1 allele that could not be phosphorylated at these positions. Homozygotes of the Dab1 5F displayed a *reeler*-like phenotype, demonstrating that these five tyrosines are essential for Dab1 function, likely because of their role as phosphorylation substrates.

7.1 Genetics of Fyn and Src Deficiency in Cortical Development

From the time of its discovery, Dab1 was known to bind nonreceptor tyrosine kinases including Src, Fyn, and Abl (Howell *et al.*, 1997a), and thus these kinases became likely candidates for mediating Reelin-dependent Dab1 phosphorylation.

Consistent with this notion, broad inhibition of Src family kinases, using the pharmacological agent PP2, both blocked Reelin-dependent Dab1 phosphorylation (Arnaud *et al.*, 2003b; Bock and Herz, 2003; Jossin *et al.*, 2003) and caused *reeler*-like malformations in cerebral cortical slices (Jossin *et al.*, 2003). Pharmacological blockade of SFKs specifically reduced Dab1 phosphorylation at tyrosines 198 and 220, the major sites of Reelin-dependent tyrosine phosphorylation (Keshvara *et al.*, 2001). Inhibition of Abl, however, does not suppress Dab1 phosphorylation (Arnaud *et al.*, 2003b). Mice singly deficient for Src (Soriano *et al.*, 1991), Fyn (Stein *et al.*, 1992), and Yes (Stein *et al.*, 1994), however, do not reveal a *reeler* mouse phenotype, implying that multiple members of the SFK family contribute to normal Reelin signaling. This hypothesis was dramatically confirmed when a double knockout of Src and Fyn was created that recapitulated major histological features of the *reeler* phenotype, including deficiency in preplate splitting, an inversion of cortical layers, and ectopic clusters of Purkinje cells in the developing cerebellum (Kuo *et al.*, 2005).

7.2 *Dab1 Clustering Activates SFKs*

One of the most important insights emerging from the studies of Dab1–SFK interaction is that simple clustering of Dab1 may be sufficient to stimulate its phosphorylation by SFKs (Fig. 7.1). Reelin is secreted as a multimer (Kubo *et al.*, 2002) and likely acts as a multivalent ligand for the Reelin receptors (Strasser *et al.*, 2004). Therefore, Reelin binding to the Reelin receptors likely causes the local clustering of both Reelin receptors and, by extension, Dab1 bound to the cytoplasmic tail of the receptors. In fact, Dab1 dimerization, independent of the Reelin receptors, is sufficient to induce Dab1 phosphorylation on Y198 and Y220 by SFKs (Strasser *et al.*, 2004). Similarly, simple coexpression of Dab1 and SFKs (either Src or Fyn) in 293 kidney cells leads to a dramatic increase in SFK activity (Bock and Herz, 2003). This amplification of SFK activity is apparently attenuated by Dab1 protein degradation mediated by the polyubiquination and proteasomal degradation (Arnaud *et al.*, 2003a; Bock *et al.*, 2004). The observed overexpression of the Dab1 5F allele suggests that proteolysis of Dab1 is dependent on Dab1 phosphorylation (Howell *et al.*, 2000; Arnaud *et al.*, 2003a). Thus, Dab1 phosphorylation initially amplifies Fyn and Src activity minutes after Reelin exposure, followed by a presumed slower time-scale dampening of Reelin signaling ($t_{1/2}$=160 minutes), subsequent to Dab1 degradation (Arnaud *et al.*, 2003a; Bock *et al.*, 2004).

8 Protein Interactions Subsequent to Dab1 Phosphorylation

Given the essential nature of Dab1 tyrosine phosphorylation, the effort to decode Reelin signaling has focused primarily on protein interactions that occur subsequent to Dab1 phosphorylation (Fig. 7.1G). One branch of the Reelin signaling pathway has the potential to regulate the actin cytoskeleton through N-WASP (Suetsugu *et al.*, 2004), as well as through Nckβ (Pramatarova *et al.*, 2003) and the Crk family of adapters (Ballif *et al.*, 2004; Chen *et al.*, 2004). Nckβ and Crk family adapters contain both SH2 domains that preferentially bind phosphorylated tyrosine motifs and SH3 domains that bind specific proline-containing sequences, such as XPXXP (Pawson and Scott, 1997). Nckβ appears to relocalize with phospho-Dab1 in neuronal growth cones in response to Reelin application, and this localization is correlated with regions of remodeled actin cytoskeleton (Pramatarova *et al.*, 2003). Similarly, the Crk family of adapter molecules also associate with phosphorylated Dab1. Crk family adapters have been implicated in a wide range of cellular processes, including migration, endocytosis, and morphological reorganization (Chen *et al.*, 2004, and references therein). A functional linkage between phospho-Dab1 and cell migration is suggested by the observation that phosphorylated Dab1 reduces migration rates in heterologous NBT-II (tumor cells), possibly by depletion of Crk protein from the focal adhesion contacts (Chen *et al.*, 2004).

Other effectors of Reelin signaling have the potential to modify the microtubule cytoskeleton. Phospho-Dab1 binds p85, the regulatory subunit of PI3 kinase (Beffert *et al.*, 2002; Ballif *et al.*, 2003; Bock *et al.*, 2003). The association of p85 with Dab1 liberates and activates the p110 catalytic subunit of PI3 kinase, leading to the subsequent activation of the serine threonine kinase Akt (also known as Protein Kinase B) and the subsequent inhibition of glycogen synthetase kinase 3β (GSK3β) (Beffert *et al.*, 2002). The microtubule-associated protein (MAP) tau is a GSK3β target and is hyperphosphorylated in some strains of *reeler* mice (Hiesberger *et al.*, 1999), possibly altering microtubule stability in these animals. The interaction of phospho-Dab1 with the lissencephaly gene Lis1 provides another route toward modification of microtubule dynamics (Assadi *et al.*, 2003). Lis1 is encoded by the gene underlying autosomal dominant Miller-Dieker lissencephaly, a severe form of lissencephaly that corresponds to nearly complete absence of sulci and gyri in the human cerebral cortex (Reiner *et al.*, 1993). Miller-Dieker lissencephaly is more severe than the lissencephaly with cerebellar hypoplasia (LCH) associated with Reelin deficiency (Hong *et al.*, 2000). Compound mutants of Lis1- and Dab1-deficient mice show enhanced histological disorganization in the hippocampus, suggesting a genetic interaction between these loci. Lis1 is known also to bind the microtubule motor cytoplasmic dynein, and via this interaction regulates cytoskeletal dynamics, including the positioning of the centrosome (Wynshaw-Boris and Gambello, 2001; Tanaka *et al.*, 2004). The interaction between phospho-Dab1 and Lis1 therefore has the potential to mediate complex regulation of microtubule dynamics and cell migration (Wynshaw-Boris and Gambello, 2001).

9 Dab1 Independent Reelin Signaling

Although interactions with phosphorylated Dab1 are clearly essential for normal Reelin signaling, Dab1 phosphorylation may not be sufficient for triggering all downstream aspects of Reelin signaling. As mentioned above, the secreted form of Reelin is now known to be multivalent and to crosslink the Reelin receptors (Kubo *et al.*, 2002). This crosslinking leads to the phosphorylation of Dab1 by SFKs. Dab1 phosphorylation can also be stimulated by the crosslinking of Reelin receptors using antibodies that recognize the extracellular domains of VLDLR and ApoER2. However, this antibody-dependent crosslinking apparently does not rescue the *reeler* phenotype in cortical slice cultures (Jossin *et al.*, 2004), whereas exogenous Reelin application will provide rescue. This finding implies the existence of a Reelin signaling pathway that works in parallel with phospho-Dab1.

10 Promotion of Receptor Expression by Dab1

Additionally, Dab1 may have an upstream role in promoting Reelin receptor presence at the cell surface (Utsunomiya-Tate *et al.*, 2000). Cultured neurons from *yotari* embryos that lack Dab1, bind approximately half the exogenously applied Reelin compared to wild-type. This deficiency in Reelin binding appears to be due to a reduction of the amount of Reelin receptors present on the cell surface and implies that experiments that rely on the analysis of Dab1-deficient or Dab1-suppressed cells may identify both the effect of deficient receptor expression and deficient downstream signaling.

11 Dab1 Expression During Development

In all areas of the CNS affected by the Reelin mutation, cells expressing Dab1 are found within a few cell diameters of cells expressing Reelin (Sheldon *et al.*, 1997; Rice *et al.*, 1998). In the mouse cerebral cortex, Dab1 expression is observed as a series of bands across the cerebral wall, including a band within the proliferative ventricular zone (Bar *et al.*, 2003). The apparent highest levels of Dab1 expression are associated with cells settling beneath the marginal zone (MZ), the location of Reelin-secreting Cajal-Retzius (CR) cells. Similar spatial relationships are observed elsewhere in the developing brain. In the hippocampus, for example, Dab1-expressing pyramidal neurons settle underneath Reelin-expressing CR cells, and in the cerebellum Dab1 is expressed by migrating Purkinje cells, as they settle underneath the Reelin-expressing cells of the external granule layer (EGL) (Rice *et al.*, 1998). In humans, Dab1 is first strongly detected at the seventh gestational week (GW) during the initial phases of cortical plate development (Meyer *et al.*, 2003). As cortical development proceeds, Dab1 expression is greatest at the upper tier of the cortical plate underneath the MZ, similar to the pattern of Dab1 expression observed in rodents. A close spatial relationship between Dab1- and Reelin-expressing cells continues through the neuraxis into the spinal cord

(Yip *et al.*, 2004) and underscores the tight functional relationship between Reelin and Dab1.

12 Cellular Response to Reelin/Dab1 Signaling

There is evidence that Reelin–Dab1 signaling modifies three possibly interrelated cellular activities during cortical development: cell adhesion (Pinto Lord *et al.*, 1982; Hoffarth *et al.*, 1995; Dulabon *et al.*, 2000; Hack *et al.*, 2002; Sanada *et al.*, 2004), somal movement and positioning (Caviness and Sidman, 1973; Super *et al.*, 2000; Magdaleno *et al.*, 2002; Bock *et al.*, 2003; Jossin *et al.*, 2003; Olson *et al.*, 2006), and dendritic or glial process growth (Pinto Lord *et al.*, 1982; Forster *et al.*, 2002; Hartfuss *et al.*, 2003; Niu *et al.*, 2004; Olson *et al.*, 2006). In principle, modification of any one of these cellular activities could affect neuronal migration, and it is not clear if any single cellular activity represents the "primary response" to Reelin, while the other activities are derived or secondary. Defining the possible function(s) of Reelin signaling is facilitated by examination of cell autonomous Dab1 deficiency. In these experiments a subset of cells that are "blinded to Reelin" due to Dab1 deficiency are allowed to develop within an otherwise normal cortex. In this way, the cellular response to Reelin can be separated from cellular responses to the gross disorganization of the cortex caused by Reelin signaling deficiency.

12.1 Dab1-Dependent Cellular Positioning

Suppression of Dab1 in a subset of developing cortical cells by RNAi (RNA interference) causes migrating neurons to arrest a few cell diameters (~40–50 μm) short of their destination at the boundary of the cortical plate (CP) and marginal zone (MZ) (Fig. 7.1A,B) (Olson *et al.*, 2006). An arrest of Dab1-deficient cells below Dab1-expressing cells is also observed in the cortices of chimeric mice composed of Dab1 –/– and Dab +/+ cells. In these animals, Dab1-deficient cells are found underneath a "supercortex" of wild-type cells (Hammond *et al.*, 2001), suggesting that Dab1 has an important, cell autonomous role in promoting somal movement and migration through the cortex. Other studies have also argued for a role of Reelin signaling in the promotion of migration (Super *et al.*, 2000; Magdaleno *et al.*, 2002; Bock *et al.*, 2003; Jossin *et al.*, 2003). In a mutant *Drosophila* eye model, induced expression of mammalian Dab1 can rescue cellular positioning defects of photoreceptors (Pramatarova *et al.*, 2006), suggesting a relatively direct interaction between Dab1 and the cellular positioning machinery. It is important to note, however, that Dab1-suppressed cells in mice can migrate appropriately for hundreds of micrometers through the intermediate zone (IZ) and lower CP (Olson *et al.*, 2006), indicating that deficits in neuronal migration associated with Reelin signaling deficiency might be relatively specific to the final ~40–50 μm of migration.

12.2 Dab1-Dependent Cellular Adhesion

Reelin–Dab1 signaling is also associated with adherent interactions between developing neurons and glia. Whereas wild-type neurons appear to detach from the glial process at the end of their migration (Pinto Lord *et al.*, 1982; Nadarajah *et al.*, 2001; Sanada *et al.*, 2004), neurons deficient in Reelin–Dab1 signaling maintain inappropriate glial adhesion (Pinto Lord *et al.*, 1982; Sanada *et al.*, 2004). This detachment at the end of migration is, however, a complex event at the molecular level, possibly involving modulation of α3β1 integrin-mediated adhesion (Dulabon *et al.*, 2000; Sanada *et al.*, 2004; Schmid *et al.*, 2005). However, mice deficient in β1 integrin are unable to form functional α3β1 heterodimeric receptors, yet show histological malformations that are more consistent with a basal lamina defect rather than cortical migration abnormalities (Graus-Porta *et al.*, 2001). Nevertheless, a number of empirical findings indicate possible cell adhesion abnormalities in the *reeler* phenotype, particularly at the end of migration, and in this regard integrin interactions remain an intriguing aspect of Dab1 function.

12.3 Dab1-Dependent Process Outgrowth

Finally, Reelin–Dab1 signaling performs a major role in stimulating process outgrowth and branching in radial glia (Hartfuss *et al.*, 2003), migrating neurons (Olson *et al.*, 2006), and postmigratory neurons (Tabata and Nakajima, 2002; Niu *et al.*, 2004; Olson *et al.*, 2006). Dab1 suppression causes a cell autonomous reduction of the size and complexity of cerebral pyramidal neuron dendrites (Fig. 7.1A–D) (Olson *et al.*, 2006). Similar dendritic reductions are also observed in cerebellar Purkinje cells (Trommsdorff *et al.*, 1999) and hippocampal CA field pyramidal neurons (Niu *et al.*, 2004) in *reeler* mice. Importantly, Reelin application appears to directly stimulate the dendritic growth of cultured hippocampal neurons (Niu *et al.*, 2004) and the phosphorylation of the dendritic protein MAP1b (Gonzalez-Billault *et al.*, 2005). In contrast, axonal growth is not affected by deficiency in Reelin signaling (Jossin and Goffinet, 2001), indicating that Reelin signaling selectively enhances dendritogenesis.

13 Models of Reelin/Dab1 Function

Reelin signaling deficiency may cause persistent and inappropriate neuron–glial adhesion at the end of migration, and this, in turn, may lead to altered cellular positioning. We propose a modified model that focuses on Reelin's demonstrated ability to stimulate dendritic growth (Niu *et al.*, 2004) and leading process transformation at the MZ (Olson *et al.*, 2006). In this new model, the leading process of the migrating neuron has a central role in directing cellular positioning (Fig. 7.1E,F). Reelin signaling modifies the leading process in such a way to permit glial-independent leading process growth. Our data indicate that Reelin signaling stimulates the formation of a branched leading process in the MZ. Since the presence of this branched leading process correlates with movement of the cell soma through the last 50 μm, this leading process may well be essential for cellular movement and positioning. The first stable branch point formed by the leading process anticipates the arrest position of the moving nucleus, and thereby defines the arrest point of the cell body. In this way, cellular positioning (i.e., lamination) in the developing cortex is the indirect consequence of the Reelin-dependent transformation of the leading process into a maturing, branched dendrite. Emerging studies suggest that inhibition of PI3 kinase affects both Reelin-dependent cellular positioning and dendritogenesis, while inhibition of Akt affects dendritogenesis specifically (Jossin and Goffinet, 2006). This study raises hope that the multiple effects of Reelin signaling on developing neurons can be dissected at the molecular level through investigations of Dab1-dependent interactions. Further studies of this important signaling pathway are likely to provide new insights and additional surprises regarding essential aspects of cortical development.

Acknowledgments C.A.W. is supported by grants from the NINDS (NS035129, NS032457, NS040043), as well as by the James and Marilyn Simons Foundation and the NLM Family Foundation. C.A.W. is an Investigator of the Howard Hughes Medical Institute. E.C.O. is supported by SUNY Upstate Medical University Departmental Funds. We thank David Cameron for valuable comments on the manuscript.

References

Arnaud, L., Ballif, B. A., and Cooper, J. A. (2003a). Regulation of protein tyrosine kinase signaling by substrate degradation during brain development. *Mol. Cell. Biol.* 23:9293–9302.

Arnaud, L., Ballif, B. A., Forster, E., and Cooper, J. A. (2003b). Fyn tyrosine kinase is a critical regulator of disabled-1 during brain development. *Curr. Biol.* 13:9–17.

Assadi, A. H., Zhang, G., Beffert, U., McNeil, R. S., Renfro, A. L., Niu, S., Quattrocchi, C. C., Antalffy, B. A., Sheldon, M., Armstrong, D. D., Wynshaw-Boris, A., Herz, J., D'Arcangelo, G., and Clark, G. D. (2003). Interaction of reelin signaling and Lis1 in brain development. *Nature Genet.* 35:270–276.

Ballif, B. A., Arnaud, L., and Cooper, J. A. (2003). Tyrosine phosphorylation of disabled-1 is essential for reelin-stimulated activation of Akt and Src family kinases. *Brain Res. Mol. Brain Res.* 117:152–159.

Ballif, B. A., Arnaud, L., Arthur, W. T., Guris, D., Imamoto, A., and Cooper, J. A. (2004). Activation of a Dab1/CrkL/C3G/Rap1 pathway in reelin-stimulated neurons. *Curr. Biol.* 14:606–610.

Bar, I., Tissir, F., Lambert de Rouvroit, C., De Backer, O., and Goffinet, A. M. (2003). The gene encoding disabled-1 (DAB1), the intracellular adaptor of the reelin pathway, reveals unusual complexity in human and mouse. *J. Biol. Chem.* 278:5802–5812.

Beffert, U., Morfini, G., Bock, H. H., Reyna, H., Brady, S. T., and Herz, J. (2002). Reelin-mediated signaling locally regulates protein kinase B/Akt and glycogen synthase kinase 3beta. *J. Biol. Chem.* 277:49958–49964.

Blaikie, P., Immanuel, D., Wu, J., Li, N., Yajnik, V., and Margolis, B. (1994). A region in Shc distinct from the SH2 domain can bind tyrosine-phosphorylated growth factor receptors. *J. Biol. Chem.* 269:32031–32034.

Bock, H. H., and Herz, J. (2003). Reelin activates SRC family tyrosine kinases in neurons. *Curr. Biol.* 13:18–26.

Bock, H. H., Jossin, Y., Liu, P., Forster, E., May, P., Goffinet, A. M., and Herz, J. (2003). Phosphatidylinositol 3-kinase interacts with the adaptor protein Dab1 in response to reelin signaling and is required for normal cortical lamination. *J. Biol. Chem.* 278:38772–38779.

Bock, H. H., Jossin, Y., May, P., Bergner, O., and Herz, J. (2004). Apolipoprotein E receptors are required for reelin-induced proteasomal degradation of the neuronal adapter protein disabled-1. *J. Biol. Chem.* 279:33471–33479.

Boycott, K. M., Flavelle, S., Bureau, A., Glass, H. C., Fujiwara, T. M., Wirrell, E., Davey, K., Chudley, A. E., Scott, J. N., McLeod, D. R., and Parboosingh, J. S. (2005). Homozygous deletion of the very low density lipoprotein receptor gene causes autosomal recessive cerebellar hypoplasia with cerebral gyral simplification. *Am. J. Hum. Genet.* 77:477–483.

Caviness, V. S., Jr., and Sidman, R. L. (1973). Time of origin or corresponding cell classes in the cerebral cortex of normal and *reeler* mutant mice: an autoradiographic analysis. *J. Comp. Neurol.* 148:141–151.

Chang, B. S., Duzcan, F., Kim, S., Cinbis, M., Aggarwal, A., Apse, K. A., Ozdel, O., Atmaca, M., Zencir, S., Bagci, H., and Walsh, C. A. (2006). The role of RELN in lissencephaly and neuropsychiatric disease. *Am. J. Med. Genet. B Neuropsychiatr. Genet.* 144:58–63.

Chen, K., Ochalski, P. G., Tran, T. S., Sahir, N., Schubert, M., Pramatarova, A., and Howell, B. W. (2004). Interaction between Dab1 and CrkII is promoted by reelin signaling. *J. Cell Sci.* 117:4527–4536.

D'Arcangelo, G. (2006). Reelin mouse mutants as models of cortical development disorders. *Epilepsy Behav.* 8:81–90.

D'Arcangelo, G., Miao, G. G., Chen, S. C., Soares, H. D., Morgan, J. I., and Curran, T. (1995). A protein related to extracellular matrix proteins deleted in the mouse mutant *reeler. Nature* 374:719–723.

D'Arcangelo, G., Homayouni, R., Keshvara, L., Rice, D. S., Sheldon, M., and Curran, T. (1999). Reelin is a ligand for lipoprotein receptors. *Neuron* 24:471–479.

Dulabon, L., Olson, E. C., Taglienti, M. G., Eisenhuth, S., McGrath, B., Walsh, C. A., Kreidberg, J. A., and Anton, E. S. (2000). Reelin binds alpha3beta1 integrin and inhibits neuronal migration. *Neuron* 27:33–44.

Falconer, D. (1951). Two new mutants, "trembler" and "*reeler*," with neurological actions in the house mouse. *J. Genet.* 13:192–201.

Forster, E., Tielsch, A., Saum, B., Weiss, K. H., Johanssen, C., Graus-Porta, D., Muller, U., and Frotscher, M. (2002). Reelin, disabled 1, and beta 1 integrins are required for the formation of the radial glial scaffold in the hippocampus. *Proc. Natl. Acad. Sci. USA* 99:13178–13183.

Goldowitz, D., Cushing, R. C., Laywell, E., D'Arcangelo, G., Sheldon, M., Sweet, H. O., Davisson, M., Steindler, D., and Curran, T. (1997). Cerebellar disorganization characteristic of *reeler* in *scrambler* mutant mice despite presence of reelin. *J. Neurosci.* 17:8767–8777.

Gonzalez, J. L., Russo, C. J., Goldowitz, D., Sweet, H. O., Davisson, M. T., and Walsh, C. A. (1997). Birthdate and cell marker analysis of *scrambler*: a novel mutation affecting cortical development with a *reeler*-like phenotype. *J. Neurosci.* 17:9204–9211.

Gonzalez-Billault, C., Del Rio, J. A., Urena, J. M., Jimenez-Mateos, E. M., Barallobre, M. J., Pascual, M., Pujadas, L., Simo, S., Torre, A. L., Gavin, R., Wandosell, F., Soriano, E., and Avila, J. (2005). A role of MAP1B in reelin-dependent neuronal migration. *Cereb. Cortex* 15:1134–1145.

Graus-Porta, D., Blaess, S., Senften, M., Littlewood-Evans, A., Damsky, C., Huang, Z., Orban, P., Klein, R., Schittny, J. C., and Muller, U. (2001). Beta1-class integrins regulate the development of laminae and folia in the cerebral and cerebellar cortex. *Neuron* 31:367–379.

Hack, I., Bancila, M., Loulier, K., Carroll, P., and Cremer, H. (2002). Reelin is a detachment signal in tangential chain-migration during postnatal neurogenesis. *Nature Neurosci.* 5:939–945.

Hammond, V., Howell, B., Godinho, L., and Tan, S. S. (2001). Disabled-1 functions cell autonomously during radial migration and cortical layering of pyramidal neurons. *J. Neurosci.* 21:8798–8808.

Hartfuss, E., Forster, E., Bock, H. H., Hack, M. A., Leprince, P., Luque, J. M., Herz, J., Frotscher, M., and Gotz, M. (2003). Reelin signaling directly affects radial glia morphology and biochemical maturation. *Development* 130:4597–4609.

Herrick, T. M., and Cooper, J. A. (2002). A hypomorphic allele of dab1 reveals regional differences in reelin–Dab1 signaling during brain development. *Development* 129:787–796.
Hiesberger, T., Trommsdorff, M., Howell, B. W., Goffinet, A., Mumby, M. C., Cooper, J. A., and Herz, J. (1999). Direct binding of reelin to VLDL receptor and ApoE receptor 2 induces tyrosine phosphorylation of disabled-1 and modulates tau phosphorylation. *Neuron* 24:481–489.
Hoffarth, R. M., Johnston, J. G., Krushel, L. A., and van der Kooy, D. (1995). The mouse mutation *reeler* causes increased adhesion within a subpopulation of early postmitotic cortical neurons. *J. Neurosci.* 15:4838–4850.
Hong, S. E., Shugart, Y. Y., Huang, D. T., Shahwan, S. A., Grant, P. E., Hourihane, J. O., Martin, N. D., and Walsh, C. A. (2000). Autosomal recessive lissencephaly with cerebellar hypoplasia is associated with human RELN mutations. *Nature Genet.* 26:93–96.
Howell, B. W., Gertler, F. B., and Cooper, J. A. (1997a). Mouse disabled (mDab1): a Src binding protein implicated in neuronal development. *EMBO J.* 16:121–132.
Howell, B. W., Hawkes, R., Soriano, P., and Cooper, J. A. (1997b). Neuronal position in the developing brain is regulated by mouse disabled-1. *Nature* 389:733–737.
Howell, B. W., Herrick, T. M., and Cooper, J. A. (1999a). Reelin-induced tryosine phosphorylation of disabled 1 during neuronal positioning. *Genes Dev.* 13:643–648.
Howell, B. W., Lanier, L. M., Frank, R., Gertler, F. B., and Cooper, J. A. (1999b). The disabled 1 phosphotyrosine-binding domain binds to the internalization signals of transmembrane glycoproteins and to phospholipids. *Mol. Cell. Biol.* 19:5179–5188.
Howell, B. W., Herrick, T. M., Hildebrand, J. D., Zhang, Y., and Cooper, J. A. (2000). Dab1 tyrosine phosphorylation sites relay positional signals during mouse brain development. *Curr. Biol.* 10:877–885.
Huang, Y., Shah, V., Liu, T., and Keshvara, L. (2005). Signaling through disabled 1 requires phosphoinositide binding. *Biochem. Biophys. Res. Commun.* 331:1460–1468.
Jossin, Y., and Goffinet, A. M. (2001). Reelin does not directly influence axonal growth. *J. Neurosci.* 21:RC183.
Jossin, Y., and Goffinet, A. M. (2006). The PI3K/Akt/mTor pathway controls brain development in reelin-dependent and independent manners. *Society for Neuroscience Abstract* 514.8.
Jossin, Y., Ogawa, M., Metin, C., Tissir, F., and Goffinet, A. M. (2003). Inhibition of SRC family kinases and non-classical protein kinases C induce a *reeler*-like malformation of cortical plate development. *J. Neurosci.* 23:9953–9959.
Jossin, Y., Ignatova, N., Hiesberger, T., Herz, J., Lambert de Rouvroit, C., and Goffinet, A. M. (2004). The central fragment of reelin, generated by proteolytic processing *in vivo*, is critical to its function during cortical plate development. *J. Neurosci.* 24:514–521.
Kavanaugh, W. M., and Williams, L. T. (1994). An alternative to SH2 domains for binding tyrosine-phosphorylated proteins. *Science* 266:1862–1865.
Kavanaugh, W. M., Turck, C.W., and Williams, L. T. (1995). PTB domain binding to signaling proteins through a sequence motif containing phosphotyrosine. *Science* 268:1177–1179.
Keshvara, L., Benhayon, D., Magdaleno, S., and Curran, T. (2001). Identification of reelin-induced sites of tyrosyl phosphorylation on disabled 1. *J. Biol. Chem.* 276:16008–16014.
Kubo, K., Mikoshiba, K., and Nakajima, K. (2002). Secreted reelin molecules form homodimers. *Neurosci. Res.* 43:381–388.
Kuo, G., Arnaud, L., Kronstad-O'Brien, P., and Cooper, J. A. (2005). Absence of Fyn and Src causes a *reeler*-like phenotype. *J. Neurosci.* 25:8578–8586.
Magdaleno, S., Keshvara, L., and Curran, T. (2002). Rescue of ataxia and preplate splitting by ectopic expression of reelin in *reeler* mice. *Neuron* 33:573–586.
Margolis, B. (1999). The PTB domain: The name doesn't say it all. *Trends Endocrinol. Metab.* 10:262–267.
Meyer, G., Lambert de Rouvroit, C., Goffinet, A. M., and Wahle, P. (2003). Disabled-1 mRNA and protein expression in developing human cortex. *Eur. J. Neurosci.* 17:517–525.
Nadarajah, B., Brunstrom, J. E., Grutzendler, J., Wong, R. O., and Pearlman, A. L. (2001). Two modes of radial migration in early development of the cerebral cortex. *Nature Neurosci.* 4:143–150.

Niu, S., Renfro, A., Quattrocchi, C. C., Sheldon, M., and D'Arcangelo, G. (2004). Reelin promotes hippocampal dendrite development through the VLDLR/ApoER2-Dab1 pathway. *Neuron* 41:71–84.

O'Bryan, J. P., Lambert, Q. T., and Der, C. J. (1998). The src homology 2 and phosphotyrosine binding domains of the ShcC adaptor protein function as inhibitors of mitogenic signaling by the epidermal growth factor receptor. *J. Biol. Chem.* 273:20431–20437.

Olson, E. C., Kim, S., and Walsh, C. A. (2006). Impaired neuronal positioning and dendritogenesis in the neocortex after cell-autonomous Dab1 suppression. *J. Neurosci.* 26:1767–1775.

Pawson, T., and Scott, J. D. (1997). Signaling through scaffold, anchoring, and adaptor proteins. *Science* 278:2075–2080.

Pinto Lord, M. C., Evrard, P., and Caviness, V. S., Jr. (1982). Obstructed neuronal migration along radial glial fibers in the neocortex of the *reeler* mouse: a Golgi-EM analysis. *Brain Res.* 256:379–393.

Pramatarova, A., Ochalski, P. G., Chen, K., Gropman, A., Myers, S., Min, K. T., and Howell, B. W. (2003). Nck beta interacts with tyrosine-phosphorylated disabled 1 and redistributes in reelin-stimulated neurons. *Mol. Cell. Biol.* 23:7210–7221.

Pramatarova, A., Ochalski, P. G., Lee, C. H., and Howell, B. W. (2006). Mouse disabled 1 regulates the nuclear position of neurons in a Drosophila eye model. *Mol. Cell. Biol.* 26:1510–1517.

Reiner, O., Carrozzo, R., Shen, Y., Wehnert, M., Faustinella, F., Dobyns, W. B., Caskey, C. T., and Ledbetter, D. H. (1993). Isolation of a Miller-Dieker lissencephaly gene containing G protein beta-subunit-like repeats. *Nature* 364:717–721.

Rice, D. S., Sheldon, M., D'Arcangelo, G., Nakajima, K., Goldowitz, D., and Curran, T. (1998). Disabled-1 acts downstream of reelin in a signaling pathway that controls laminar organization in the mammalian brain. *Development* 125:3719–3729.

Sanada, K., Gupta, A., and Tsai, L. H. (2004). Disabled-1-regulated adhesion of migrating neurons to radial glial fiber contributes to neuronal positioning during early corticogenesis. *Neuron* 42:197–211.

Schmid, R. S., Jo, R., Shelton, S., Kreidberg, J. A., and Anton, E. S. (2005). Reelin, integrin and Dab1 Interactions during embryonic cerebral cortical development. *Cereb. Cortex.* 10: 1632–1636.

Sheldon, M., Rice, D. S., D'Arcangelo, G., Yoneshima, H., Nakajima, K., Mikoshiba, K., Howell, B. W., Cooper, J. A., Goldowitz, D., and Curran, T. (1997). *Scrambler* and *yotari* disrupt the disabled gene and produce a *reeler*-like phenotype in mice. *Nature* 389:730–733.

Sheppard, A. M., and Pearlman, A. L. (1997). Abnormal reorganization of preplate neurons and their associated extracellular matrix: an early manifestation of altered neocortical development in the *reeler* mutant mouse. *J. Comp. Neurol.* 378:173–179.

Songyang, Z., Margolis, B., Chaudhuri, M., Shoelson, S. E., and Cantley, L. C. (1995). The phosphotyrosine interaction domain of SHC recognizes tyrosine-phosphorylated NPXY motif. *J. Biol. Chem.* 270:14863–14866.

Soriano, P., Montgomery, C., Geske, R., and Bradley, A. (1991). Targeted disruption of the c-src proto-oncogene leads to osteopetrosis in mice. *Cell* 64:693–702.

Stein, P. L., Lee, H. M., Rich, S., and Soriano, P. (1992). pp59fyn mutant mice display differential signaling in thymocytes and peripheral T cells. *Cell* 70:741–750.

Stein, P. L., Vogel, H., and Soriano, P. (1994). Combined deficiencies of Src, Fyn, and Yes tyrosine kinases in mutant mice. *Genes Dev.* 8:1999–2007.

Stolt, P. C., Jeon, H., Song, H. K., Herz, J., Eck, M. J., and Blacklow, S. C. (2003). Origins of peptide selectivity and phosphoinositide binding revealed by structures of disabled-1 PTB domain complexes. *Structure (Camb.)* 11:569–579.

Stolt, P. C., Vardar, D., and Blacklow, S. C. (2004). The dual-function disabled-1 PTB domain exhibits site independence in binding phosphoinositide and peptide ligands. *Biochemistry* 43:10979–10987.

Strasser, V., Fasching, D., Hauser, C., Mayer, H., Bock, H. H., Hiesberger, T., Herz, J., Weeber, E. J., Sweatt, J. D., Pramatarova, A., Howell, B., Schneider, W. J., and Nimpf, J. (2004). Receptor clustering is involved in reelin signaling. *Mol. Cell. Biol.* 24:1378–1386.

Suetsugu, S., Tezuka, T., Morimura, T., Hattori, M., Mikoshiba, K., Yamamoto, T., and Takenawa, T. (2004). Regulation of actin cytoskeleton by mDab1 through N-WASP and ubiquitination of mDab1. *Biochem. J.* 384–1-8.

Super, H., Del Rio, J. A., Martinez, A., Perez-Sust, P., and Soriano, E. (2000). Disruption of neuronal migration and radial glia in the developing cerebral cortex following ablation of Cajal-Retzius cells. *Cereb. Cortex* 10:602–613.

Sweet, H. O., Bronson, R. T., Johnson, K. R., Cook, S. A., and Davisson, M. T. (1996). *Scrambler*, a new neurological mutation of the mouse with abnormalities of neuronal migration. *Mamm. Genome* 7:798–802.

Tabata, H., and Nakajima, K. (2002). Neurons tend to stop migration and differentiate along the cortical internal plexiform zones in the reelin signal-deficient mice. *J. Neurosci. Res.* 69:723–730.

Tanaka, T., Serneo, F. F., Higgins, C., Gambello, M. J., Wynshaw-Boris, A., and Gleeson, J. G. (2004). Lis1 and doublecortin function with dynein to mediate coupling of the nucleus to the centrosome in neuronal migration. *J. Cell Biol.* 165:709–721.

Trommsdorff, M., Borg, J. P., Margolis, B., and Herz, J. (1998). Interaction of cytosolic adaptor proteins with neuronal apolipoprotein E receptors and the amyloid precursor protein. *J. Biol. Chem.* 273:33556–33560.

Trommsdorff, M., Gotthardt, M., Hiesberger, T., Shelton, J., Stockinger, W., Nimpf, J., Hammer, R. E., Richardson, J. A., and Herz, J. (1999). *Reeler*/disabled-like disruption of neuronal migration in knockout mice lacking the VLDL receptor and ApoE receptor 2. *Cell* 97:689–701.

Uhlik, M. T., Temple, B., Bencharit, S., Kimple, A. J., Siderovski, D. P., and Johnson, G. L. (2005). Structural and evolutionary division of phosphotyrosine binding (PTB) domains. *J. Mol. Biol.* 345:1–20.

Utsunomiya-Tate, N., Kubo, K., Tate, S., Kainosho, M., Katayama, E., Nakajima, K., and Mikoshiba, K. (2000). Reelin molecules assemble together to form a large protein complex, which is inhibited by the function-blocking CR-50 antibody. *Proc. Natl. Acad. Sci. USA* 97:9729–9734.

Ware, M. L., Fox, J. W., Gonzalez, J. L., Davis, N. M., Lambert de Rouvroit, C., Russo, C. J., Chua, S. C., Jr., Goffinet, A. M., and Walsh, C. A. (1997). Aberrant splicing of a mouse disabled homolog, mdab1, in the *scrambler* mouse. *Neuron* 19:239–249.

Wynshaw-Boris, A., and Gambello, M. J. (2001). LIS1 and dynein motor function in neuronal migration and development. *Genes Dev.* 15:639–651.

Xu, M., Arnaud, L., and Cooper, J. A. (2005). Both the phosphoinositide and receptor binding activities of Dab1 are required for reelin-stimulated Dab1 tyrosine phosphorylation. *Brain Res. Mol. Brain Res.* 139:300–305.

Yip, Y. P., Capriotti, C., Magdaleno, S., Benhayon, D., Curran, T., Nakajima, K., and Yip, J. W. (2004). Components of the reelin signaling pathway are expressed in the spinal cord. *J. Comp. Neurol.* 470:210–219.

Yun, M., Keshvara, L., Park, C. G., Zhang, Y. M., Dickerson, J. B., Zheng, J., Rock, C. O., Curran, T., and Park, H. W. (2003). Crystal structures of the Dab homology domains of mouse disabled 1 and 2. *J. Biol. Chem.* 278:36572–36581.

Chapter 8
Ultrastructural Localization of Reelin

Rosalinda C. Roberts and Emma Perez-Costas

Contents

Different chapters of the present book provide background about the molecular, genetic, and biochemical features of the reelin gene, its transcript, and final protein forms. It is not, therefore, our intention to provide an extensive background about the features of the reelin gene and protein, but to emphasize some relevant features that will be helpful for a better understanding of the data reviewed in the present chapter.

1 Reelin Expression and Distribution in Vertebrates

As far as we know, reelin is expressed throughout the vertebrate scale from the most ancient living vertebrate group to humans (for reviews see Lambert de Rouvroit and Goffinet, 1998; Tissir and Goffinet, 2003; Herz and Chen, 2006). The reelin gene has an 87.2% sequence homology between mice and humans, and the protein encoded by this gene presents an even higher homology (94.2%) between these two

R. C. Roberts
Department of Psychiatry and Behavioral Neurobiology, University of Alabama at Birmingham, SPARKS Center (SC 865D), 1720 7th Avenue South, Birmingham, AL 35294-0017
e-mail: rcusidor@uab.edu

E. Perez-Costas
Department of Psychiatry and Behavioral Neurobiology, University of Alabama at Birmingham, SPARKS Center (SC 865D), 1720 7th Avenue South, Birmingham, AL 35294-0017
e-mail: epcostas@uab.edu

species (DeSilva *et al.*, 1997). In addition, reelin is also present in other vertebrates, for which partial sequences of the gene have been obtained (Bernier *et al.*, 1999, 2000; Goffinet *et al.*, 1999; Costagli *et al.*, 2002), or for which reelin mRNA and/or protein have been detected by different techniques (Pesold *et al.*, 1998, 1999; Bernier *et al.*, 1999, 2000; Goffinet *et al.*, 1999; Rodriguez *et al.*, 2000; Perez-Garcia *et al.*, 2001; Costagli *et al.*, 2002; Martinez-Cerdeno and Clasca, 2002; Martinez-Cerdeno *et al.*, 2002, 2003; Perez-Costas *et al.*, 2002, 2004; Tissir *et al.*, 2003; Candal *et al.*, 2005; Ramos-Moreno *et al.*, 2006).

During embryonic development in mice, reelin is expressed by different neuronal groups in the brain and spinal cord (D'Arcangelo *et al.*, 1995; Ikeda and Terashima, 1997; Schiffman *et al.*, 1997; Alcantara *et al.*, 1998; Kubasak *et al.*, 2004). Reelin is also widely expressed in different neuronal groups during brain development in other vertebrates, including humans (Bernier *et al.*, 1999, 2000; Goffinet *et al.*, 1999; Costagli *et al.*, 2002; Meyer and Goffinet, 1998; Meyer *et al.*, 2002; Perez-Costas *et al.*, 2002; Tissir *et al.*, 2003; Abraham *et al.*, 2004; Candal *et al.*, 2005; Roberts *et al.*, 2005). During embryonic development of mammals, reelin is synthesized and secreted in the cortex by Cajal-Retzius cells and in the cerebellum by cells in the external granular layer. In both cases, reelin secreted into the extracellular matrix is involved in the signaling for the correct positioning of postmitotic neurons that migrate radially from proliferative zones (for reviews see D'Arcangelo and Curran, 1998; Lambert de Rouvroit and Goffinet, 1998; Rice and Curran, 2001; Tissir and Goffinet, 2003). In addition, a possible role for reelin in neuronal positioning has also been reported in non-cortical areas of the mammalian brain, such as in the rhombencephalic motor nuclei, sympathetic preganglionic neurons, and even in autonomic neurons in the spinal cord (Yip *et al.*, 2000; Ohshima *et al.*, 2002; Phelps *et al.*, 2002). Finally, several research groups have provided cumulative evidence indicating that reelin may be involved directly or indirectly in other important functions in the developing brain of vertebrates such as synaptogenesis and proper dendritic development (Del Rio *et al.*, 1997; Miyata *et al.*, 1997; Forster *et al.*, 1998; Borrell *et al.*, 1999; Ohshima *et al.*, 2001; Rice *et al.*, 2001; Niu *et al.*, 2004; Yabut *et al.*, 2007).

Even though early studies associated reelin expression almost exclusively with developmental stages of the brain and it was initially considered as a "developmental molecule," an increasing number of studies have shown that reelin is widely expressed in the adult vertebrate brain as well (Perez-Garcia *et al.*, 2001; Costagli *et al.*, 2002; Martinez-Cerdeno and Clasca, 2002; Martinez-Cerdeno *et al.*, 2002, 2003; Perez-Costas *et al.*, 2004; Roberts *et al.*, 2005; Ramos-Moreno *et al.*, 2006). This robust presence of reelin in adult brain contrasts with early works in which the presence of reelin in the adult brain was considered an almost residual expression that remained after development (e.g., see, the early descriptions of reelin expression in the mouse by Ikeda and Terashima, 1997, and Alcantara *et al.*, 1998). It has been hypothesized that reelin in the adult brain is involved in synaptic function, in regulating neuronal plasticity, and could have a role in the modulation of learning and memory processes (Impagnatiello *et al.*, 1998; Pesold *et al.*, 1998, 1999;

Guidotti *et al.*, 2000; Rodriguez *et al.*, 2000, 2002; Weeber *et al.*, 2002; Perez-Costas *et al.*, 2004; Roberts *et al.*, 2005). This hypothesis is supported by evidence showing that reelin plays a role in controlling synaptic plasticity in the adult brain (Weeber *et al.*, 2002; Beffert *et al.*, 2005; Chen *et al.*, 2005; Qiu *et al.*, 2006).

Some of the findings that have brought more attention to the study of reelin are the downregulation of reelin protein levels in several neurological and psychiatric diseases, including schizophrenia, bipolar disorder, major depression, autism, and Alzheimer's disease (Impagnatiello *et al.*, 1998; Fatemi *et al.*, 2000, 2001, 2002, 2005; Guidotti *et al.*, 2000). However, none of these disorders are linked with direct genetic defects of the reelin gene. In fact, the only human disease that has been linked with defects in the reelin gene is a severe autosomal recessive form of lissencephaly with cerebellar hypoplasia, which is accompanied by severe delays in cognitive and motor development (Hong *et al.*, 2000). Thus, the continued study of reelin is important not only to learn more about its normal function, but also to determine how abnormalities in reelin may contribute to the neuropathology of the aforementioned diseases.

In summary, reelin is a highly ubiquitous molecule in the vertebrate brain and is widely present in both the developing and adult brain. Currently, the mechanism of action of this protein is not completely understood in the developing brain and is even less clear in the adult. The high conservation of the reelin protein sequence throughout the vertebrate scale, as well as its high expression during development and in the adult brain, indicate that this protein may be involved in many roles, ranging from the most studied ones (such as its involvement in neuronal migration and proper neuronal positioning) to other possible functions, such as a role in adult neural plasticity related to learning and memory processes.

An important contribution toward a better knowledge of reelin function is provided by detailed studies of the cellular and subcellular localization of this protein. Detailed light microscopy can identify cellular types that express reelin and also perform a gross examination of its intracellular localization, but electron microscopy studies are necessary to discern where the protein is localized at the subcellular level. Detailed knowledge of the ultrastructural localization of this protein will help to shed light on some of the questions that are still open about reelin storage, transport, and secretion, as well as to identify cellular types that express reelin in the vertebrate brain.

2 Types of Cells that Express Reelin

The expression of reelin is not only widespread in different brain regions and virtually present in all vertebrates (Fig. 8.1 shows examples of reelin labeling in lamprey, rat, and human brain), but also is seen in a variety of cell types as detected by *in situ* hybridization, and/or antibodies against different epitopes of the reelin protein. As with other aspects of the study of reelin, early studies using mRNA *in situ* hybridization techniques reported a quite limited expression of reelin, largely confined to

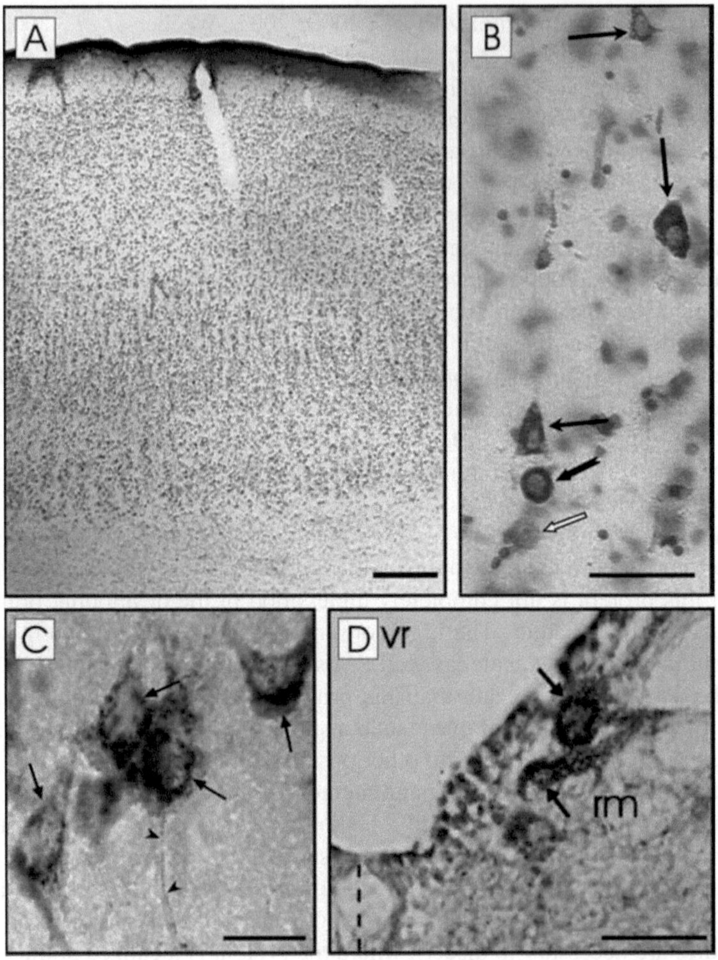

Fig. 8.1 Reelin-labeled neurons in the vertebrate brain. (**A**) Low magnification image of reelin labeling in the adult human cortex (BA39) demonstrating the abundant presence of reelin-labeled cells in all layers of the cortex (brown-stained cells). The section is counterstained with cresyl violet. (**B**) High magnification of the same cortical area as in **A** showing reelin-labeled pyramidal (plain black arrows) and nonpyramidal (notched arrow) cells. An unlabeled pyramidal cell is indicated with a white arrow. (**C**) Reelin-labeled cells of the adult rat entorhinal cortex. Arrows indicate the particle reelin labeling present in the cytoplasm, while arrowheads indicate reelin-labeled processes. (**D**) Reelin-labeled cells of the reticular rhombencephalic nucleus of the lamprey. Note the high similarity of the intracytoplasmic staining of these cells with the staining shown in **C**. vr, rhombencephalic ventricle; rm, nucleus reticularis medius. Scale bars: 500 μm (**A**); 50 μm (**B**); 15 μm (**C**); 150 μm (**D**). [**A, B** extracted from Roberts *et al.* (2005) *J. Comp. Neurol*. 482:294–308; **C** extracted from Perez-Costas (2002) Doctoral Thesis, p. 143; **D** extracted from Perez-Costas *et al.* (2004) *J. Chem. Neuroanat*. 27:7–21] (*See Color Plates*)

the cortex, cerebellum, and other laminar structures of the brain (Ikeda and Terashima, 1997; Alcantara *et al.*, 1998). Initially, the expression of reelin in the adult cortex was attributed almost exclusively to GABAergic interneurons, and the presence of reelin-expressing neurons in other areas of the brain was timidly reported (Ikeda and Terashima, 1997; Alcantara *et al.*, 1998). Subsequently, the improvement in reelin antibodies allowed more detailed studies that revealed that reelin is also widely expressed in noncortical structures of the vertebrate brain (Costagli *et al.*, 2002; Martinez-Cerdeno *et al.*, 2002, 2003; Perez-Costas *et al.*, 2002, 2004; Ramos-Moreno *et al.*, 2006). Moreover, the idea that reelin is almost exclusively present in a subset of GABAergic interneurons in the adult vertebrate cortex (Pesold *et al.*, 1998, 1999) has been strongly challenged by several studies pointing out a much higher variety of cell types that express reelin in the cortex of adult vertebrates (Fig. 8.1; Martinez-Cerdeno and Clasca, 2002; Martinez-Cerdeno *et al.*, 2002, 2003; Roberts *et al.*, 2005, Ramos-Moreno *et al.*, 2006). In fact, the high amount and variety of reelin-labeled cells observed in the adult human cortex (Fig. 8.1; Roberts *et al.*, 2005) is comparable with what has been reported in the cortex of other adult mammals (Martinez-Cerdeno and Clasca, 2002; Martinez-Cerdeno *et al.*, 2002, 2003; Deguchi *et al.*, 2003; Ramos-Moreno *et al.*, 2006).

In addition to the variety of neuronal types that express reelin, some studies indicate that reelin can also be present in glial cells (Perez-Garcia *et al.*, 2001; Roberts *et al.*, 2005) but at a much lower rate than in neurons, being only clearly distinguishable at the electron microscopic level (Roberts *et al.*, 2005).

3 Subcellular and Ultrastructural Localization of Reelin

Currently, a considerable number of research works have provided data about reelin distribution in different species, and some of them include detailed studies using light microscopy. On the contrary, a surprisingly small number of studies have been done at the electron microscopic level. High magnification lenses in light microscopy allow an overall study of the presence or absence of reelin in a specific part of a cell (for example, in the soma or in the processes of the cell), making it possible to discern the "gross" subcellular localization of reelin. However, only the electron microscope provides the appropriate resolution to analyze which organelles or ultrastructural components of the cell may be taking part in the synthesis, storage, or transport of reelin.

3.1 Where Is Reelin Located at the Subcellular Level?

The subcellular localization of reelin is almost as ubiquitous as its regional distribution in the brain. Antibodies against reelin have allowed a thorough analysis of reelin localization at the subcellular level. In all vertebrates studied, reelin consistently appears located in the soma, axonal processes, and dendrites of reelin-containing cells. Another consistent pattern is the abundance of extracellular matrix reelin labeling in developmental stages and a dramatic decrease of this kind of labeling in the adult brain (Perez-Garcia *et al.*, 2001; Martinez-Cerdeno and Clasca, 2002; Martinez-Cerdeno *et al.*, 2002, 2003; Perez-Costas *et al.*, 2002, 2004; Tissir and Goffinet, 2003; Candal *et al.*, 2005; Ramos-Moreno *et al.*, 2006). One of the best-known examples is the presence of intense extracellular matrix labeling for reelin in cortical areas of mammals during development that is replaced by a predominantly intracellular staining in the adult brain (Tissir and Goffinet, 2003; Caruncho *et al.*, 2004; Roberts *et al.*, 2005). Another less-known phenomenon, but also consistent throughout the vertebrate scale, is the presence of a transient expression of reelin in some specific major fiber tracts, coinciding in time with the moment of their development and/or maturation. Examples of this are the transient labeling for reelin in the afferent tract of the habenula, or the transient expression of reelin in the optic tract (Fig. 8 2; Perez-Costas *et al.*, 2002).

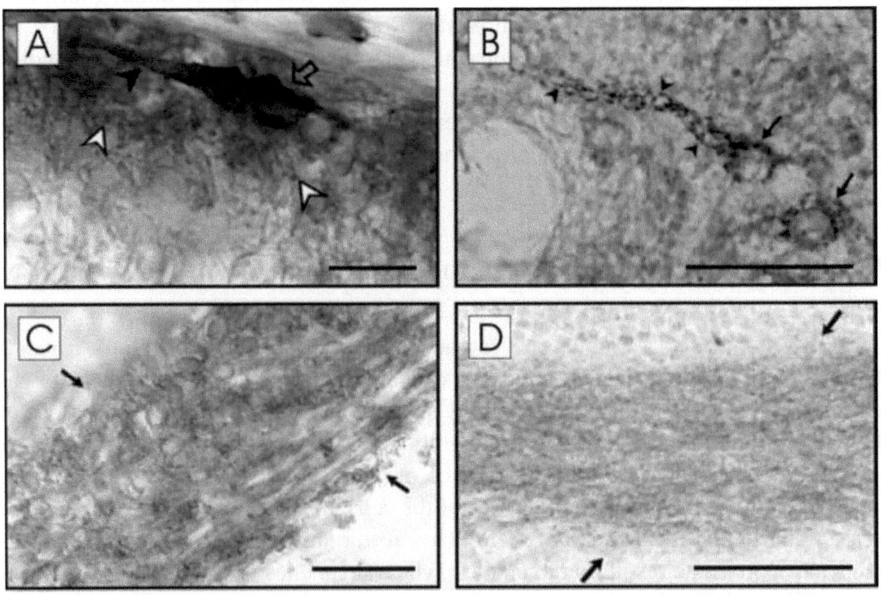

Fig. 8.2 Subcellular localization of reelin in vertebrates at the light microscopic level. (**A**) Reelin-labeled Cajal-Retzius cell in the cortex of a 17-day-old rat embryo. The labeled cell is surrounded by strong extracellular reelin labeling (white arrowheads). Note that the cell soma presents a solid reelin labeling, and the dendrite of the Cajal-Retzius cell is also labeled (black arrowhead). (**B**) Reelin-labeled cells of the retinopetal mesencephalic nucleus in the adult lamprey brain. Neuronal somata (arrows) present conspicuous particulate labeling that is also present in the initial segment of a process (arrowheads). (**C**) Transient reelin labeling in the afferent habenular tract (stria medullaris) of a 17-day-old rat embryo. (**D**) Transient reelin labeling in the optic tract of the larval lamprey brain. Scale bars: 10μm (**A**); 150μm (**B**); 25μm (**C**); 50μm (**D**). [**A**, **C**, and **D** extracted and modified from Perez-Costas *et al.* (2002) *J. Chem. Neuroanat.* 23:211–221; **B** extracted and modified from Perez-Costas *et al.* (2004) *J. Chem. Neuroanat.* 27:7–21]

An excellent paper by Martinez-Cerdeno *et al.* (2002) has shown the presence of two distinct patterns of staining for neuronal somata at the light microscopic level in different neuronal populations of the adult primate brain. One type of staining consisted in labeled particles located in the cell soma, and a second type consisted in a solid staining throughout the soma, as well as primary and secondary dendrites, the cell nuclei being devoid of reelin labeling in both cases. A review of the literature demonstrates that both types of staining are also present in other vertebrates (Fig. 8.2; see also Perez-Garcia *et al.*, 2001; Martinez-Cerdeno *et al.*, 2003; Perez-Costas *et al.*, 2004; Ramos-Moreno *et al.*, 2006). The possible biological meaning of the presence of these two different patterns of subcellular reelin labeling in the neuronal somata can be answered, in part, by electron microscopy. Ultrastructural analysis of the human cortex suggests that the different patterns of staining result from the labeling of specific organelles or other intracellular components, such as ribosomes (Fig. 8.3; Roberts *et al.*, 2005). Within the somata of neurons, round membrane-bound profiles (probably endosomes) (Fig. 8.3A), and labeling deposited in outpockets of the nuclear membrane (Fig. 8.3B) may account for the particulate staining observed at the light microscopic level. The more diffuse staining observed at the light microscopic level in somata and proximal dendrites is probably due to the labeling of reelin on ribosomes and rough endoplasmic reticulum (Fig. 8.3A, 8.4A).

3.2 *Ultrastructural Localization of Reelin*

Studies of reelin at the electron microscopic level have been performed only in mammalian species, with detailed studies available in the mouse (Pappas *et al.*, 2001) and human brain (Roberts *et al.*, 2005). Reelin labeling at the electron microscopic level is consistent with what has been described using light microscopy, identifying reelin-labeled structures in neuronal somata, axons, and dendrites, as well as extracellular matrix labeling (Pesold *et al.*, 1998; Rodriguez *et al.*, 2000; Derer *et al.*, 2001; Pappas *et al.*, 2001, 2003; Martinez-Cerdeno *et al.*, 2002; Roberts *et al.*, 2005). In addition, the high resolution of the electron microscope has demonstrated that reelin can also be present in glial cells but in very low levels, compared with its presence in neurons (Roberts *et al.*, 2005). Although most of the studies agree on the consistency of this pattern, some discrepancies are present.

3.2.1 Reelin in Neuronal Somata

Within the somata of human cortical neurons, reelin labeling is present in the euchromatin of the nucleus, ribosomes on the outer nuclear membrane, rough endoplasmic reticulum, and polyribosome rosettes (Fig. 8.3, 8.4). In these structures, the labeling is very discretely deposited (Fig. 8.4, 8.5). For example, spherical membrane-bound cytoplasmic organelles are labeled fairly regularly (Fig. 8.3A). In many instances, an outpocketing of the nuclear outer membrane appears filled

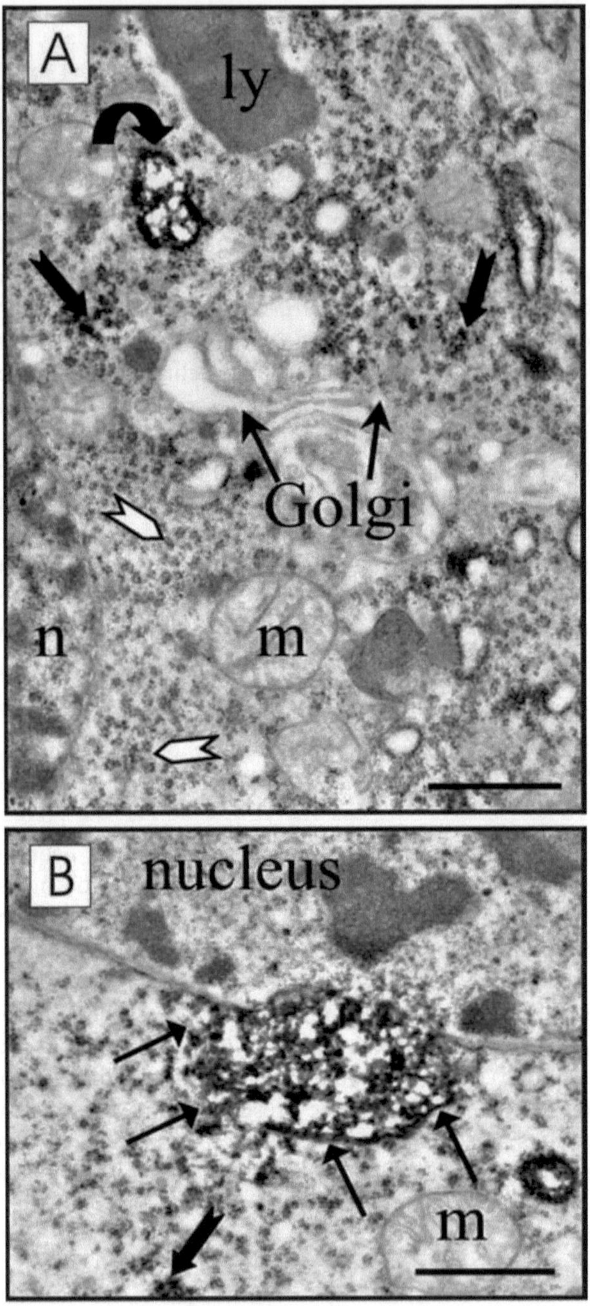

Fig. 8.3 Ultrastructural localization of reelin labeling in neuronal somata. (**A**) Many ribosomes are labeled (black arrows), while some are unlabeled (white arrows). Labeled oval organelles (curved arrow), which may be endoplasmic reticulum or endosomes, are present in some cell bodies. The Golgi apparatus is unlabeled as are mitochondria (m) and lysosomes (ly). n, nucleus. (**B**) An example of labeling in the space between the inner and outer nuclear membrane (plain black arrows). Labeled ribosomes (arrow) and unlabeled mitochondrion are shown. Scale bars: 1 μm. [Figures extracted and modified from Roberts *et al.* (2005) *J. Comp. Neurol.* 482:294–308]

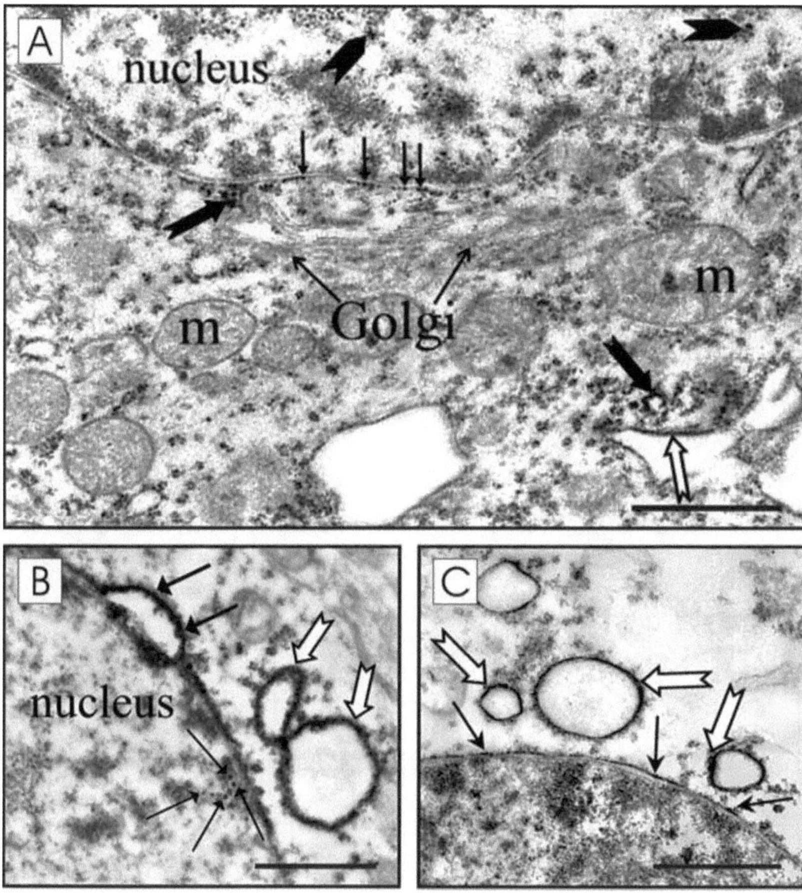

Fig. 8.4 Ultrastructural localization of reelin labeling in neuronal somata. (**A**, **B**) Reelin labeling in adult human cortex. (**A**) Reelin appears in ribosomes (thick arrows), rough endoplasmic reticulum and ribosomes of the outer nuclear membrane (thin arrows). Note also some particulate reelin labeling in the nucleus (arrowheads). (**B**) More detailed image from a different neuron showing reelin labeling in ribosomes on the outer nuclear membrane (arrows), as well as in the rough endoplasmic reticulum (white arrows). Note also the presence of reelin-labeled vesicles in the nucleus (arrows). (**C**) Reelin labeling in fetal human cortex. Reelin labeling appears in the rough endoplasmic reticulum (white arrows) as well as in ribosomes of the outer nuclear membrane (thin arrows). Note the similarity of this staining with that of the adult brain shown in **A** and **B**. Scale bars: 1 μm in A-C. [Figures extracted and modified from Roberts *et al.* (2005) *J. Comp. Neurol.* 482:294–308]

with reelin-labeled material (Fig. 8.3B). Ribosomes on the outer nuclear membrane and rough endoplasmic reticulum are robustly labeled (Fig. 8.4A,B), as well as vesicles within the nucleus at the nuclear membrane (Fig. 8.4B). In developing human brain (Fig. 8.4C), reelin is located on the euchromatin within the nucleus, in the innermost part of the nuclear membrane, on rough endoplasmic reticulum, and on ribosomes along the outer nuclear membrane (Roberts *et al.*, 2005). Presence of reelin on ribosomes, polyribosome rosettes, rough endoplasmic reticulum, and

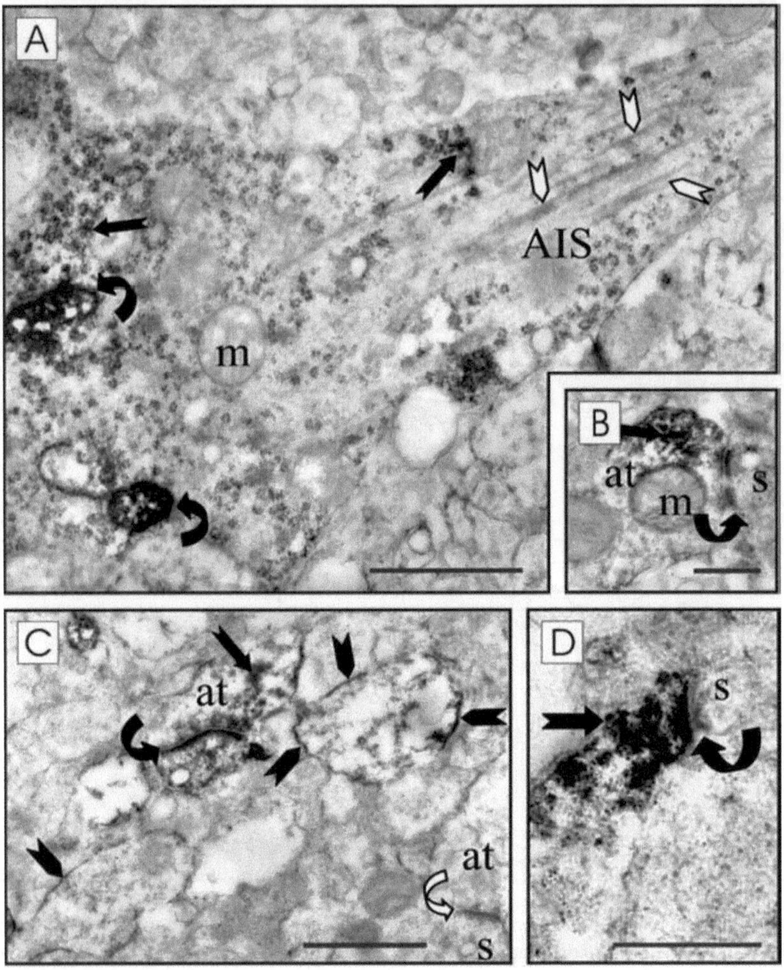

Fig. 8.5 Ultrastructural localization of reelin labeling in neuronal processes and synapses. (**A–C**) Reelin labeling in the adult human cortex. (**A**) Reelin-labeled axon initial segment (AIS) in which the axon hillock contains numerous reelin-labeled ribosomes (straight black arrows), as well as labeled oval organelles (curved arrows). (**B**) Labeled axon terminal (straight arrow) forming a synapse (curved arrow) with a spine (s). (**C**) Neuropil showing a labeled axon terminal (straight black arrow) forming a synapse (curved black arrow) with a labeled spine (s). (**D**) Reelin labeling in the human fetal cortex. A labeled axon terminal (straight arrow) forms a synapse (curved arrow) with a spine (s). Scale bars: 1 μm (**A**, **C**); 0.5 μm (**B**, **D**). [Figures extracted and modified from Roberts *et al.* (2005) *J. Comp. Neurol.* 482:294–308]

outer nuclear membrane of cells in the human cortex (Roberts *et al.*, 2005) is in general agreement with what has been observed in rodent and nonhuman primates (Pappas *et al.*, 2001; Martinez-Cerdeno *et al.*, 2002).

Reelin labeling in adult human cortex is conspicuously absent from heterochromatin, most mitochondria, and lysosomes (Fig. 8.3A, 8.4A; Roberts *et al.*, 2005).

The Golgi apparatus appears almost devoid of reelin (Fig. 8.3A, 8.4A), while the extracellular matrix (Fig. 8.5C) and multivesicular bodies (Fig. 8.6A) appear occasionally labeled (Roberts *et al.*, 2005). A major difference in the results found in human and those in other species for which electron microscopy was utilized is the paucity of reelin in the adult human extracellular space and the subcellular machinery in which secreted proteins are made and taken back up into the cell. In contrast, in the adult mouse, reelin has been found in the Golgi apparatus and extracellular space (Pappas *et al.*, 2001) and is secreted in cell cultures of murine neurons (Lacor *et al.*, 2000). In macaques, Martinez-Cerdeno *et al.* (2002) reported the presence of reelin in secretory organelles, but the most robust staining was present in the rough endoplasmic reticulum, rather than the smooth endoplasmic reticulum or the Golgi

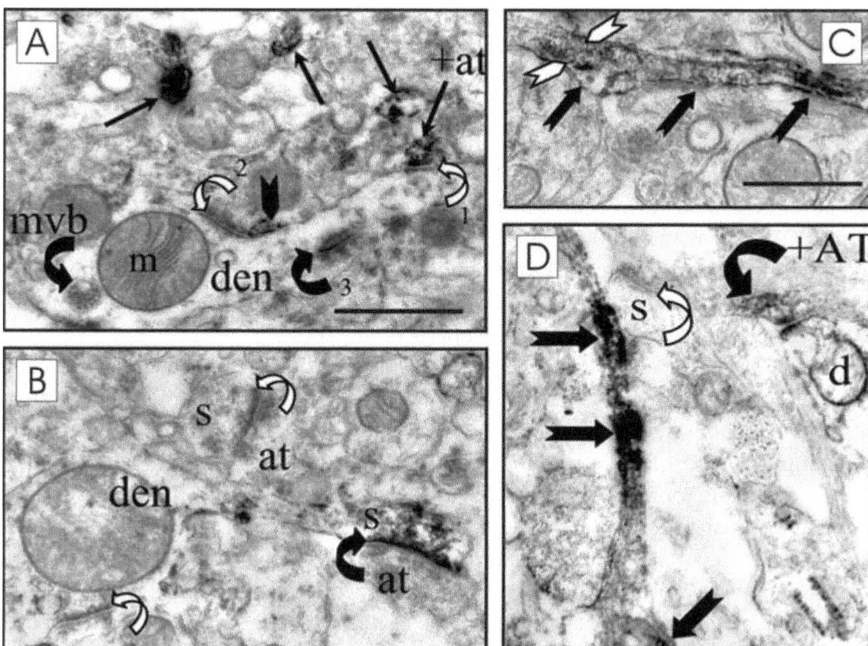

Fig. 8.6 Ultrastructural localization of reelin labeling in neuronal processes and synapses. (**A–C**) Reelin labeling in the neuropil of the adult human cortex. (**A**) Diffuse reelin labeling in profiles (straight arrows) that could correspond with small dendrites or spine necks. A symmetric synapse (curved black arrow) is formed by a labeled terminal (+at) but the postsynaptic density is not labeled (white arrow 1). Another labeled synapse (curved black arrow 3) is formed by an unlabeled terminal and a reelin-labeled postsynaptic density. Note also the presence of a labeled glial process (black arrowhead). (**B**) A long labeled spine (s) receives an asymmetric synapse (black arrow) from an unlabeled terminal (at). (**C**) Diffusely labeled unmyelinated axon (straight arrows). (**D**) In this image of the fetal human cortex, reelin labeling appears in long thin neuronal processes (black arrows) and in an axon terminal (+AT) forming a synapse with a dendrite (d). Scale bars: 1 μm (**A–C**); 0.5 μm (**D**). [Figures extracted and modified from Roberts *et al.* (2005) *J. Comp. Neurol.* 482:294–308]

apparatus. This group reported also that all reelin labeling was intracellular in the adult nonhuman primate brain (Martinez-Cerdeno *et al.*, 2002).

3.2.2 Reelin in Neuronal Processes and Synapses (Neuropil)

The presence of abundant neuropil staining has been described at the light microscopic level in different vertebrates (Martinez-Cerdeno *et al.*, 2002, 2003; Perez-Costas *et al.*, 2002; Ramos-Moreno *et al.*, 2006). Neuropil includes both dendritic and axonal processes and electron microscopy is necessary to identify the labeled profiles. Reelin labeling in the human cortical neuropil is richly distributed among axons, terminals, dendritic shafts, and spines (Roberts *et al.*, 2005).

Reelin labeling in the adult human appears throughout the axon, including the axon hillock and initial segment, and axon terminals (Fig. 8.5B,C, 8.6A). In the axon hillock and axon initial segment, reelin labeling is deposited on polyribosome rosettes, and present diffusely in spherical membrane-bound cytoplasmic organelles (Fig. 8.5A). In addition, in small unmyelinated axons reelin labeling is diffusely deposited inside the axon (Fig. 8.6C). Reelin axonal labeling is also present in fetal human brain (Fig. 8.5D, 8.6D; Roberts *et al.*, 2005). The ultrastructural localization of reelin-labeled axonal processes has also been described in rodents and nonhuman primates (Derer *et al.*, 2001; Pappas *et al.*, 2001; Martinez-Cerdeno *et al.*, 2002). The presence of reelin labeling throughout the extent of axonal processes strongly suggests that reelin can be transported long distances through axonal tracts (Derer *et al.*, 2001; Perez-Costas *et al.*, 2002; Martinez-Cerdeno *et al.*, 2003; Roberts *et al.*, 2005).

In some axon terminals, it is possible to discern that reelin labeling is located around and between synaptic vesicles (Roberts *et al.*, 2005). The labeled axon terminals formed both asymmetric (Fig. 8.5B,C) and symmetric synapses (Fig. 8.6A), indicating that reelin can be present in inhibitory and excitatory terminals. Labeling in axon terminals forming symmetric synapses is consistent with the localization of reelin in cortical interneurons, which form this type of synapse (DeFelipe *et al.*, 2002). The location of reelin in pyramidal cells and in axon terminals that form asymmetric synapses suggests that reelin is involved in corticocortical signaling (DeFelipe *et al.*, 2002). Thalamocortical projections represent another potential source of reelin-labeled terminals forming asymmetric synapses in the cortex (Peters, 2002).

Throughout the neuropil, reelin labeling in the human cortex also appears in dendrites and spines (Roberts *et al.*, 2005). In dendritic shafts, reelin labeling appears in the rough endoplasmic reticulum and polyribosome rosettes, as well as in some of the postsynaptic densities (Fig. 8.6A). Dendritic spines contain diffuse labeling in the head and neck (Fig. 8.6B) and/or prominent labeling in the postsynaptic density (Fig. 8.5C, 8.6A). The postsynaptic densities that reveal reelin labeling in spines and dendritic shafts are associated with both asymmetric (Fig. 8.5C) and symmetric synapses (Fig. 8.6A). These results in human are consistent with the localization of reelin labeling in spines and/or the postsynaptic densities in adult mice (Pappas *et al.*, 2001) and nonhuman primates (Rodriguez *et al.*, 2000; Martinez-Cerdeno *et al.*, 2002). Moreover, recent studies have shown that reelin

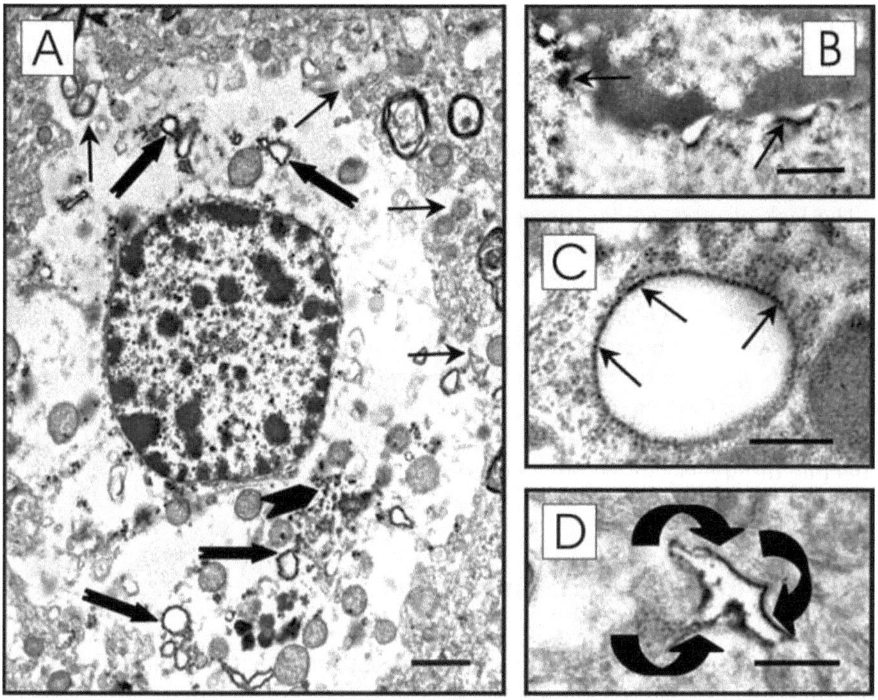

Fig. 8.7 Localization of reelin labeling in glial cells in the human cortex. (**A**) Astrocyte showing reelin labeling in the rough endoplasmic reticulum (thick arrows) as well as in free ribosomes (arrowheads). (**B**) Reelin labeling in the outer nuclear membrane of a glial cell (arrows). (**C**) Detail of the rough endoplasmic reticulum of a glial cell showing labeled ribosomes (arrows). (**D**) A small irregularly shaped labeled profile (curved arrows) that probably corresponds to an astrocytic process. Scale bars: 1 µm (**A**); 0.5 µm (**B–D**). [Figures extracted and modified from Roberts *et al.* (2005) *J. Comp. Neurol.* 482:294–308]

and its receptors can be pivotal regulators of *N*-methyl-D-aspartate receptor function and could also regulate GABA circuits (Herz and Chen, 2006).

3.2.3 Reelin Labeling in Glial Cells

At the electron microscopic level, reelin labeling is also found in glial cells of the human cortex (Fig. 8.7A) but in lower levels than in neurons. In cell bodies, the labeling appears in ribosomes attached to the outer nuclear membrane (Fig. 8.7B), and as discrete labeling on the rough endoplasmic reticulum and in polyribosome rosettes (Fig. 8.7A,C). Small labeled astrocytic processes are observed throughout the neuropil often associated with synapses (Fig. 8.7D; Roberts *et al.*, 2005). The presence and extent of reelin labeling in glial cells in nonhuman primates and other species is somewhat unclear, and, if present, is probably less abundant than in humans. Ultrastructural studies in nonhuman primates (Martinez-Cerdeno *et al.*,

2002) and rodents (Pappas *et al.*, 2001) did not report the presence of reelin labeling in glial cells. However, a study by Rodriguez *et al.* (2000) presenting electron microscopy data on reelin labeling in a nonhuman primate shows an image of a small irregularly shaped reelin-labeled process remarkably similar in morphology to that of an astroglial process, though the profile was interpreted by the authors to be a dendritic spine.

The lack of other studies describing reelin labeling in glial cells could reflect several technical factors, including that the low level of reelin labeling in glial cells could surpass the detection limits of light microscopy, as well as possible differences in tissue processing and, perhaps, misinterpretation of electron microscope images. On the other hand, it is quite possible that there are real differences in reelin content in glial cells among different vertebrate species, as occurs for other structures in the brain (Martinez-Cerdeno and Clasca, 2002; Deguchi *et al.*, 2003; Martinez-Cerdeno *et al.*, 2003). Thus, reelin labeling in human glial cells may represent a pattern reflecting differences in the phylogenetic scale.

4 Reelin Function and Malfunction

During neocortical development, reelin is present in the extracellular matrix and Cajal-Retzius cells and functions to guide developing neurons (Meyer and Goffinet, 1998; Meyer *et al.*, 2002). However, once neocortical neuronal migration ceases during the first half of gestation (Rakic and Sidman, 1968; Rakic, 1978) and the degeneration of Cajal-Retzius cells occurs by the end of the second trimester (Meyer and Goffinet, 1998), the major function of reelin in human must change. At midgestation in human, the localization of reelin has both fetal and adult patterns (Meyer *et al.*, 2002; Roberts *et al.*, 2005). Thus, the second trimester of human development appears to represent the transition in location and function of reelin. This pattern of change in location across development is consistent with the pattern of other proteins and molecules whose function is critical during development but that continue to be expressed in adulthood.

At the synapse, reelin labeling has been found associated with presynaptic and postsynaptic elements (Rodriguez *et al.*, 2000; Pappas *et al.*, 2001; Roberts *et al.*, 2005). The location of reelin in axon terminals, dendritic spines, the postsynaptic density, and astroglial processes in adult brain is consistent with the apparent role of reelin in synaptic plasticity (Rice *et al.*, 2001; Weeber *et al.*, 2002), which is a function of the tripartite synapse (spine, terminal, and astroglial process). In adult human cortex, reelin is present in 53% of synaptic complexes, where labeling is present in the axon terminal, the postsynaptic density, or both, suggesting an important role for reelin in synaptic processes (Roberts *et al.*, 2005). The abundance of reelin associated with cortical synapses suggests its involvement in the function of many cortical connections.

In summary, the ultrastructural localization of reelin indicates that this protein can be synthesized by different types of neurons in the vertebrate brain, transported even long distances through neuronal processes, and is abundantly present in synapses, both in the developing and adult brain. The prevalence of this protein in the adult brain also indicates that reelin's functions go far beyond its role in neuronal migration and correct positioning of neurons in the cortex. In fact, the remarkable abundance of reelin labeling in axon terminals, as well as dendrites and dendritic spines in adult brain is consistent with a role for reelin in normal mature synaptic function, including synaptic plasticity and appropriate remodeling of synaptic contacts. Altogether, these findings suggest that proper levels of reelin and its receptors could be crucial for the proper functioning of synaptic transmission in the adult brain.

An abundance of evidence suggests that the malfunction of reelin may have a role in several diseases of the brain. Reelin abnormalities are found in several brain regions in schizophrenia, bipolar disease, depression (Impagnatiello *et al.*, 1998; Fatemi *et al.*, 2000, 2001; Guidotti *et al.*, 2000; Eastwood and Harrison, 2002; Caruncho *et al.*, 2004), autism (Persico *et al.*, 2001; Reichelt *et al.*, 2001; Fatemi *et al.*, 2002, 2005), and lissencephaly (Hong *et al.*, 2000). The decrease in reelin mRNA in schizophrenia (Impagnatiello *et al.*, 1998; Guidotti *et al.*, 2000) is particularly interesting, considering the evidence from postmortem studies that schizophrenia is associated with impaired cell migration in the neocortex (Akbarian *et al.*, 1996; Kirkpatrick *et al.*, 1999, 2003), and the observation that reelin is present in dendritic spines and terminals, which are affected in this disease (Roberts *et al.*, 1996; Glantz and Lewis, 2000). Moreover, it has been reported that heterozygous reeler mice (that have reduced levels of reelin in the brain) present anatomical and behavioral deficits that resemble some of the behavioral and anatomical deficits present in schizophrenia (Impagnatiello *et al.*, 1998; Tueting *et al.*, 1999; Guidotti *et al.*, 2000; Ballmaier *et al.*, 2002). Whether reelin is affected during development, adulthood, or in both of these stages in brain disease remains to be determined. In any case, reelin and/or molecules in its signaling pathway may be potential therapeutic targets for several brain disorders.

Acknowledgments The authors thank Dr. Brian Kirkpatrick for his participation in the study of reelin expression in the human brain, as well as Joy K. Roche for her technical assistance. We also thank Dr. Miguel Melendez-Ferro for his critical review and participation in the editing of this manuscript.

References

Abraham, H., Perez-Garcia, C. G., and Meyer, G. (2004). p73 and reelin in Cajal-Retzius cells of the developing human hippocampal formation. *Cereb. Cortex* 14:484–495.

Akbarian, S., Kim, J. J., Potkin, S. G., Hetrick, W. P., Bunney, W. E., Jr., and Jones, E. G. (1996). Maldistribution of interstitial neurons in prefrontal white matter of the brains of schizophrenic patients. *Arch. Gen. Psychiatry* 53:425–436.

Alcantara, S., Ruiz, M., D'Arcangelo, G., Ezan, F., de Lecea, L., Curran, T., Sotelo, C., and Soriano, E. (1998). Regional and cellular patterns of reelin mRNA expression in the forebrain of the developing and adult mouse. *J. Neurosci.* 17:7779–7799.

Ballmaier, M., Zoli, M., Leo, G., Agnati, L. F., and Spano, P. (2002). Preferential alterations in the mesolimbic dopamine pathway of heterozygous reeler mice: an emerging animal-based model of schizophrenia. *Eur. J. Neurosci.* 15:1197–1205.

Beffert, U., Weeber, E. J., Durudas, A., Quiu, S., Masiulis, I., Sweatt, J. D., Li, W. P., Adelmann, G., Frotscher, M., Hammer, R. E., and Herz, J. (2005). Modulation of synaptic plasticity and memory by reelin involves differential splicing of the lipoprotein receptor Apoer2. *Neuron* 47:567–579.

Bernier, B., Bar, I., Pieau, C., Lambert de Rouvroit, C., and Goffinet, A. M. (1999). Reelin mRNA expression during embryonic brain development in the turtle *Emys orbicularis. J. Comp. Neurol.* 413:463–479.

Bernier, B., Bar, I., D'Arcangelo, G., Curran, T., and Goffinet, A. M. (2000). Reelin mRNA expression during embryonic brain development in the chick. *J. Comp. Neurol.* 422:448–463.

Borrell, V., Del Rio, J. A., Alcantara, S., Derer, M., Martinez, A., D'Arcangelo, G., Nakajima, K., Mikoshiba, K., Derer, P., Curran, T., and Soriano, E. (1999). Reelin regulates the development and synaptogenesis of the layer-specific entorhino-hippocampal connections. *J. Neurosci.* 19:1345–1358.

Candal, E. M., Caruncho, H. J., Sueiro, C., Anadon, R., and Rodriguez-Moldes, I. (2005). Reelin expression in the retina and optic tectum of developing common brown trout. *Brain Res. Dev. Brain Res.* 154:187–197.

Caruncho, H. J., Dopeso-Reyes, I. G., Loza, M. I., and Rodriguez, M. A. (2004). GABA, reelin, and the neurodevelopmental hypothesis of schizophrenia. *Crit. Rev. Neurobiol.* 16:25–32.

Chen, Y., Beffert, U., Ertunc, M., Tanq, T. S., Kavalali, E. T., Bezprovanny, I., and Herz, J. (2005). Reelin modulates NMDA receptor activity in cortical neurons. *J. Neurosci.* 25: 8209–8216.

Costagli, A., Kapsimali, M., Wilson, S. W., and Mione, M. (2002). Conserved and divergent patterns of reelin expression in the zebrafish central nervous system. *J. Comp. Neurol.* 450:73–93.

D'Arcangelo, G., and Curran, T. (1998). Reeler: new tales on an old mutant mouse. *BioEssays* 20:235–244.

D'Arcangelo, G., Miao, G. G., Chen, S. C., Soares, H, D., Morgan, J. I., and Curran, T. (1995). A protein related to extracellular matrix proteins deleted in the mouse mutant reeler. *Nature* 374:719–723.

DeFelipe, J., Alonso-Nanclares, L., and Arellano, J. I. (2002). Microstructure of the neocortex: comparative aspects. *J. Neurocytol.* 31:299–316.

Deguchi, K., Inoue, K., Avila, W. E., Lopez-Terrada, D., Antalffy, B. A., Quattrocchi, C. C., Sheldon, M., Mikoshiba, K., D'Arcangelo, G., and Armstrong, D. L. (2003). Reelin and disabled-1 expression in developing and mature human cortical neurons. *J. Neuropathol. Exp. Neurol.* 62:676–684.

Del Rio, J. A., Heimrich, B., Borrell, V., Forster, E., Drakew, A., Alcantara, S., Nakajima, K., Miyata, T., Ogawa, M., Mikoshiba, K., Derer, P., Frotscher, M., and Soriano, E. (1997). A role for Cajal-Retzius cells and reelin in the development of hippocampal connections. *Nature* 385:70–74.

Derer, P., Derer, M., and Goffinet, A. M. (2001). Axonal secretion of reelin by Cajal-Retzius cells: evidence from comparison of normal and $Reln^{Orl}$ mutant mice. *J. Comp. Neurol.* 440:136–143.

DeSilva, U., D'Arcangelo, G., Braden, V. V., Chen, J., Miao, G. G., Curran, T., and Green, E. D. (1997). The human reelin gene: isolation, sequencing, and, mapping on chromosome 7. *Genome Res.* 7:157–164.

Eastwood, S. L., and Harrison, P. J. (2003). Interstitial white matter neurons express less reelin and are abnormally distributed in schizophrenia: towards an integration of molecular and morphologic aspects of the neurodevelopmental hypothesis. *Mol. Psychiatry* 8:821–831.

Fatemi, S. H., Earle, J. A., and McMenomy, T. (2000). Reduction in reelin immunoreactivity in hippocampus of subjects with schizophrenia, bipolar disorder and major depression. *Mol. Psychiatry.* 5:654–663.

Fatemi, S. H., Kroll, J. L., and Stary, J. M. (2001). Altered levels of reelin and its isoforms in schizophrenia and mood disorders. *Neuroreport* 12:3209–3215.

Fatemi, S. H., Stary, J. M., and Egan, E. A. (2002). Reduced blood levels of reelin as a vulnerability factor in pathophysiology of autistic disorder. *Cell. Mol. Neurobiol.* 22:139–152.

Fatemi, S. H., Snow, A. V., Stary, J. M., Araghi-Niknam, M., Reutiman, T. J., Lee, S., Brooks, A. I., and Pearce D. A. (2005). Reelin signaling is impaired in autism. *Biol. Psychiatry* 57:777–787.

Forster, E., Kaltschmidt, C., Deng, J., Cremer, H., Deller, T., and Frotscher, M. (1998). Lamina-specific cell adhesion on living slices of hippocampus. *Development* 125:3399–3410.

Glantz, L. A., and Lewis, D. A. (2000). Decreased dendritic spine density on prefrontal cortical pyramidal neurons in schizophrenia. *Arch. Gen. Psychiatry* 57:65–73.

Goffinet, A. M., Bar, I., Bernier, B., Trujillo, C., Raynaud, A., and Meyer, G. (1999). Reelin expression during embryonic brain development in lazertidian lizards. *J. Comp. Neurol.* 414:533–550.

Guidotti, A., Auta, J., Davis, J. M., DiGiorgi-Generini, V., Dwivedi, Y., Grayson, D. R., Impagnatiello, F., Pandey, G. N., Pesold, C., Sharma, R. F., Uzunov, D. P., and Costa, E. (2000). Decreased reelin and glutamic acid decarboxylase 67 (GAD67) expression in schizophrenia and bipolar disorders. *Arch. Gen. Psychiatry* 57:1061–1069.

Herz, J., and Chen, Y. (2006). Reelin, lipoprotein receptors and synaptic plasticity. *Nature Rev. Neurosci.* 7:850–859.

Hong, S. E., Shugart, Y. Y., Huang, D. T., Shahwan, S. A., Grant, P. E., Hourihane, J. O., Martin, N. D., and Walsh, C. A. (2000). Autosomal recessive lissencephaly with cerebellar hypoplasia is associated with human RELN mutations. *Nature Genet.* 26:93–96.

Ikeda, Y., and Terashima, T. (1997). Expression of reelin, the gene responsible for the reeler mutation, in embryonic development and adulthood in the mouse. *Dev. Dyn.* 210:157–172.

Impagnatiello, F., Guidotti, A., Pesold, C., Dwivedi, Y., Caruncho, H. J., Pisu, M. G., Uzunov, D. P., Smalheiser, N. R., Davis, J. M., Pandey, G. N., Pappas, G. D., Tueting, P., Sharma, R. P., and Costa, E. (1998). A decrease of reelin expression as a putative vulnerability factor in schizophrenia. *Proc. Natl. Acad. Sci. USA* 95:15718–15723.

Kirkpatrick, B., Conley, R. C., Kakoyannis, A., Reep, R. L., and Roberts, R. C. (1999). Interstitial cells of the white matter in the inferior parietal cortex in schizophrenia: An unbiased cell-counting study. *Synapse* 34:95–102.

Kirkpatrick, B., Messias, N. C., Conley, R. R., and Roberts, R. C. (2003). Interstitial cells of the white matter in the dorsolateral prefrontal cortex in deficit and nondeficit schizophrenia. *J. Nerv. Ment. Dis.* 191:563–567.

Kubasak, M. D., Brooks, R., Chen, S., Villeda, S. A., and Phelps, P. E. (2004). Developmental distribution of reelin-positive cells and their secreted production in the rodent spinal cord. *J. Comp. Neurol.* 468:165–178.

Lacor, P. N., Grayson, D. R., Auta, J., Sugaya, I., Costa, E., and Guidotti, A. (2000). Reelin secretion from glutamatergic neurons in culture is independent from neurotransmitter regulation. *Proc. Natl. Acad. Sci. USA* 97:3556–3561.

Lambert de Rouvroit, C., and Goffinet A. M. (1998). The reeler mouse as a model of brain development. *Adv. Anat. Embryol.* 150:1–106.

Martinez-Cerdeno, V., and Clasca, F. (2002). Reelin immunoreactivity in the adult neocortex: a comparative study in rodents, carnivores and non-human primates. *Brain Res. Bull.* 57:485–488.

Martinez-Cerdeno, V., Galazo, M. J., Cavada, C., and Clasca, F. (2002). Reelin immunoreactivity in the adult primate brain: intracellular localization in projecting and local circuit neurons of the cerebral cortex, hippocampus and subcortical regions. *Cereb. Cortex* 12:1298–1311.

Martinez-Cerdeno, V., Galazo, M. J., and Clasca, F. (2003). Reelin immunoreactive neurons, axons, and neuropil in the adult ferret brain: evidence for axonal secretion of reelin in long pathways. *J. Comp. Neurol.* 463:92–116.

Meyer, G., and Goffinet, A. M. (1998). Prenatal development of reelin-immunoreactive neurons in the human neocortex. *J. Comp. Neurol.* 397:29–40.

Meyer, G., Perez-Garcia, C. G., Abraham, H., and Caput, D. (2002). Expression of p73 and reelin in the developing human cortex. *J. Neurosci.* 22:4973–4986.

Miyata, T., Nakajima, K., Mikoshiba, K., and Ogawa, M. (1997). Regulation of Purkinje cell alignment by reelin as revealed with CR-50 antibody. *J. Neurosci.* 17:3599–3609.

Niu, S., Renfro, A., Quattrocchi, C. C., Sheldon, M., and D'Arcangelo, G. (2004). Reelin promotes hippocampal dendrite development through the VLDLR/ApoER2-Dab1 pathway. *Neuron* 41:71–84.

Ohshima, T., Ogawa, M., Veeranna, Hirasawa, M., Longenecker, G., Ishiguro, K., Pant, H. C., Brady, R.O., Kulkarni, A. B., and Mikoshiba, K. (2001). Synergistic contributions of cyclin-dependent kinase 5/p35 and reelin/Dab 1 to the positioning of cortical neurons in the developing mouse brain. *Proc. Natl. Acad. Sci. USA* 98:2764–2769.

Ohshima, T., Ogawa, M., Takeuchi, K., Takahashi, S., Kulkarni, A. B., and Mikoshiba, K. (2002). Cyclin-dependent kinase 57p35 contributes synergistically with reelin/dab1 to the positioning of facial branchiomotor and inferior olive neurons in the developing mouse hindbrain. *J. Neurosci.* 22:4036–4044.

Pappas, G. D., Kriho, V., and Pesold, C. (2001). Reelin in the extracellular matrix and dendritic spines of the cortex and hippocampus: a comparison between wild type and heterozygous reeler mice by electron microscopy. *J. Neurocytol.* 30:413–425.

Pappas, G. D., Kriho, V., Liu, W. S., Tremolizzo, L., Lugli, G., and Larson, J. (2003). Immunocytochemical localization of reelin in the olfactory bulb of the heterozygous reeler mouse: an animal model for schizophrenia. *Neurobiol. Res.* 25:819–830.

Perez-Costas, E. (2002). *Expresion y distribucion de reelina en el sistema nervioso central de la lamprea.* Doctoral thesis. University of Santiago de Compostela, Spain.

Perez-Costas, E., Melendez-Ferro, M., Santos, Y., Anadon, R., Rodicio, M. C., and Caruncho, H. J. (2002). Reelin immunoreactivity in the larval sea lamprey brain. *J. Chem. Neuroanat.* 23:211–221.

Perez-Costas, E., Melendez-Ferro, M., Perez-Garcia, C. G., Caruncho, H. J., and Rodicio, M. C. (2004). Reelin immunoreactivity in the adult sea lamprey brain. *J. Chem. Neuroanat.* 27:7–21.

Perez-Garcia, C. G., Gonzalez-Delgado, F. J., Suarez-Sola, M. L., Castro-Fuentes, R., Martin-Trujillo, J. M., Ferres-Torres, R., and Meyer, G. (2001). Reelin-immunoreactive neurons in the adult vertebrate pallium. *J. Chem. Neuroanat.* 21:41–51.

Persico, A. M., D'Agruma, L., Maiorano, N., Totaro, A., Militerni, R., Bravaccio, C., Wassink, T. H., Schneider, C., Melmed, R., Trillo, S., Montecchi, F., Palermo, M., Pascucci, T., Puglisi-Allegra, S., Reichelt, K. L., Conciatori, M., Marino, R., Quattrocchi, C. C., Baldi, A., Zelante, L., Gasparini, P., and Keller, F. (2001). Reelin gene alleles and haplotypes as a factor predisposing to autistic disorder. *Mol. Psychiatry* 6:150–159.

Pesold, C., Impagnatiello, F., Pisu, M. G., Uzunov, D. P., Costa, E., Guidotti, A., and Caruncho, H. J. (1998). Reelin is preferentially expressed in neurons synthesizing γ-aminobutyric acid in cortex and hippocampus of adult rats. *Proc. Natl. Acad. Sci. USA* 95:3221–3226.

Pesold, C., Liu, W. S., Guidotti, A., Costa, E., and Caruncho, H. J. (1999). Cortical bitufted, horizontal, and Martinotti cells preferentially express and secrete reelin into perineuronal nets, nonsynaptically modulating gene expression. *Proc. Natl. Acad. Sci. USA* 96:3127–3222.

Peters, A. (2002). Examining neocortical circuits: some background and facts. *J. Neurocytol.* 31:183–193.

Phelps, P. E., Rich, R., Dupuy-Davies, S., Rios, Y., and Wong, T. (2002). Evidence for a cell-specific action of reelin in the spinal cord. *Dev. Biol.* 244:180–198.

Qiu, S., Korwek, K. M., Pratt-Davis, A. R., Peters, M., Bergman, M. Y., and Weeber, E. J. (2006). Cognitive disruption and altered hippocampus synaptic function in reelin haploinsufficient mice. *Neurobiol. Learn. Memory* 85:228–242.

Rakic, P. (1978). Neuronal migration and contact guidance in the primate telencephalon. *Postgrad. Med. J.* 1:25–40.

Rakic, P., and Sidman, R. L. (1968). Supravital DNA synthesis in the developing human and mouse brain. *J. Neuropathol. Exp. Neurol.* 27:246–276.

Ramos-Moreno, T., Galazo, M. J., Porrero, C., Martinez-Cerdeno, V., and Clasca, F. (2006). Extracellular matrix molecules and synaptic plasticity: immunomapping of intracellular and secreted reelin in the adult rat brain. *Eur. J. Neurosci.* 23:401–422.
Reichelt, K. L., Conciatori, M., Marino, R., Quattrocchi, C.C., Baldi, A., Zelante, L., Gasparini, P., and Keller, F. (2001). Reelin gene alleles and haplotypes as a factor predisposing to autistic disorder. *Mol. Psychiatry* 6:150–159.
Rice, D. S., and Curran, T. (2001). Role of reelin signalling pathway in central nervous system development. *Annu. Rev. Neurosci.* 24:1005–1039.
Rice, D. S., Nusinowitz, S., Azimi, A. M., Martinez, A., Soriano, E., and Curran, T. (2001). The reelin pathway modulates the structure and function of retinal synaptic circuitry. *Neuron* 31:929–941.
Roberts, R.C., Conley, R., Kung, L., Peretti, F. J., and Chute, D. J. (1996). Reduced striatal spine size in schizophrenia: a postmortem ultrastructural study. *Neuroreport* 7:1214–1218.
Roberts, R. C., Xu, L., Roche, J. K., and Kirkpatrick, B. (2005). Ultrastructural localization of reelin in the cortex in postmortem human brain. *J. Comp. Neurol.* 482:294–308.
Rodriguez, M. A., Pesold, C., Liu, W. S., Kriho, V., Guidotti, A., Pappas, G. D., and Costa, E. (2000). Colocalization of integrin receptors and reelin in dendritic spine postsynaptic densities of adult nonhuman primate brain. *Proc. Natl. Acad. Sci. USA* 97:3550–3555.
Rodriguez, M. A., Caruncho, H. J., Costa, E., Pesold, C., Liu, W. S., and Guidotti, A. (2002). In Patas monkey, glutamic acid decarboxylase-67 and reelin mRNA coexpression varies in a manner dependent on layers and cortical areas. *J. Comp. Neurol.* 45:279–288.
Schiffmann, S. N., Bernier, B., and Goffinet, A. M. (1997). Reelin mRNA expression during mouse brain development. *Eur. J. Neurosci.* 9:1055–1071.
Tissir, F., and Goffinet, A. M. (2003). Reelin and brain development. *Nature Rev. Neurosci.* 4:496–505.
Tissir, F., Lambert de Rouvroit, C., Sire, J. Y., Meyer, G., and Goffinet, A. M. (2003). Reelin expression during embryonic brain development in *Crocodylus niloticus*. *J. Comp. Neurol.* 457:250–262.
Tueting, P., Costa, E., Dwivedi, Y., Guidotti, A., Impagnatiello, F., Manev, R., and Pesold, C. (1999). The phenotypic characteristics of heterozygous reeler mouse. *Neuroreport* 10:1329–1334.
Weeber, E. J., Beffert, U., Jones, C., Christian, J. M., Forster, E., Sweatt, J. D., and Herz, J. (2002). Reelin and ApoE receptors cooperate to enhance hippocampal synaptic plasticity and learning. *J. Biol. Chem.* 277:39944–39952.
Yabut, O., Renfro, A., Niu, S., Swann, J. W., Marin, O., and D'Arcangelo, G. (2007). Abnormal laminar position and dendrite development of interneurons in the reeler forebrain. *Brain Res.* 1140:75–83.
Yip, J., Yip, Y. P., Nakajima, K., and Capriotti, C. (2000). Reelin controls position of autonomic neurons in the spinal cord. *Proc. Natl. Acad. Sci. USA* 97:8612–8616.

Chapter 9
Reelin and Cyclin-Dependent Kinase 5

Toshio Ohshima

Contents

1 Introduction

Reelin, an extracellular signaling molecule, and cyclin-dependent kinase 5 (Cdk5), a cytoplasmic kinase, are key regulators of normal brain development, including establishment of the complex brain structure. Recent studies have indicated that both Reelin signaling and Cdk5 are also involved in synaptic plasticity and neurodegeneration. In this chapter, I shall describe the functions of Cdk5 in neuronal migration during brain development and present an overview of the relationship of Cdk5 with Reelin signaling based on analyses of mutant mouse models. I shall also

T. Ohshima
Laboratory for Developmental Neurobiology, RIKEN Brain Science Institute, 2-1 Hirosawa, Wako City, Saitama 351-0198, Japan and Laboratory for Molecular Brain Science, Department of Life Science and Medical Bio-Science, Waseda University, Sinjuku, Tokyo 169-8555, Japan
e-mail: ohshima@brain.riken.go.jp

S. H. Fatemi (ed.), *Reelin Glycoprotein: Structure, Biology and Roles in Health and Disease.*

refer to the functions of Reelin signaling and Cdk5 in dendrite development, synaptic plasticity, and neurodegeneration.

2 What Is Cdk5?

2.1 Cdk5 Is a Neuron-Specific Serine/Threonine Kinase

Cdk5 is a homologue of the Cdk protein family of serine/threonine (Ser/Thr) kinases (Meyerson *et al.*, 1992; Hellmich *et al.*, 1992). While the other Cdks are active in proliferating cells, Cdk5 is mainly active in postmitotic neurons (Hellmich *et al.*, 1992; Tsai *et al.*, 1993). The activity of Cdk5 is regulated by its binding with a neuron-specific regulatory subunit, either p35 (Lew *et al.*, 1994; Tsai *et al.*, 1994) or its isoform p39 (Tang *et al.*, 1995); its activity is, therefore, correlated with the expression of p35 and p39. Expressions of p35 and p39 have been detected throughout the central nervous system (CNS), as well as in peripheral neurons such as the dorsal root ganglia (DRG) neurons (Zheng *et al.*, 1998; Delalle *et al.*, 1997). Cdk5 activity in the CNS has been shown to reach a peak from the late embryonic stage to the first week of the postnatal period, with later decline in the adult stage (Wu *et al.*, 2000; Takahashi *et al.*, 2003). However, some areas of the brain, including the hippocampus, have been shown to retain a relatively high expression level of p35 and high activity of Cdk5 (Wu *et al.*, 2000; Takahashi *et al.*, 2003).

2.2 Role of Cdk5 in Neuronal Positioning

Based on studies of the phenotypes of KO mice, it is considered that Cdk5 and p35 are critical for neuronal migration and positioning. Cdk5 KO mice exhibit perinatal lethality with disruption of the cortical laminar structures in the cerebral cortex, olfactory bulb, hippocampus, and cerebellum (Ohshima *et al.*, 1996). p35 KO mice showed a milder phenotype than Cdk5 KO mice owing to the redundancy of p39 (Chae *et al.*, 1997; Ohshima *et al.*, 2001). Although p39 KO mice show no phenotype, double-null mice for p35 and p39 display a phenotype identical to that of the Cdk5 KO mice (Ko *et al.*, 2001), confirming redundancy in these subunits. Birthdate labeling studies using BrdU confirmed that the migration defects of the neuronal subsets cause an abnormal laminar structure of the cerebral cortex in Cdk5 KO mice (Gilmore *et al.*, 1998). In addition to cortical neurons, the formation of some nuclei, including the facial motor nucleus and inferior olive, in the hindbrain, is defective in Cdk5 KO mice (Ohshima *et al.*, 2002).

How does Cdk5 regulate neuronal migration and positioning? Cdk5 modulates the actin cytoskeleton dynamics through phosphorylation of Pak1 (Nikolic *et al.*, 1998; Rashid *et al.*, 2001) and filamin 1 (Fox *et al.*, 1998). Cdk5 also modulates the microtubule dynamics through phosphorylation of microtubule-associated proteins, including tau (Kobayashi *et al.*, 1993), MAP1b (Paglini *et al.*, 1998), doublecortin (Tanaka *et al.*,

2004), Nudel (Sasaki *et al.*, 2000; Niethammer *et al.*, 2000), and Collapsin Response Mediator Proteins (CRMPs) (Uchida *et al.*, 2005). Among these substrates, defect of filamin 1 causes human periventricular heterotopia (Fox *et al.*, 1998), and defects of Lis1 and doublecortin cause human lissencephaly type 1 (Reiner *et al.*, 1993; des Portes *et al.*, 1998; Gleeson *et al.*, 1998). These findings indicate that Cdk5 regulates cytoskeletal dynamics that determines the speed of migration, extension of the leading processes, and cell-soma propulsion in migrating neurons. Cdk5 may also regulate cellular adhesion in neuronal–glial interaction through phosphorylation of β-catenin to regulate its interaction with N-cadherin (Kwon *et al.*, 2000), and this Cdk5-mediated adhesion may also be important for the neuronal migration.

3 Relationship Between Reelin Signaling and Cdk5

3.1 Relationship in Neuronal Migration and Positioning

3.1.1 Similarities and Differences: Defects of Reelin Signaling Versus Defects of Cdk5/p35

During cortical development, the preplate splits into a marginal zone and the subplate in wild-type mice. However, in the *reeler* mice, the preplate remains undivided as a superficial layer (the "superplate"), and cohorts of cortical pyramidal neurons accumulate underneath the superplate (Caviness, 1982; Goffinet, 1984; Sheppard and Pearlman, 1997). In the case of Cdk5 KO mice, the preplate is formed normally, with the earliest-generated pyramidal neurons being positioned within the preplate. However, the subsequent cohorts of pyramidal neurons accumulate underneath the subplate (Gilmore *et al.*, 1998), as shown schematically in Fig. 9.1.

In addition to those in the cerebral cortex, the *reeler* phenotype and Cdk5/p35 mice also share similarities in abnormalities of neuronal positioning in other areas of the CNS (Ohshima and Mikoshiba, 2002). For example, the cerebellum is another typical region of the CNS in which the *reeler* phenotype and Cdk5/p35 deficiency share a similar phenotype, although the similarity was only noted at first glance, and detailed analysis revealed significant differences. Both Reelin signaling and Cdk5/p35 are required for proper positioning of Purkinje cells (Mariani *et al.*, 1977; Ohshima *et al.*, 1996, 1999). Cdk5/p35 deficiency also results in failure of complete migration of the granule cells from the external granule cell layer (EGL) to the internal granule cell layer (IGL) (Chae *et al.*, 1997; Ohshima *et al.*, 1999). This inward migration of granule cells, however, occurs normally in the *reeler* mice. As summarized in Table 9.1, there is considerable overlap in the affected neuronal population between *reeler* mice and Cdk5 KO mice, with only some exceptions. For example, positioning of mitral cells in the olfactory bulb is abnormal in the Cdk5 KO mice, and positioning of the granule cells in the dentate gyrus is abnormal in the *reeler* mice (Stanfield and Cowan, 1979; Drakew *et al.*, 2002);

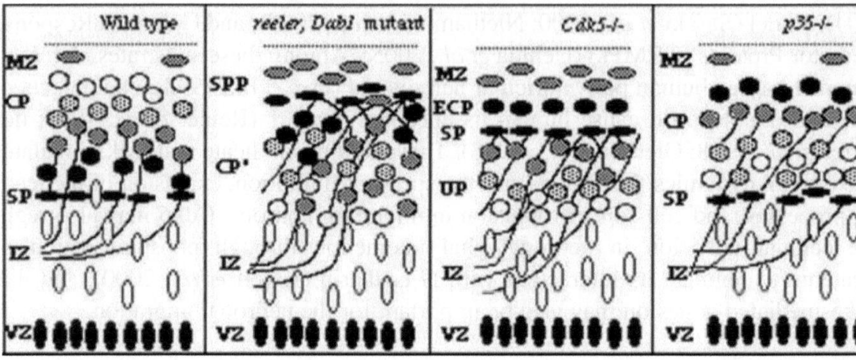

Fig. 9.1 Schematic representation of the cerebral cortex in the wild-type and mutant mice. In the wild-type mice, migrating neurons split preplate into a marginal zone (MZ) and subplate (SP), to form the cortical plate (CP). In *reeler* and Dab1 mutant mice, the preplate is not split and remains as a superplate (SPP) with mutant neurons stacked up in an inverse order. In the Cdk5-/- mice, the initial wave of migrating neurons (indicated by black circles) splits the preplate to form a narrow ectopic cortical plate (ECP), but later-born neurons (light gray and white circles) stack up *reeler*-like under the subplate in an inverted fashion as an underplate (UP). In the p35-/- mice, the later compensatory effects of Cdk5/p39 result in normal positioning of the subplate neurons. (Figure adapted with permission from Ohshima and Mikoshiba, 2002)

Table 9.1 Comparison of affected brain regions or neuronal subtypes in *reeler*/Dab1 mutant, Cdk5 KO, and p35 KO mice

Structure or neuronal type in CNS		*reeler*/Dab1 mutant	Cdk5 KO	p35 KO
Olfactory bulb	Mitral cells	–	+	–
Cerebral cortex	Subplate neurons	+	–	–
	Cortical neurons	+	+	+
Hippocampus	Pyramidal cell layer	+	+	±
	Dentate gyrus	+	±	±
Midbrain	Dopamine neurons in SN	+	+	–
Cerebellum	Purkinje cells	+	+	±
	Granule cells (inward)	–	n.d.*	+
Brainstem	Facial motor nucleus	±	+	–
	Inferior olive	±	+	–

+, affected; ±, mildly affected; –, unaffected. CNS, central nervous system; SN, substantia nigra.
*n.d., could not be determined because of perinatal lethality. This type of migration of granule cells is Cdk5-dependent (Ohshima *et al.*, 1999).

on the other hand, only mild abnormality in the positioning of these neurons is observed in the p35 KO mice (Wenzel *et al.*, 2001).

3.1.2 Genetic Interaction Between Reelin Signaling and Cdk5/p35 in Neuronal Positioning

To study the relationship between Reelin signaling and Cdk5/p35, double mutant mice with respect to these two signaling proteins were generated and their phenotypes analyzed (Ohshima *et al.*, 2001; Beffert *et al.*, 2004). Because all of the Cdk5 KO mice die perinatally, the genetic interaction was examined by comparing the phenotype of those double knockout mice for p35 and Reelin (Ohshima *et al.*, 2001), Dab1 (Ohshima *et al.*, 2001), ApoER2 and VLDLR (Beffert *et al.*, 2004) versus that in the respective single KO mice. Exaggerated neuronal migration defects in the hippocampus, as compared with that in the respective single KO mice, were typically observed in the double KO mice for p35/Dab1, p35/Reelin, p35/ApoER2, and p35/VLDLR (Ohshima *et al.*, 2001; Beffert *et al.*, 2004). Invasion of cells into layer I of the cerebral cortex is known to be a typical feature in the *reeler* and Dab1 mutant mice; this phenotype was observed in the double KO mice for p35/ApoER2 and p35/VLDLR, but not in the single KO mice for p35, ApoER2, or VLDLR (Beffert *et al.*, 2004). Exaggerated defects of Purkinje cell migration in the cerebellum have also been reported in double KO mice for p35/Reelin and p35/Dab1 (Ohshima *et al.*, 2001). These findings indicate that Reelin signaling and Cdk5 function together in a parallel manner to effect proper neuronal migration and positioning in the developing brain (Ohshima and Mikoshiba, 2002; Beffert *et al.*, 2004).

3.2 Roles of Reelin Signaling and Cdk5 in Dendrite Development

Impairment of Reelin signaling during neuronal development has been shown to result in a reduced complexity of the dendritic tree (Borrell *et al.*, 1999) and in lowering the synaptic complexity of the hippocampus (Del Rio *et al.*, 1997). Another study, using cultured hippocampal neurons, has shown the important role of Reelin in the regulation of dendritic branching and demonstrated that this effect of Reelin was inhibited by the addition of a Dab1 phosphorylation inhibitor (Niu *et al.*, 2004). The results of a recent Dab1 RNAi study in the developing cortex lent support to these findings (Olson *et al.*, 2006). Cdk5 has also been shown to be involved in neurite outgrowth. Reduction of Cdk5 kinase activity by expression of the dominant-negative form of Cdk5 (Nikolic *et al.*, 1996), or the addition of antisense oligonucleotides of Cdk5, p35, and p39 in a primary culture of neurons was shown to inhibit neurite outgrowth (Xiong *et al.*, 1997). The synergistic effects of Reelin signaling and Cdk5 in dendritic development remain to be elucidated.

3.3 Roles of Reelin Signaling and Cdk5 in Synaptic Plasticity

Cdk5 and Reelin have both been shown to be involved in synaptic function. Cdk5 phosphorylates presynaptic proteins and may be involved in exocytosis (Matsubara *et al.*, 1996; Shuang *et al.*, 1998; Tomizawa *et al.*, 2002) and endocytosis at presynaptic sites (Tan *et al.*, 2003; Tomizawa *et al.*, 2003). Cdk5 also phosphorylates postsynaptic proteins including NMDA receptors (Li *et al.*, 2001) and PSD-95 (Morabito *et al.*, 2004), and may be involved in the regulation of synaptic plasticity. Both ApoER2 KO mice and VLDLR KO mice show defects of LTP induction and fear-conditioned associative learning (Weeber *et al.*, 2002). p35 KO mice were shown to have defects of LTD induction and spatial learning (Ohshima *et al.*, 2005). Interestingly, Reelin has been shown to enhance LTP of Schaffer collateral synapse at CA1 of hippocampus (Weeber *et al.*, 2002). Such enhancement was, however, not observed in the hippocampus of the p35 KO mice, indicating that Reelin-dependent enhancement of LTP in the hippocampus may be dependent on Cdk5 activity (Beffert *et al.*, 2004). The molecular mechanism underlying this relation would serve as an excellent subject for future study.

3.4 Roles of Cdk5 and Reelin Signaling in Alzheimer's Disease: Phosphorylation of Tau and Regulation of APP

Possible involvement of Cdk5 in Alzheimer's disease (AD) has been discussed. Accumulation of p25, which is a cleavage product of p35, and a stable protein, and elevated Cdk5 activity have been reported in the brains of AD patients (Patrick *et al.*, 1999), although these observations have not been consistent among investigators (Takashima *et al.,* 2001;Yoo and Lubec, 2001). The pathological hallmarks of AD brains are amyloid plaques, which is caused by the abnormal processing of APP, and neurofibrillary tangles that result from abnormal phosphorylation and aggregation of tau protein. Increased tau phosphorylation has been demonstrated in the brains of *reeler,* Dab1 mutant, and ApoER2/VLDLR KO mice (Hiesberger *et al.*, 1999; Brich *et al.*, 2003). Dab1 can physically interact with intracellular domains of APP and APP-like proteins (Homayouni *et al.*, 1999; Howell *et al.*, 1999) and regulate their trafficking and proteolytic processing (Hoe *et al.*, 2006). Interestingly, Cdk5 phosphorylates intracellular domain of APP at Thr668 (Iijima *et al.*, 2000) and affects its intracellular trafficking and proteolytic processing (Lee *et al.*, 2003).

4 Possible Molecular Mechanisms Underlying the Relation Between Reelin Signaling and Cdk5

4.1 Through the Phosphorylation of Dab1 by Cdk5

It has been reported that intracellular adapter protein Dab1 is phosphorylated at tyrosine sites, as well as Ser/Thr sites *in vivo* (Arnaud *et al.*, 2003; Ohshima *et al.*, 2007). *In vitro*, Cdk5 has been shown to phosphorylate Dab1 at multiple Ser/Thr sites in the carboxyl terminus (Ohshima *et al.*, 2007). Among them, phosphorylation of Dab1 at Ser491 by Cdk5 *in vivo* has been shown using phospho-specific antibody (Keshvara *et al.*, 2002). However, functional significance of Cdk5-mediated Dab1 phosphorylation in Reelin signaling remains to be elucidated.

4.2 Sharing of Common Downstream Targets

Reelin signaling and Cdk5 may share downstream targets, such as the microtubule-associated protein, tau (Fig. 9.2). It had been shown that Reelin signaling induces Akt activation along with inactivation of GSK3β, a known major tau kinase (Beffert *et al.*, 2002). Cdk5 is also known as tau kinase (Kobayashi *et al.*, 1993; Paudel *et al.*, 1993) and phosphorylates tau at several sites (Baumann *et al.*, 1993). Interestingly, inhibition of Cdk5 can lead to increased phosphorylation of several neuronal proteins by GSK3β, including neurofilaments and kinesin light chain (Morfini *et al.*, 2004).

Sharing of other downstream targets of Reelin signaling and Cdk5 may explain the synergic function of these two pathways (Fig. 9.2). Reelin signaling facilitates interaction between Dab1 and Lis1, and Lis1 is also associated with Nudel, which is a Cdk5 substrate as described above (Sasaki *et al.*, 2000; Niethammer *et al.*,

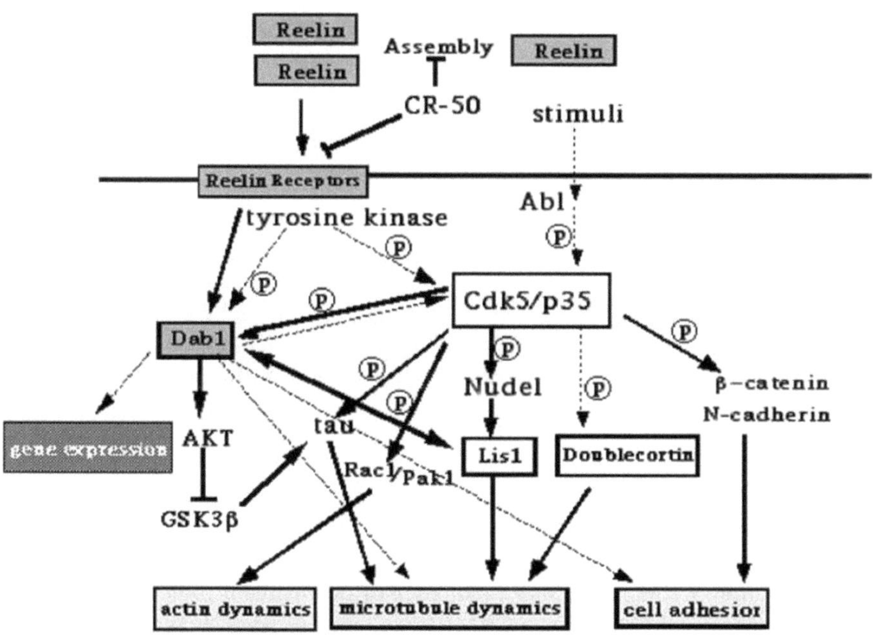

Fig. 9.2 Model of the signaling pathway of Reelin and Cdk5 in the control of neuronal positioning. Reelin and Cdk5 function in a parallel fashion. (Figure adapted with permission from Ohshima and Mikoshiba, 2002, with modification)

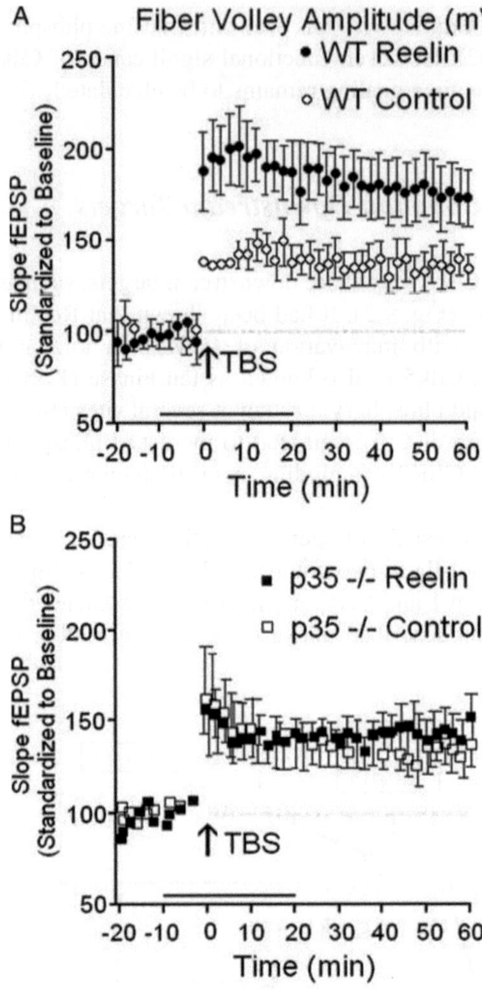

Fig. 9.3 Reelin treatment potentiates LTP in hippocampal slices in the wild-type (WT) mice (**A**). This potentiation is not observed in the p35-/- slice (**B**) indicating Cdk5/p35-dependency. TBS, theta burst stimulation. (Figures adapted with permission from Beffert *et al.*, 2004)

2000; Assadi *et al.*, 2003). Cdk5-mediated phosphorylation of Nudel might be important for the interaction of these proteins to activate nucleokinesis during neuronal migration. CRMPs are also candidates of downstream molecules shared by Reelin signaling and Cdk5, since CRMP1 has recently been shown to be involved in Reelin signaling, and it is a substrate of Cdk5 (Uchida *et al.*, 2005; Yamashita *et al.*, 2006).

5 Closing Remarks

As described above, Reelin signaling and Cdk5 are involved together in many aspects of brain development and functions. Genetic studies using double mutant mice indicate synergistic function of Reelin signaling and Cdk5 in brain development. The relationships between Reelin signaling and Cdk5 in other functions of the brain, synaptic plasticity, and neurological diseases remain to be precisely elucidated in future studies. Reelin seems to enhance LTP in the hippocampus and this enhancement appears to be Cdk5-dependent (Fig. 9.3). This is a good example indicating the close relationship between Reelin signaling and Cdk5 in the execution of brain function.

References

Arnaud, L., Ballif, B. A., and Cooper, J. A. (2003). Regulation of protein tyrosine kinase signaling by substrate degradation during brain development. *Mol. Cell. Biol.* 23:9293–9302.

Assadi, A. H., Zhang, G., Beffert, U., McNeil, R. S., Renfro, A. L., Niu, S., Quattrocchi, C. C., Antalffy, B. A., Sheldon, M., Armstrong, D. D., Wynshaw-Boris, A., Herz, J., D'Arcangelo, G.., and Clark G.. D. (2003). Interaction of reelin signaling and Lis1 in brain development. *Nature Genet.* 35:270–276.

Baumann, K., Mandelkow, E. M., Biernat, J., Piwnica-Worms, H., and Mandelkow, E. (1993). Abnormal Alzheimer-like phosphorylation of Tau-protein by cyclin-dependent kinases Cdk2 and Cdk5. *FEBS Lett.* 336:417–424.

Beffert, U., Morfini, G., Bock H. H., Reyna, H., Brady, S. T., and Herz, J. (2002). Reelin-mediated signaling locally regulates protein kinase B/Akt and glycogen synthase kinase 3. *J. Biol. Chem.* 277:49958–49964.

Beffert, U., Weeber E. J., Morfini, G., Ko, J., Brady, S. T., Tsai, L.H., Sweatt, D., and Herz J. (2004). Reelin and cyclin-dependent kinase 5-dependent signals cooperate in regulating neuronal migration and synaptic transmission. *J. Neurosci.* 24:1897–1906.

Bibb, J. A., Snyder, G. L., Nishi, A., Zhen, Y., Meijer, L., Fienberg, A. A., Tsai, L. H., Kwon, Y. T., Girault, J. A., Czernik, A. J., Huganir, R. L., Hemmings, H. C., Jr., Nairn, A. C., and Greengard, P. (1999). Protein kinase and phosphatase control by distinct phosphorylation sites within a single regulatory protein. *Nature* 402:669–671.

Borrell, V., Del Rio, J. A., Alcantara, S., Derer, M., Martinez, A., D'Arcangelo, G., Nakajima, K., Mikoshiba, K., Derer, P., Curran, T., and Soriano, E. (1999). Reelin regulates the development and synaptogenesis of the layer-specific entorhino-hippocampal connections. *J. Neurosci.* 19:1345–1358.

Brich, J., Shie, F. S., Howell, B. W., Li, R., Tus, K., Wakeland, E. K., Jin, L. W., Mumby, M., Churchill, G., Herz, J., and Cooper, J. A. (2003). Genetic modulation of phosphorylation in the mouse. *J. Neurosci.* 23:187–192.

Caviness, V. S. (1982). Neocortical histogenesis in normal and reeler mice: a developmental study based upon [3H] thymidine autoradiography. *Brain Res.* 256:293–302.

Chae, T., Kwon, Y. T., Bronson, R., Dikkes, P., Li, E., and Tsai, L. H. (1997). Mice lacking p35, a neuronal specific activator of Cdk5, display cortical lamination defects, seizures and adult lethality. *Neuron* 18:29–42.

Delalle, I., Bhide, P. G., Caviness, V. S. J., and Tsai, L. H. (1997). Temporal and spatial patterns of expression of p35, a regulatory subunit of cyclin-dependent kinase 5, in the nervous system of the mouse. *J. Neurocytol.* 26:283–296.

Del Rio, J. A., Heimrich, B., Borrell, V., Forster, E., Drakew, A., Alcantara, S., Nakajima, K., Miyata, T., Ogawa, M., Mikoshiba, K., Derer, P., Frotscher, M., and Soriano, E. (1997). A role for Cajal–Retzius cells and reelin in the development of hippocampal connections. *Nature* 385:70–74.

des Portes, V., Pinard, J. M., Billuart, P., Vinet, M. C., Koulakoff, A., Carrie, A., Gelot, A., Dupuis, E., Motte, J., Berwald-Netter, Y., Catala, M., Kahn, A., Beldjord, C., and Chelly, J. (1998). A novel CNS gene required for neuronal migration and involved in X-linked subcortical laminar heterotopia and lissencephaly syndrome. *Cell* 92:51–61.

Drakew, A., Deller, T., Heimrich, B., Gebhardt, C., Del Turco, D., Tielsch, A., Forster, E., Herz, J., and Frotscher, M. (2002). Dentate granule cells in reeler mutants and VLDLR and ApoER2 knockout mice. *Exp. Neurol.* 176:12–24.

Fox, J. W., Lamperti, E. D., Eksioglu, Y. Z., Hong, S. E., Feng, Y., Graham, D. A., Scheffer, I. E., Dobyns, W. B., Hirsch, B. A., Radtke, R. A., Berkovic, S. F., Huttenlocher, P. R., and Walsh, C. A. (1998). Mutations in filamin 1 prevent migration of cerebral cortical neurons in human periventricular heterotopia. *Neuron* 21:1315–1325.

Gilmore, E. C., Ohshima, T., Goffinet, A. M., Kulkarni, A. B., and Herrup, K. (1998). Cyclin-dependent kinase 5-deficient mice demonstrate novel developmental arrest in cerebral cortex. *J. Neurosci.* 18:6370–6377.

Gleeson, J. G., Allen, K. M., Fox, J. W., Lamperti, E. D., Berkovic, S., Scheffer, I., Cooper, E. C., Dobyns, W. B., Minnerath, S. R., Ross, M. E., and Walsh, C. A. (1998). Doublecortin, a brain-specific gene mutated in human X-linked lissencephaly and double cortex syndrome, encodes a putative signaling protein. *Cell* 92:63–72.

Goffinet, A. M. (1984). Events governing organization of postmigratory neurons: studies on brain development in normal and reeler mice. *Brain Res. Rev.* 7:261–296.

Hellmich, M. R., Pant, H. C., Wada, E., and Battey, J. F. (1992). Neuronal cdc2-like kinase: a CDC2-related protein kinase with predominantly neuronal expression. *Proc. Natl. Acad. Sci. USA* 89:10867–10871.

Hiesberger, T., Trommsdorff, M., Howell, B. W., Goffinet, A., Mumby, M. C., Cooper, J. A., and Herz, J. (1999). Direct binding of reelin to VLDL receptor and ApoE receptor 2 induces tyrosine phosphorylation of disabled-1 and modulates tau phosphorylation. *Neuron* 24:481–489.

Hoe, H. S., Tran, T. S., Matsuoka, Y., Howell, B. W., and Rebeck, G. W. (2006). Dab1 and reelin effects on APP and ApoEr2 trafficking and processing. *J. Biol. Chem.* 281:35176–35185.

Homayouni, R., Rice, D. S., Sheldon, M., and Curran, T. (1999). Disabled-1 binds to the cytoplasmic domain of amyloid precursor-like protein 1. *J. Neurosci.* 19:7507–7515.

Howell, B. W., Herrick, T. M., and Cooper, J. A. (1999). Reelin-induced tyrosine phosphorylation of disabled 1 during neuronal positioning. *Genes Dev.* 13:643–648.

Iijima, K., Ando, K., Takeda, S., Satoh, Y., Seki, T., Itohara, S., Greengard, P., Kirino, Y., Nairn, A. C., and Suzuki, T. (2000). Neuron-specific phosphorylation of Alzheimer's -amyloid precursor protein by cyclin-dependent kinase 5. *J. Neurochem.* 75:1085–1091.

Keshvara, L., Magdaleno, S., Benhayon, D., and Curran, T. (2002). Cyclin-dependent kinase 5 phosphorylates disabled 1 independently of reelin signaling. *J. Neurosci.* 22:4869–4877.

Ko, J., Humbert, S., Brorson, T., Takahashi, S., Kulkarni, A. B., Li, E., and Tsai, L. H. (2001). p35 and p39 are essential for Cdk5 function during neurodevelopment. *J. Neurosci.* 21:6758–6771.

Kobayashi, S., Ishiguro, K., Omori, A., Takamatsu, M., Arioka, M., Imahori, K., and Uchida, T. (1993). Cdc2-related kinase PSSALRE/Cdk5 is homologous with the 30 kDa subunit of tau protein kinase II, a proline-directed protein kinase associated with microtubule. *FEBS Lett.* 335:171–175.

Kwon, Y. T., Gupta, A., Zhou, Y., Nikolic, M., and Tsai, L. H. (2000). Regulation of the N-cadherin-mediated adhesion by the p35/Cdk5 kinase. *Curr. Biol.* 10:363–372.

Lee, M. S., Kao, S. C., Lemere, C. A., Xia, W., Tseng, H. C., Zhou, Y., Neve, R., Ahlijanian, M. K., and Tsai, L. H. (2003). APP processing is regulated by cytoplasmic phosphorylation. *J. Cell. Biol.* 163:83–95.

Lew, J., Huang, Q. Q., Qi, Z., Winkfein, R. J., Aebersold, R., Hunt, T., and Wang J. H. (1994). Neuronal cdc2-like kinase is a complex of cyclin-dependent kinase 5 and a novel brain-specific regulatory subunit. *Nature* 371:423–425.

Li, B. S., Sun, M. K., Zhang, L., Takahashi, S., Ma, W., Vinade, L., Kulkarni, A. B., Brady, R. O., and Pant, H. C. (2001). Regulation of NMDA receptors by cyclin-dependent kinase-5. *Proc. Natl. Acad. Sci. USA* 98:12742–12747.

Mariani, J., Crepel, F., Mikoshiba, K., Changeux, J. P., and Sotelo, C. (1977). Anatomical, physiological and biochemical studies of the cerebellum from Reeler mutant mouse. *Philos. Trans. R. Soc. London Ser. B Biol. Sci.* 281:1–28.

Matsubara, M., Kusubata, M., Ishiguro, K., Uchida, T., Titani, K., and Taniguchi, H. (1996). Site-specific phosphorylation of synapsin I by mitogen-activated protein kinase and Cdk5 and its effects on physiological functions. *J. Biol. Chem.* 271:21108–21113.

Meyerson, M., Enders, G. H., Wu, C. L., Su, L. K., Gorka, C., Nelson, C., Harlow, E., and Tsai, L. H. (1992). A family of human CDC2-related protein kinases. *EMBO J.* 11:2909–2917.

Morabito, M. A., Sheng, M., and Tsai, L. H. (2004). Cyclin-dependent kinase 5 phosphorylates the N-terminal domain of the postsynaptic density protein PSD-95 in neurons. *J. Neurosci.* 24:865–876.

Morfini, G., Szebenyim, G., Brown, H., Pant, H. C., Pigino, G., DeBoer, S., Beffert, U., and Brady, S. T. (2004). A novel CDK5-dependent pathway for regulating GSK3 activity and kinesin-driven motility in neurons. *EMBO J.* 23:2235–2245.

Niethammer, M., Smith, D. S., Ayala, R., Peng, J., Ko, J., Lee, M. S., Morabito, M., and Tsai, L. H. (2000). NUDEL is a novel Cdk5 substrate that associates with LIS1 and cytoplasmic dynein. *Neuron* 28:697–711.

Nikolic, M., Dudek, H., Kwon, Y. T., Ramos, Y. F. M., and Tsai, L. H. (1996). The Cdk5/p35 kinase is essential for neurite outgrowth during neuronal differentiation. *Genes Dev.* 10:816–825.

Nikolic, M., Chou, M. M., Lu, W., Mayer, B. J., and Tsai, L. H. (1998). The p35/Cdk5 kinase is a neuron-specific Rac effector that inhibits Pak1 activity. *Nature* 395:194–198.

Niu, S., Renfro, A., Quattrocchi, C. C., Sheldon, M., and D'Arcangelo, G. (2004). Reelin promotes hippocampal dendrite development through the VLDLR/ApoER2-Dab1 pathway. *Neuron* 41:71–84.

Ohshima, T., and Mikoshiba, K. (2002). Reelin signaling and Cdk5 in the control of neuronal positioning. *Mol. Neurobiol.* 26:153–166.

Ohshima, T., Ward, J. M., Huh, C. G., Longenecker, G., Veeranna, Pant, H. C., Brady, R. O., Martin, L. J., and Kulkarni, A. B. (1996). Targeted disruption of the cyclin-dependent kinase 5 gene results in abnormal corticogenesis, neuronal pathology and perinatal death. *Proc. Natl. Acad. Sci. USA* 93:11173–11178.

Ohshima, T., Gilmore, E. C., Longenecker, G., Jacobowitz, D. M., Brady, R. O., Herrup, K., and Kulkarni, A. B. (1999). Migration defects of Cdk5(-/-) neurons in the developing cerebellum is cell autonomous. *J. Neurosci.* 19:6017–6026.

Ohshima, T., Ogawa, M., Veeranna, Hirasawa, M., Longenecker, G., Ishiguro, K., Pant, H. C., Brady, R. O., Kulkarni, A. B., and Mikoshiba, K. (2001). Synergistic contributions of cyclin-dependent kinase 5/p35 and reelin/Dab1 to the positioning of cortical neurons in the developing mouse brain. *Proc. Natl. Acad. Sci. USA* 98:2764–2769.

Ohshima, T., Ogawa, M., Takeuchi, K., Takahashi, S., Kulkarni, A. B., and Mikoshiba, K. (2002). Cyclin-dependent kinase 5/p35 contributes synergistically with reelin/Dab1 to the positioning of facial branchiomotor and inferior olive neurons in the developing mouse hindbrain. *J. Neurosci.* 22:4036–4044.

Ohshima, T., Ogura, H., Tomizawa, K., Hayashi, K., Suzuki, H., Saito, T., Kamei, H., Nishi, A., Bibb, J. A., Hisanaga, S., Matsui, H., and Mikoshiba, K. (2005). Impairment of hippocampal long-term depression and defective spatial learning and memory in p35-/- mice. *J. Neurochem.* 94:917–925.

Ohshima, T., Suzuki, H., Morimura, T., Ogawa, M., and Mikoshiba, K. (2007). Modulation of reelin signaling by cyclin-dependent kinase 5. *Brain Res.* 1140:84–95.

Olson, E C., Kim, S., and Walsh, C. A. (2006). Impaired neuronal positioning and dendritogenesis in the neocortex after cell-autonomous Dab1 suppression. *J. Neurosci.* 26:1767–1775.

Paglini, G., Pigino, G., Kunda, P., Morfini, G., Maccioni, R., Quiroga, S., Ferreira, A., and Caceres, A. (1998). Evidence for the participation of the neuron-specific CDK5 activator P35 during laminin-enhanced axonal growth. *J. Neurosci.* 18:9858–9869.

Patrick, G. N., Zukerberg, L., Nikolic, M., de la Monte, S., Dikkes, P., and Tsai, L. H. (1999). Conversion of p35 to p25 de-regulates Cdk5 activity and promotes neurodegeneration. *Nature* 402:615–622.

Paudel, H. K., Lew, J., Ali, Z., and Wang, J. H. (1993). Brain proline-directed protein kinase phosphorylates tau on sites that are abnormally phosphorylated in tau associated with Alzheimer's paired helical filaments. *J. Biol. Chem.* 268:23512–23518.

Rashid, T., Banerjee, M., and Nikolic, M. (2001). Phosphorylation of Pak1 by the p35/Cdk5 kinase affects neuronal morphology. *J. Biol. Chem.* 276:49043–49052.

Reiner, O., Carrozzo, R., Shen, Y., Wehnert, M., Faustinella, F., Dobyns, W. B., Caskey, C. T., and Ledbetter, D. H. (1993). Isolation of a Miller-Dieker lissencephaly gene containing G-protein-subunit-like repeats. *Nature* 364:717–721.

Sasaki, S., Shionoya, A., Ishida, M., Gambello, M. J., Yingling, J., Wynshaw-Boris, A., and Hirotsune, S. (2000). A LIS1/NUDEL/cytoplasmic dynein heavy chain complex in the developing and adult nervous system. *Neuron* 28:681–696.

Sheppard, A. M., and Pearlman, A. L. (1997). Abnormal reorganization of preplate neurons and their associated extracellular matrix: an early manifestation of altered neocortical development in the *reeler* mutant mouse. *J. Comp. Neurol.* 378:173–179.

Shuang, R., Zhang, L., Fletcher, A., Groblewski, G. E., Pevsner, J., and Stuenkel, E. L. (1998). Regulation of Munc18/syntaxin 1A interaction by cyclin-dependent kinase 5 in nerve endings. *J. Biol. Chem.* 273:4957–4966.

Stanfield, B. B., and Cowan, W. M. (1979). The morphology of the hippocampus and dentate gyrus in normal and reeler mice. *J. Comp. Neurol.* 185:393–422.

Takahashi, S., Saito, T., Hisanaga, S., Pant, H. C., and Kulkarni, A. B. (2003). Tau phosphorylation by cyclin-dependent kinase 5/p39 during brain development reduces its affinity for microtubules. *J. Biol. Chem.* 278:10506–10515.

Takashima, A., Murayama, M., Yasutake, K., Takahashi, H., Yokoyama, M., and Ishiguro, K. (2001). Involvement of cyclin dependent kinase 5 activator p25 on tau phosphorylation in mouse brain. *Neurosci. Lett.* 306:37–40.

Tan, T. C., Valova, V. A., Malladi, C. S., Graham, M. E., Berven, L. A., Jupp, O. J., Hansra, G., McClure, S. J., Sarcevic, B., Boadle, R. A., Larsen, M. R., Cousin, M. A., and Robinson, P. J. (2003). Cdk5 is essential for synaptic vesicle endocytosis. *Nature Cell Biol.* 5:701–710.

Tanaka, T., Serneo, F. F., Tseng, H. C., Kulkarni, A. B., Tsai, L. H., and Gleeson, J. G. (2004). Cdk5 phosphorylation of doublecortin ser297 regulates its effect on neuronal migration. *Neuron* 41:215–227.

Tang, D., Yeung, J., Lee, K. Y., Matsushita, M., Matsui, H., Tomizawa, K., Hatase, O., and Wang, J. H. (1995). An isoform of the neuronal cyclin-dependent kinase 5 (cdk5) activator. *J. Biol. Chem.* 270:26897–26903.

Tomizawa, K., Ohta, J., Matsushita, M., Moriwaki, A., Li, S. T., Takei, K., and Matsui, H. (2002). Cdk5/p35 regulates neurotransmitter release through phosphorylation and downregulation of P/Q-type voltage-dependent calcium channel activity. *J. Neurosci.* 22:2590–2597.

Tomizawa, K., Sunada, S., Lu, Y. F., Oda, Y., Kinuta, M., Ohshima, T., Saito, T., Wei, F. Y., Matsushita, M., Li, S. T., Tsutsui, K., Hisanaga, S., Mikoshiba, K., Takei, K., and Matsui, H. (2003). Cophosphorylation of amphiphysin I and dynamin I by Cdk5 regulates clathrin-mediated endocytosis of synaptic vesicles. *J. Cell Biol.* 163:813–824.

Trommsdorff, M., Gotthardt, M., Hiesberger, T., Shelton, J., Stockinger, W., Nimpf, J., Hammer, R. E., Richardson, J. A., and Herz, J. (1999). Reeler/disabled-like disruption of neuronal migration in knockout mice lacking the VLDL receptor and ApoE receptor 2. *Cell* 97:689–701.

Tsai, L. H., Takahashi, T., Caviness, V. S., Jr., and Harlow, E. (1993). Activity and expression pattern of cyclin-dependent kinase 5 in the embryonic mouse nervous system. *Development* 119:1029–1040.

Tsai, L. H., Delalle, I., Caviness, V. S., Jr., Chae, T., and Harlow, E. (1994). p35 is a neural-specific regulatory subunit of cyclin-dependent kinase 5. *Nature* 371:419–423.

Uchida, Y., Ohshima, T., Sasaki, Y., Suzuki, H., Yanai, S., Yamashita, N., Nakamura, F., Takei, K., Ihara, Y., Mikoshiba, K., Kolattukudy, P., Honnorat, J., and Goshima, Y. (2005). Semaphorin3A signalling is mediated via sequential Cdk5 and GSK3β phosphorylation of CRMP2: implication of common phosphorylating mechanism underlying axon guidance and Alzheimer's disease. *Genes Cells* 10:165–179.

Weeber, E. J., Beffert, U., Jones, C., Christian, J. M., Forster, E., Sweatt, J. D., and Herz, J. (2002). Reelin and ApoE receptors cooperate to enhance hippocampal synaptic plasticity and learning. *J. Biol. Chem.* 277:39944–39952.

Wenzel, H. J., Robbins, C. A., Tsai, L. H., and Schwartzkroin, P. A. (2001). Abnormal morphological and functional organization of the hippocampus in a p35 mutant model of cortical dysplasia associated with spontaneous seizure. *J. Neurosci.* 21:983–998.

Wu, D. C., Yu, Y. P., Lee, N. T., Yu, A. C., Wang, J. H., and Han, Y. F. (2000). The expression of Cdk5, p35, p39, and Cdk5 kinase activity in developing, adult, and aged rat brains. *Neurochem. Res.* 25:923–929.

Xiong, W., Pestell, R., and Rosner, M. R. (1997). Role of cyclins in neuronal differentiation of immortalized hippocampal cells. *Mol. Cell. Biol.* 17:6585–6597.

Yamashita, N., Uchida, Y., Ohshima, T., Hirai, S. I., Nakamura, F., Taniguchi, M., Mikoshiba, K., Honnorat, J., Kolattukudy, P., Thomasset, N., Takei, K., Takahashi, T., and Goshima, Y. (2006). CRMP1 mediates reelin signaling in cortical neuronal migration. *J. Neurosci.* 26:13357–13362.

Yoo, B. C., and Lubec, G. (2001). p25 protein in neurodegeneration. *Nature* 30:135–147.

Zheng, M., Leung, C. L., and Liem, R. K. (1998). Region-specific expression of cyclin-dependent kinase 5 (Cdk5) and its activators, p35 and p39, in the developing and adult rat central nervous system. *J. Neurobiol.* 35:141–159.

Chapter 10
Reelin and the Cerebellum

Robert F. Hevner

Contents

1 Introduction

The cerebellum is a large, complex brain structure that mediates essential functions for movement, balance, cognition, and language (Ito, 2005). Development of the cerebellum critically depends on Reelin signaling. Complete deficiency of Reelin causes a severe cerebellar malformation, with extensive cellular disorganization and hypoplasia. Identical cerebellar defects are observed in mice lacking downstream components of the Reelin signaling pathway, including Reelin receptors VLDLR and ApoER2, adapter protein Dab1, or kinases Fyn and Src. The brain malformation results in ataxia and loss of balance, manifesting as a reeling gait in mice (hence the

R. F. Hevner
Department of Pathology/Neuropathology, University of Washington, HMC Box 359791, 325 Ninth Avenue, Seattle, WA 98104
e-mail: rhevner@u.washington.edu

name Reelin) and multiple neurological problems in humans. More subtle abnormalities of Reelin signaling may underlie important neurobehavioral disorders in humans. In particular, some studies have linked *RELN* gene polymorphisms and reduced Reelin expression to autism. Since cerebellar defects are frequently observed in autistic brains, an attractive hypothesis is that Reelin signaling abnormalities may cause autism by perturbing cerebellar development or plasticity.

Despite growing biomedical significance, our understanding of how Reelin regulates cerebellar morphogenesis is far from complete. Reelin, its receptors, and downstream effectors are expressed by different cohorts of cells at different time points throughout cerebellar development. Most Reelin-producing cells are located near the surface of the developing cerebellar cortex, including cells of the rostral rhombic lip migratory stream (RLS), the nuclear transitory zone (NTZ), and the external granular layer (EGL). Other Reelin-producing cells are located deeper in the cerebellum, including some neurons of the deep cerebellar nuclei (DCN) and internal granular layer. Much evidence suggests that one important function of Reelin is to promote detachment of Purkinje cells from radial glia in the mantle zone of the embryonic cerebellar cortex, thus allowing multiple Purkinje cells to migrate along the same radial glia. The migrating Purkinje cells respond to Reelin signaling by activating a signaling cascade that includes Reelin receptors (VLDLR and ApoER2), adapter protein Dab1, and kinases Src and Fyn. Besides promoting Purkinje cell detachment from radial glia, Reelin may also regulate Purkinje cell spreading and monolayer formation, radial glia morphology, granule cell proliferation, unipolar brush cell migration, DCN cytoarchitecture, axon guidance, dendrite morphology, and synaptic plasticity. These mechanisms will require further basic research.

2 Overview of Cerebellar Development

Cerebellar development involves a complex sequence of events in neurogenesis, cell migration, axon pathfinding, dendritogenesis, synaptogenesis, and other mechanisms (Sotelo, 2004). Recent studies have made substantial progress in defining the cell types that express Reelin, and in characterizing neuron sources, lineages, and migrations. New discoveries that have changed the context for understanding Reelin functions are emphasized here.

2.1 Patterning and Rotation of the Cerebellar Anlage

The cerebellum develops from the dorsal part of hindbrain rhombomere 1 (r1), which undergoes a complex process of growth, flexure, and rotation to generate the cerebellar plate and upper rhombic lip, which together represent the cerebellar anlage (Fig. 10.1). The cerebellar plate and upper rhombic lip are distinct neurogenic compartments that express different genes and produce different sets of cerebellar neurons. Interestingly, most or all Reelin-producing cells are produced in the

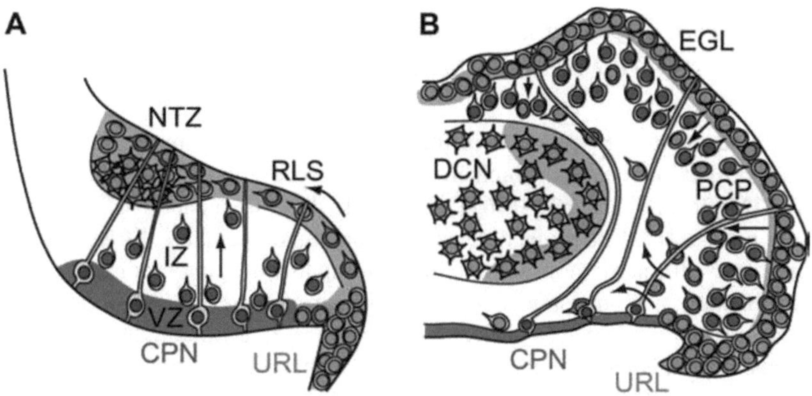

Fig. 10.1 Reelin signaling and cell migrations in cerebellar development. The diagrams show schematic views of the developing cerebellum in sagittal sections through the vermis, oriented with rostral to the left and dorsal to the top. (**A**) Early stage of cerebellar development (mouse E13.5). Cells derived from the upper rhombic lip (URL) (green nuclei) migrate nonradially (curved arrow) through the rostral rhombic lip migratory stream (RLS) to the nuclear transitory zone (NTZ). Reelin (blue) is expressed by many cells in the RLS and NTZ. At the same time, Purkinje cells (red nuclei) migrate radially (straight arrow) from the ventricular zone (VZ) of the cerebellar plate neuroepithelium (CPN) along radial glial cells (gray) through the intermediate zone (IZ), toward the RLS and NTZ. The Purkinje cells express cytoplasmic Dab1 (yellow). (**B**) Later stage of cerebellar development (mouse E17.5). The Purkinje cell plate (PCP) has formed, and the external granular layer (EGL) has replaced the RLS. Cells from the EGL migrate radially inward through the PCP (straight arrows), while unipolar brush cells migrate directly from the URL into the IZ (curved arrows). The deep cerebellar nuclei (DCN) contain neurons derived from the NTZ that have migrated radially inward toward the VZ (*See Color Plates*)

upper rhombic lip, while Reelin-responsive Purkinje cells are produced in the cerebellar plate (Fig. 10.1). Fate mapping and molecular expression studies indicate that the upper rhombic lip and cerebellar plate are initially patterned by signals from the roof plate and isthmic organizer (Zervas *et al.*, 2004; Sgaier *et al.*, 2005; Chizhikov *et al.*, 2006). Initially, in the hindbrain neural tube, the upper rhombic lip primordium is located in dorsomedial r1, adjacent to the roof plate, while the cerebellar plate primordium is located in dorsolateral r1. Their relative positions subsequently shift as a result of cerebellar plate rotation and pontine flexure, so that the cerebellar plate comes to be located rostral to the upper rhombic lip. The cerebellar plate is a conventional neuroepithelium with radially organized ventricular zone (VZ), intermediate zone (IZ), and mantle zone (Fig. 10.1A). In contrast, the upper rhombic lip is a pure progenitor compartment, producing postmitotic neurons and precursors that migrate tangentially out of the rhombic lip and into the cerebellar plate, as well as to some brainstem nuclei. Accordingly, regional subdivisions within the cerebellum (longitudinal stripes and transverse domains) appear to be specified primarily by patterning of the cerebellar plate, which occurs prior to the start of neurogenesis (Zervas *et al.*, 2004; Sgaier *et al.*, 2005; Chizhikov *et al.*, 2006).

2.2 *Different Types of Cerebellar Neurons Are Produced Sequentially*

The first neurons to be generated in the embryonic cerebellum are Purkinje cells and DCN projection neurons, which are produced from embryonic day (E) 11 to E13 in mice (Miale and Sidman, 1961). Although they are generated concurrently, Purkinje cells and DCN projection neurons are fundamentally different types of neurons. They differ by morphology, migration pathways, axon connections, and, most significantly, neurotransmitter systems. Purkinje cells release GABA, an inhibitory neurotransmitter, while DCN projection neurons release glutamate, an excitatory neurotransmitter. Studies in forebrain and spinal cord have demonstrated that transmitter phenotype is a fundamental aspect of neuron identity and is determined very early in neuronal fate specification (Schuurmans and Guillemot, 2002; Cheng *et al.*, 2005). In the forebrain, glutamatergic and GABAergic neurons are produced in separate progenitor compartments, known as the cortical neuroepithelium (pallium) and ganglionic eminences, respectively. This suggests the possibility that Purkinje cells, DCN projection neurons, and other cerebellar neuron types might likewise originate from separate glutamatergic and GABAergic progenitor compartments. This hypothesis was recently supported (Machold and Fishell, 2005; Wang *et al.*, 2005; Englund *et al.*, 2006; Fink *et al.*, 2006).

Following Purkinje cells and DCN projection neurons, the next neurons to be produced are local inhibitory interneurons (GABAergic) and unipolar brush cells (glutamatergic). Inhibitory interneurons differentiate into diverse morphological subtypes and migrate to locations throughout the cerebellar cortex and DCN. Unipolar brush cells differentiate into at least two subtypes and migrate only to the internal granular layer. The last neurons to be produced are granule cells (glutamatergic), which are the most abundant neuron type in the brain. Granule cells are produced from amplifying precursors in the EGL, which are, in turn, derived from the upper rhombic lip via migration through the RLS. Neurogenesis of granule cells covers a relatively long interval, lasting through postnatal day (P) 15 in mice (Miale and Sidman, 1961).

2.2.1 Purkinje Cells and Inhibitory Interneurons Are Derived from the Cerebellar Plate VZ

Efforts to determine the origins and lineages of cerebellar neurons have recently been enhanced by the introduction of new experimental approaches and genetic technologies. Specific progenitor compartments have been identified on the basis of molecular expression, and have been linked to the separate production of different classes of neurons. The new findings have important implications for understanding cerebellar development, and for dissecting the functions of Reelin signaling.

Purkinje cells have long been thought to arise from the cerebellar plate neuroepithelium (Altman and Bayer, 1985a,c). Recent studies have indeed confirmed and expanded this view. Purkinje cells are produced from progenitors in the cerebellar plate VZ that specifically express Ptf1a (pancreas transcription factor 1a), a basic helix–loop–helix (bHLH) transcription factor that is essential for the development of Purkinje neurons (Hoshino *et al.*, 2005). Mice that lack Ptf1a expression in the cerebellum, known as *cerebelless* mutants, completely lack cerebellar cortex due to decreased production of Purkinje cells and other GABAergic neurons, and secondary deficiency of granule neurons (Hoshino *et al.*, 2005). Interestingly, *cerebelless* mutants survive up to 2 years, indicating that the cerebellum is virtually dispensable for mice in laboratory conditions.

Inhibitory interneurons of the cerebellum likewise originate from $Ptf1a^+$ progenitor cells in the cerebellar plate VZ (Hoshino *et al.*, 2005). However, many inhibitory interneurons are not produced directly from VZ progenitor cells, but instead are generated by division of precursor cells migrating through the developing white matter (Zhang and Goldman, 1996). These precursor cells express transcription factor Pax2 (Maricich and Herrup, 1999) and require Ptf1a for their development (Hoshino *et al.*, 2005). Since Purkinje cells and inhibitory interneurons are produced from the cerebellar plate VZ while glutamatergic neuron types are not, the cerebellar plate VZ may be regarded as a distinct compartment for GABAergic neuron production.

2.2.2 DCN Projection Neurons, Unipolar Brush Cells, and Granule Neurons Are Derived from the Upper Rhombic Lip

DCN projection neurons, like Purkinje cells, were thought to arise from progenitors in the cerebellar plate VZ (Altman and Bayer, 1985a,b). But surprisingly, recent experiments have demonstrated that DCN projection neurons actually arise from progenitors in the upper rhombic lip. These progenitors express Math1 (mouse atonal homolog 1), a bHLH transcription factor and specific marker of rhombic lip lineages in the cerebellum (Machold and Fishell, 2005; Wang *et al.*, 2005). Newly generated DCN projection neurons migrate rostrally from the upper rhombic lip to the nuclear transitory zone (NTZ), a transient cell mass that is subsequently partitioned and organized to form the DCN (Fig. 10.1). The migration pathway from upper rhombic lip to NTZ, known as the rostral rhombic lip migratory stream (RLS), traverses a nonradial route along the subpial surface of the cerebellar plate (Wang *et al.*, 2005). The migrating DCN projection neurons form a thin, almost continuous sheet across the dorsal surface of the cerebellum. Importantly, many cells in the RLS express Reelin, which thus accumulates at high levels throughout the subpial zone of the early embryonic cerebellum (Miyata *et al.*, 1996). In addition, Reelin is expressed by some cells in the NTZ and later DCN, which may contribute to overall Reelin signaling in the cerebellum (Jensen *et al.*, 2002). As discussed further below, Reelin evidently acts as a signal to regulate the radial migration of Purkinje neurons from the VZ toward the cerebellar cortex.

Unipolar brush cells are a unique type of glutamatergic interneurons in the cerebellar cortex. Their origins were revealed in 2006 by a careful study of transcription factor expression patterns and cell migration in organotypic slice cultures (Englund *et al.*, 2006). As shown in that study, unipolar brush cells are generated in the upper rhombic lip and migrate tangentially through the IZ or developing white matter to the internal granular layer (Fig. 10.1B). Like DCN projection neurons, unipolar brush cells arise from a $Math1^+$ lineage, although they can be identified more specifically by high-level expression of Tbr2, a T-domain transcription factor (Englund *et al.*, 2006). The discovery that unipolar brush cells come from the upper rhombic lip supported the hypothesis that glutamatergic and GABAergic neurons are produced in separate progenitor compartments. Interestingly, many unipolar brush cells express Reelin, but the significance of this for cerebellar development is unknown.

Granule neurons are produced from proliferating precursor cells in the EGL, which are in turn derived from the upper rhombic lip. These origins were postulated many years ago on the basis of histological observations, and have been fully supported by newer approaches (Machold and Fishell, 2005; Wang *et al.*, 2005). The EGL develops by replacement of the RLS with granule neuron precursors migrating from the upper rhombic lip (Fig. 10.1). Thus, granule neurons are related to DCN projection neurons and unipolar brush cells by common origins from $Math1^+$ lineages in the upper rhombic lip. Notably, newly differentiated granule neurons express Reelin in the inner part of the EGL and in the internal granular layer (Miyata *et al.*, 1996). This source of Reelin may regulate Purkinje cell spreading during middle to late stages of cerebellar development.

2.3 Cerebellar Cell Migrations and Interactions

The five major types of cerebellar neurons (Purkinje cells, inhibitory interneurons, DCN projection neurons, unipolar brush cells, and granule neurons) undergo complex migrations and interactions during morphogenesis. The migrating cells interact with each other and with radial glia by cell–cell contacts and by production of extracellular factors that are detected by specific receptors. Reelin is one example of an extracellular factor that mediates important interactions in cell guidance (and possibly other processes as well). Another important extracellular factor is Sonic hedgehog, which is produced by Purkinje cells to stimulate the proliferation of granule cell precursors. A basic knowledge of these migrations and interactions is essential to understanding the role of Reelin signaling in cerebellar development.

2.3.1 Radial Migration of Purkinje Cells and Inhibitory Interneurons

At early stages of neurogenesis, the embryonic cerebellum lacks folia and instead exhibits a relatively flat surface topography (Fig. 10.1A). The cerebellar plate VZ appears to map directly onto the mantle (subpial) zone in a point-to-point manner, defined by radial glial cells whose processes span the distance from ventricular to pial surface (Fig. 10.1A). This cellular organization resembles that in the embryonic neocortex, where radial glial cells organize the cortex into columnar units and guide radial migration of pyramidal neurons from VZ to cortical plate (Rakic, 1988, 1995). Similarly, radial glia in the embryonic cerebellum guide radial migration of Purkinje cells from the VZ to the mantle zone, where they form the Purkinje cell plate (Goffinet, 1983; Yuasa *et al.*, 1993, 1996) directly beneath Reelin-expressing cells in the RLS and EGL (Miyata *et al.*, 1996, 1997). Initially, the Purkinje cell plate is several cell layers thick, but the Purkinje cells subsequently spread out to form a monolayer, concurrent with the entry of granule cells from the external granular layer (Altman and Bayer, 1985c). Reelin signaling may play an important role in the tangential spread of Purkinje cells and monolayer formation (Miyata *et al.*, 1997).

The radial phase of Purkinje cell migration appears to be guided mainly by contacts with radial glia. Electron microscopy has revealed adherens junctions (puncta and macula adhaerentia) between Purkinje cells and radial glia in the IZ (Yuasa *et al.*, 1996). During this phase, the Purkinje cells express high levels of Reelin receptors as well as Dab1 mRNA and protein, which are downstream mediators of Reelin signaling (Trommsdorff *et al.*, 1999; Rice and Curran, 2001). As the Purkinje cells enter the mantle zone and accumulate in the Purkinje cell plate, they lose adherens junctions to radial glia, suggesting that the Purkinje cells release their hold on radial glia in order to facilitate spreading (Yuasa *et al.*, 1993, 1996).

Like Purkinje cells, inhibitory interneurons initially migrate radially from the cerebellar plate VZ into the IZ (Maricich and Herrup, 1999; Hoshino *et al.*, 2005). However, the inhibitory interneuron precursors continue to proliferate as they migrate through cerebellar white matter, undergoing widespread tangential as well as radial dispersion. The inhibitory interneurons eventually settle in the DCN and all layers of cerebellar cortex (Zhang and Goldman, 1996; Maricich and Herrup, 1999). The signals that guide interneurons are unknown.

2.3.2 Tangential (Nonradial) Migration of Rhombic Lip Derivatives

The upper rhombic lip, also known as the germinal trigone of the cerebellum, gives rise to a massive efflux of neurons and neural progenitor cells that migrate rostrally into the cerebellar plate and brainstem. Genetic lineage tracing with *Math1* reporter mice revealed that cells exit the rhombic lip in a subpial stream, the RLS (later replaced by EGL), which delivers DCN projection neurons and granule cells to the cerebellum, and many additional neurons to diverse brainstem nuclei (Machold and Fishell, 2005; Wang *et al.*, 2005; Fink *et al.*, 2006). The subpial migration of cells in the RLS/EGL is regulated in part by the chemoattractant activity of stromal cell derived factor 1 (SDF-1), a chemokine secreted from the meninges (Zhu *et al.*, 2002). The SDF-1 is detected by chemokine (C-X-C motif) receptor 4 (CXCR4) on RLS/EGL cells (Zou *et al.*, 1998). The signals that control partition of RLS cells into DCN, the granular layer, and brainstem nuclei are not known.

The RLS is not the only pathway out of the rhombic lip. Unipolar brush cells migrate through a distinct narrow channel between the Purkinje cell plate and the cerebellar plate VZ into the cerebellar plate IZ (Englund *et al.*, 2006). Having entered the IZ, the unipolar brush cells disperse widely in the developing white matter and enter the internal granular layer from below. The signals that guide this migration have not been studied.

2.3.3 Inward Radial Migration of Granule Neurons

Much evidence indicates that granule neurons use radial glia as guides to migrate from the EGL though the Purkinje cell plate to the internal granular layer (Hatten, 1999). Notably, this migration proceeds in the opposite direction as the earlier radial migration of Purkinje cells, i.e., inward rather than outward. The inward migration of granule neurons may be directed by repellent factors from the meninges (Zhu *et al.*, 2002), presumably associated with downregulation of SDF-1 in the meninges and/or CXCR4 in granule neurons. Interestingly, the radial glia that mediate granule cell migration display a specialized phenotype with processes that extend to the pial surface but not the ventricular surface. These unique radial glia in the cerebellum are known as Bergmann glia (Hatten, 1999).

2.3.4 Sonic Hedgehog from Purkinje Cells Promotes Granule Cell Neurogenesis and Cerebellar Foliation

Adequate production of granule neurons critically depends on an interaction with Purkinje cells, which produce factors that stimulate proliferation of granule cell precursors in the EGL. This interaction presumably serves to optimally regulate the ratio of granule neurons to Purkinje cells, which is approximately 1000:1 in humans (Nolte, 1999). The main proliferative signal produced by Purkinje cells is Sonic hedgehog, which strongly stimulates granule cell proliferation (reviewed by Ruiz i Altaba *et al.*, 2002). The proliferative effect of Sonic hedgehog is further potentiated by other factors (Mills *et al.*, 2006). In turn, the proliferation of granule neurons determines the number of cerebellar folia, as demonstrated by experiments in which Sonic hedgehog signaling was manipulated genetically (Corrales *et al.*, 2006). The interaction between Purkinje cells and granule cell precursors is severely impaired by Reelin deficiency, as explained in subsequent sections of this chapter.

3 Reelin Signaling in the Cerebellum

The Reelin signaling pathway involves the same receptors and downstream mediators in the developing cerebellum as in other regions such as developing cerebral cortex (reviewed in Rice and Curran, 2001; Förster *et al.*, 2006). The receptors for

Reelin are ApoER2, VLDLR, and α3β1 integrin, and downstream signaling mediators include Dab1 adapter protein and Src family tyrosine kinases (Rice and Curran, 2001; Kuo *et al.*, 2005; Förster *et al.*, 2006). The expression of Reelin and other signaling components is extremely complex in the developing cerebellum, and not yet fully understood. For example, Miyata *et al.* (1996) detected Reelin protein at various expression levels in different cell types and extracellular zones throughout cerebellar development. The importance of such subtle gradations of Reelin expression and accumulation is unknown. Nevertheless, the broad outlines of Reelin expression are clear enough to support the hypothesis that Reelin signaling regulates the migration of Purkinje cells, and probably other aspects of cerebellar development as well.

3.1 Purkinje Cell Radial Migration from VZ to Mantle Zone

The first cells to express Reelin during cerebellar development (~E13 in mouse) are located along the RLS and scattered in the NTZ (Miyata *et al.*, 1996). These cells are derived from the rhombic lip and differentially express transcription factors Zic1, Pax6, Tbr2, and Tbr1 (Miyata *et al.*, 1996; Fink *et al.*, 2006). Since the RLS defines the subpial surface of the embryonic cerebellum (Fig. 10.1A), the early Reelin$^+$ cells are well positioned to signal Purkinje cells that they should cease migration and remain in the mantle zone to form the Purkinje cell plate. Consistent with this interpretation, much evidence indicates that migrating Purkinje cells express Reelin receptors and downstream mediators, including VLDLR, ApoER2, and Dab1 (Trommsdorff *et al.*, 1999; Rice and Curran, 2001; Perez-Garcia *et al.*, 2004). The role of Reelin signaling from the NTZ is somewhat unclear and problematic. As the NTZ grows and partitions to form the DCN, its cells migrate radially inward, toward the VZ (in the opposite direction to Purkinje cells) and ultimately come to reside near the roof of the fourth ventricle (Altman and Bayer, 1985a,b). It is unclear whether Purkinje cells migrate around or between these NTZ/DCN cells to reach overlying regions of cerebellar cortex. It is likewise unclear whether radial glia pass through the NTZ/DCN or curve around them (Fig. 10.1B).

The mechanisms of Purkinje cell migration that are regulated by Reelin have been studied very little. The available evidence suggests that Reelin may promote detachment of Purkinje cells from radial glial fibers. In *reeler* mice, which lack Reelin protein due to a genetic mutation, there are increased numbers of adherens junctions (puncta adhaerentia) between migrating Purkinje cells and radial glia and between pairs of migrating Purkinje cells (Yuasa *et al.*, 1993). The prolonged attachment between Purkinje cells and radial glia may obstruct the arrival of following Purkinje cell cohorts migrating on the same radial glial fibers, and thus block formation of the Purkinje cell plate (Goffinet, 1983). Additional support for this hypothesis comes from studies of the cerebral cortex, where evidence indicates that Reelin/Dab1 signaling promotes detachment from radial glia and inhibits neuron migration (Dulabon *et al.*, 2000; Sanada *et al.*, 2004).

It has also been suggested that Reelin might regulate Purkinje cell migrations indirectly, by effects on radial glia morphology (Yuasa *et al.*, 1993). This hypothesis was based on the observation that radial glia have abnormally curved and disorganized processes in *reeler* mice (Yuasa *et al.*, 1993). On the other hand, it is equally possible that the radial glia abnormalities may be secondary effects of prolonged Purkinje cell attachment and impaired migration.

3.2 Purkinje Cell Spreading and Monolayer Formation

Once Purkinje cells have completed their radial migration into the mantle zone (by the day of birth in mice), they coalesce in the Purkinje cell plate, a continuous cellular stratum that is ~2–4 cells thick in newborn mice. During the subsequent prolonged period of granule cell neurogenesis and migration, which continue during the first 3 postnatal weeks in mice (Miale and Sidman, 1961), the Purkinje cells spread into a highly convoluted monolayer that defines the contours of the cerebellar folia (Goffinet, 1983). The signals that regulate Purkinje cell spreading and monolayer formation are unknown. Interestingly, Reelin protein is expressed by numerous EGL cells and accumulates in the extracellular matrix of the EGL during this period, and then is downregulated after monolayer formation is complete (Miyata *et al.*, 1996). Thus, Reelin is expressed in appropriate patterns to guide Purkinje cell spreading. However, this aspect of cerebellar development has been studied very little, and the proposed function of Reelin signaling during this stage has not yet been confirmed experimentally. In mutant animals, such as *reeler*, the early defects of Purkinje cell radial migration are so severe that no useful information can be obtained about the role of Reelin in Purkinje cell spreading.

3.3 Regulation of Reelin Expression in Cerebellum

Reelin is expressed by a limited set of cerebellar neuron types, implying that expression is stringently regulated by upstream factors. Interestingly, all Reelin$^+$ cell types in the developing cerebellum share common origins from the upper rhombic lip. Cells in the upper rhombic lip lineage express numerous specific transcription factors, including Math1, Zic1, Pax6, Tbr2, and Tbr1. Of these, Zic1, Pax6, Tbr2, and Tbr1 have been co-localized with Reelin in subsets of rhombic lip-derived cells (Miyata *et al.*, 1996; Englund *et al.*, 2006; Fink *et al.*, 2006). In the developing cerebral cortex, Reelin appears to be regulated by Tbr1. *Tbr1* null mutant mice have a severe cortical malformation with marked reduction of Reelin mRNA and protein (Hevner *et al.*, 2001). In the cerebellum, however, *Tbr1* inactivation has no apparent effect on Reelin expression (Fink *et al.*, 2006). Thus, despite similar roles played by Reelin signaling in controlling radial migration of neurons, the upstream regulation of Reelin evidently differs between cerebral cortex and cerebellum.

4 Cerebellar Malformations Caused by Reelin Deficiency

Under conditions of complete Reelin deficiency, as first described in *reeler* mutant mice, many brain structures develop abnormally. The most severe malformation is observed in the cerebellum, which is affected by both neuronal disorganization and marked hypoplasia. The cerebral cortex is also malformed, but unlike the cerebellum, shows no significant hypoplasia. The *reeler* brain malformations are beginning to be understood at a mechanistic level (Rice and Curran, 2001; Förster *et al.*, 2006). Here, I focus on the mechanisms of cerebellar defects caused by Reelin deficiency.

4.1 Neurological Defects Caused by Cerebellar Malformations

The cerebellum is primarily a motor center, although some cognitive and language functions are also mediated by cerebellar circuits (Ito, 2005). In humans, cerebellar lesions cause mainly ataxia (incoordination) and vertigo (inappropriate sensation of movement or spinning). Since the cerebellum is small and malformed in rodents and humans with Reelin deficiency, one would expect to observe ataxia in affected individuals. Indeed, ataxia is a prominent symptom in Reelin-deficient mice (*reeler*) and rats (*SRK*), along with tremor or "shaking" of the body (Aikawa *et al.*, 1988). Affected rodents generally die young, usually around the end of the first postnatal month (Mariani *et al.*, 1977; Aikawa *et al.*, 1988). In humans with *RELN* mutations, ataxia is not so prominent. Rather, the major symptoms are neurodevelopmental delay, hypotonia ("floppy baby"), language deficit, and generalized seizures (Hong *et al.*, 2000; Chang *et al.*, 2007). Presumably, ataxia would be more apparent if not masked by severe hypotonia. With appropriate medical care, affected humans can survive for many years, despite the severe neurological problems.

4.2 Cerebellar Hypoplasia and Disorganization in reeler and SRK

SRK rats and *reeler* mice exhibit virtually identical cerebellar phenotypes (Kikkawa *et al.*, 2003). In *reeler* mice, the first defect to appear (by E14) is absent formation of the Purkinje cell plate (Goffinet, 1983). By E17, when normal embryos begin to show evidence of foliation, *reeler* mutants exhibit not only defective Purkinje cell migrations, but also reduced tangential growth of the cerebellar surface and absent foliation. These defects progress, and by the time cerebellar development is complete (~3 postnatal weeks in mice), the *reeler* cerebellum is only 25–33% of normal size, contains only 12–14% of the normal DNA amount, has no foliation, and has severe neuronal disorganization throughout most regions of cortex (Fig. 10.2). The disproportionate decrease of DNA relative to cerebellar mass in *reeler* is attributed

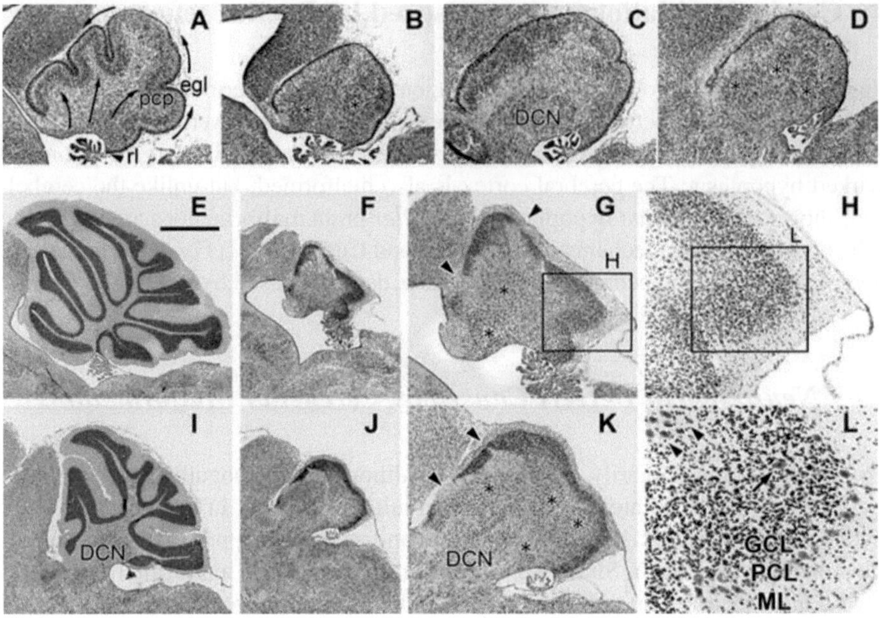

Fig. 10.2 Cerebellar histology in control and *reeler* mice. Sagittal sections through the cerebellar vermis (**A**, **B**, **E–H**, **L**) or hemisphere (**C**, **D**, **I–K**) of control and *reeler* (**B**, **D**, **F–H**, **J–L**) mice were stained with cresyl violet on P0.5 (**A–D**) or P22 (**E–L**). The boxed area in **G** is enlarged in **H**, and the boxed area in **H** is enlarged in **L**. In P0.5 controls, Purkinje cells had migrated to the Purkinje cell plate (pcp), and folia were developing by migration and proliferation of cells in the external granular layer (egl). In P0.5 *reeler* mice, the cerebellum was hypoplastic, no folia were developing, and Purkinje cells formed large, centrally located ectopic clusters (asterisks). The hypoplasia and defective foliation of the *reeler* cerebellum became even more obvious by P22. Most Purkinje cells in the P22 *reeler* cerebellum are located in the large central clusters, although some are isolated ectopically in the granule cell layer (GCL), and others form a nearly normal Purkinje cell layer (PCL) below the molecular layer (ML). In **L**, arrowheads indicate Purkinje cells in deep ectopia, and the arrow indicates a Purkinje cell in the GCL. The GCL in *reeler* consistently shows gaps (arrowheads in **G**, **K**), which may be related to the presumptive locations of fissures (Goldowitz *et al.*, 1997). The deep cerebellar nuclei (DCN) in *reeler* are located near the normal location, but somewhat distorted by the Purkinje cell ectopia (Goffinet, 1983; Goffinet *et al.*, 1984). Sections oriented as described for Figure 1. Scale bar (in **E**): **A–D**, 400 μm; **E**, **F**, **I**, **J**, 1000 μm; **G**, **K**, 500 μm; **H**, 200 μm; **L**, 100 μm (*See Color Plates*)

to selective depletion of granule neurons (Mariani *et al.*, 1977). Despite the absence of foliation, some small regions of the cerebellar surface display essentially normal cortical cytoarchitecture, including formation of a monolayer by ~7% of remaining Purkinje cells (Mariani *et al.*, 1977; Goffinet *et al.*, 1984; Goldowitz *et al.*, 1997).

The DCN develop relatively normally in *reeler* mice, although the organization of the lateral (dentate) nucleus is perturbed (Goffinet, 1983), and the medial (fastigial) nucleus is displaced laterally (Goffinet *et al.*, 1984). Cerebellar patterning into longitudinal and transverse compartments, and cerebellar axon connections likewise show only mild abnormalities, presumably secondary consequences of distortion caused by altered Purkinje cell migrations and hypoplasia (Mariani *et al.*, 1977; Goffinet *et al.*, 1984). Mice with deficiencies of downstream Reelin signaling molecules including VLDLR/ApoER2, Dab1, and Src/Fyn exhibit essentially identical cerebellar malformations as in *reeler* (Rice and Curran, 2001; Kuo *et al.*, 2005).

The primary defect of cerebellar morphogenesis in Reelin deficiency is thought to be abnormal migration of Purkinje cells. This conclusion is supported by several observations. First, a defect of Purkinje cell migration (agenesis of the Purkinje cell plate) is the earliest morphological abnormality to appear in *reeler* mice (Goffinet, 1983). Second, Dab1 protein is upregulated in *reeler* Purkinje cells, indicating a biochemical response to decreased Reelin signaling (Sheldon *et al.*, 1997; Rice *et al.*, 1998). Third, the impairment of granule cell proliferation, and consequent hypoplasia and lack of foliation, can be adequately explained as a secondary consequence of abnormal Purkinje cell positioning. Granule cell precursors in the EGL require Sonic hedgehog, which is produced by Purkinje cells, to stimulate mitosis and neurogenesis (Ruiz i Altaba *et al.*, 2002). Any decrease of Sonic hedgehog signaling reduces the neurogenesis of granule cells, which in turn leads to reduced foliation (Corrales *et al.*, 2006). Since most Purkinje cells in *reeler* are located in abnormally deep positions far from the EGL, granule cell precursors are presumably exposed to lower concentrations of Sonic hedgehog and thus proliferate less. The abnormal organization of neuronal processes, such as Purkinje cell dendrites, may also impair transport of Sonic hedgehog to the EGL (Ruiz i Altaba *et al.*, 2002).

Other cerebellar defects that have been reported in *reeler* include disorganized arrangement of radial glia processes and cell bodies (Yuasa *et al.*, 1993), reduced numbers of unipolar brush cells (Ilijic *et al.*, 2005; Englund *et al.*, 2006), altered synaptic organization and physiological responses of some cerebellar neurons (Mariani *et al.*, 1977), and mild defects of axon connections (Goffinet *et al.*, 1984). Some or all of these abnormalities may be secondary to malpositioning of Purkinje cells. On the other hand, Reelin signaling regulates axon growth and branching independently of cell migration in the hippocampal formation, and thus might do so in the cerebellum as well (Del Río *et al.*, 1997; Borrell *et al.*, 1999).

4.3 Human Reelin Deficiency: Lissencephaly with Cerebellar Hypoplasia

Genetic studies have demonstrated that Reelin deficiency in humans causes a severe brain malformation involving the cerebellum, brainstem, and cerebral cortex. The malformation is classified as a form of lissencephaly (reduced number of cerebral cortical gyri) with cerebellar hypoplasia (Hong *et al.*, 2000; Chang *et al.*, 2006). The malformation syndrome is inherited in an autosomal recessive pattern and is caused by mutations affecting both copies of the *RELN* gene, located on chromosome 7q22. On neuroimaging scans, the cerebellum appears severely hypoplastic and lacks folia (Hong *et al.*, 2000; Chang *et al.*, 2007). In addition, the cerebral cortex exhibits a simplified gyral pattern; cortical thickness is increased; the hippocampus appears malrotated and flat; and the brainstem is hypoplastic. The brain phenotype caused by Reelin deficiency is unique and can easily be distinguished from other forms of lissencephaly or cerebellar hypoplasia by neuroimaging (Hong *et al.*, 2000; Chang *et al.*, 2007). No histologic studies of the human brain phenotype have been reported.

Humans with a heterozygous mutation affecting only one copy of the *RELN* gene show no apparent neuropsychiatric abnormalities (Hong *et al.*, 2000; Chang *et al.*, 2007). Thus, the proposed role of Reelin signaling in diseases, such as schizophrenia and autism (Fatemi, 2005; Fatemi *et al.*, 2005), may involve changes that are more complex than a partial reduction of mRNA and protein levels due to gene dosage effect.

4.4 Autism, Reelin, and the Cerebellum

Several studies have suggested that alterations of Reelin signaling may contribute to the pathogenesis of neurobehavioral disorders, especially autism (Fatemi, 2005; Fatemi *et al.*, 2005). Genetic studies have linked autism to specific polymorphisms of the *RELN* gene, and decreased levels of Reelin mRNA and protein have been found in autistic brains relative to controls, particularly in the cerebellum (reviewed in Fatemi, 2005). Intriguingly, abnormalities of cerebellar structure are among the most consistent neuropathologic findings in autism (Kemper and Bauman, 1998; Palmen *et al.*, 2004; Bauman and Kemper, 2005; Pickett and London, 2005). Specifically, the autistic cerebellum is smaller, contains fewer Purkinje cells (~41% loss), has atrophic Purkinje cells (~24% decrease of cell size), shows variable gliosis, and expresses increased levels of glial fibrillary acidic protein (Kemper and Bauman, 1998; Carper and Courchesne, 2000; Fatemi *et al.*, 2002; Palmen *et al.*, 2004; Laurence and Fatemi, 2005; Pickett and London, 2005). Defects of cerebellar development have been further implicated in autism by linkage to the *EN2* gene, which regulates embryonic patterning of the cerebellum (Kuemerle *et al.*, 2007). These associations support the hypothesis that some symptoms in autism may be caused by cerebellar defects, sometimes arising from perturbations of Reelin signaling. Further studies will be necessary to clarify how abnormal Reelin signaling in humans may cause a reduction in the number of Purkinje cells, versus ectopic migration as observed in *reeler* mice and *SRK* rats.

5 Conclusion

The role of Reelin signaling in cerebellar development and disease is not yet fully understood. Existing data strongly suggest that Reelin signaling regulates radial migration of Purkinje cells by modulating their adhesive properties and contacts with radial glia. Some data also suggest that other cell types and developmental mechanisms may be regulated by Reelin signaling, although the current evidence is inconclusive. For example, DCN neurons reportedly express Dab1 mRNA and protein during development, which would imply that they can respond to Reelin (Rice *et al.*, 1998), but the role (if any) of Reelin signaling in regulating DCN cell migrations remains unclear. Molecular expression patterns further suggest that Reelin signaling may regulate Purkinje cell spreading into a monolayer. Likewise, the possible role of Reelin signaling in regulating axon pathfinding of cerebellar efferent and afferent axons deserves more study. Future studies along these lines would benefit from more sophisticated genetic models, such as conditional *reelin* knockout mice, to investigate the role of Reelin signaling in different cell types and at different stages of development. Such analysis will be essential to fully understand how Reelin signaling affects all aspects of cerebellar development, and ultimately lay the scientific foundation for interpreting links between Reelin signaling, cerebellar development, and autism.

References

Aikawa, H., Nonaka, I., Woo, M., Tsugane, T., and Esaki, K. (1988). Shaking rat Kawasaki (SRK): a new neurological mutant in the Wistar strain. *Acta Neuropathol. (Berl.)* 76:366–372.

Altman, J., and Bayer, S. A. (1985a). Embryonic development of the rat cerebellum. I. Delineation of the cerebellar primordium and early cell movements. *J. Comp. Neurol.* 231:1–26.

Altman, J., and Bayer, S. A. (1985b). Embryonic development of the rat cerebellum. II. Translocation and regional distribution of the deep neurons. *J. Comp. Neurol.* 231:27–41.

Altman, J., and Bayer, S. A. (1985c). Embryonic development of the rat cerebellum. III. Regional differences in the time of origin, migration, and settling of Purkinje cells. *J. Comp. Neurol.* 231:42–65.

Bauman, M. L., and Kemper, T. L. (2005). Neuroanatomic observations of the brain in autism: a review and future directions. *Int. J. Del. Neurosci.* 23:183–187.

Borrell, V., Del Río, J. A., Alcántara, S., Derer, M., Martínez, A., D'Arcangelo, G., Nakajima, K., Mikoshiba, K., Derer, P., Curran, T., and Soriano, E. (1999). Reelin regulates the development and synaptogenesis of the layer-specific entorhino-hippocampal connections. *J. Neurosci.* 19:1345–1358.

Carper, R. A., and Courchesne, E. (2000). Inverse correlation between frontal lobe and cerebellum sizes in children with autism. *Brain* 123:836–844.

Chang, B. S., Duzcan, F., Kim, S., Cinbis, M., Aggarwal, A., Apse, K. A., Ozdel, O., Atmaca, M., Zencir, S., Bagci, H., and Walsh, C. A. (2007). The role of *RELN* in lissencephaly and neuropsychiatric disease. *Am. J. Med. Genet. B Neuropsychiatr. Genet.* 144:58–63.

Cheng, L., Samad, O. A., Xu, Y., Mizuguchi, R., Luo, P., Shirasawa, S., Goulding, M., and Ma, Q. (2005). Lbx1 and Tlx3 are opposing switches in determining GABAergic versus glutamatergic transmitter phenotypes. *Nature Neurosci.* 8:1510–1515.

Chizhikov, V. V., Lindgren, A. G., Currie, D. S., Rose, M. F., Monuki, E. S., and Millen, K. J. (2006). The roof plate regulates cerebellar cell-type specification and proliferation. *Development* 133:2793–2804.

Corrales, J. D., Blaess, S., Mahoney, E. M., and Joyner, A. L. (2006). The level of sonic hedgehog signaling regulates the complexity of cerebellar foliation. *Development* 133:1811–1821.

Del Río, J. A., Heimrich, B., Borrell, V., Förster, E., Drakew, A., Alcántara, S., Nakajima, K., Miyata, T., Ogawa, M., Mikoshiba, K., Derer, P., Frotscher, M., and Soriano, E. (1997). A role for Cajal-Retzius cells and *reelin* in the development of hippocampal connections. *Nature* 385:70–74.

Dulabon, L., Olson, E. C., Taglienti, M. G., Eisenhuth, S., McGrath, B., Walsh, C. A., Kreidberg, J. A., and Anton, E. S. (2000). Reelin binds $\alpha 3\beta 1$ integrin and inhibits neuronal migration. *Neuron* 27:33–44.

Englund, C., Kowalczyk, T., Daza, R. A. M., Dagan, A., Lau, C., Rose, M. F., and Hevner, R. F. (2006). Unipolar brush cells of the cerebellum are produced in the rhombic lip and migrate through developing white matter. *J. Neurosci.* 26:9184–9195.

Fatemi, S. H. (2005). Reelin glycoprotein: structure, biology and roles in health and disease. *Mol. Psychiatry* 10:251–257.

Fatemi, S. H., Halt, A. R., Realmuto, G., Earle, J., Kist, D. A., and Merz, A. (2002). Purkinje cell size is reduced in cerebellum of patients with autism. *Cell. Mol. Neurobiol.* 22:171–175.

Fatemi, S. H., Snow, A. V., Stary, J. M., Araghi-Niknam, M., Reutiman, T. J., Lee, S., Brooks, A. I., and Pearce, D. A. (2005). Reelin signaling is impaired in autism. *Biol. Psychiatry* 57:777–787.

Fink, A. J., Englund, C., Daza, R. A. M., Pham, D., Lau, C., Nivison, M., Kowalczyk, T., and Hevner, R. F. (2006). Development of the deep cerebellar nuclei: transcription factors and cell migration from the rhombic lip. *J. Neurosci.* 26:3066–3076.

Förster, E., Jossin, Y., Zhao, S., Chai, X., Frotscher, M., and Goffinet, A. M. (2006). Recent progress in understanding the role of reelin in radial neuronal migration, with specific emphasis on the dentate gyrus. *Eur. J. Neurosci.* 23:901–909.

Goffinet, A. M. (1983). The embryonic development of the cerebellum in normal and reeler mutant mice. *Anat. Embryol.* 168:73–86.

Goffinet, A. M., So, K.-F., Yamamoto, M., Edwards, M., and Caviness, V. S., Jr. (1984). Architectonic and hodological organization of the cerebellum in reeler mutant mice. *Brain Res. Dev. Brain Res.* 16:263–276.

Goldowitz, D., Cushing, R. C., Laywell, E., D'Arcangelo, G., Sheldon, M., Sweet, H. O., Davisson, M., Steindler, D., and Curran, T. (1997). Cerebellar disorganization characteristic of reeler in scrambler mutant mice despite presence of reelin. *J. Neurosci.* 17:8767–8777.

Hatten, M. E. (1999). Central nervous system neuronal migration. *Annu. Rev. Neurosci.* 22:511–539.

Hevner, R. F., Shi, L., Justice, N., Hsueh, Y.-P., Sheng, M., Smiga, S., Bulfone, A., Goffinet, A. M., Campagnoni, A. T., and Rubenstein, J. L. R. (2001). Tbr1 regulates differentiation of the preplate and layer 6. *Neuron* 29:353–366.

Hong, S. E., Shugart, Y. Y., Huang, D. T., Al Shahwan, S., Grant, P. E., Hourihane, J. O'B., Martin, N. D. T., and Walsh, C. A. (2000). Autosomal recessive lissencephaly with cerebellar hypoplasia is associated with human *RELN* mutations. *Nature Genet.* 26:93–96.

Hoshino, M., Nakamura, S., Mori, K., Kawauchi, T., Terao, M., Nishimura, Y. V., Fukuda, A., Fuse, T., Matsuo, N., Sone, M., Watanabe, M., Bito, H., Terashima, T., Wright, C. V. E., Kawaguchi, Y., Nakao, K., and Nabeshima, Y. (2005). *Ptf1a*, a bHLH transcription gene, defines GABAergic neuronal fates in the cerebellum. *Neuron* 47:201–213.

Ilijic, E., Guidotti, A., and Mugnaini, E. (2005). Moving up or moving down? Malpositioned cerebellar unipolar brush cells in reeler mouse. *Neuroscience* 136:633–647.

Ito, M. (2005). Bases and implications of learning in the cerebellum—adaptive control and internal model mechanism. *Prog. Brain Res.* 148:95–109.

Jensen, P., Zoghbi, H. Y., and Goldowitz, D. (2002). Dissection of the cellular and molecular events that position cerebellar Purkinje cells: a study of the *math1* null-mutant mouse. *J. Neurosci.* 22:8110–8116.

Kemper, T. L., and Bauman, M. (1998). Neuropathology of infantile autism. *J. Neuropathol. Exp. Neurol.* 57:645–652.

Kikkawa, S., Yamamoto, T., Misaki, K., Ikeda, Y., Okado, H., Ogawa, M., Woodhams, P. L., and Terashima, T. (2003). Missplicing resulting from a short deletion in the *reelin* gene causes *reeler*-like neuronal disorders in the mutant Shaking Rat Kawasaki. *J. Comp. Neurol.* 463:303–315.

Kuemerle, B., Gulden, F., Cherosky, N., Williams, E., and Herrup, K. (2007). The mouse Engrailed genes: a window into autism. *Behav. Brain Res.* 176:121–132.

Kuo, G., Arnaud, L., Kronstad-O'Brien, P., and Cooper, J. A. (2005). Absence of Fyn and Src causes a reeler-like phenotype. *J. Neurosci.* 25:8578–8586.

Laurence, J. A., and Fatemi, S. H. (2005). Glial fibrillary acidic protein is elevated in superior frontal, parietal and cerebellar cortices of autistic subjects. *Cerebellum* 4:206–210.

Machold, R., and Fishell, G. (2005). Math1 is expressed in temporally discrete pools of cerebellar rhombic-lip neural progenitors. *Neuron* 48:17–24.

Mariani, J., Crepel, F., Mikoshiba, K., Changeux, J. P., and Sotelo, C. (1977). Anatomical, physiological and biochemical studies of the cerebellum from *reeler* mutant mouse. *Philos. Trans. R. Soc. London Ser. B Biol. Sci.* 281:1–28.

Maricich, S. M., and Herrup, K. (1999). Pax-2 expression defines a subset of GABAergic interneurons and their precursors in the developing murine cerebellum. *J. Neurobiol.* 41:281–294.

Miale, I. L., and Sidman, R. L. (1961). An autoradiographic analysis of histogenesis in the mouse cerebellum. *Exp. Neurol.* 4:277–296.

Mills, J., Niewmierzycka, A., Oloumi, A., Rico, B., St-Arnaud, R., Mackenzie, I. R., Mawji, N. M., Wilson, J., Reichardt, L. F., and Dedhar, S. (2006). Critical role of integrin-linked kinase in granule cell precursor proliferation and cerebellar development. *J. Neurosci.* 26:830–840.

Miyata, T., Nakajima, K., Aruga, J., Takahashi, S., Ikenaka, K., Mikoshiba, K., and Ogawa, M. (1996). Distribution of a reeler gene-related antigen in the developing cerebellum: an immunohistochemical study with an allogeneic antibody CR-50 on normal and reeler mice. *J. Comp. Neurol.* 372:215–228.

Miyata, T., Nakajima, K., Mikoshiba, K., and Ogawa, M. (1997). Regulation of Purkinje cell alignment by reelin as revealed with CR-50 antibody. *J. Neurosci.* 17:3599–3609.

Nolte, J. (1999). *The Human Brain: An Introduction to Its Functional Anatomy*, 4th ed. Mosby, St. Louis.

Palmen, S. J. M. C., van Engeland, H., Hof, P. R., and Schmitz, C. (2004). Neuropathological findings in autism. *Brain* 127:2572–2583.

Perez-Garcia, C. G., Tissir, F., Goffinet, A. M., and Meyer, G. (2004). Reelin receptors in developing laminated brain structures of mouse and human. *Eur. J. Neurosci.* 20:2827–2832.

Pickett, J., and London, E. (2005). The neuropathology of autism: a review. *J. Neuropathol. Exp. Neurol.* 64:925–935.

Rakic, P. (1988). Specification of cerebral cortical areas. *Science* 241:170–176.

Rakic, P. (1995). A small step for the cell, a giant leap for mankind: a hypothesis of neocortical expansion during evolution. *Trends Neurosci.* 18:383–388.

Rice, D. S., and Curran, T. (2001). Role of the reelin signaling pathway in central nervous system development. *Annu. Rev. Neurosci.* 24:1005–1039.

Rice, D. S., Sheldon, M., D'Arcangelo, G., Nakajima, K., Goldowitz, D., and Curran, T. (1998). *Disabled-1* acts downstream of *Reelin* in a signaling pathway that controls laminar organization in the mammalian brain. *Development* 125:3719–3729.

Ruiz i Altaba, A., Palma, V., and Dahmane, N. (2002). Hedgehog-Gli signaling and the growth of the brain. *Nature Rev. Neurosci.* 3:24–33.

Sanada, K., Gupta, A., and Tsai, L.-H. (2004). Disabled-1-regulated adhesion of migrating neurons to radial glial fiber contributes to neuronal positioning during early corticogenesis. *Neuron* 42:197–211.

Schuurmans, C., and Guillemot, F. (2002). Molecular mechanisms underlying cell fate specification in the developing telencephalon. *Curr. Opin. Neurobiol.* 12:26–34.

Sgaier, S. K., Millet, S., Villanueva, M. P., Berenshteyn, F., Song, C., and Joyner, A. L. (2005). Morphogenetic and cellular movements that shape the mouse cerebellum: insights from genetic fate mapping. *Neuron* 45:27–40.

Sheldon, M., Rice, D. S., D'Arcangelo, G., Yoneshima, H., Nakajima, K., Mikoshiba, K., Howell, B. W., Cooper, J. A., Goldowitz, D., and Curran, T. (1997). *Scrambler* and *yotari* disrupt the *disabled* gene and produce a *reeler*-like phenotype in mice. *Nature* 389:730–733.

Sotelo, C. (2004). Cellular and genetic regulation of the development of the cerebellar system. *Prog. Neurobiol.* 72:295–339.

Trommsdorff, M., Gotthardt, M., Hiesberger, T., Shelton, J., Stockinger, W., Nimpf, J., Hammer, R. E., Richardson, J. A., and Herz, J. (1999). Reeler/disabled-like disruption of neuronal migration in knockout mice lacking the VLDL receptor and ApoE receptor 2. *Cell* 97:689–701.

Wang, V. Y., Rose, M. F., and Zoghbi, H. Y. (2005). *Math1* expression redefines the rhombic lip derivatives and reveals novel lineages within the brainstem and cerebellum. *Neuron* 48:31–43.

Yuasa, S., Kitoh, J., Oda, S., and Kawamura, K. (1993). Obstructed migration of Purkinje cells in the developing cerebellum of the reeler mutant mouse. *Anat. Embryol.* 188:317–329.

Yuasa, S., Kawamura, K., Kuwano, R., and Ono, K. (1996). Neuron–glia interrelations during migration of Purkinje cells in the mouse embryonic cerebellum. *Int. J. Dev. Neurosci.* 14:429–438.

Zervas, M., Millet, S., Ahn, S., and Joyner, A. L. (2004). Cell behaviors and genetic lineages of the mesencephalon and rhombomere 1. *Neuron* 43:345–357.

Zhang, L., and Goldman, J. E. (1996). Generation of cerebellar interneurons from dividing progenitors in white matter. *Neuron* 16:47–54.

Zhu, Y., Yu, T., Zhang, X.-C., Nagasawa, T., Wu, J. Y., and Rao, Y. (2002). Role of the chemokine SDF-1 as the meningeal attractant for embryonic cerebellar neurons. *Nature Neurosci.* 5:719–720.

Zou, Y. R., Kottmann, A. H., Kuroda, M., Taniuchi, I., and Littman, D. R. (1998). Function of the chemokine CXCR4 in haematopoiesis and in cerebellar development. *Nature* 393:595–599.

Chapter 11
Reelin and Radial Glial Cells

Eckart Förster, Shanting Zhao, and Michael Frotscher

Contents

1 Introduction

Defects of the radial glial scaffold in reeler mice were detected and characterized after the radial neuronal migration defects in this mutant had been described (Caviness and Rakic, 1978; Caviness *et al.*, 1988; Pinto-Lord *et al.*, 1982). Based on these findings, it has been hypothesized that radial glial defects contribute to the malpositioning of radially migrating neurons. Experimental evidence that Reelin may directly influence the development of radial glial cells is quite recent (Förster *et al.*, 2002; Weiss *et al.*, 2003; Hartfuss *et al.*, 2003; Luque *et al.*, 2003). The question as to why Reelin should simultaneously act on two different cell types, neurons

E. Förster
Institut für Anatomie I, Zelluläre Neurobiologie, Universität Hamburg, Martinistrasse 52, D-20246 Hamburg, Germany

S. Zhao
Institut für Anatomie und Zellbiologie, Abteilung für Neuroanatomie, Universität Freiburg, Albertstrasse 17, D-79104 Freiburg, Germany

M. Frotscher
Institut für Anatomie und Zellbiologie, Abteilung für Neuroanatomie, Universität Freiburg, Albertstrasse 17, D-79104 Freiburg, Germany
e-mail: Michael.Frotscher@anat.uni-freiburg.de

S. H. Fatemi (ed.), *Reelin Glycoprotein: Structure, Biology and Roles in Health and Disease.*

and radial glial cells, turned out to be a semantic problem when radial glial cells were shown to be precursors of radially migrating neurons (Malatesta *et al.*, 2000; Noctor *et al.*, 2001; Miyata *et al.*, 2001). Thus, when discussing a role of Reelin in the formation of the radial glial scaffold, the changing view of radial glial cell function in cortical development has to be taken into account.

2 Definition of Radial Glial Cells: Changing Views in History

By the end of the nineteenth century, different hypotheses existed on the nature of radial glial cells. Wilhelm His (1889) was the first to show that glial cells are generated in the anlage of the central nervous system and are not of mesenchymal origin, as suggested by Virchow (in: Jacobson, 1991). His and Ramón y Cajal (1911) assumed that glial cells and neurons were derived from different precursor populations. In contrast, Magini (1888a,b) and Kölliker (1896) had shown that the majority of neuroepithelial cells form long radially oriented processes toward the pial surface. The predominance of these cells already suggested that they might be neuronal precursors. Correspondingly, this cell type was termed "radial neuroglia" by Magini (1888a). Nonetheless, for more than 100 years, the generally accepted view of the nature of neuroepithelial cells was based on His's idea that glial cells and neurons were derived from different progenitor cells. Only recently, independently obtained results from different laboratories clearly demonstrated that neurons in the neuroepithelium originate from dividing radial glial cells (Malatesta *et al.*, 2000; Noctor *et al.*, 2001; Miyata *et al.*, 2001). Thus, asymmetric division of a radial glial cell generates a neuron that migrates along the radial glial process of its own precursor cell (Alvarez-Buylla *et al.*, 2001; Nadarajah and Parnavelas, 2002; Noctor *et al.*, 2002). In view of these findings, the hypothesis of the different origins of radial glial cells and neurons in the ventricular zone had to be revised and the term "radial neuroglia," proposed more than 100 years ago by Magini (1888a), came to late honors. Radial glial cells were shown to transform into astrocytes when their role in neuronal migration is accomplished (Schmechel and Rakic, 1979; Mission *et al.*, 1991).

3 Role of Radial Glial Cells in Neuronal Migration: Different Models

How do radial glial cells guide neuronal migration? In the current chapter, different models are discussed, thereby providing a base for the interpretation of Reelin's role in organizing the radial glial scaffold.

The earliest morphological evidence, suggesting that newly generated neurons could migrate along radial glial processes, was provided by Magini (1888b), and numerous later studies support this model (Rakic, 1971, 1972, 1988; Sidman and Rakic, 1973; Noctor *et al.*, 2001).

By contrast, a different migration mode was proposed by Berry and Rogers (1965). After nuclear division within the perikaryon of the neuronal precursor cell in the neuroepithelium, one nucleus migrates within the radial process of the cell toward the cortical plate, whereas the second nucleus remains in the cell body close to the ventricular zone. A similar migration mode had been suggested by Morest (1970). According to Morest's model, the newly generated neuron first loses its contact to the ventricular surface, the cell body then translocates, being "pulled" by its radial process, toward the cortical plate.

Only recently, the existence of both modes of migration, i.e., migration of the neuron along a radial glial process, as well as translocation of the cell body connected to a radial glial process, could be confirmed by using video microscopy to monitor the migration of fluorescently labeled neurons in living slices of embryonic cortex. Thus, Noctor *et al.* (2001) found the generation of neurons by asymmetric division of radial glial cells and migration of the newly generated neuron along the radial glial process of its mother radial glial cell. These observations confirm Magini's interpretation (1888b), and the model suggested by Rakic (1972). By contrast, Miyata *et al.* (2001) could show that a radial glial cell divides asymmetrically and generates a neuron that keeps the existing radial process, whereas the perikaryon detaches from the ventricular zone. These observations are in line with the model suggested by Berry and Rogers (1965) and Morest (1970). Translocation of the cell body was documented for ontogenetically earlier generated neurons, and migration along radial glial processes was shown predominantly for neurons that were generated late in ontogenesis (Nadarajah and Parnavelas, 2002).

4 Reelin-Secreting Cajal-Retzius Cells Organize the Radial Glial Scaffold

Reelin is expressed by Cajal-Retzius (CR) cells in the marginal zone of the developing cortex (Fig. 11.1). Several studies have shown that CR cells play an important role in the organization of radial glial cells. Thus, application of 6-hydroxydopamine (6-OHDA), a toxin that causes CR cells to degenerate (Del Rio *et al.*, 1996, 1997; Supér *et al.*, 1997), induces disorganization of radial glial cells and astrocytes in the dentate gyrus (Hartmann *et al.*, 1992). A dramatic decrease in the number of radial glial apical processes was observed after ablation of CR cells with domoic acid (Supér *et al.*, 2000), indicating that CR cell-specific factors are required to maintain the radial glial phenotype and to anchor radial glial processes to the marginal zone.

In the homozygous reeler mutant, CR cells are present in the marginal zone but do not express Reelin. Absence of Reelin causes, in addition to malpositioning of neurons, an impaired development of radial glial cells (Caviness and Rakic, 1978; Pinto-Lord *et al.*, 1982; Yuasa *et al.*, 1993; Caviness *et al.*, 1988; Hunter-Schaedle, 1997). Alterations include defasciculation of radial glial fibers (Caviness *et al.*, 1988) and malformation of their apical processes (Pinto-Lord *et al.*, 1982).

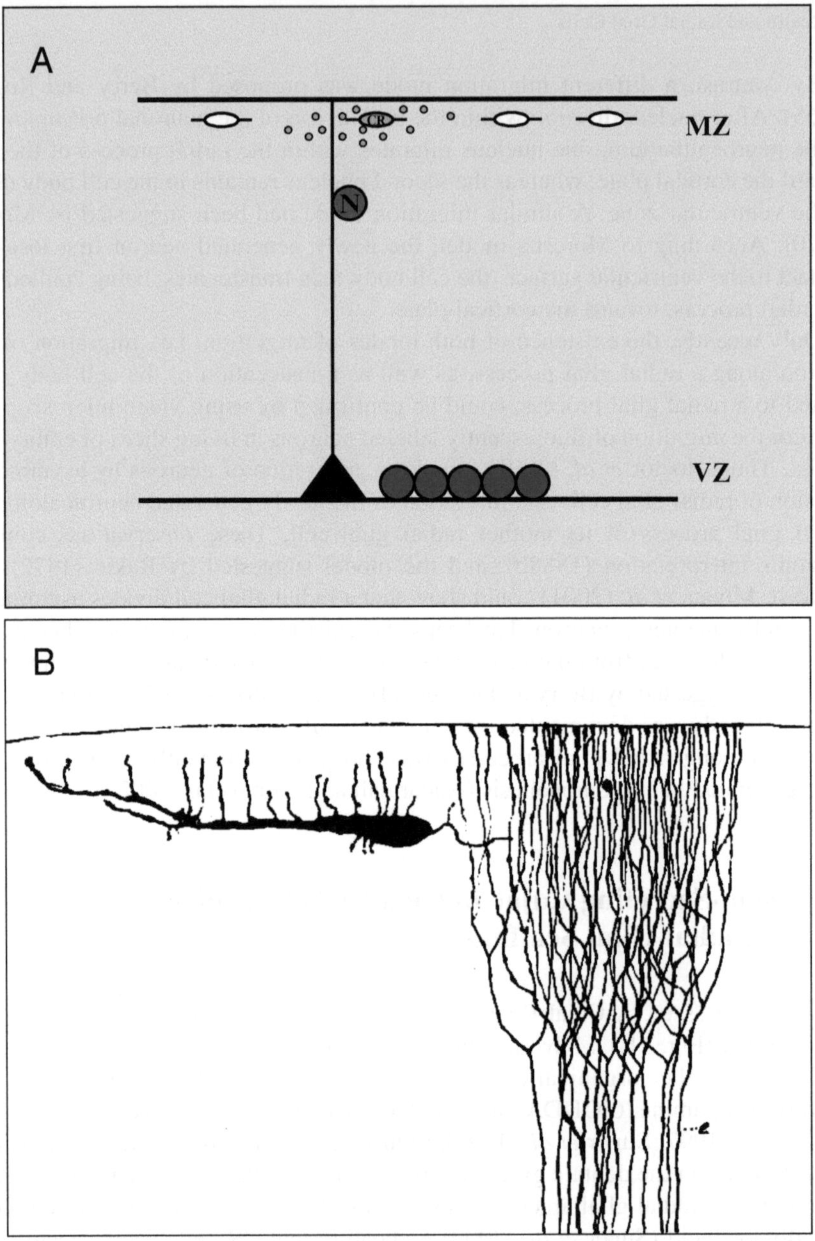

Fig. 11.1 (**A**) Schematic view of the developing cortex. A radial glial cell (black) is shown, extending a radial process from its perikaryon in the ventricular zone (VZ) toward the marginal zone (MZ). Neurons (red) in the ventricular zone are generated by asymmetric division of radial glial cells. A newly generated neuron (N) migrates along the radial glial process toward the marginal zone. Cajal-Retzius cells (CR; green) located in the marginal zone, secrete the glycoprotein Reelin (green dots) into the extracellular matrix. Reelin controls the positioning of radially migrating neurons by acting on both radial glial cells and migrating neurons (*See Color Plates*). (**B**) Detail of a drawing by Gustav Retzius from a silver-stained preparation of developing cortex (modified, Retzius, 1893). The drawing shows a Cajal-Retzius cell (horizontal cell, left) in the marginal zone below the pial surface and radial glial processes (right) that reach the pial surface. Note that radial glial fibers branch when entering the marginal zone

An intrinsic glial cell defect has been proposed to contribute to the radial glial malformations in the reeler mutant (Hunter-Schaedle, 1997).

Migrating neurons detach from radial glial fibers before they reach the most apical portions of these processes. In wild-type mice, apical radial glial processes give rise to multiple terminal branches when they reach the marginal zone of the neocortex (Fig. 11.1B; Retzius, 1893; Pinto-Lord *et al.*, 1982) or the dentate gyrus (Rickmann *et al.*, 1987). In contrast, apical radial glial processes in the reeler mutant do not branch and often do not reach the marginal zone (Pinto-Lord *et al.*, 1982). An oblique or horizontal course of radial glial apical processes has also been described in the cerebellum of reeler mutants (Yuasa *et al.*, 1993). These malformations indicate that Reelin is recognized by apical radial glial processes, induces their branching, and is required to anchor the ramified processes in the marginal zone. In most cortical regions, alterations of the radial glial scaffold in the reeler mutant are subtle when compared to neuronal migration defects.

Where may radial glial cells encounter Reelin? Immunocytochemistry with antibodies against Reelin suggests that Reelin is predominantly localized in the marginal zone (Ogawa *et al.*, 1995). Furthermore, the finding that Reelin molecules assemble to form a large protein complex (Utsunomiya-Tate *et al.*, 2000) suggests that Reelin exerts its action locally, i.e., in the marginal zone rather than as a freely diffusible factor. Thus, a direct action of Reelin on radial glial cells, particularly on the branching of their terminals and on the anchorage of their endfeet to the pial surface, could be a means to indirectly influence migrating neurons. The severe radial glial malformations in the dentate gyrus of mutants with defects in the Reelin signaling pathway suggest that this mode of Reelin action may be particularly important in the dentate gyrus (Förster *et al.*, 2002; Weiss *et al.*, 2003). However, phenotypic morphological studies could not answer the question as to whether Reelin may directly act on radial glial cells.

5 Reelin Signaling via Reelin Receptors and Dab1 Controls Radial Glial Differentiation in the Dentate Gyrus

Due to its simple cytoarchitecture, the dentate gyrus of the hippocampal formation is a particularly suitable model to study the formation of neuronal layers in the cerebral cortex (Stanfield and Cowan, 1979a,b; Cowan *et al.*, 1980, 1981; Förster *et al.*, 2006). Newly generated dentate granule cells were shown to migrate along radial glial fibers from the ventricular zone toward the dentate anlage (Rickmann *et al.*, 1987), reminiscent of radial migration of neurons in the neocortex. Granule cells that are later generated from precursors in the secondary proliferation zone of the hilus were suggested to migrate along a specialized radial glial scaffold to their final positions in the dentate gyrus (Rickmann *et al.*, 1987).

In rodents, the majority of dentate granule cells are born postnatally (Angevine, 1965; Schlessinger *et al.*, 1975; Bayer, 1980), and radial glial cells in the dentate gyrus of rodents are known to persist after birth and to express GFAP (Woodhams *et al.*, 1981; Levitt and Rakic, 1980; Eckenhoff and Rakic, 1984; Rickmann *et al.*, 1987).

In contrast to neocortical cell layers, the dentate granule cell layer is not formed in an inside-out gradient, but later-generated granule cells are apposed beneath the earlier-formed granule cells (Stanfield and Cowan, 1979a). In the reeler mutant, dentate granule cells fail to migrate and accumulate in the hilar region. Granule cells are scattered in the hilar region of VLDLR/ApoER2-deficient mice, similar to the Reeler dentate gyrus (Caviness and Sidman, 1973; Stanfield and Cowan, 1979b; Deller *et al.*, 1999; Drakew *et al.*, 2002), demonstrating that the Reelin signaling cascade is required for the correct positioning of dentate granule cells. Dab1 has been shown to function downstream of Reelin. Dab1-deficient mice also develop a reeler-like phenotype (Sweet *et al.*, 1996; Howell *et al.*, 1997; Sheldon *et al.*, 1997; Ware *et al.*, 1997). Dab1 mRNA is predominantly expressed in neurons; however, GFAP-positive cells in the hippocampus may also express Dab1 mRNA (Förster *et al.*, 2002). Like in reeler mutants, a regular radial glial scaffold does not develop in mutants deficient in Dab1 (Förster *et al.*, 2002; Frotscher *et al.*, 2003; Zhao *et al.*, 2004, 2006), suggesting that Reelin may directly act on radial glial cells via Dab1 interaction. Mutant mice deficient in the Reelin receptors ApoER2 and VLDLR display phenotypically similar defects of the dentate radial glial scaffold. A gradual expression of the radial glial scaffold defects is seen in the dentate gyrus of mice deficient in only ApoER2 or VLDLR, in accordance with the gradual expression of granule cell migration defects in these mutants (Drakew *et al.*, 2002; Gebhardt *et al.*, 2002; Weiss *et al.*, 2003). Thus, the Reelin signaling pathway is required for the formation of the dentate radial glial scaffold. Dentate radial glial defects in mutants lacking Reelin, Dab1, or ApoER2 and VLDLR, are likely to contribute to the granule cell migration defects seen in these mutants.

In the stripe choice assay, originally developed to study axonal repulsion or attraction, hippocampal GFAP-positive glial cells preferred Reelin-coated stripes to control stripes not containing Reelin (Förster *et al.*, 2002; Frotscher *et al.*, 2003). Moreover, long GFAP-positive processes, supposedly radial glial processes, branched significantly more often on the Reelin stripes than on control stripes (Förster *et al.*, 2002).

These findings further support Reelin signaling in radial glial cells via ApoER2, VLDLR, and Dab1. There is also evidence for an additional involvement of β1-integrins, putative Reelin receptors (Dulabon *et al.*, 2000), in radial glial cell differentiation in the hippocampus. In β1-integrin-deficient mice, subtle malformation of the radial glial scaffold in the dentate gyrus and similar migrational defects of the granule cells were described in the dentate gyrus (Förster *et al.*, 2002).

6 Rescue of the Reeler Radial Glial Scaffold in Slice Cultures

How does Reelin act on radial glial cells? By using hippocampal slice cultures as a model, the question could be addressed as to whether Reelin has to be secreted in the marginal zone to exert its role in granule cell positioning. Zhao *et al.* (2004) have added recombinant Reelin to the incubation medium of reeler hippocampal slices not expressing Reelin. In these Reelin-treated cultures, the length of GFAP-expressing glial fibers was significantly increased when compared to untreated reeler slices. However, the elongated GFAP-positive glial fibers did not form the characteristic radially oriented glial scaffold of the dentate gyrus. In line with this, recombinant Reelin in the medium did not rescue the formation of a compact dentate granule cell layer, and granule cells remained scattered over the dentate gyrus (Zhao *et al.*, 2004).

The result of the experiment was different when a reeler hippocampal slice was co-cultured with a wild-type hippocampal slice, such that the Reelin-containing dentate marginal zone was closely apposed to the reeler slice. In this co-culture, Reelin secreted by the wild-type slice induced a growth of radial glial fibers in the reeler dentate gyrus that was oriented toward the source of Reelin, namely, the wild-type marginal zone. Moreover, in parallel with the rescued radial glial scaffold in reeler slices co-cultured with wild-type, a dense granule cell layer had formed in the reeler slice (Zhao *et al.*, 2004; Förster *et al.*, 2006). These findings suggest that Reelin has to be present in a specific topographic position, i.e., in the marginal zone, in order to exert its effect on granule cell positioning.

7 Reelin Acts Directly on Radial Glial Cells in the Developing Neocortex

Radial glial cell defects in the reeler neocortex are less severe than in the hippocampus. This raises the question as to whether a similar Reelin action, as in radial glial cells in the dentate gyrus, may also be attributed to the radial glial cells in the neocortex.

When labeling radial glial cells in reeler with the lipophilic dye DiI from the ventricular surface, the number of ventricular zone cells with long radial processes was shown to be significantly reduced compared to wild-type (Hartfuss *et al.*, 2003). This reduction in process length was accompanied by a reduced expression of brain lipid binding protein (BLBP), a radial glial marker protein. Interestingly, only radial glial cells of the dorsal telencephalon displayed these defects, whereas radial glial cells of the ventral telencephalon, or the ganglionic eminence, were not altered in reeler mutants, pointing to regional differences in radial glial cell populations (Hartfuss *et al.*, 2003).

In vitro, Reelin treatment increased both the BLBP content and process extension of isolated cortical radial glial cells, but not of dissociated radial glial cells isolated from the ventral telencephalon, thereby confirming the *in situ* observations. This Reelin effect was dependent on the presence of Dab1, suggesting that the Reelin signaling cascade is also active in radial glial cells and mediates these effects. Along this line, isolated radial glial cells from embryonic cortex (E14) expressed mRNA coding for the Reelin receptors. In addition, these cells were immunopositive for ApoER2 and VLDLR (Hartfuss *et al.*, 2003). Localization of components of the Reelin signaling cascade in radial glial cells has also been shown by Luque *et al.* (2003), thereby confirming the above interpretations.

BLBP expression in the E12 to E16 reeler cortex was shown to remain constant, whereas an increase in the number of BLBP-expressing cells was observed in the wild-type cortex until almost all precursor cells contained BLBP. Thus, maturation of radial glial cells in the dorsal telencephalon, but not the ventral telencephalon, seems to depend on the presence of Reelin.

By contrast, immunoreactivity for the radial glial cell marker RC2 was similar in wild-type and reeler, suggesting that RC2 expression is not regulated by Reelin (Hartfuss *et al.*, 2003).

In line with these findings, Magdaleno *et al.* (2002) have shown that ventricular zone cells in the neocortex are competent to respond to Reelin, likely by their long radial fibers extending toward the marginal zone, and that the Reelin signaling pathway is already activated in these cells before they start to migrate. Thus, Reelin expression in ventricular zone precursor cells was sufficient to rescue some of the migrational defects seen in the reeler mutant (Magdaleno *et al.*, 2002). These findings support the idea that Reelin may directly act on radial glial cells.

As radial glial cells can transform into neurons that keep the radial fiber (Miyata *et al.*, 2001), Dab1-mediated Reelin signaling in neurons may be required to maintain the radial orientation of the radial fiber toward the marginal zone. In fact, phosphorylation of Dab1 by Reelin binding to VLDLR and ApoER2 modulates cytoskeletal proteins (Hiesberger *et al.*, 1999). Similar mechanisms may operate while these cells are still neuronal precursors, i.e., radial glial cells.

References

Alvarez-Buylla, A., Garcia-Verdugo, J. M., and Tramóntin, A. D. (2001). A unified hypothesis on the lineage of neural stem cells. *Nature Rev. Neurosci.* 2:287–293.

Angevine, J. B. (1965). Time of origin in the hippocampal region. An autoradiographic study in the mouse. *Exp. Neurol. Suppl.* 2:1–70.

Bayer, S. A. (1980). Development of the hippocampal region in the rat. I. Neurogenesis examined with (^{3}H) thymidine autoradiography. *J. Comp. Neurol.* 190:87–114.

Berry, M., and Rogers, A. W. (1965). The migration of neuroblasts in the developing cerebral cortex. *J. Anat.* 99:691–709.

Caviness, V. S., and Sidman, R. L. (1973). Time of origin of corresponding cell classes in the cerebral cortex of normal and reeler mutant mice: an autoradiographic analysis. *J. Comp. Neurol.* 148:141–151.

Caviness, V. S., Jr., and Rakic, P. (1978). Mechanisms of cortical development: a view from mutations in mice. *Annu. Rev. Neurosci.* 1:297–326.

Caviness, V.S., Crandall, J. E., and Edwards, M. A. (1988). The reeler malformation. Implications for neocortical histogenesis. *Cereb. Cortex* 7:59–89.

Cowan, W. M., Stanfield, B., and Kishi, K. (1980). The development of the dentate gyrus. *Curr. Top. Dev. Biol.* 15 Part 1:103–157.

Cowan, W. M., Stanfield, B. B., and Amaral, D. G. (1981). Further observations on the development of the dentate gyrus. In Cowan, W.M. (ed.), *Studies in Developmental Neurobiology.* Oxford University Press, New York, pp. 395–435.

Deller, T., Drakew, A., Heimrich, B., Förster, E., Tielsch, A., and Frotscher, M. (1999). The hippocampus of the reeler mutant mouse: fiber segregation in area CA1 depends on the position of the postsynaptic target cells. *Exp. Neurol.* 156:254–267.

Del Rio, J. A., Heimrich, B., Supér, H., Borrell, V., Frotscher, M., and Soriano, E. (1996). Differential survival of Cajal-Retzius cells in organotypic cultures of hippocampus and neocortex. *J. Neurosci.* 16:6896–6907.

Del Rio, J. A., Heimrich, B., Borrell, V., Förster, E., Drakew, A., Alcántara, S., Nakajima, K., Miyata, T., Ogawa, M., Mikoshiba, K., Derer, P., Frotscher, M., and Soriano, E. (1997). A role for Cajal-Retzius cells and reelin in the development of hippocampal connections. *Nature* 385:70–74.

Drakew, A., Deller, T., Heimrich, B., Gebhardt, C., Del Turco, D., Tielsch, A., Förster, E., Herz, J., and Frotscher, M. (2002). Dentate granule cells in reeler mutants and VLDLR and ApoER2 knockout mice. *Exp. Neurol.* 176:12–24.

Dulabon, L., Olson, E. C., Taglienti, M. G., Eisenhuth, S., McGrath, B., Walsh, C. A., Kreidberg, J. A., and Anton, E. S. (2000). Reelin binds $\alpha 3\beta 1$ integrin and inhibits neuronal migration. *Neuron* 27:33–44.

Eckenhoff, M. F., and Rakic, P. (1984). Radial organization of the hippocampal dentate gyrus: A Golgi, ultrastructural, and immunocytochemical analysis in the developing rhesus monkey. *J. Comp. Neurol.* 223:1–21.

Förster, E., Tielsch, A., Saum, B., Weiss, K. H., Johanssen, C., Graus-Porta, D., Müller, U., and Frotscher, M. (2002). Reelin, disabled 1, and $\beta 1$-integrins are required for the formation of the radial glial scaffold in the hippocampus. *Proc. Natl. Acad. Sci. USA* 99:13178–13183.

Förster, E., Zhao, S., and Frotscher, M. (2006). Laminating the hippocampus. *Nature Rev. Neurosci.* 7:259–267.

Frotscher, M., Haas, C., and Förster, E. (2003). Reelin controls granule cell migration in the dentate gyrus by acting on the radial glial scaffold. *Cereb. Cortex* 13:634–640.

Gebhardt, C., del Turco, D., Drakew, A., Tielsch, A., Herz, J., Frotscher, M., and Deller, T. (2002). Abnormal positioning of granule cells alters afferent fiber distribution in the mouse fascia dentata: morphologic evidence from reeler, apolipoprotein E receptor 2-, and very low density lipoprotein receptor knockout mice. *J. Comp. Neurol.* 445:278–292.

Hartfuss, E., Förster, E., Bock, H. H., Hack, M. A., Leprince, P., Luque, J. M., Herz, J., Frotscher, M., and Götz, M. (2003). Reelin signaling directly affects radial glial morphology and biochemical maturation. *Development* 130:4597–4609.

Hartmann, D., Sievers, J., Pehlemann, F. W., and Berry, M. (1992). Destruction of meningeal cells over the medial cerebral hemisphere of newborn hamsters prevents the formation of the infrapyramidal blade of the dentate gyrus. *J. Comp. Neurol.* 320:33–61.

Hiesberger, T., Trommsdorff, M., Howell, B. W., Goffinet, A., Mumby, M. C., Cooper, J. A., and Herz, J. (1999). Direct binding of reelin to VLDL receptor and ApoE receptor 2 induces tyrosine phosphorylation of disabled-1 and modulates tau phosphorylation. *Neuron* 24:481–489.

His, W. (1889). Die Neuroblasten und deren Entstehung im embryonalen Marke. *Abh. Math. Phys. Cl. Kgl. Sächs. Ges. Wiss.* 15:313–372.

Howell, B. W., Hawkes, R., Soriano, P., and Cooper, J. A. (1997). Neuronal positioning in the developing brain is regulated by mouse disabled-1. *Nature* 389:733–737.

Hunter-Schaedle, K. E. (1997). Radial glial cell development and transformation are disturbed in the reeler forebrain. *J. Neurobiol.* 33:459–472.

Jacobson, M. (1991). *Developmental Neurobiology*, 2nd ed. Plenum Press, New York.

Kölliker, A. (1896). *Handbuch der Gewebelehre des Menschen*, 6th ed. W. Engelmann, Leipzig.

Levitt, P., and Rakic, P. (1980). Immunoperoxidase localization of glial fibrillary acidic protein in radial glial cells and astrocytes of the developing rhesus monkey. *J. Comp. Neurol.* 193:815–840.

Luque, J. M., Morante-Oria, J., and Fairen, A. (2003). Localization of ApoER2, VLDLR and Dab1 in radial glia: groundwork for a new model of reelin action during cortical development. *Dev. Brain Res.* 140:195–203.

Magdaleno, S., Keshvara, L., and Curran, T. (2002). Rescue of ataxia and preplate splitting by ectopic expression of reelin in reeler mice. *Neuron* 33:573–586.

Magini, J. (1888a). Sur la névroglie et les cellules nerveuses cerebrales chez les fœtus. *Arch. Ital. Biol.* 9:59–60.

Magini, J. (1888b). Nouvelles recherches histologiques sur le cerveau du foetus. *Arch. Ital. Biol.* 10:384–387.
Malatesta, P., Hartfuss, E., and Götz, M. (2000). Isolation of radial glial cells by fluorescent-activated cell sorting reveals a neuronal lineage. *Development* 127:5253–5263.
Mission, J. P., Takahashi, T., and Caviness, V. S. (1991). Ontogeny of radial and other astroglial cells in the murine cerebral cortex. *Glia* 4:138–148.
Miyata, T., Kawaguchi, A., Okano, H., and Ogawa, M. (2001). Asymmetric inheritance of radial glial fibers by cortical neurons. *Neuron* 31:727–741.
Morest, D. K. (1970). A study of neurogenesis in the forebrain of opossum pouch young. *Z. Anat. Entwicklungsgesch.* 130:265–305.
Nadarajah, B., and Parnavelas, J. G. (2002). Modes of neuronal migration in the developing cerebral cortex. *Nature Rev. Neurosci.* 3:423–432.
Noctor, S. C., Flint, A. C., Weissman, T. A., Dammerman, R. S., and Kriegstein, A. R. (2001). Neurons derived from radial units in neocortex. *Nature* 409:714–720.
Noctor, S. C., Flint, A. C., Weissman, T. A., Wong, W. S., Clinton, B. K., and Kriegstein, A. R. (2002). Dividing precursor cells of the embryonic cortical ventricular zone have morphological and molecular characteristics of radial glia. *J. Neurosci.* 22:3161–3173.
Ogawa, M., Miyata, T., Nakajima, K., Yagyu, K., Seike, M., Ikenaka, K., Yamamoto, H., and Mikoshiba, K. (1995). The reeler gene-associated antigen on Cajal-Retzius neurons is a crucial molecule for laminar organization of cortical neurons. *Neuron* 14:899–912.
Pinto-Lord, M. C., Evrard, P., and Caviness, V. S. (1982). Obstructed neuronal migration along radial glia fibers in the neocortex of the reeler mouse: a Golgi-EM analysis. *Dev. Brain Res.* 4:379–393.
Rakic, P. (1971). Neuron–glia relationship during granule cell migration in the developing cerebellar cortex. A Golgi and electron microscopic study in Macacus rhesus. *J. Comp. Neurol.* 141:283–312.
Rakic, P. (1972). Mode of cell migration to the superficial layers of fetal monkey neocortex. *J. Comp. Neurol.* 145:61–84.
Rakic, P. (1988). Specification of cerebral cortical areas. *Science* 241:170–176.
Ramón y Cajal, S. (1911). *Histologie du Système Nerveux de l'Homme et des Vertébrés*, Vol. 2. Maloine, Paris.
Retzius, G. (1893). Die Cajalschen Zellen der Grosshirnrinde beim Menschen und bei Säugetieren. *Biol. Unters.* 5:1–9.
Rickmann, M., Amaral, D. G., and Cowan, M. (1987). Organization of radial glia cells during the development of the rat dentate gyrus. *J. Comp. Neurol.* 264:449–479.
Schlessinger, A. R., Cowan, W. M., and Gottlieb, D. I. (1975). An autoradiographic study of the time of origin and the pattern of granule cell migration in the dentate gyrus of the rat. *J. Comp. Neurol.* 159:149–176.
Schmechel, S. E., and Rakic, P. (1979). A Golgi study of radial glial cells in developing monkey telencephalon: morphogenesis and transformation into astrocytes. *Anat. Embryol.* 156:115–152.
Sheldon, M., Rice, D. S., d'Arcangelo, G., Yoneshima, H., Nakajima, M., Mikoshiba, K., Howell, B. W., Cooper, J. A., Goldowitz, D., and Curran, T. (1997). Scrambler and yotari disrupt the disabled gene and produce a reeler-like phenotype in mice. *Nature* 389:730–733.
Sidman, R. L., and Rakic, P. (1973). Neuronal migration with special reference to developing human brain. *Brain Res.* 62:1–35.
Stanfield, B. B., and Cowan, W. M. (1979a). The development of the hippocampus and dentate gyrus in normal and reeler mice. *J. Comp. Neurol.* 185:423–460.
Stanfield, B. B., and Cowan, W. M. (1979b). The morphology of the hippocampus and dentate gyrus in normal and reeler mice. *J. Comp. Neurol.* 185:393–422.
Supér, H., Martinez, A., and Soriano, E. (1997). Degeneration of Cajal-Retzius cells in the developing cortex of the mouse after ablation of meningeal cells by 6-hydroxydopamine. *Dev. Brain Res.* 98:15–20.
Supér, H., Del Rio, J. A., Martinez, A., Perez-Sust, P., and Soriano, E. (2000). Disruption of neuronal migration and radial glia in the developing cerebral cortex following ablation of Cajal-Retzius cells. *Cereb. Cortex* 10:602–613.

Sweet, H. O., Bronson, R. T., Johnson, K. R., Cook, S. A., and Davisson, M. T. (1996). Scrambler, a new neurological mutation of the mouse with abnormalities of neuronal migration. *Mamm. Genome* 7:798–802.

Utsunomiya-Tate, N., Kubo, K., Tate, S., Kainosho, M., Katayama, E., Nakajima, K., and Mikoshiba, K. (2000). Reelin molecules assemble together to form a large protein complex, which is inhibited by the function-blocking CR-50 antibody. *Proc. Natl. Acad. Sci. USA* 97:9729–9734.

Ware, M. L., Fox, J. W., Gonzalez, J. L., Davis, N. M., Lambert de Rouvroit, C., Russo, C. J., Chua, S. C., Jr., Goffinet, A. M., and Walsh, C. A. (1997). Aberrant splicing of a mouse disabled homolog, mdab1, in the scrambler mouse. *Neuron* 19:239–249.

Weiss, K.-H., Johanssen, C., Tielsch, A., Saum, B., Frotscher, M., and Förster, E. (2003). Malformation of the radial glial scaffold in the dentate gyrus of reeler mice, scrambler mice and ApoER2/VLDLR deficient mice. *J. Comp. Neurol.* 460:56–65.

Woodhams, E. B., Basco, E., Hajos, F., Csillag, A., and Balazs, R. (1981). Radial glia in the developing mouse cerebral cortex and hippocampus. *Anat. Embryol.* 163:331–343.

Yuasa, S., Kitoh, J., Oda, S., and Kawamura, K. (1993). Obstructed migration of Purkinje cells in the developing cerebellum of the reeler mutant mouse. *Anat. Embryol.* 188:317–329.

Zhao, S., Chai, X., Förster, E., and Frotscher, M. (2004). Reelin is a positional signal for dentate granule cells. *Development* 131:5117–5125.

Zhao, S., Chai, X., Bock, H., Brunne, B., Förster, E., and Frotscher, M. (2006). Rescue of the reeler phenotype in the dentate gyrus by wildtype coculture is mediated by lipoprotein receptors for reelin and disabled1. *J. Comp. Neurol.* 495:1–9.

Chapter 12
Reelin and Cognition

Shenfeng Qiu and Edwin John Weeber

Contents

1 Introduction

The cognitive function of the human brain refers to the mental manipulation of all the information acquired by the sensory system. A more physiological definition of cognition by Ulric Neisser states, "...the term cognition refers to all processes by which the sensory input is transformed, reduced, elaborated, stored, recovered, and used." It is our unique cognitive ability acquired from our distinct experience that ultimately makes us who we are; as Rene Descartes said, "I think, therefore I am." The authors have no intention or expertise to embark on a philosophical discussion of cognition in this chapter; we rather safely rely on our assertion that normal cognition requires proper function of the central nervous system, which constitutes hundreds of millions of neurons and supporting cells and is correctly "wired" during development and, equally as important, "rewired" in response to our postnatal experiences.

S. Qiu
754 Robinson Research Building, 23rd Avenue South and Pierce Avenue, Department of Molecular Physiology and Biophysics, Vanderbilt University Medical Center, Nashville, TN37232-0615

E. J. Weeber, Department of Molecular Pharmacology and Physiology, University of South Florida, 12901 Bruce B. Downs Blvd, MDC Room 3159, Tampa, FL 33612
email: eweeber@health.usf.edu

S. H. Fatemi (ed.), *Reelin Glycoprotein: Structure, Biology and Roles in Health and Disease.*

The correct construction of neuronal circuits of the embryonic brain provides a structural basis for the CNS to carry out its normal function throughout life. This process requires a myriad of proteins encoded by numerous genes that are properly expressed spatiotemporally during development. For example, the expression of the extracellular matrix protein Reelin by specialized Cajal-Retzius cells located in the marginal zone is required for the ventricular migrating neurons to form a highly laminated structure in the neocortex, hippocampus, and cerebellum, which is indispensable for the formation of a functional network and activity-driven synaptogenesis (D'Arcangelo *et al.*, 1995; Aguilo *et al.*, 1999; Soda *et al.*, 2003). Other proteins, whether they are located in the same [e.g., apolipoprotein E receptor 2 (ApoER2) and very-low-density lipoprotein receptor (VLDLR), disabled-1 (Dab1)] or distinct pathways (e.g., cyclin-dependent kinase 5 and its activators p35 and p39), are also required for the correct lamination (Ohshima *et al.*, 1996; Howell *et al.*, 1997; Trommsdorff *et al.*, 1999; Ko *et al.*, 2001). This typical gross laminar arrangement of neurons is necessary, albeit not sufficient, to carry out the normal flow of information, which is achieved by numerous chemical and electrical synapses and the electrical pulses generated there. It is perceivable that disruption of cortical, hippocampal, and cerebellar lamination, as observed in the Reeler mouse, would result in profound cognitive deficits that are similar to those observed in human lissencephaly patients bearing a *RELN* gene mutation (Hong *et al.*, 2000). However, the severe motor deficits in these mutant mice preclude the performance of meaningful cognitive analyses on these animals.

Although much of our knowledge on the actions of Reelin comes from investigation of embryonic CNS development, recent studies have indicated that Reelin signaling also plays an important role in synaptic function in the postnatal brain (Weeber *et al.*, 2002; Beffert *et al.*, 2005; Ramos-Moreno *et al.*, 2006). Reelin-expressing neurons are widely distributed in the adult brain (Pesold *et al.*, 1998; Martinez-Cerdeno *et al.*, 2002; Abraham and Meyer, 2003; Roberts *et al.*, 2005; Ramos-Moreno *et al.*, 2006), at a period long after the decrease in number of Cajal-Retzius cells (Marin-Padilla, 1998; Meyer *et al.*, 1998; Sarnat and Flores-Sarnat, 2002). The functional significance of the persistent expression of Reelin after the completion of cell migration remains enigmatic in most brain regions. However, in the hippocampus, Reelin and its lipoprotein receptors are required for proper function of synaptic transmission and plasticity, and this signaling system profoundly modifies mammalian learning and memory behavior (Weeber *et al.*, 2002; Beffert *et al.*, 2005; D'Arcangelo, 2005). This chapter will focus on existing experimental evidence supporting the requirement of Reelin for normal synaptic function in the hippocampus as well as mammalian cognitive ability.

2 Reelin in the Adult Hippocampus

2.1 Expression in the Hippocampus

The hippocampus is a structure within the brain's limbic system and is one of the most comprehensively studied regions in the CNS due to the fundamental role it plays in many forms of learning and memory (Scoville and Milner, 1957), and its distinctive laminar structure that exhibits the readily identifiable "trisynaptic" neuronal circuitry (Anderson *et al.*, 1971). The developmental lamination of hippocampal principal neurons and the correct formation of innervating fibers require Reelin (Del Rio *et al.*, 1997; Forster *et al.*, 2002). In the postnatal mouse hippocampus, Reelin is expressed primarily by a group of interneurons (Alcantara *et al.*, 1998; Pesold *et al.*, 1998). Immunohistochemical labeling using the monoclonal G10 antibody demonstrates that Reelin-expressing neurons are located primarily in the hilar region of dentate gyrus and the stratum lacunosum-moleculare layer of the hippocampus proper. Reelin-positive cells can also be found in stratum oriens and stratum radiatum of the CA1 and CA3 regions (Fig. 12.1). Reelin is also required for dentate granule cells to form a compact layer during embryonic development (Drakew *et al.*, 2002; Frotscher *et al.*, 2003; Zhao *et al.*, 2004). In the postnatal dentate gyrus, Reelin seems to promote radial glial fiber differentiation and may support the neurogenesis of granule cells in the hilus region and their migration and integration into the granule cell layer (Forster *et al.*, 2002; Frotscher *et al.*, 2003). In agreement, a recent study has shown that Reeler mice exhibit impaired neurogenesis in the dentate gyrus (Won *et al.*, 2006). However, it is currently not known

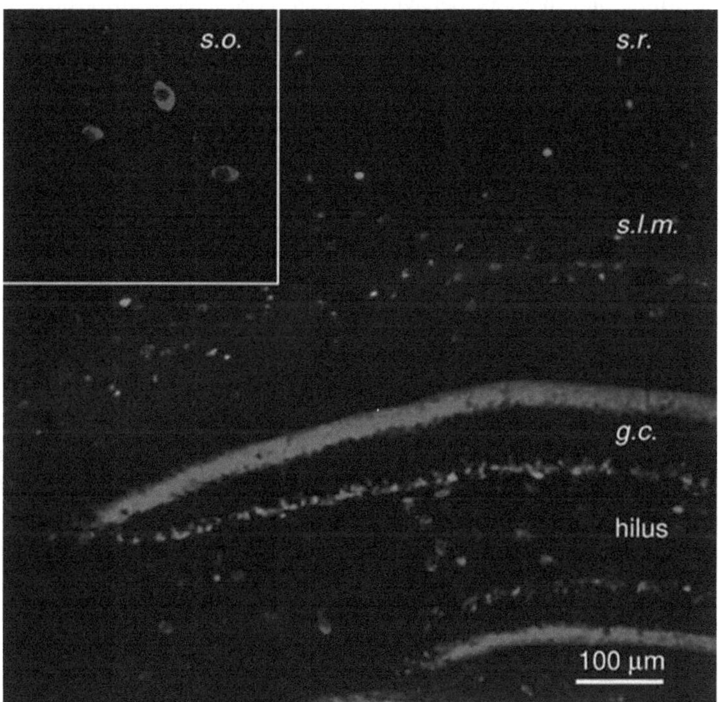

Fig. 12.1 Reelin-expressing cells in adult mouse hippocampus. Double immunofluorescent staining of a hippocampus cryosection obtained from a 6-week-old wild-type mouse. Note that Reelin-containing cells (red) were primarily distributed in the dentate hilar region (hilus) and stratum lacunosum-moleculare (*s.l.m.*) but also can be found in stratum oriens (*s.o.*) and stratum radiatum (*s.r.*) of CA1 region. Immunostaining of the calcium-binding protein calretinin (green) was used to visualize the dentate gyrus layers (*See Color Plates*)

whether Reelin signaling contributes to the synaptic function in the dentate gyrus, which receives heavy innervation from the perforant fibers and sends its mossy fibers to synapse on CA3 pyramidal cells.

2.2 Electrophysiology

The initial effort made to identify a putative role for Reelin in plasticity and memory was rather straightforward by using electrophysiological and behavioral studies in mice to identify and measure the effects of different components in the Reelin signaling pathway. The electrophysiological studies were carried out on acute hippocampal slices prepared from 4- to 6-week-old adult mice. Synaptic transmission and plasticity were tested in the Schaffer collateral-CA1 synapses, primarily because this area of the hippocampus is the most well characterized in the murine brain and due to the remarkable property of activity-dependent synaptic plasticity, such as long-term potentiation (LTP) and long-term depression (LTD) (Bliss and Lomo, 1973; Nicoll and Malenka, 1995). Both LTP and LTD are thought to underlie the cellular mechanisms of learning and memory behavior and are correlated with positive results as measured by cognitive tests in animals (Giese *et al.*, 1998; Lee *et al.*, 2003).

In electrophysiological studies using hippocampal slices, a single electric stimulus evokes action potentials on the presynaptic fibers and causes release of glutamate neurotransmitter at axon terminals. The postsynaptic glutamate receptors, which are ligand-gated ion channels, open in response to glutamate to allow ionic influx and efflux (Dingledine *et al.*, 1999). One major type of ionotropic glutamate receptors, named α-amino-5-hydroxy-3-methyl-4-isoxazole propionic acid receptors (AMPARs), mediates fast synaptic transmission. Another major type of ionotropic glutamate receptors, called the *N*-methyl-D-aspartic acid receptors (NMDARs), serve as co-incidence detectors and play a major role in the induction of synaptic plasticity. The opening of NMDAR-type ion channels requires both glutamate binding and membrane depolarization as a result of AMPAR activation. These receptors also have high permeability for calcium, whose intracellular dynamics affect the polarity of synaptic plasticity. Some NMDAR subunits, such as NR1, NR2A, and NR2B, are also subjected to modulatory phosphorylation at serine/threonine or tyrosine residues. These sites of phosphorylation not only affect channel kinetics but also control their trafficking to synaptic sites (Wang and Salter, 1994; Lau and Huganir, 1995; Yu *et al.*, 1997; Chung *et al.*, 2004). The most well characterized form of LTP in area CA1 induced by high-frequency stimulation relies on NMDA receptor activation and the subsequent modification of AMPA receptors. AMPAR modification can be achieved either by increased subunit phosphorylation (Barria *et al.*, 1997; Banke *et al.*, 2000; Lee *et al.*, 2003) or by increased subunit synthesis and trafficking to synaptic sites (Shi *et al.*, 1999; Hayashi *et al.*, 2000), leading to long-lasting changes in synaptic strength. This artificially induced change in synaptic strength is thought to incorporate similar biochemical and structural changes as those that underlie the mechanisms involved in memory formation *in vivo*.

In order to investigate the potential effect of Reelin on synaptic transmission and plasticity, slices were perfused with recombinant Reelin isolated from HEK293 cells stably transfected with full-length Reelin cDNA (D'Arcangelo *et al.*, 1997). Synaptically evoked field excitatory postsynaptic potentials (fEPSP) were recorded and the slope of fEPSP was used to quantify the strength of synaptic transmission. The input–output curve was then constructed by using fEPSP slope as a function of presynaptic fiber volley. It was observed that a short perfusion of Reelin medium did not change the baseline response, nor did it change the input–output curve (proportional increase in the fEPSP with increased stimulus intensity). However, it dramatically elevated the magnitude of LTP induced by two trains of 1-sec high-frequency stimulation (Fig. 12.2E). These results revealed that Reelin perfusion "preconditions" the CA1 synapses to favor LTP induction and maintenance. The mechanisms were largely speculative at the time of observation. It was proposed that Reelin activation of its lipoprotein receptors, ApoER2 and VLDLR, leads to Dab1 and Src kinase phosphorylation, and subsequently Src kinases phosphorylate NMDA receptors on tyrosine residues (Weeber *et al.*, 2002). Therefore, the increased tyrosine phosphorylation of NMDA subunits leads to increased receptor conductance and enhanced calcium influx during LTP induction and thus leads to elevated LTP magnitude or lowered LTP threshold.

2.3 Reelin Signaling and Glutamate Receptors

This hypothesis was supported by two recent functional studies at cellular levels (Beffert *et al.*, 2005; Chen *et al.*, 2005). The enhanced NMDAR-mediated response in CA1 neurons following Reelin perfusion was confirmed in subsequent studies (Beffert *et al.*, 2005). Isolated NMDAR-mediated whole cell currents were recorded in CA1 pyramidal neurons in the presence of bicuculline to block GABA responses and 6-cyano-7-nitroquinoxaline-2,3-dione (CNQX) to block AMPAR responses, and the NMDAR currents were evoked at +30 mV to remove the voltage-dependent NMDAR magnesium block. It was observed that perfusion of Reelin significantly increased NMDAR-mediated charge transfer, indicating that there is a functional connection between Reelin signaling and enhanced glutamatergic response in CA1. Moreover, in NR2A-immunoprecipitated CA1 tissues, Reelin perfusion significantly increased the level of phosphotyrosine, indicating that there was enhanced NMDAR subunit tyrosine phosphorylation. Therefore, increased tyrosine phosphorylation and unitary conductance at least partially account for the enhanced NMDAR-mediated responses, although it is likely that phosphorylation-dependent trafficking of NMDAR subunits (Hayashi and Huganir, 2004) may also play a role. In a complementary study, Chen *et al.* (2005) have shown that increased tyrosine phosphorylation of another NMDAR subunit NR2B is increased. Moreover, in the presence of Reelin, glutamate-induced calcium influx through NMDA receptors is markedly enhanced, an effect that is dependent on lipoprotein receptors and Src family tyrosine kinases. These studies provide

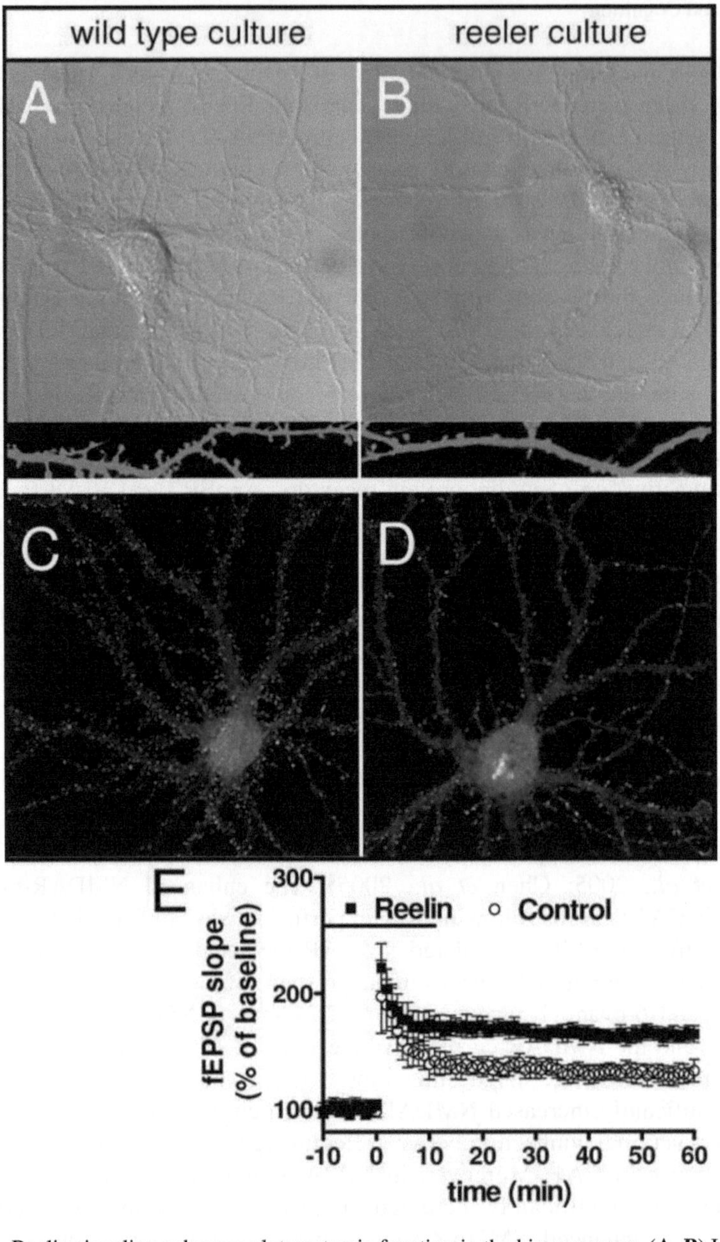

Fig. 12.2 Reelin signaling enhances glutamatergic function in the hippocampus. (**A**, **B**) In cultured embryonic mouse hippocampal neurons derived from homozygous Reeler embryos, stunted neurite growth and fewer neurite ramifications are seen; in addition, when neurons were filled with fluorophores to reveal dendritic spines, it was observed that neurons from wild-type cultures show significantly more spines in their primary dendrites. (**C**, **D**) Neurons from both wild-type and Reeler embryos are cultured for 2 weeks and then immunostained with NMDA receptor subunit NR1 and AMPA receptor subunit (GluR1) antibodies. A larger number of puncta that are positive for both NR1 and GluR1 were observed in wild-type cultures compared with Reeler cultures. (**E**) Long-term potentiation experiments using acute hippocampal slices prepared from 6-week-old mice. A 20-min perfusion of Reelin dramatically elevated the magnitude of tetanus-induced LTP (*See Color Plates*)

functional evidence that Reelin signaling in the adult hippocampus is connected with enhanced glutamatergic function.

Recent studies reveal that Reelin signaling not only enhances NMDA receptor function but also increases AMPA receptor-mediated synaptic responses through distinct mechanisms. Specifically, perfusion of purified recombinant Reelin for an extended time increased whole cell synaptic responses when CA1 neurons were voltage-clamped at −70 mV, indicating that increased AMPA receptor response was not secondary to enhanced NMDAR function. Moreover, this increase was not associated with changes of subunit phosphorylation. Further nonstationary fluctuation analysis of AMPA receptor-mediated miniature excitatory postsynaptic currents (mEPSCs) revealed that this increase was due to increased receptor numbers at synaptic sites. Therefore, elevated Reelin levels in the hippocampus would result in a profound enhancement of glutamatergic function at CA1 synapses. During *in vivo* conditions, this elevated level of Reelin may be achieved through increased activity of Reelin-expressing interneurons. However, obtaining a selective activation of these interneurons and dissecting the effects of endogenous Reelin on synaptic activity present a major technical challenge.

3 Reelin and Postnatal Hippocampus Development

Dendritic spines are small protrusions that cover the surface of dendrites and bear the postsynaptic structures that form excitatory synapses. Abnormally shaped or reduced numbers of dendritic spines are found in a number of cognitive diseases, such as fragile X syndrome, Williams syndrome, Rett syndrome, Down's syndrome, Angelman syndrome, and autism (Irwin *et al.*, 2000; Kaufmann and Moser, 2000; Barnes and Milgram, 2002; Weeber *et al.*, 2003). In cultured hippocampal neurons, Reelin signaling is required for normal development of dendritic structures. In the absence of Reelin or the intracellular adapter protein Dab1, neurons show stunted dendritic growth and fewer number of branches, similar to neurons of double knockout mice lacking both ApoER2 and VLDLR (Niu *et al.*, 2004). This can also be seen using a patch electrode filled with fluorescent dyes to dialyze cultured neurons and analyze dendritic spine density. This technique reveals that hippocampal neurons cultured from homozygous Reeler embryos had significantly fewer dendritic spines, a phenotype that can be rescued by adding exogenous recombinant Reelin to the culture (Fig. 12.2A,B; Qiu *et al.*, unpublished observations). The reduced number of dendritic spines indicates that a normal constitutive level of Reelin-lipoprotein receptor-mediated signaling is required for development of dendritic structures, which are crucial for intensive information processing by the neurons. This notion is in agreement with studies showing that heterozygote Reeler mice exhibit reduced dendritic spine densities and impaired performance in certain learning and memory behaviors (Tueting *et al.*, 1999; Liu *et al.*, 2001; Pappas *et al.*, 2001; Qiu *et al.*, 2006), which will be discussed later in this chapter.

Two major forms of NR2 subunits in adult hippocampus, NR2A and NR2B, are developmentally regulated (Monyer *et al.*, 1994; Sheng *et al.*, 1994; Chavis and Westbrook, 2001), subjected to activity- and behavior-induced changes (Carmignoto and Vicini, 1992; Quinlan *et al.*, 2004; Barria and Malinow, 2005), and also help to determine the polarity of synaptic plasticity under certain circumstances (Liu *et al.*, 2004; Massey *et al.*, 2004). Recently, it has been shown that Reelin signaling is required for maturation of glutamate receptor function during postnatal development. For example, Reelin is required for a normal developmental switch from NR2B to NR2A of somatic NMDA receptor subunits in cultured hippocampal neurons (Sinagra *et al.*, 2005), a process that is dependent on ApoER2, VLDLR, and the intracellular Dab1 and Src. NMDAR function is critically dependent on the composition of its subunits. Because both NR2A and NR2B can be tyrosine phosphorylated in response to synaptic activity (Lau and Huganir, 1995; Yu *et al.*, 1997; Lu *et al.*, 1998; Salter and Kalia, 2004) and Reelin signaling (Beffert *et al.*, 2005; Chen *et al.*, 2005), changes of NMDAR activity through developmental alteration of subunit composition and/or tyrosine phosphorylation by Reelin could have a dramatic impact on synaptic strength and plasticity. The study by Sinagra *et al.* (2005) investigated the maturation of somatic NMDA receptors in cultured neurons and found that Reelin signaling is required for normal developmental switch of somatic NMDAR subunits from NR2B to NR2A in cultured hippocampal neurons. However, a caveat should be noted that the contribution of Reelin signaling in neuronal maturation in the hippocampus during *in vivo* conditions may differ, due to the fact that *in vivo* maturation and subunit switch of NMDA receptors is critically dependent on neuronal activity that cannot be mimicked under culture conditions (Monyer *et al.*, 1994; Quinlan *et al.*, 1999; Lu and Constantine-Paton, 2004; Nakayama *et al.*, 2005). In contrast to using neuronal culture, examination of the effect of Reelin on glutamatergic maturation was performed using *in vitro* cultured hippocampal slices obtained from 6- to 7-day-old mice. When slice cultures are prepared from young rodents, neural circuitries of the hippocampal formation are known to mature and to maintain a surprising three-dimensional, organotypic organization for many weeks *in vitro* (Zimmer and Gahwiler, 1984; Gahwiler *et al.*, 1997). Using this preparation, it was found that Reelin signaling also facilitates a functional switch of synaptic NMDA receptor subunits from NR2B to NR2A. Additionally, Reelin promotes AMPA receptor surface expression and reduces the number of silent synapses in cultured hippocampal slices (Qiu and Weeber, unpublished observations). Therefore, Reelin-mediated signaling may play a previously unappreciated role to facilitate maturation of glutamate receptors at early postnatal stages.

4 Reelin Receptors and Synaptic Function

4.1 ApoER2 and VLDLR

Genetic ablation of both ApoER2 and VLDLR recapitulates the Reeler phenotype and precludes the ability to perform any meaningful cognitive tests on these animals (Trommsdorff *et al.*, 1999). In comparison, knockout of either ApoER2 or

VLDLR does not produce the severe neuronal migration defects of the Reeler mouse, allowing behavioral tests that rely on normal locomotor ability for the tests (Trommsdorff *et al.*, 1999; Weeber *et al.*, 2002). It was found that mice lacking either ApoER2 or VLDLR showed hippocampus-dependent contextual fear conditioning deficits. In addition, these mice had impaired LTP and diminished responses to the Reelin enhancement of LTP in the hippocampal slices. Therefore, it seemed that both ApoER2 and VLDLR are required to carry out the normal synaptic function in the hippocampus. It must be noted that both ApoER2 and VLDLR knockout mice exhibited some degree of anatomical abnormalities, such as a distinct split of area CA1 pyramidal neurons and a less compact granule cell layer in dentate gyrus, which suggests that the contextual fear conditioning deficits in these animals may still be attributed to defects in developmental lamination and neuronal network connections. However, the observation that receptor-associated protein (RAP), a functional antagonist to LDLR family member, blocked hippocampal LTP favors the hypothesis that ApoER2- and VLDLR-mediated signaling *per se* is required for LTP induction and normal cognitive function (Zhuo *et al.*, 2000; Weeber *et al.*, 2002).

The importance of one of the Reelin receptors, ApoER2, in cognitive function was further underscored by two recent studies from Joachim Herz's research group. Using a homologous knockin gene replacement approach, Beffert *et al.* (2005) have shown that an alternatively spliced intracellular domain of ApoER2 plays an important role in normal cognition and Reelin modulation of synaptic plasticity. This intracellular domain, encoded by exon 19, is required for Reelin enhancement of LTP, maintaining normal associative and spatial learning, but not required for developmental lamination of cortical structures. Moreover, the tyrosine kinase activation does not require this alternatively spliced insert, because the NPXY-containing Dab1 binding motif is not affected by splicing. The connection of ApoER2 with cognitive function was supported by several lines of additional evidence. For example, the alternative splicing of exon 19 causes changes in neuronal activity and behavior. Additionally, ApoER2 and NMDAR are physically associated and coimmunoprecipitate in heterologous expression cell lines and colocalize to CA1 postsynaptic densities (Beffert *et al.*, 2005; Hoe *et al.*, 2006). Moreover, like NMDARs, ApoER2 also binds to PDZ domains of PSD-95 (Hoe *et al.*, 2006), and therefore may recruit Src through postsynaptic scaffolding proteins and result in NMDAR tyrosine phosphorylation. Lastly, in mice carrying the mutations in the Dab1 binding sites on ApoER2, where the normal NFDNPVY motif was replaced by EIGNPVY, disrupted Dab1 binding leads to moderate lamination defects and inability to maintain late-phase LTP (Beffert *et al.*, 2006). Therefore, there seems to be a mechanistic divergence downstream of Reelin binding to ApoER2, whereby one mechanism is coupled with proper lamination and the other coupled with glutamate receptor-mediated function and cognition. These seemingly divergent mechanisms may be coherently connected as more signaling components downstream to lipoprotein receptors are revealed, such as those that enlist MAP kinases (Senokuchi *et al.*, 2004), tyrosine kinases (Arnaud *et al.*, 2003; Ballif *et al.*, 2003; Bock and Herz, 2003), lipid kinases (Beffert *et al.*, 2002; Bock *et al.*, 2003), and ligand-gated ion channels (Beffert *et al.*, 2005; Chen *et al.*, 2005; Sinagra *et al.*, 2005).

4.2 Reelin Receptors and ApoE

In addition to being the major receptor for Reelin, ApoER2 also serves as a receptor for apolipoprotein E (apoE) in the CNS (Herz and Bock, 2002). ApoE is a small (34-kDa) secreted glycoprotein that associates with lipoproteins and mediates uptake of these particles into target cells via receptor-mediated endocytosis by the low-density lipoprotein (LDL) receptor family. Three commonly occurring isoforms have been identified in the human population due to single nucleotide polymorphisms on the APOE gene on chromosome 19. The apoE3 isoform (Cys112, Arg158) occurs at the highest frequency, followed by apoE4 (Arg112, Arg158) and apoE2 (Cys112, Cys158).

There exists a genetic connection between apoE isoform expression and Alzheimer's disease. ApoE4 allele inheritance increases the risk of early onset Alzheimer's disease, while apoE2 allele inheritance decreases the risk (Strittmatter *et al.*, 1993; West *et al.*, 1994). ApoE is known to play an important role in neural development, regeneration, and synapse remodeling. Consistently, apoE knockout mice show severe deficits in learning and memory (Oitzl *et al.*, 1997), and mice with targeted replacement of the human apoE isoforms under the control of the mouse apoE promoter produce differential spatial memory performance depending on isoform expression (Grootendorst *et al.*, 2005). Moreover, strong experimental evidence supports different neurotoxicity properties of distinct isoform-specific apoE microdomains (Brecht *et al.*, 2004; Chang *et al.*, 2005; Ye *et al.*, 2005; Mahley *et al.*, 2006). It is not clear whether these distinct learning and memory effects or susceptibility to cytotoxicity can be attributed to different signaling pathways mediated by apoE receptors in the CNS. However, evidence exists that relates apoE signaling to neuronal maturation. For example, apoE-mediated glia-derived cholesterol transport is required for the functional maturation of cultured central neurons (Mauch *et al.*, 2001). This role of cholesterol can be attributed to neurons' limited capability for cholesterol synthesis and the fact that cholesterol directly participates in neuronal remodeling and synaptogenesis (Goritz *et al.*, 2005). Again, it cannot be ruled out that apoE initiates signaling through ApoER2 and/or other CNS lipoprotein receptors that induce neuronal growth and functional maturation. These potential signaling effects by apoE may also profoundly modulate Reelin signaling, due to the fact that apoE affects the Reelin signaling through binding to ApoER2 and VLDLR and imposes a steric hindrance for Reelin binding. Because apoE4 preferentially associates with larger lipoprotein particles, it may have the greatest inhibition on Reelin signaling (Ohkubo *et al.*, 2003).

Reelin-mediated signaling may contribute to the pathogenesis of Alzheimer's disease through the interaction with apoE receptors and intracellular adapter proteins. Mice lacking either Reelin, or both ApoER2 and VLDLR, exhibit hyperphosphorylation of microtubule-stabilizing protein tau (Hiesberger *et al.*, 1999). One of the histopathological hallmarks of Alzheimer's disease, i.e., neurofibrillary tangles, contains hyperphosphorylated tau as the basic structural component. The tau protein is associated with the stabilization of microtubules and phosphorylation of tau resulting in the destabilization of microtubules and allowing cytostructural remodeling. The hyperphosphorylated tau protein disrupts neuronal axonal transport and cytoarchitectural integrity and leads to impaired neuronal synaptic function and neurodegeneration. A putative pathway for Reelin-mediated control of tau phosphorylation is via Dab1 and Src activation, which leads to protein kinase B/AKT activation and inhibition of GSK3β, a key kinase involved in tau phosphorylation (Beffert *et al.*, 2002). Therefore, changes in Reelin signaling can, in turn, lead to altered GSK3β activity and subsequent increased tau phosphorylation. The observation of a Reelin–tau connection reveals an interesting possibility that Reelin signaling may represent a new therapeutic target for regulating tau phosphorylation in Alzheimer's disease. In agreement with a potential role of Reelin signaling in Alzheimer's disease, a recent study has shown that Reelin level is elevated, and its glycosylation pattern is altered in the brains of Alzheimer's patients, which nicely correlates with changes in tau protein phosphorylation in the CSF of these patients (Botella-Lopez *et al.*, 2006).

4.3 Integrins

In addition to the well-characterized ApoER2 and VLDLR, Reelin also binds to integrins. The functional significance of Reelin binding to these receptors is poorly understood, especially in the postnatal brain. However, it is well established that integrin-mediated signaling plays an important role in synaptic function. Integrins constitute a large family of heterodimeric membrane-spanning receptors for some RGD-motif-containing extracellular matrix proteins. Their interaction activates a myriad of intracellular signaling cascades, leading to reorganization of cytoskeleton actin filaments (Tozer *et al.*, 1996; Milner and Campbell, 2002). This signaling mediated by integrins is also crucial for lamination of cortical structures (Graus-Porta *et al.*, 2001), and together with Reelin signaling, is required for the formation of the radial glial scaffold in the hippocampus (Forster *et al.*, 2002). The correct lamination effects of Reelin may be partly attributed to inhibitory effects on neuronal migration upon Reelin binding to integrins (Dulabon *et al.*, 2000). Notably, multiple studies implicate integrins in cognitive ability and synaptic function. For example, integrins can modulate NMDA receptor maturation and function and play an important role in LTP and spatial memory function (Chavis and Westbrook, 2001; Chan *et al.*, 2003, 2006). In addition, the levels of integrin-associated protein (IAP) mRNA correlate with performance in the inhibitory avoidance learning task, which is a hippocampus-dependent learning paradigm (Lee *et al.*, 2000). Moreover, mice deficient in α3, α4, and α8 integrins exhibit impaired LTP in hippocampal CA1 field and impaired spatial memory performance (Chan *et al.*, 2003). Additionally, in conditional β1 integrin knockout mice, with forebrain excitatory neuron-specific knockout of β1 integrin, they exhibit impaired synaptic transmission through AMPA receptors and diminished NMDA receptor-dependent LTP. These mutant mice also displayed normal hippocampus-dependent spatial and contextual memory but showed impairment in a hippocampus-dependent working memory task (Chan *et al.*, 2006). Therefore, β1-integrins may function as potential regulators of synaptic glutamate receptor function and working memory.

It is currently not clear how Reelin may activate integrin-mediated signaling in the postnatal brain. However, Reelin signaling through ApoER2/VLDLR may be overlapping with the integrin pathways. Integrins and Reelin receptors are both found at the dendritic spine and postsynaptic densities in CA1 neurons (Bi *et al.*, 2001; Beffert *et al.*, 2005), and like Reelin and apolipoprotein receptors, integrin-mediated signaling is also required for the maturation of cultured hippocampal neurons. For example, Chavis and Westbrook (2001) have shown that chronic blockade of integrins with RGD peptide prevents the developmental switch of NMDA receptor subunit NR2B to NR2A; similarly, chronic blockade of Reelin signaling with Src inhibitor or LDL receptor antagonist also delays the functional switch of somatic NMDA receptors (Sinagra *et al.*, 2005). Moreover, the formation of dendritic spines during neuronal development and their structural plasticity also require integrin. In cultured hippocampal neurons, integrins are required for activity-dependent spine remodeling through mechanisms involving NMDA receptors and calcium–calmodulin-dependent kinase II (Shi and Ethell, 2006). For example, reduced dendritic spine numbers and hypoplasticity were correlated with decreased cortical Reelin expression in human schizophrenic patients (Eastwood and Harrison, 2006). Integrin-mediated signaling has also been shown to lead to tyrosine phosphorylation of Crk family adapter proteins, resulting in cytoskeletal rearrangement through recruitment of guanine-nucleotide exchange factors and small GTPases. Crk proteins bind to Dab1 only when Dab1 is tyrosine phosphorylated at Y232 and Y220. Reelin application leads to C3G tyrosine phosphorylation and the downstream small GTPase Rap1 activation in Dab1-dependent manner (Ballif *et al.*, 2004). This pathway employed by Reelin signaling involving C3G and Rap1 is also operating to regulate integrin-dependent cell adhesion and motility. Therefore, it is possible that Reelin and integrin affect synaptic function through discrete but interacting pathways involving ApoER2/VLDLR and other extracellular matrix proteins, and these two signaling systems may overlap and cooperate in dictating synaptic function and plasticity.

5 The Heterozygote Reeler Mouse

The role of Reelin in cognition is best supported by research showing that gene dosage of Reelin is correlated with learning and memory in mice (Qiu *et al.*, 2006). In contrast to the Reeler phenotype, a single allele inactivation of the *RELN* gene, as happens with the heterozygote Reeler mouse (HRM), results in some subtle yet distinct neuroanatomical, neurochemical, and behavioral features compared to their wild-type counterparts. The HRM exhibits ~50% overall Reelin reduction in the brain. Phenotypically, the HRM shows reduced dendritic spine density in the parietal-frontal cortex pyramidal neurons and the basal dendrites of CA1 pyramidal neurons, reduced cortical thickness, and decreased number of glutamic acid decarboxylase 67-kDa-positive cells. Moreover, the HRM also exhibits a reduced density of nicotinamide adenine dinucleotide phosphate-diaphorase (NADPH) -positive neurons in cortical gray matter and a decreased prepulse inhibition behavior (Tueting *et al.*, 1999; Liu *et al.*, 2001; Salinger *et al.*, 2003; Qiu *et al.*, 2006). These unique features of the HRM are also partially shared by human schizophrenic patients and other cognitive disorders, such as lissencephaly, bipolar disorder, and autism (Fatemi *et al.*, 2000, 2001, 2005a,b; Hong *et al.*, 2000; Assadi *et al.*, 2003), suggesting a potential role of Reelin in modulating cognitive function in the postnatal brain and a reduced Reelin signaling causing deficiency in synaptic function.

A single functional *RELN* allele in the HRM does not seem to change the gross anatomy in the mouse hippocampus, nor does it change the overall levels of glutamate receptor subunits including GluR1, GluR2/3, NR1, NR2A, and NR2B (Qiu *et al.*, 2006). The HRM also does not differ from wild-type mice in rotorod, open field testing, hot plate nociception, light/dark preference, elevated plus platform tests, startle response thresholds, and water maze tests. However, the HRM exhibits a pronounced deficit in prepulse inhibition, which is a measure of sensory-motor gating of the nervous system (Tueting *et al.*, 1999; Qiu *et al.*, 2006). The HRM also exhibits contextual fear conditioning deficit, a hippocampus-dependent functional test. Examination of the synaptic transmission and plasticity in CA1 revealed a surprising plethora of phenotypic differences: the HRM exhibited decreased input–output responses, reduced paired pulse facilitation, and impaired LTP and LTD in CA1 (Qiu *et al.*, 2006). The impaired LTP was reproducible at the single-cell level when using a theta burst stimulation protocol. Strikingly, there was a disrupted excitatory–inhibitory network balance in CA1 in the HRM. Although CA1 pyramidal neurons receive similar excitatory input, the inhibitory postsynaptic currents onto these neurons are greatly reduced. These phenotypes of the HRM strengthen the similarity of the HRM and human schizophrenic patients, in that the latter show diminished inhibition (Escobar *et al.*, 2002; Lewis *et al.*, 2005). The reduced inhibition of the HRM may reflect the failure of inhibitory interneurons to reach their final destination and form functional connections with pyramidal neurons, or failure of release of GABA neurotransmitters due to less excitatory drive onto these neurons. Although these phenotypes manifested by the HRM could be caused by the subtle developmental effects resulting from Reelin haploinsufficiency, a reduced signaling of Reelin through ApoER2/VLDLR may also account for at least some of these deficits.

6 Summary

The expression of Reelin protein persists in the postnatal brain at a stage when neuronal migration is long completed. The functional significance of this continued presence of Reelin is underscored by recent findings that Reelin- and ApoER2/VLDLR-mediated signaling play a novel role in synaptic function in the hippocampus (Fig. 12.3). This signaling system is coupled with enhanced glutamatergic function, facilitates glutamatergic maturation at postnatal stages, modulates synaptic plasticity,

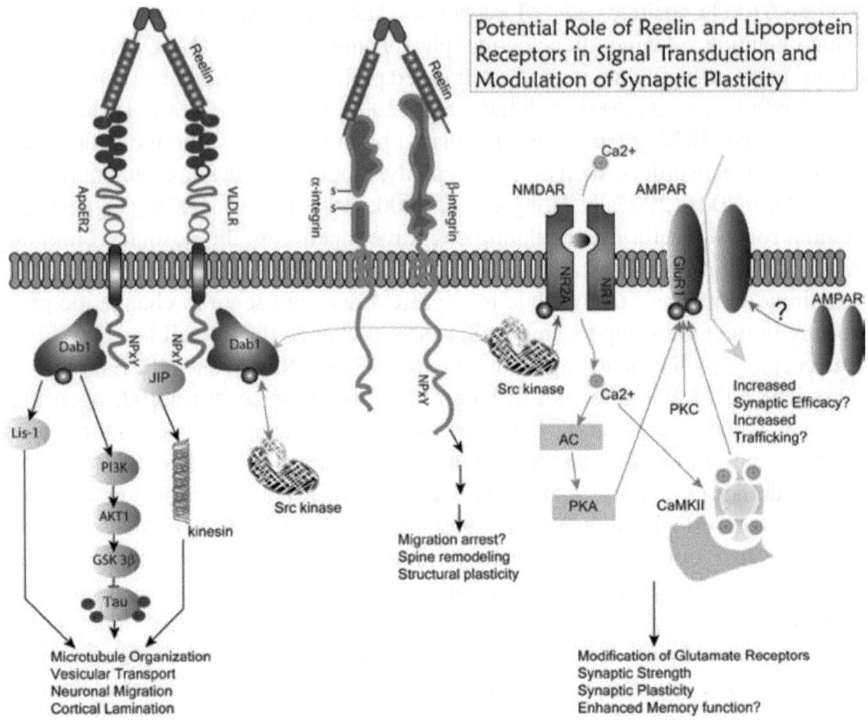

Fig. 12.3 Schematic representation of Reelin signaling and the subsequent enhancement of synaptic function in the adult hippocampus. Reelin binds and activates ApoER2/VLDLR and leads to tyrosine phosphorylation and activation of Dab1 and Src family protein tyrosine kinases. Src kinases phosphorylate NMDA receptor subunits and lead to enhanced channel conductance, augmented Ca^{2+} influx during activation, and increased synaptic plasticity. This increased synaptic plasticity may involve changes of AMPA receptor phosphorylation and trafficking as well. In response to Reelin signaling, PI3K and PKB/AKT can be activated as well, resulting in inhibition of tau phosphorylation. In addition to ApoER2/VLDLR, Reelin also activates integrins (*See Color Plates*)

and enhances memory performance in experimental animals. Considering the presence of Reelin-expressing cells in other parts of brain, it is possible that Reelin signaling is operating in various CNS structures with similar effects on the glutamatergic synapses. Because much of our current knowledge on the postnatal role of Reelin is acquired through perfusion of recombinant Reelin on the hippocampal slices, a selective activation of Reelin-expressing neurons in the hippocampus and subsequent study of adjacent synapses would more faithfully reflect the *in vivo* roles of Reelin signaling, although currently this is not technically feasible. Alternatively, a selective ablation of Reelin-expressing neurons (i.e., conditional knockout) in the postnatal hippocampus without the pandemic developmental disruption of overall hippocampal synaptic connection would be more informative than the promiscuous pharmacological blockade of Reelin receptors to address the consequences of reduced Reelin signaling in the adult brain. Given that aberrant Reelin signaling is implicated in many human cognitive diseases, the authors consider these novel approaches to represent promising future directions to further understanding of the role of Reelin signaling in cognitive function.

References

Abraham, H., and Meyer, G. (2003). Reelin-expressing neurons in the postnatal and adult human hippocampal formation. *Hippocampus* 13:715–727.

Aguilo, A., Schwartz, T. H., Kumar, V. S., Peterlin, Z. A., Tsiola, A., Soriano, E., and Yuste, R. (1999). Involvement of Cajal-Retzius neurons in spontaneous correlated activity of embryonic and postnatal layer 1 from wild-type and reeler mice. *J. Neurosci.* 19:10856–10868.

Alcantara, S., Ruiz, M., D'Arcangelo, G., Ezan, F., de Lecea, L., Curran, T., Sotelo, C., and Soriano, E. (1998). Regional and cellular patterns of reelin mRNA expression in the forebrain of the developing and adult mouse. *J. Neurosci.* 18:7779–7799.

Anderson, P., Bliss, T. V., and Skrede, K. K. (1971). Lamellar organization of hippocampal pathways. *Exp. Brain Res.* 13:222–238.

Arnaud, L., Ballif, B. A., Forster, E., and Cooper, J. A. (2003). Fyn tyrosine kinase is a critical regulator of disabled-1 during brain development. *Curr. Biol.* 13:9–17.

Assadi, A. H., Zhang, G., Beffert, U., McNeil, R. S., Renfro, A. L., Niu, S., Quattrocchi, C. C., Antalffy, B. A., Sheldon, M., Armstrong, D. D., Wynshaw-Boris, A., Herz, J., D'Arcangelo, G., and Clark, G. D. (2003). Interaction of reelin signaling and Lis1 in brain development. *Nature Genet.* 35:270–276.

Ballif, B. A., Arnaud, L., and Cooper, J. A. (2003). Tyrosine phosphorylation of disabled-1 is essential for reelin-stimulated activation of Akt and Src family kinases. *Brain Res. Mol. Brain Res.* 117:152–159.

Ballif, B. A., Arnaud, L., Arthur, W. T., Guris, D., Imamoto, A., and Cooper, J. A. (2004). Activation of a Dab1/CrkL/C3G/Rap1 pathway in reelin-stimulated neurons. *Curr. Biol.* 14:606–610.

Banke, T. G., Bowie, D., Lee, H., Huganir, R. L., Schousboe, A., and Traynelis, S. F. (2000). Control of GluR1 AMPA receptor function by cAMP-dependent protein kinase. *J. Neurosci.* 20:89–102.

Barnes, A. P., and Milgram, S. L. (2002). Signals from the X: signal transduction and X-linked mental retardation. *Int. J. Dev. Neurosci.* 20:397–406.

Barria, A., and Malinow, R. (2005). NMDA receptor subunit composition controls synaptic plasticity by regulating binding to CaMKII. *Neuron* 48:289–301.

Barria, A., Muller, D., Derkach, V., Griffith, L. C., and Soderling, T. R. (1997). Regulatory phosphorylation of AMPA-type glutamate receptors by CaM-KII during long-term potentiation. *Science* 276:2042–2045.

Beffert, U., Morfini, G., Bock, H. H., Reyna, H., Brady, S. T., and Herz, J. (2002). Reelin-mediated signaling locally regulates protein kinase B/Akt and glycogen synthase kinase 3beta. *J. Biol. Chem.* 277:49958–49964.

Beffert, U., Weeber, E. J., Durudas, A., Qiu, S., Masiulis, I., Sweatt, J. D., Li, W. P., Adelmann, G., Frotscher, M., Hammer, R. E., and Herz, J. (2005). Modulation of synaptic plasticity and memory by reelin involves differential splicing of the lipoprotein receptor Apoer2. *Neuron* 47:567–579.

Beffert, U., Durudas, A., Weeber, E. J., Stolt, P. C., Giehl, K. M., Sweatt, J. D., Hammer, R. E., and Herz, J. (2006). Functional dissection of reelin signaling by site-directed disruption of disabled-1 adaptor binding to apolipoprotein E receptor 2: distinct roles in development and synaptic plasticity. *J. Neurosci.* 26:2041–2052.

Bi, X., Lynch, G., Zhou, J., and Gall, C. M. (2001). Polarized distribution of alpha5 integrin in dendrites of hippocampal and cortical neurons. *J. Comp. Neurol.* 435:184–193.

Bliss, T. V., and Lomo, T. (1973). Long-lasting potentiation of synaptic transmission in the dentate area of the anaesthetized rabbit following stimulation of the perforant path. *J. Physiol. (London)* 232:331–356.

Bock, H. H., and Herz, J. (2003). Reelin activates SRC family tyrosine kinases in neurons. *Curr. Biol.* 13:18–26.

Bock, H. H., Jossin, Y., Liu, P., Forster, E., May, P., Goffinet, A. M., and Herz, J. (2003). Phosphatidylinositol 3-kinase interacts with the adaptor protein Dab1 in response to reelin signaling and is required for normal cortical lamination. *J. Biol. Chem.* 278: 38772–38779.

Botella-Lopez, A., Burgaya, F., Gavin, R., Garcia-Ayllon, M. S., Gomez-Tortosa, E., Pena-Casanova, J., Urena, J. M., Del Rio, J. A., Blesa, R., Soriano, E., and Saez-Valero, J. (2006). Reelin expression and glycosylation patterns are altered in Alzheimer's disease. *Proc. Natl. Acad. Sci. USA* 103:5573–5578.

Brecht, W. J., Harris, F. M., Chang, S., Tesseur, I., Yu, G. Q., Xu, Q., Dee Fish, J., Wyss-Coray, T., Buttini, M., Mucke, L., Mahley, R. W., and Huang, Y. (2004). Neuron-specific apolipoprotein e4 proteolysis is associated with increased tau phosphorylation in brains of transgenic mice. *J. Neurosci.* 24:2527–2534.

Carmignoto, G., and Vicini, S. (1992). Activity-dependent decrease in NMDA receptor responses during development of the visual cortex. *Science* 258:1007–1011.

Chan, C. S., Weeber, E. J., Kurup, S., Sweatt, J. D., and Davis, R. L. (2003). Integrin requirement for hippocampal synaptic plasticity and spatial memory. *J. Neurosci.* 23:7107–7116.

Chan, C. S., Weeber, E. J., Zong, L., Fuchs, E., Sweatt, J. D., and Davis, R. L. (2006). Beta 1-integrins are required for hippocampal AMPA receptor-dependent synaptic transmission, synaptic plasticity, and working memory. *J. Neurosci.* 26:223–232.

Chang, S., ran Ma, T., Miranda, R. D., Balestra, M. E., Mahley, R. W., and Huang, Y. (2005). Lipid- and receptor-binding regions of apolipoprotein E4 fragments act in concert to cause mitochondrial dysfunction and neurotoxicity. *Proc. Natl. Acad. Sci. USA* 102: 18694–18699.

Chavis, P., and Westbrook, G. (2001). Integrins mediate functional pre- and postsynaptic maturation at a hippocampal synapse. *Nature* 411:317–321.

Chen, Y., Beffert, U., Ertunc, M., Tang, T. S., Kavalali, E. T., Bezprozvanny, I., and Herz, J. (2005). Reelin modulates NMDA receptor activity in cortical neurons. *J. Neurosci.* 25:8209–8216.

Chung, H. J., Huang, Y. H., Lau, L. F., and Huganir, R. L. (2004). Regulation of the NMDA receptor complex and trafficking by activity-dependent phosphorylation of the NR2B subunit PDZ ligand. *J. Neurosci.* 24:10248–10259.

D'Arcangelo, G. (2005). Apoer2: a reelin receptor to remember. *Neuron* 47:471–473.

D'Arcangelo, G., Miao, G. G., Chen, S. C., Soares, H. D., Morgan, J. I., and Curran, T. (1995). A protein related to extracellular matrix proteins deleted in the mouse mutant reeler. *Nature* 374:719–723.

D'Arcangelo, G., Nakajima, K., Miyata, T., Ogawa, M., Mikoshiba, K., and Curran, T. (1997). Reelin is a secreted glycoprotein recognized by the CR-50 monoclonal antibody. *J. Neurosci.* 17:23–31.

Del Rio, J. A., Heimrich, B., Borrell, V., Forster, E., Drakew, A., Alcantara, S., Nakajima, K., Miyata, T., Ogawa, M., Mikoshiba, K., Derer, P., Frotscher, M., and Soriano, E. (1997). A role for Cajal-Retzius cells and reelin in the development of hippocampal connections. *Nature* 385:70–74.

Dingledine, R., Borges, K., Bowie, D., and Traynelis, S. F. (1999). The glutamate receptor ion channels. *Pharmacol. Rev.* 51:7–61.

Drakew, A., Deller, T., Heimrich, B., Gebhardt, C., Del Turco, D., Tielsch, A., Forster, E., Herz, J., and Frotscher, M. (2002). Dentate granule cells in reeler mutants and VLDLR and ApoER2 knockout mice. *Exp. Neurol.* 176:12–24.

Dulabon, L., Olson, E. C., Taglienti, M. G., Eisenhuth, S., McGrath, B., Walsh, C. A., Kreidberg, J. A., and Anton, E. S. (2000). Reelin binds alpha3beta1 integrin and inhibits neuronal migration. *Neuron* 27:33–44.

Eastwood, S. L., and Harrison, P. J. (2006). Cellular basis of reduced cortical reelin expression in schizophrenia. *Am. J. Psychiatry* 163:540–542.

Escobar, M., Oberling, P., and Miller, R. R. (2002). Associative deficit accounts of disrupted latent inhibition and blocking in schizophrenia. *Neurosci. Biobehav. Rev.* 26:203–216.

Fatemi, S. H., Earle, J. A., and McMenomy, T. (2000). Reduction in reelin immunoreactivity in hippocampus of subjects with schizophrenia, bipolar disorder and major depression. *Mol. Psychiatry* 5:654–663.

Fatemi, S. H., Kroll, J. L., and Stary, J. M. (2001). Altered levels of reelin and its isoforms in schizophrenia and mood disorders. *Neuroreport* 12:3209–3215.

Fatemi, S. H., Stary, J. M., Earle, J. A., Araghi-Niknam, M., and Eagan, E. (2005a). GABAergic dysfunction in schizophrenia and mood disorders as reflected by decreased levels of glutamic acid decarboxylase 65 and 67 kDa and reelin proteins in cerebellum. *Schizophr. Res.* 72:109–122.

Fatemi, S. H., Snow, A. V., Stary, J. M., Araghi-Niknam, M., Reutiman, T. J., Lee, S., Brooks, A. I., and Pearce, D. A. (2005b). Reelin signaling is impaired in autism. *Biol. Psychiatry* 57:777–787.

Forster, E., Tielsch, A., Saum, B., Weiss, K. H., Johanssen, C., Graus-Porta, D., Muller, U., and Frotscher, M. (2002). Reelin, disabled 1, and beta 1 integrins are required for the formation of the radial glial scaffold in the hippocampus. *Proc. Natl. Acad. Sci. US.A* 99:13178–13183.

Frotscher, M., Haas, C. A., and Forster, E. (2003). Reelin controls granule cell migration in the dentate gyrus by acting on the radial glial scaffold. *Cereb. Cortex* 13:634–640.

Gahwiler, B. H., Capogna, M., Debanne, D., McKinney, R. A., and Thompson, S. M. (1997). Organotypic slice cultures: a technique has come of age. *Trends Neurosci.* 20:471–477.

Giese, K. P., Fedorov, N. B., Filipkowski, R. K., and Silva, A. J. (1998). Autophosphorylation at Thr286 of the alpha calcium-calmodulin kinase II in LTP and learning. *Science* 279:870–873.

Goritz, C., Mauch, D. H., and Pfrieger, F. W. (2005). Multiple mechanisms mediate cholesterol-induced synaptogenesis in a CNS neuron. *Mol. Cell. Neurosci.* 29:190–201.

Graus-Porta, D., Blaess, S., Senften, M., Littlewood-Evans, A., Damsky, C., Huang, Z., Orban, P., Klein, R., Schittny, J. C., and Muller, U. (2001). Beta1-class integrins regulate the development of laminae and folia in the cerebral and cerebellar cortex. *Neuron* 31:367–379.

Grootendorst, J., Bour, A., Vogel, E., Kelche, C., Sullivan, P. M., Dodart, J. C., Bales, K., and Mathis, C. (2005). Human apoE targeted replacement mouse lines: h-apoE4 and h-apoE3 mice differ on spatial memory performance and avoidance behavior. *Behav. Brain Res.* 159:1–14.

Hayashi, T., and Huganir, R. L. (2004). Tyrosine phosphorylation and regulation of the AMPA receptor by SRC family tyrosine kinases. *J. Neurosci.* 24:6152–6160.

Hayashi, Y., Shi, S. H., Esteban, J. A., Piccini, A., Poncer, J. C., and Malinow, R. (2000). Driving AMPA receptors into synapses by LTP and CaMKII: requirement for GluR1 and PDZ domain interaction. *Science* 287:2262–2267.

Herz, J., and Bock, H. H. (2002). Lipoprotein receptors in the nervous system. *Annu. Rev. Biochem.* 71:405–434.

Hiesberger, T., Trommsdorff, M., Howell, B. W., Goffinet, A., Mumby, M. C., Cooper, J. A., and Herz, J. (1999). Direct binding of reelin to VLDL receptor and ApoE receptor 2 induces tyrosine phosphorylation of disabled-1 and modulates tau phosphorylation. *Neuron* 24:481–489.

Hoe, H. S., Pocivavsek, A., Chakraborty, G., Fu, Z., Vicini, S., Ehlers, M. D., and Rebeck, G. W. (2006). Apolipoprotein E receptor 2 interactions with the N-methyl-D-aspartate receptor. *J. Biol. Chem.* 281:3425–3431.

Hong, S. E., Shugart, Y. Y., Huang, D. T., Shahwan, S. A., Grant, P. E., Hourihane, J. O., Martin, N. D., and Walsh, C. A. (2000). Autosomal recessive lissencephaly with cerebellar hypoplasia is associated with human RELN mutations. *Nature Genet.* 26:93–96.

Howell, B. W., Hawkes, R., Soriano, P., and Cooper, J. A. (1997). Neuronal position in the developing brain is regulated by mouse disabled-1. *Nature* 389:733–737.

Irwin, S. A., Galvez, R., and Greenough, W. T. (2000). Dendritic spine structural anomalies in fragile-X mental retardation syndrome. *Cereb. Cortex* 10:1038–1044.

Kaufmann, W. E., and Moser, H. W. (2000). Dendritic anomalies in disorders associated with mental retardation. *Cereb. Cortex* 10:981–991.

Ko, J., Humbert, S., Bronson, R. T., Takahashi, S., Kulkarni, A. B., Li, E., and Tsai, L. H. (2001). p35 and p39 are essential for cyclin-dependent kinase 5 function during neurodevelopment. *J. Neurosci.* 21:6758–6771.

Lau, L. F., and Huganir, R. L. (1995). Differential tyrosine phosphorylation of N-methyl-D-aspartate receptor subunits. *J. Biol. Chem.* 270:20036–20041.

Lee, E. H., Hsieh, Y. P., Yang, C. L., Tsai, K. J., and Liu, C. H. (2000). Induction of integrin-associated protein (IAP) mRNA expression during memory consolidation in rat hippocampus. *Eur. J. Neurosci.* 12:1105–1112.

Lee, H. K., Takamiya, K., Han, J. S., Man, H., Kim, C. H., Rumbaugh, G., Yu, S., Ding, L., He, C., Petralia, R. S., Wenthold, R. J., Gallagher, M., and Huganir, R. L. (2003). Phosphorylation of the AMPA receptor GluR1 subunit is required for synaptic plasticity and retention of spatial memory. *Cell* 112:631–643.

Lewis, D. A., Hashimoto, T., and Volk, D. W. (2005). Cortical inhibitory neurons and schizophrenia. *Nature Rev. Neurosci.* 6:312–324.

Liu, L., Wong, T. P., Pozza, M. F., Lingenhoehl, K., Wang, Y., Sheng, M., Auberson, Y. P., and Wang, Y. T. (2004). Role of NMDA receptor subtypes in governing the direction of hippocampal synaptic plasticity. *Science* 304:1021–1024.

Liu, W. S., Pesold, C., Rodriguez, M. A., Carboni, G., Auta, J., Lacor, P., Larson, J., Condie, B. G., Guidotti, A., and Costa, E. (2001). Down-regulation of dendritic spine and glutamic acid decarboxylase 67 expressions in the reelin haploinsufficient heterozygous reeler mouse. *Proc. Natl. Acad. Sci. USA* 98:3477–3482.

Lu, W., and Constantine-Paton, M. (2004). Eye opening rapidly induces synaptic potentiation and refinement. *Neuron* 43:237–249.

Lu, Y. M., Roder, J. C., Davidow, J., and Salter, M. W. (1998). Src activation in the induction of long-term potentiation in CA1 hippocampal neurons. *Science* 279:1363–1367.

Mahley, R. W., Weisgraber, K. H., and Huang, Y. (2006). Apolipoprotein E4: a causative factor and therapeutic target in neuropathology, including Alzheimer's disease. *Proc. Natl. Acad. Sci. USA* 103:5644–5651.

Marin-Padilla, M. (1998). Cajal-Retzius cells and the development of the neocortex. *Trends Neurosci.* 21:64–71.

Martinez-Cerdeno, V., Galazo, M. J., Cavada, C., and Clasca, F. (2002). Reelin immunoreactivity in the adult primate brain: intracellular localization in projecting and local circuit neurons of the cerebral cortex, hippocampus and subcortical regions. *Cereb. Cortex* 12:1298–1311.

Massey, P. V., Johnson, B. E., Moult, P. R., Auberson, Y. P., Brown, M. W., Molnar, E., Collingridge, G. L., and Bashir, Z. I. (2004). Differential roles of NR2A and NR2B-containing NMDA receptors in cortical long-term potentiation and long-term depression. *J. Neurosci.* 24:7821–7828.

Mauch, D. H., Nagler, K., Schumacher, S., Goritz, C., Muller, E. C., Otto, A., and Pfrieger, F. W. (2001). CNS synaptogenesis promoted by glia-derived cholesterol. *Science* 294:1354–1357.

Meyer, G., Soria, J. M., Martinez-Galan, J. R., Martin-Clemente, B., and Fairen, A. (1998). Different origins and developmental histories of transient neurons in the marginal zone of the fetal and neonatal rat cortex. *J. Comp. Neurol.* 397:493–518.

Milner, R., and Campbell, I. L. (2002). The integrin family of cell adhesion molecules has multiple functions within the CNS. *J. Neurosci. Res.* 69:286–291.

Monyer, H., Burnashev, N., Laurie, D. J., Sakmann, B., and Seeburg, P. H. (1994). Developmental and regional expression in the rat brain and functional properties of four NMDA receptors. *Neuron* 12:529–540.

Nakayama, K., Kiyosue, K., and Taguchi, T. (2005). Diminished neuronal activity increases neuron-neuron connectivity underlying silent synapse formation and the rapid conversion of silent to functional synapses. *J. Neurosci.* 25:4040–4051.

Nicoll, R. A., and Malenka, R. C. (1995). Contrasting properties of two forms of long-term potentiation in the hippocampus. *Nature* 377:115–118.

Niu, S., Renfro, A., Quattrocchi, C. C., Sheldon, M., and D'Arcangelo, G. (2004). Reelin promotes hippocampal dendrite development through the VLDLR/ApoER2-Dab1 pathway. *Neuron* 41:71–84.

Ohkubo, N., Lee, Y. D., Morishima, A., Terashima, T., Kikkawa, S., Tohyama, M., Sakanaka, M., Tanaka, J., Maeda, N., Vitek, M. P., and Mitsuda, N. (2003). Apolipoprotein E and reelin ligands modulate tau phosphorylation through an apolipoprotein E receptor/disabled-1/glycogen synthase kinase-3beta cascade. *FASEB J.* 17:295–297.

Ohshima, T., Ward, J. M., Huh, C. G., Longenecker, G., Veeranna, Pant, H. C., Brady, R. O., Martin, L. J., and Kulkarni, A. B. (1996). Targeted disruption of the cyclin-dependent kinase 5 gene results in abnormal corticogenesis, neuronal pathology and perinatal death. *Proc. Natl. Acad. Sci. USA* 93:11173–11178.

Oitzl, M. S., Mulder, M., Lucassen, P. J., Havekes, L. M., Grootendorst, J., and de Kloet, E. R. (1997). Severe learning deficits in apolipoprotein E-knockout mice in a water maze task. *Brain Res.* 752:189–196.

Pappas, G. D., Kriho, V., and Pesold, C. (2001). Reelin in the extracellular matrix and dendritic spines of the cortex and hippocampus: a comparison between wild type and heterozygous reeler mice by immunoelectron microscopy. *J. Neurocytol.* 30:413–425.

Pesold, C., Impagnatiello, F., Pisu, M. G., Uzunov, D. P., Costa, E., Guidotti, A., and Caruncho, H. J. (1998). Reelin is preferentially expressed in neurons synthesizing gamma-aminobutyric acid in cortex and hippocampus of adult rats. *Proc. Natl. Acad. Sci. US.A* 95:3221–3226.

Qiu, S., Korwek, K. M., Pratt-Davis, A. R., Peters, M., Bergman, M. Y., and Weeber, E. J. (2006). Cognitive disruption and altered hippocampus synaptic function in reelin haploinsufficient mice. *Neurobiol. Learn. Mem.* 85:228–242.

Quinlan, E. M., Philpot, B. D., Huganir, R. L., and Bear, M. F. (1999). Rapid, experience-dependent expression of synaptic NMDA receptors in visual cortex in vivo. *Nature Neurosci.* 2:352–357.

Quinlan, E. M., Lebel, D., Brosh, I., and Barkai, E. (2004). A molecular mechanism for stabilization of learning-induced synaptic modifications. *Neuron* 41:185–192.

Ramos-Moreno, T., Galazo, M. J., Porrero, C., Martinez-Cerdeno, V., and Clasca, F. (2006). Extracellular matrix molecules and synaptic plasticity: immunomapping of intracellular and secreted reelin in the adult rat brain. *Eur. J. Neurosci.* 23:401–422.

Roberts, R. C., Xu, L., Roche, J. K., and Kirkpatrick, B. (2005). Ultrastructural localization of reelin in the cortex in post-mortem human brain. *J. Comp. Neurol.* 482:294–308.

Salinger, W. L., Ladrow, P., and Wheeler, C. (2003). Behavioral phenotype of the reeler mutant mouse: effects of RELN gene dosage and social isolation. *Behav. Neurosci.* 117:1257–1275.

Salter, M. W., and Kalia, L. V. (2004). Src kinases: a hub for NMDA receptor regulation. *Nature Rev. Neurosci.* 5:317–328.

Sarnat, H. B., and Flores-Sarnat, L. (2002). Cajal-Retzius and subplate neurons: their role in cortical development. *Eur. J. Paediatr. Neurol.* 6:91–97.

Scoville, W. B., and Milner, B. (1957). Loss of recent memory after bilateral hippocampal lesions. *J. Neurochem.* 20:11–21.

Senokuchi, T., Matsumura, T., Sakai, M., Matsuo, T., Yano, M., Kiritoshi, S., Sonoda, K., Kukidome, D., Nishikawa, T., and Araki, E. (2004). Extracellular signal-regulated kinase and p38 mitogen-activated protein kinase mediate macrophage proliferation induced by oxidized low-density lipoprotein. *Atherosclerosis* 176:233–245.

Sheng, M., Cummings, J., Roldan, L. A., Jan, Y. N., and Jan, L. Y. (1994). Changing subunit composition of heteromeric NMDA receptors during development of rat cortex. *Nature* 368:144–147.

Shi, S. H., Hayashi, Y., Petralia, R. S., Zaman, S. H., Wenthold, R. J., Svoboda, K., and Malinow, R. (1999). Rapid spine delivery and redistribution of AMPA receptors after synaptic NMDA receptor activation. *Science* 284:1811–1816.

Shi, Y., and Ethell, I. M. (2006). Integrins control dendritic spine plasticity in hippocampal neurons through NMDA receptor and Ca^{2+}/calmodulin-dependent protein kinase II-mediated actin reorganization. *J. Neurosci.* 26:1813–1822.

Sinagra, M., Verrier, D., Frankova, D., Korwek, K. M., Blahos, J., Weeber, E. J., Manzoni, O. J., and Chavis, P. (2005). Reelin, very-low-density lipoprotein receptor, and apolipoprotein E receptor 2 control somatic NMDA receptor composition during hippocampal maturation in vitro. *J. Neurosci.* 25:6127–6136.

Soda, T., Nakashima, R., Watanabe, D., Nakajima, K., Pastan, I., and Nakanishi, S. (2003). Segregation and coactivation of developing neocortical layer 1 neurons. *J. Neurosci.* 23:6272–6279.

Strittmatter, W. J., Saunders, A. M., Schmechel, D., Pericak-Vance, M., Enghild, J., Salvesen, G. S., and Roses, A. D. (1993). Apolipoprotein E: high-avidity binding to beta-amyloid and increased frequency of type 4 allele in late-onset familial Alzheimer disease. *Proc. Natl. Acad. Sci. USA* 90:1977–1981.

Tozer, E. C., Hughes, P. E., and Loftus, J. C. (1996). Ligand binding and affinity modulation of integrins. *Biochem. Cell. Biol.* 74:785–798.

Trommsdorff, M., Gotthardt, M., Hiesberger, T., Shelton, J., Stockinger, W., Nimpf, J., Hammer, R. E., Richardson, J. A., and Herz, J. (1999). Reeler/disabled-like disruption of neuronal migration in knockout mice lacking the VLDL receptor and ApoE receptor 2. *Cell* 97:689–701.

Tueting, P., Costa, E., Dwivedi, Y., Guidotti, A., Impagnatiello, F., Manev, R., and Pesold, C. (1999). The phenotypic characteristics of heterozygous reeler mouse. *Neuroreport* 10:1329–1334.

Wang, Y. T., and Salter, M. W. (1994). Regulation of NMDA receptors by tyrosine kinases and phosphatases. *Nature* 369:233–235.

Weeber, E. J., Beffert, U., Jones, C., Christian, J. M., Forster, E., Sweatt, J. D., and Herz, J. (2002). Reelin and ApoE receptors cooperate to enhance hippocampal synaptic plasticity and learning. *J. Biol. Chem.* 277:39944–39952.

Weeber, E. J., Jiang, Y. H., Elgersma, Y., Varga, A. W., Carrasquillo, Y., Brown, S. E., Christian, J. M., Mirnikjoo, B., Silva, A., Beaudet, A. L., and Sweatt, J. D. (2003). Derangements of hippocampal calcium/calmodulin-dependent protein kinase II in a mouse model for Angelman mental retardation syndrome. *J. Neurosci.* 23:2634–2644.

West, H. L., Rebeck, G. W., and Hyman, B. T. (1994). Frequency of the apolipoprotein E epsilon 2 allele is diminished in sporadic Alzheimer disease. *Neurosci. Lett.* 175:46–48.

Won, S. J., Kim, S. H., Xie, L., Wang, Y., Mao, X. O., Jin, K., and Greenberg, D. A. (2006). Reelin-deficient mice show impaired neurogenesis and increased stroke size. *Exp. Neurol.* 198:250–259.

Ye, S., Huang, Y., Mullendorff, K., Dong, L., Giedt, G., Meng, E. C., Cohen, F. E., Kuntz, I. D., Weisgraber, K. H., and Mahley, R. W. (2005). Apolipoprotein (apo) E4 enhances amyloid beta peptide production in cultured neuronal cells: apoE structure as a potential therapeutic target. *Proc. Natl. Acad. Sci. USA* 102:18700–18705.

Yu, X.M., Askalan, R., Keil, G. J., 2nd, and Salter, M. W. (1997). NMDA channel regulation by channel-associated protein tyrosine kinase Src. *Science* 275:674–678.

Zhao, S., Chai, X., Forster, E., and Frotscher, M. (2004). Reelin is a positional signal for the lamination of dentate granule cells. *Development* 131:5117–5125.

Zhuo, M., Holtzman, D. M., Li, Y., Osaka, H., DeMaro, J., Jacquin, M., and Bu, G. (2000). Role of tissue plasminogen activator receptor LRP in hippocampal long-term potentiation. *J. Neurosci.* 20:542–549.

Zimmer, J., and Gahwiler, B. H. (1984). Cellular and connective organization of slice cultures of the rat hippocampus and fascia dentata. *J. Comp. Neurol.* 228:432–446.

Chapter 13
Protein Kinases and Signaling Pathways that Are Activated by Reelin

Jonathan A. Cooper, Nathaniel S. Allen, and Libing Feng

Contents

1 Introduction

Defects in the cortex, hippocampus, inferior olive, and cerebellum of Reeler mutant mice were first detected many decades ago (Caviness and Rakic, 1978; Rice and Curran, 2001). Recently, a plethora of other developmental and adult phenotypes have been detected in mutant mice, including misplacement of olfactory

J. A. Cooper
Division of Basic Sciences, Fred Hutchinson Cancer Research Center, 1100 Fairview Avenue N, Seattle, WA 98109
e-mail: jcooper@fhcrc.org

N. S. Allen
Division of Basic Sciences, Fred Hutchinson Cancer Research Center, 1100 Fairview Avenue N, Seattle, WA 98109

L. Feng
Division of Basic Sciences, Fred Hutchinson Cancer Research Center, 1100 Fairview Avenue N, Seattle, WA 98109

S. H. Fatemi (ed.), *Reelin Glycoprotein: Structure, Biology and Roles in Health and Disease.*

interneurons (Hack *et al.*, 2002), facial motor neurons (FMNs) (Ohshima *et al.*, 2002; Rossel *et al.*, 2005), sympathetic preganglionic neurons (SPNs) (Yip *et al.*, 2000), and gonadotropin-releasing hormone (GnRH) neurons (Cariboni *et al.*, 2005), reduced dendrite outgrowth in the hippocampus (Niu *et al.*, 2004), and defective long-term potentiation (LTP) and memory (Weeber *et al.*, 2002). In some genetic backgrounds, the Reeler mutation also causes neurodegeneration and early death, but these phenotypes are not detected in other backgrounds and are likely to be indirect (Brich *et al.*, 2003; Goffinet, 1990).How Reelin, the Reeler gene product, creates these different phenotypes is still incompletely understood.

The discovery that Reelin is secreted (D'Arcangelo *et al.*, 1995; Ogawa *et al.*, 1995) and acts nonautonomously, as shown by chimera experiments (Terashima *et al.*, 1986; Yoshiki and Kusakabe, 1998), led to the idea it might be a positional signal. The discoveries of other mutations that recapitulate the Reeler phenotype (Sweet *et al.*, 1996; Trommsdorff *et al.*, 1999; Yoneshima *et al.*, 1997) but act (more or less) cell autonomously (Hammond *et al.*, 2001; Yang *et al.*, 2002), led to the identification of proteins that are required in cells receiving the Reelin signal. Biochemical experiments in cultured neurons and brain slices, backed up by *in vivo* observations, provide a reasonable model (described in Section 2) for how these components fit together to form a Reelin-activated signaling pathway (reviewed by Gupta *et al.*, 2002; Jossin, 2004; Lambert de Rouvroit and Goffinet, 2001; Rice and Curran, 2001). This pathway seems to be needed for almost all of the known Reelin functions—LTP and dendrite outgrowth, as well as positioning of most classes of neurons. However, there are three outstanding questions:

- When and where does Reelin act during development of the cortex and cerebellum? Both migrating neurons and the radial glia are affected by Reelin signaling (Dulabon *et al.*, 2000; Forster *et al.*, 2002; Hartfuss *et al.*, 2003; Hunter-Schaedle, 1997; Olson *et al.*, 2006; Pinto-Lord *et al.*, 1982; Tabata and Nakajima, 2002; Zhao *et al.*, 2004), but it is not clear whether abnormal lamination in the Reeler mutants results from defects in the neurons, radial glia, or both.
- At the cellular level, what is the effect of Reelin? Is it a change in radial glia differentiation or branching or attachment at the pial surface? Or does Reelin regulate the leading edge, nucleokinesis, or cell–cell adhesion of migrating neurons? Or all of the above?
- What other genes and proteins link the known intracellular signaling components to the Reelin-regulated cellular responses that result in a normal brain structure?

The purpose of this chapter is to briefly review the core components and signaling mechanism of the Reelin pathway, and then to present evidence on possible downstream components. Issues related to the first two questions, the timing and site of Reelin action and the possible changes in cell biology, are left for other chapters.

2 Core Components of the Reelin Signaling Pathway

The central core of the Reelin response is shown in Fig. 13.1 This pathway operates in all known Reelin-induced events, including neuron positioning, dendrite outgrowth, and LTP, with the conspicuous exception of the positioning of GnRH neurons (Cariboni *et al.*, 2005). In the core pathway, Reelin binding to either or both of the Reelin receptors, VLDLR and ApoER2, induces receptor clustering, which is translated, on the inside of the membrane, into the mutual activation of the tyrosine kinases Src and Fyn and the phosphorylation on tyrosine of the adapter protein Dab1. Evidence supporting this core pathway has come from a wealth of experiments, described below, that have made use of mouse mutants, cultured neuronal progenitor cells, differentiating neurons, brain slices, and reconstituted systems.

2.1 The Reelin Receptors

VLDLR and ApoER2 are two transmembrane apolipoprotein E (ApoE) receptors that are expressed in the developing CNS. ApoER2 is expressed at higher levels in the cortex and hippocampus, and VLDLR at higher levels in the developing cerebellum (Trommsdorff *et al.*, 1999), but it appears that, in most respects, their roles are highly overlapping and essentially redundant. An important exception, a nonredundant role of a specific splice form of ApoER2 in Reelin-regulated LTP, will be discussed briefly in Section 3.1.1.

Trommsdorff *et al.* (1999) first showed that *apoer2-/-* mice have subtle developmental defects in the cortex and hippocampus, while *vldlr-/-* have a defective cerebellum, but the double mutant develops the Reeler phenotype. The ectodomains of ApoER2 and VLDLR bind to recombinant Reelin—specifically the central four repeats (R3–7) (Jossin *et al.*, 2004)—with similar affinity (D'Arcangelo *et al.*, 1999; Hiesberger *et al.*, 1999). Reelin may bind to other cell surface proteins, but functional

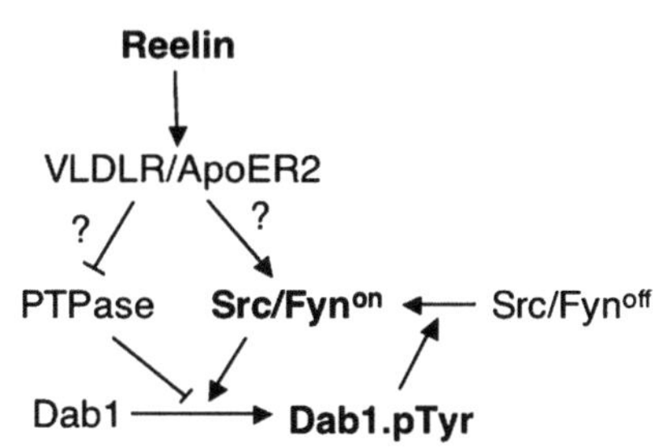

Fig. 13.1 Core components of the Reelin signaling pathway. This figure shows proteins and causality relationships demonstrated by study of biochemical events and phenotypes of mutant mice and neurons

blockade or knockout of VLDLR and ApoER2 is sufficient to block Reelin-induced activation of Dab1 (Benhayon *et al.*, 2003; D'Arcangelo *et al.*, 1999; Hiesberger *et al.*, 1999), and expression of either receptor on NIH3T3 cells allows coexpressed Dab1 to be tyrosine phosphorylated in response to added Reelin (Mayer *et al.*, 2006). Furthermore, antibody-induced cross-linking of either VLDLR or ApoER2 on cultured neurons activates Dab1 (Jossin *et al.*, 2004; Strasser *et al.*, 2004). The cytoplasmic tails of VLDLR and ApoER2 contain canonical endocytosis motifs found in all of the LDLR family, FXNPXY, and these motifs bind to the Dab1 adapter protein (Homayouni *et al.*, 1999; Howell *et al.*, 1999b). A mutant of ApoER2, in which the NFDNPVY stretch was replaced with an EIGNPVY signal, does not bind Dab1 and shows classic Reeler-like defects in the cortex, hippocampus, cerebellum, and LTP (Beffert *et al.*, 2006). These results demonstrate that VLDLR and ApoER2 are necessary for Reelin-dependent brain development *in vivo* and are necessary and sufficient for Reelin-dependent Dab1 phosphorylation in neurons.

2.2 Src-Family Kinases

Src and Fyn are two Src-family kinases (SFKs), which contain an N-terminal myristoylation site for membrane association, SH3 and SH2 domains for intramolecular regulation and for protein–protein interactions when activated, a tyrosine kinase domain, and a C-terminal inhibitory phosphorylation site (Sicheri and Kuriyan, 1997). SFKs undergo conformation switching between an inactive, C-terminally phosphorylated state and a fully activated state in which the C-terminus is dephosphorylated and the activation loop of the kinase domain is phosphorylated. Both Src and Fyn are highly expressed in the developing CNS. *Src-/-* mice have no obvious brain defects (Soriano *et al.*, 1991). *Fyn-/-* mice have an undulating instead of straight layer of pyramidal neurons in the CA3 region of the hippocampus (Grant *et al.*, 1992), which does not resemble the Reeler phenotype. Also, *fyn-/-* layer V projection neurons have misaligned dendritic arbors (Sasaki *et al.*, 2002). Interestingly, this is exacerbated in semaphorin 3a (*sema3a-/-*) mutants, and projection neurons of all layers are misoriented in Reeler mutants. Late-generated neurons of *fyn-/-* cortex undermigrate (Yuasa *et al.*, 2004), resembling Reeler. However, when both Src and Fyn are mutants, a more obvious Reeler phenotype results, with inefficient preplate splitting, inverted cortical plate, and undermigration of Purkinje cells (Kuo *et al.*, 2005). Unfortunately, it is not known whether a complete Reeler phenotype results, because double mutant pups die perinatally.

Neurons from *src-/- fyn-/-* embryos have greatly reduced tyrosine phosphorylation of Dab1 when stimulated with Reelin (Kuo *et al.*, 2005). Moreover, pharmacological inhibition of SFKs impairs Dab1 phosphorylation in neurons (Arnaud *et al.*, 2003b; Bock and Herz, 2003). Application of SFK inhibitors to cortical plate cultures prevents preplate splitting *in vitro* (Jossin *et al.*, 2003b). Reelin stimulates SFK activity when added to neurons (Arnaud *et al.*, 2003b; Bock and Herz, 2003). This activation is rather slight, suggesting that a small proportion of the total kinase population is activated at any one time. The activation depends on both Reelin and Dab1, suggesting that Dab1 is needed for SFK activation, as well as vice versa (Ballif *et al.*, 2003). There is thus a mutual interdependence (positive feedback loop), linking SFKs and Dab1 as dual outcomes of Reelin receptor clustering on the surface.

As is the case for VLDLR and ApoER2, Src and Fyn do not have completely overlapping roles. Overall, the absence of Fyn has a bigger effect than absence of Src, and Src only plays a detectable role when Fyn is absent (Arnaud *et al.*, 2003b; Bock and Herz, 2003). However, Src and Fyn may play different roles in different types of neurons, and, given that Fyn is more closely associated with lipid rafts than Src, they may contribute differently to Reelin signaling in different membrane microdomains.

2.3 Dab1

Dab1 is a cytoplasmic protein that lacks a catalytic domain and appears to act as an adapter, to create signaling complexes via protein–protein interactions (Howell *et al.*, 1997a). Dab1 gene deletion, or reduced expression due to retrotransposon insertion, causes the Reeler phenotype (Howell *et al.*, 1997b; Sheldon *et al.*, 1997; Ware *et al.*, 1997). At the protein level, Dab1 is activated by tyrosine phosphorylation in Reelin-stimulated cultures, and the stoichiometry of Dab1 phosphorylation is decreased in embryonic brains of Reeler mutants in which Reelin is missing (Howell *et al.*, 1999a). Dab1 tyrosine phosphorylation is also completely, or nearly completely, abolished in *vldlr-/- apoer2-/-* and in *src-/- fyn-/-* neurons and brains (Benhayon *et al.*, 2003; Kuo *et al.*, 2005) or by SFK inhibitors *in vitro* (Arnaud *et al.*, 2003b; Bock and Herz, 2003). Dab1 is an *in vitro* substrate for Src and Fyn, binds to Src and Fyn SH2 domains after it is tyrosine phosphorylated, and activates Src and Fyn when coexpressed (Bock and Herz, 2003; Howell *et al.*, 1997a), although very little if any Src or Fyn forms a stable complex with Dab1 in Reelin-stimulated neurons (Bock *et al.*, 2003). Finally, a mutant form of Dab1, Dab1 5F, in which the potential phosphorylation sites are mutated to phenylalanine, is not phosphorylated *in vivo* and cannot support normal brain development (Howell *et al.*, 2000). Dab1 5F homozygous animals exhibit all the phenotypes of Reeler mutants that have been studied, proving that Dab1 phosphorylation is required for its Reelin-induced functions.

Dab1 contains an N-terminal PTB-related domain which can bind to PIP_2 (Howell *et al.*, 1999b; Stolt *et al.*, 2003; Yun *et al.*, 2003). Indeed, even though much of the Dab1 in neurons seems to be cytoplasmic, a fraction partitions to the membrane. The Dab1 PTB domain has a second binding site for the FXNPXY signal found in the cytoplasmic tails of the Reelin receptors VLDLR and ApoER2 and in proteins related to amyloid precursor protein (APP) (Homayouni *et al.*, 1999; Howell *et al.*, 1999b). Dab1 binds independently and noncooperatively to PIP_2 and NPXY (Howell *et al.*, 1999b; Stolt *et al.*, 2003, 2004). PIP_2 binding promotes basal (Reelin-independent) Dab1 tyrosine phosphorylation (Huang *et al.*, 2005), but both binding sites are needed for Reelin-stimulated Dab1 tyrosine phosphorylation in neurons (Stolt *et al.*, 2005; Xu *et al.*, 2005). Even though Dab1 activation is driven by ApoER2/VLDLR clustering (Strasser *et al.*, 2004), Dab1 binding to the receptors in neurons is barely detectable by coimmunoprecipitation (Bock *et al.*, 2003), and it is not known whether Dab1 binding to receptors increases if they are clustered. However, small regions of Dab1 immunoreactivity are transiently detected adjacent to sites of immunoreactive Reelin on the surface of Reelin-stimulated neurons (Morimura *et al.*, 2005), suggesting mutual coclustering. Why this makes clustered Dab1 a better substrate for SFKs [or a worse substrate for phosphotyrosine phosphatases (PTPs)] remains unclear. Conformational changes may be involved, or an as yet unidentified tyrosine kinase may initiate the process.

One question raised by the PIP_2-binding activity of the Dab1 PTB domain is whether signaling occurs preferentially from lipid-ordered membrane microdomains (lipid rafts) which are enriched for cholesterol and PIP_2. ApoER2 partitions to lipid rafts (Riddell *et al.*, 2001), as does Fyn and some Src (Mukherjee *et al.*, 2003). Cholesterol depletion of neurons reduces Reelin-induced Dab1 tyrosine phosphorylation *in vitro* (Bock *et al.*, 2003). However, VLDLR is excluded from lipid rafts (Mayer *et al.*, 2006) but can activate Dab1 in the absence of ApoER2, leaving open the role of lipid rafts.

Potential tyrosine phosphorylation sites in Dab1 were mapped by mutation to four tyrosine residues (a, Y185; b, Y198; c, Y220; d, Y232) contained in four repeated exons. These residues come in pairs (a and b; c and d) of very similar local sequence (Fig. 13.2). Cotransfected Src causes phosphorylation of Dab1 at multiple sites, and mutants lacking various combinations of sites show progressive loss of phosphotyrosine (Howell *et al.*, 2000). Src can phosphorylate all four sites, as detected by specific phosphopeptide antibodies (Keshvara *et al.*, 2001) and, for sites b and c, by mass spectrometry (Howell *et al.*, 2000). In neurons, Reelin stimulates phosphorylation of sites b and c, detected using phosphopeptide antibodies (Keshvara *et al.*, 2001), and site d, detected using phosphopeptide mapping (Ballif *et al.*, 2004) and mass spectrometry (B. Ballif, personal communication). It is not known whether site a is phosphorylated in response to Reelin. There has been one attempt to functionally test the importance of individual phosphorylation sites for Dab1 function *in vivo*: Sanada *et al.* (2004) used *in utero* microinjection to express wild-type or point mutant Dab1 proteins in *dab1-/-* brain and followed the location of the altered neurons using green fluorescent protein. They found that wild-type Dab1 rescued the ability of neurons to enter the cortical plate and to detach from their sister radial glia fiber. Surprisingly, mutation of the b site had no effect on rescue, while mutation of sites c or d abolished rescue. Reciprocally, expression of wild-type Dab1 or b mutant Dab1 had no deleterious effects on neurons in a normal brain, while the c and d mutants acted as dominant negatives. These results suggest essential roles for the c and d sites, and a nonessential (or redundant) role for the b site.

Dab1 is alternatively spliced (Howell *et al.*, 1997a). One form (555*) is relatively highly expressed in the VZ compared to the canonical form (555, or p80) (Jossin *et al.*, 2003a). Knockin mutants demonstrate that p80 can fulfill all of the

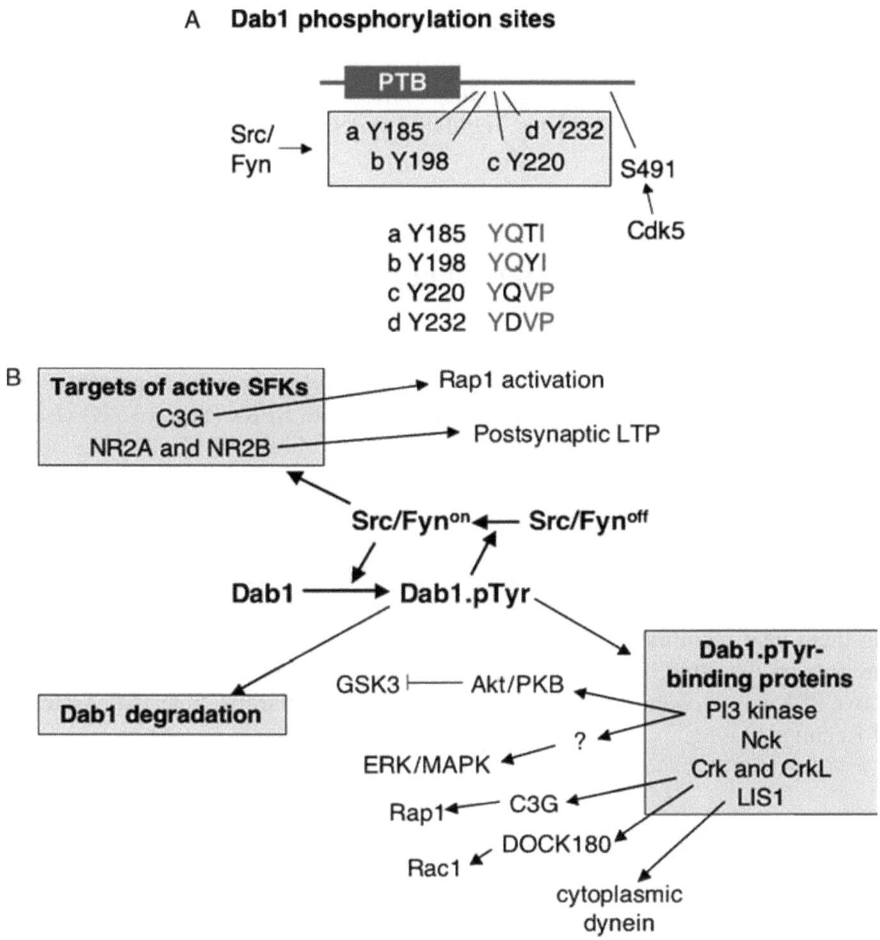

Fig. 13.2 Events downstream of Dab1 phosphorylation and SFK activation. (**A**) Phosphorylation sites in Dab1 that are phosphorylated by SFKs and Cdk5. (**B**) Events that may be important in Reelin signaling are shown separated into two categories: those triggered by active SFKs and those dependent directly on Dab1 phosphorylation (*See Color Plates*)

functions of Dab1 (Howell *et al.*, 2000). Another splice form, p45, which lacks most sequence C-terminal to the PTB domain and tyrosine phosphorylation sites, is expressed at very low level in normal brain but at increased level in *cdk5-/-* and *reelin-/-* brain (Ohshima *et al.*, 2007). p45 can substitute when p80 is absent, but does so inefficiently, such that a single copy of the p45 allele causes a phenotype if p80 is absent (Herrick and Cooper, 2002). Other splice forms, which lack two or more tyrosine phosphorylation sites, were detected in cDNA libraries and may correspond to small proteins detected with certain Dab1 antisera in embryonic brain (Howell *et al.*, 1997a). However, in chicken retina, expression of a short (two site) form of Dab1 is detected in early, proliferating, retinal cells and the long (four site) form of Dab1 in later, differentiating cells (Katyal and Godbout, 2004). Forced expression of the long form induces differentiation in cultured retinal cells (Katyal and Godbout, 2004). Thus, regulated splicing of Dab1 may permit different signaling events.

3 Possible Downstream Signaling Pathways

It is not known how Dab1 phosphorylation leads to downstream signaling. However, there are two obvious mechanisms, either or both of which may be important. First, Dab1 tyrosine phosphorylation is needed to activate SFKs (Fig. 13.2B) (Ballif *et al.*, 2003; Bock and Herz, 2003). Other proteins phosphorylated by the active SFKs may transmit downstream responses. Second, phosphorylated Dab1 may directly regulate other proteins by binding to them. Phosphorylation of proteins such as Dab1 allows them to bind to other proteins that contain SH2, PTB, or (rarely) C2 domains (Sondermann and Kuriyan, 2005). Phosphorylated Dab1 could influence the activity of bound proteins by inducing a conformation change, holding them near membranes, or bringing two bound proteins into proximity so that one acts on the other. In the following sections, we review several ways in which active SFKs or phosphorylated Dab1 may regulate cellular machinery downstream of Reelin signaling.

3.1 Kinase Pathways

3.1.1 SFK Activation and Possible Other Substrates: C3G and NMDAR

Once SFKs are stimulated by Reelin in the presence of phosphorylated Dab1, they may phosphorylate other cell proteins to relay the signal. However, simple examination of Western blots probed with antibodies to phosphotyrosine shows that tyrosine phosphorylation events elicited by Reelin do not occur wholesale, and phosphorylation of many proteins does not change detectably. Indeed, apart from Dab1 itself and the activation loop tyrosine residues of Src and Fyn, few other tyrosine phosphorylation events have been detected.

When the adapter proteins Crk and Crk-like (CrkL) were found to bind to phosphorylated Dab1 (see Section 3.2.2), Ballif *et al.* (2004) investigated whether Reelin might regulate another Crk/CrkL-binding protein, C3G, also known as Rapgef1. Indeed, when immunoprecipitated from Reelin-treated neurons, C3G contains increased levels of phosphotyrosine, while phosphorylation of another Crk-binding protein, p130Cas, did not change (Ballif *et al.*, 2004). C3G tyrosine phosphorylation was also increased in developing normal embryo brain compared with Dab1 5F mutant brain (Ballif *et al.*, 2004). This suggests that C3G tyrosine phosphorylation is increased by Reelin, dependent on Dab1 tyrosine phosphorylation. One possible downstream consequence of this phosphorylation is increased activity of C3G on its substrate, Rap1. C3G is a guanine nucleotide exchange factor, or GEF, for the small GTPase Rap1 (Tanaka *et al.*, 1994). Human C3G activity was previously reported to be stimulated by phosphorylation on a specific tyrosine residue (de Jong *et al.*, 1998; Ichiba *et al.*, 1999). This tyrosine residue is conserved in mouse C3G but its local environment is altered, and it is not known if this same tyrosine is phosphorylated in response to Reelin. However, slight but significant increases in Rap1 activity were detected in Reelin-stimulated neurons (Ballif *et al.*, 2004). Whether this occurs *in vivo* is not clear, but Rap1 has been implicated in regulating the actin cytoskeleton via another small GTPase, Rac1, and in regulating adhesion (Ohba *et al.*, 2001). The importance of C3G or Rap1 in the Reeler phenotype has not been assessed.

Another protein whose phosphorylation is regulated by Reelin is the NMDA receptor. This is important for the effect of Reelin on LTP (Beffert *et al.*, 2005; Chen *et al.*, 2005). The same phosphorylation may or may not occur during Reelin-regulated neuron migrations, but it is unlikely to be important for neuron positioning because it depends on a specific splice form of one of the Reelin receptors, ApoER2, and this splice form (exon 19+) is not needed for normal cortical lamination or cerebellar development (Beffert *et al.*, 2005). Reelin treatment of hippocampal slices sensitizes neurons to glutamate, and this sensitization requires the Reelin receptors, as well as Dab1 and SFK activity (Beffert *et al.*, 2005; Chen *et al.*, 2005). Reelin stimulates the phosphorylation of the NR2A and NR2B subunits of the NMDA receptor (Chen *et al.*, 2005), which are known to regulate channel function (Yu *et al.*, 1997). Both splice forms of ApoER2 bind to NMDA receptors, but only the exon 19+ form of ApoER2 binds to PSD95 (Beffert *et al.*, 2005; Gotthardt *et al.*, 2000), which also binds to NMDA receptors and promotes their phosphorylation by SFKs (Tezuka *et al.*, 1999) and localizes to postsynaptic density (Beffert *et al.*, 2005). These results provide strong evidence that Reelin regulates tyrosine phosphorylation of NR2B via Dab1 and SFKs, and thus regulates LTP. However, the results also prove that this signaling is not needed during Reelin-regulated neuron positioning events.

3.1.2 PI3 Kinase, Akt, and GSK3

Reelin stimulates the protein kinase Akt (also called PKB), leading to phosphorylation and inhibition of another kinase, GSK3, in cultured neurons (Ballif *et al.*, 2003; Beffert *et al.*, 2002). Beffert *et al.* (2002) also detected a decrease in tau phosphorylation, consistent with GSK3 inhibition. Consistent with this pathway operating *in vivo*, mutation of Reelin causes an increased level and activity of GSK3 and tau phosphorylation (Ohkubo *et al.*, 2003). This effect may be related to the observed changes in dendrite morphology in Reeler mutant brain and to the postnatal accumulation of phosphorylated tau and associated neurotoxicity (Brich *et al.*, 2003; Hiesberger *et al.*, 1999). In other cell types, Akt is activated by 3′-phosphoinositides generated through the activity of PI3 kinase (Vanhaesebroeck and Alessi,

2000). A role for PI3 kinase in Reelin activation of Akt was inferred from the effects of various PI3 kinase inhibitors on phosphorylation of Akt and GSK3 but not Dab1 (Beffert *et al.*, 2002; Bock *et al.*, 2003), and small amounts of PI3 kinase were subsequently found to coimmunoprecipitate with Dab1 from Reelin-stimulated neurons (Bock *et al.*, 2003). Whether this binding is involved in or required for PI3 kinase activation remains unknown. An important role for PI3 kinase in cortical plate formation has been inferred from slice culture experiments, in which PI3 kinase inhibitors prevent normal cortical plate formation (Bock *et al.*, 2003), but the relevance *in vivo*, or the roles of Akt and GSK3, are unclear.

3.1.3 MAP Kinase Pathways

Recently, Simo *et al.* (2007) detected an increase in MAP kinase (ERK) activity in Reelin-stimulated cortical neuron cultures. The effect was specific in that two other MAP kinases, p38 and JNK, were not activated, it required SFKs, and was not detected in *dab1-/-* neurons. The level of activity was less than that induced by a growth factor (BDNF) and may have been overlooked by other investigators (Ballif *et al.*, 2003). The mechanism of activation is unclear—in other cell types, ERK is activated via the small GTPase Ras and the upstream kinase Raf1, but in Reelin-stimulated neurons, Ras activity does not appear to increase and, unexpectedly, ERK activation is prevented by PI3 kinase inhibitors (Simo *et al.*, 2007). While Ras-independent ERK activation has been detected in other systems, a role for PI3 kinase independent of Ras and upstream of ERK is unusual but not unprecedented (Schmidt *et al.*, 2004; Takeda *et al.*, 1999; York *et al.*, 1998). ERK activation was also detected by immunofluorescence in chain-migrating subventricular zone neurons stimulated with Reelin, and exposure of these cultures to an ERK inhibitor prevented Reelin-induced scattering of the chain-migrating cells (Simo *et al.*, 2007). Moreover, Reelin was found to induce expression of an early response gene, Egr1, and this induction was inhibited when ERK activation was blocked (Simo *et al.*, 2007). Therefore, ERK activation is involved in at least some aspects of the Reelin response.

3.2 Dab1-Binding Proteins

Some Dab1-binding proteins, including the Reelin receptors (Section 2.1), PI3 kinase (Section 3.1.2), and Lis1 (Section 3.4.1), are discussed in other sections. Here we discuss proteins that bind the PTB domain and phosphotyrosines of Dab1.

3.2.1 Nck

Various methods have been used to identify proteins that might bind to Dab1 after it is phosphorylated. Pramatarova *et al.* (2003) used the yeast two-hybrid system, in the presence of Src, and identified Nckβ (also called Nck2 or Grb4) as a phosphorylation-dependent Dab1-binding partner. In tissue culture, Nckβ but not Nckα (Nck1) could bind to phosphorylated Dab1. Both proteins contain three SH3 domains and one SH2 domain. SH2 domains bind to specific tyrosine-phosphorylated sequences, and SH3 domains to proline-rich peptides (Pawson, 1995). Even though the SH2 domains of Nckα and Nckβ bind comparable sequences *in vitro* (Frese *et al.*, 2006), there are reports that they have different activities in cells (Chen *et al.*, 2000). To test whether Dab1 and Nckβ might affect nonneuronal cells, they were coexpressed in fibroblasts and developing fruitfly eye. In both assays, the effects of combined Dab1 and Nck were worse than with either protein on its own (Pramatarova *et al.*, 2003).

To test whether Nckβ might be involved in Reelin signaling in neurons, Pramatarova *et al.* (2003) showed that Nckβ is expressed in Reelin-responsive Purkinje cells and cortical neurons, and that Nckβ relocalizes with Dab1 to growing neurites when early differentiating neurons are exposed to Reelin. Unfortunately, complexes between phosphorylated Dab1 and Nckβ were not detected in Reelin-stimulated neurons or developing brain, and it is possible that the Reelin-induced relocalization of Nckβ and Dab1 is indirect.

Nck family proteins are involved in axon pathfinding in fruit flies and mice (Cowan and Henkemeyer, 2001; Garrity *et al.*, 1996) and in regulating actin dynamics in mouse fibroblasts and kidney cells (Bladt *et al.*, 2003; Jones *et al.*, 2006), by binding to transmembrane proteins that contain phosphorylated pYDXV sequences (Frese *et al.*, 2006; Jones *et al.*, 2006). No site in Dab1 is a perfect candidate to bind Nck, but sites c and d are implicated (Pramatarova *et al.*, 2003). Nck SH3 domains bind a variety of proteins, of which N-WASP and PAK1 are good candidates to regulate the actin cytoskeleton downstream of Nck *in vivo* (Eden *et al.*, 2002; Li *et al.*, 2001; O'Sullivan *et al.*, 1999; Rohatgi *et al.*, 2001). However, it remains unclear whether Nck is directly regulated by Dab1 during Reelin signaling *in vivo*.

3.2.2 Crk and CrkL

The closely related Crk and CrkL adapter proteins were identified as binding to Dab1 by use of various affinity purification methods (Ballif *et al.*, 2004; Chen *et al.*, 2004; Huang *et al.*, 2004). Importantly, Crk and CrkL form a complex with Dab1 following Reelin stimulation of neurons and in embryonic brain. Indeed, a large fraction of the tyrosine phosphorylated Dab1 can be immunoprecipitated from neurons with antibodies against Crk and CrkL (Ballif *et al.*, 2004). Crk and CrkL both contain a single SH2 domain and one or two SH3 domains (Feller, 2001). The SH2 domain of Crk binds to sequences containing pY(D/k/n)(H/f/r)P (Songyang *et al.*, 1993), in good agreement with the sequences of Dab1 sites c (pYQVP) and d (pYDVP). Indeed, mutation of either or both sites c and d abolishes binding of phosphorylated Dab1 to the Crk or CrkL SH2 domain *in vitro* (Ballif *et al.*, 2004; Huang *et al.*, 2004). Crk/CrkL SH3 domains bind a variety of proteins, including the guanine nucleotide exchange factors DOCK180 (also called DOCK1) and C3G (Hasegawa *et al.*, 1996; Tanaka *et al.*, 1994). Evidence that Reelin stimulates C3G tyrosine phosphorylation and activates its substrate, Rap1, was reported in Section 3.1.1. Evidence that DOCK180 is regulated by Reelin in neurons is lacking, but abnormal fruit-fly eye development caused by ectopic murine Dab1 expression is ameliorated by mutation of the fruit-fly DOCK180 homologue, suggesting that Dab1 can activate endogenous DOCK180 under some conditions.

The importance of Crk/CrkL for Reelin signaling *in vivo* awaits genetic testing. Mutation of the *crkl* gene reportedly causes either no phenotype (Hemmeryckx *et al.*, 2002) or perinatal death (Guris *et al.*, 2001; Hemmeryckx *et al.*, 2002), depending on mouse strain background. A Reeler-like phenotype has not been reported. Crk is more highly expressed than CrkL in the developing cortex. Knockout of the *crk* gene was recently reported to cause late gestation or perinatal lethality (Park *et al.*, 2006), so that conditional mutations in the *crk* or *crkl* or both genes will likely be needed to assess their importance in brain development.

3.2.3 Proteins that Bind to the PTB Domain

In addition to the Reelin receptors VLDLR and ApoER2, several proteins have been described that bind to the Dab1 PTB domain, at least *in vitro* or in tissue culture. These proteins include APP (Homayouni *et al.*, 1999; Howell *et al.*, 1999b), Dab2IP (Homayouni *et al.*, 2003), and N-WASP (Suetsugu *et al.*, 2004). However, even if these interactions occur in neurons, it seems unlikely that binding would be affected by Dab1 tyrosine phosphorylation state, which is essential for downstream signaling (Howell *et al.*, 2000). Therefore, these proteins are more likely to be involved in events upstream of Dab1 activation rather than downstream. For example, they could act as endogenous competitors of the Reelin receptor–Dab1 interaction, and thus increase the threshold level of Reelin needed to activate Dab1.

3.3 Adhesion Molecules

One effect of the Reeler mutation is increased interaction between neurons and radial glia *in vivo* (Pinto-Lord *et al.*, 1982) and abnormal reaggregation of neurons in culture (DeLong and Sidman, 1970; Ogawa *et al.*, 1995). This suggests changes in adhesive properties of neurons. Anton *et al.* (1999) noted that antibodies to integrin α3β1 inhibited migration *in vitro*, and that integrin *α3-/-* cortex showed abnormal layering. Indeed, *α3-/-* neurons migrate abnormally in culture systems (Schmid *et al.*, 2004). A surprisingly direct link between integrin α3β1 and the Reelin pathway came with the report that Reelin, α3β1, and Dab1 could all be coprecipitated from neurons (Dulabon *et al.*, 2000), suggesting that α3β1 is a Reelin receptor. Dab1 binds to integrin tails *in vitro* (Calderwood *et al.*, 2003). However, α3β1 was not needed for Reelin-induced Dab1 tyrosine phosphorylation (Dulabon *et al.*, 2000), and α3β1 binds to the nonessential N-terminus of Reelin (Jossin *et al.*, 2004; Schmid *et al.*, 2005), suggesting that α3β1 may be a nonessential coreceptor. More recently, Sanada *et al.* (2004) showed that α3β1 levels are elevated in *dab1-/-* brain relative to normal brain, and α3β1 levels were locally increased in marked clones of cells in which Dab1 was mutant. RNAi-mediated knockdown of α3 rescued positioning defects caused by inhibiting Dab1 function. These results led to a model where α3β1 may be downregulated following receipt of a Reelin signal, permitting detachment of migrating neurons from radial glia.

These models were questioned when it was found that although targeted knockout of the *β1* gene caused brain abnormalities, the phenotype did not resemble Reeler and was likely due to a requirement for β1 in the basement membrane (Graus-Porta *et al.*, 2001) and defects in radial glia (Forster *et al.*, 2002). However, gene deletion and RNAi knockdown may cause different effects if integrin subunits can switch between different partners with distinct specificities.

3.4 Other Genes Involved in Neuron Migrations: Lis1 and Cdk5

3.4.1 Lis1

The gene encoding Lis1 (Pafah1b1) is mutated in a human lissencephaly syndrome (Reiner *et al.*, 1993). Lis1 biochemistry and biology is complex: it is a component of at least two biochemical complexes, one that catalyzes breakdown of platelet-activating factor, an intracellular second messenger that is present in brain cells, and another that interacts with and regulates cytoplasmic dynein/dynactin complexes (Feng and Walsh, 2001; Reiner, 2000). Mutations of other components of the second complex also cause neuron migration defects—typically slower migration and uncoupling of the centrosome from the nucleus of the migrating cell (Shu *et al.*, 2004). In addition, mutation of Lis1 and Lis1-binding components can affect neurogenesis, possibly by affecting mitotic spindle function (Feng and Walsh, 2004; Sheen *et al.*, 2006). The cellular phenotypes are not obviously related to those caused by mutation of Reelin pathway components (Gupta *et al.*, 2002).

Nevertheless, Assadi *et al.* (2003) tested for genetic and biochemical interactions between the Reelin pathway and Lis1. They made use of *lis1* and *dab1* mutants to test for genetic interactions. The most conspicuous phenotype was hydrocephalus, possibly due to partial blockage of the aqueduct of Sylvius, which implies abnormalities in the ependymal cells and is probably not related directly to changes in Reelin signaling. However, they also detected subtle changes in layering of neurons in the cortex and hippocampus. Lis1+/– has a partial dominant phenotype, with some disorganization of the hippocampus. The hippocampus was more disrupted in *lis1+/– dab1+/–*, although many of these mice had severe hydrocephalus, complicating the interpretation. There was also some disorganization of deep layer cortical plate neurons in the double heterozygotes. It is possible that some of these effects are secondary to a change in neural progenitors in the ventricular zone, since these are the cells that give rise later to the ependyma. If so, this would be the first evidence for a role for Dab1 in the progenitors. Next, they tested for a direct interaction between Dab1 and Lis1 proteins. In tissue culture cells, Lis1 and Dab1 can coimmunoprecipitate, depending on tyrosine phosphorylation of Dab1 at sites b and c and on residues in Lis1 that are mutated in human lissencephaly patients. Strikingly, Dab1 and Lis1 also coimmunoprecipitate from embryonic brain extracts, dependent on Reelin. These results provide strong evidence that Reelin may signal via an inducible Dab1–Lis1 complex to regulate neuron migrations.

3.4.2 Cdk5

The serine/threonine protein kinase, Cdk5, and its activators p35 and p39, have essential functions in neuron migrations in the cortex, hippocampus, and cerebellum, as shown by the phenotypes of *cdk5-/-* (Chae *et al.*, 1997; Ohshima *et al.*, 1996) or double mutant *p35-/- p39-/-* (Ko *et al.*, 2001) mice. However, the *cdk5-/-* phenotype differs from the Reeler phenotype in several important respects. First, in the *cdk5-/-* or *p35-/-* cortex, preplate splitting does occur (Chae *et al.*, 1997; Ohshima *et al.*, 1996), although the splitting is abnormal (Rakic *et al.*, 2006). Second, cortical plate layers are intermixed, rather than inverted as in Reeler (Gupta *et al.*, 2002; Kwon and Tsai, 1998). Nevertheless, there are signs of interactions between the Reelin pathway and Cdk5 at both the genetic and biochemical level.

Several investigators have studied the phenotypes of mice with compound mutations in Cdk5 or p35 and Reelin, Dab1, ApoER2, or VLDLR. Cerebellar and facial motor neuron development is more severely affected in *dab1-/- p35-/-* than in either *dab1-/-* or *p35-/-* (Ohshima *et al.*, 2001, 2002). The fact that p35 plays a role even when Reelin signaling is totally ablated (by *dab1-/-*) means that p35 (and, by inference, Cdk5) functions independently of Reelin. However, Ohshima *et al.* (2001) also found that *dab1–/+ p35-/-* cerebellar Purkinje cells were more disordered than in *p35-/-*, while *dab1–/+* has no phenotype. This could be due to the aforementioned Dab1-independent role of p35, or could be due to interdependence between Dab1 and p35. Beffert *et al.* (2004) found that mutation of *p35-/-* and *apoer2-/-* singly impairs hippocampal development slightly, but double mutation caused a Reeler-like defect. However, *p39-/-* and *vldlr-/-*, each of which singly has little effect on the hippocampus, had no greater effect in combination. It is difficult to exclude that the effects in the hippocampus are not simply additive. More informatively, double mutation of *p35-/-* and *vldlr-/-* or *apoer2-/-* caused invasion of cortical plate neurons into the marginal zone, which is not seen with any single mutant.

Since neither *vldlr-/-* nor *apoer2-/-* totally blocks Reelin signaling, these results suggest that decreased Reelin signaling makes a worse phenotype when p35 is missing. There was also a requirement for p35 or p39 for Reelin-induced LTP in hippocampal cultures. Interpreting these experiments, in which pathways are partly or fully inactivated by homozygous mutation of a component, is difficult, and may mean that the pathways are parallel and converge downstream, or that the components interact.

To test whether Reelin and Cdk5 signaling might intersect directly, several investigators have used biochemical approaches. Initially, Keshvara *et al.* (2002) found that Cdk5 could phosphorylate Dab1 (p80 splice form) at one or more sites near the C-terminus, including serine 491 (S491). There was reduced tyrosine phosphorylation of Dab1 in *cdk5-/-* brain samples, but this may be due to altered access of migrating neurons to Reelin in the mutant cortex since Reelin stimulated Dab1 tyrosine phosphorylation equally in cultured *cdk5-/-* and wild-type neurons. Also, Dab1 was phosphorylated at S491 equally in control and Reeler brains. Similarly, Beffert *et al.* (2004) found no effect of *p35-/-* mutation or Cdk5 inhibitors on Reelin-induced Dab1 tyrosine phosphorylation in neurons, nor was there an effect of Reelin on phosphorylation of Cdk5 substrates. However, Ohshima *et al.* (2007) obtained somewhat different results. In transfected tissue culture cells, Cdk5 inhibited Fyn-mediated p80 tyrosine phosphorylation, and Reelin was more active in inducing Dab1 tyrosine phosphorylation in *cdk5-/-* than control neurons. It is not clear why these results differ from those of Beffert *et al.* (2004) and Keshvara *et al.* (2002), except that Ohshima *et al.* (2007) used neurons from *cdk5-/-* and *cdk5+/+* animals in a *reelin-/-* background, to remove possible effects of Reelin exposure *in vivo*.

If C-terminal phosphorylation of Dab1 by Cdk5 has any effect on development, it may be subtle. Expression of the p45 splice from of Dab1, which lacks the normal C-terminus of p80, rescues the *dab1-/-* phenotype (Herrick and Cooper, 2002). This protein is relatively highly expressed, implying increased stability, but still undergoes Reelin-stimulated tyrosine phosphorylation and degradation. The phenotypic rescue is not perfect, and phenotypes are revealed when gene dosage of the Dab1 p45 allele is reduced. Therefore, the normal C-terminus of Dab1 p80 is not required, but promotes normal brain development. The novel C-terminus of Dab1 p45 may also have a function: Ohshima *et al.* (2007) found a Cdk5 phosphorylation site in the C-terminus of p45, and provided evidence that Cdk5 may stimulate p45 tyrosine phosphorylation and its degradation. However, these results have not been confirmed in neurons.

3.5 Ubiquitin–Proteasome System

Following Reelin-induced tyrosine phosphorylation, Dab1 is targeted for degradation by the ubiquitin–proteasome system (UPS) (Arnaud *et al.*, 2003a; Bock *et al.*, 2004). This mechanism probably underlies the observed increases in Dab1 protein levels in embryonic brains that have mutations in *reelin*, both *vldlr* and *apoer2*, or both *src* and *fyn*, and in the *dab1 5F* knockin mouse (Arnaud *et al.*, 2003b; Howell *et al.*, 2000; Kuo *et al.*, 2005; Rice *et al.*, 1998; Trommsdorff *et al.*, 1999). Reelin-induced degradation in neuron cultures requires SFK activity and the Reelin receptors (Arnaud *et al.*, 2003a; Bock *et al.*, 2004). Since Dab1 tyrosine phosphorylation always correlates with increased SFK activity, the requirement of Dab1 phosphorylation for Dab1 degradation could be explained if the UPS was activated by SFKs. However, in neurons that contain both phosphorylated and nonphosphorylated Dab1, only the phosphorylated Dab1 is degraded (Arnaud *et al.*, 2003a), implying that phosphorylation tags Dab1 for recognition by the UPS.

A direct role for Dab1 ubiquitination or degradation in Reelin signaling has been suggested (Rice *et al.*, 1998). For example, codegradation of Dab1-associated molecules may relay a signal. However, this is currently a matter of conjecture. What is clear is that changes in Dab1 level due to the UPS do affect the ability of neurons to respond to Reelin challenge. Neurons from Reeler mice contain more Dab1 protein and more phosphorylated Dab1 after Reelin challenge, than controls (Howell *et al.*, 1999a). Neurons that were pretreated with Reelin have less Dab1 and less phosphorylated Dab1 after repeat Reelin challenge (Bock *et al.*, 2004). Strikingly, the UPS provides a main mechanism for removing phosphorylated Dab1 from stimulated neurons: inhibiting the UPS allows Dab1 to remain phosphorylated, and SFKs to remain active, for many hours longer than normal (Arnaud *et al.*, 2003a). This points to the UPS as a major negative regulator of Reelin signaling.

3.6 Phosphatases

Phosphotyrosine phosphatases (PTPs) are potentially involved at several steps in the Reelin response, including reversal of Dab1 tyrosine phosphorylation and activation of SFKs. When SFK inhibitors are added to Reelin-stimulated neurons, Dab1 rapidly loses its phosphotyrosine, implying the existence of PTPs that dephosphorylate Dab1 (L. Arnaud, personal communication). In addition, initial activation of SFKs by Reelin implies a switch of phosphotyrosine from the C-terminal inhibitory site to the activation loop of the SFK, again involving one or more PTPs. One candidate PTP that may be involved in SFK activation in neurons is the transmembrane receptor PTP, PTPα (also called RPTPα). This PTP was shown to activate Src by dephosphorylating its C-terminal tyrosine (Zheng *et al.*, 1992). Brains and fibroblasts of mice lacking PTPα have decreased Src and Fyn activity (Ponniah *et al.*, 1999; Su *et al.*, 1999), and close examination reveals lamination defects in the hippocampus and impaired NMDA receptor phosphorylation and decreased LTP. These results suggest that PTPα may be involved in SFK activation in response to Reelin, but this has not been studied directly.

4 Summary and Conclusions

In summary, the core signaling pathway illustrated in Fig. 13.1 mediates almost all known effects of Reelin, but the events occurring downstream of SFK activation and Dab1 tyrosine phosphorylation remain to be critically evaluated. Directed tests for genetic interactions between mutations in Reelin pathway components and Cdk5 or Lis1 components suggest ways that Reelin signaling could impinge on other pathways implicated in neuron migrations. On the other hand, biochemical approaches, some unbiased, have suggested alternative mechanisms, illustrated in Fig. 13.2. It is quite possible, even likely, that SFK-Dab1 will be a branch point in the signaling pathway, and different aspects of the Reeler phenotype may be mediated by different branches of the pathway. Critical analysis in the mouse, and hopefully in more genetically amenable vertebrates, will be needed to fully understand how Reelin regulates brain development and function.

References

Anton, E. S., Kreidberg, J. A., and Rakic, P. (1999). Distinct functions of alpha3 and alpha(v) integrin receptors in neuronal migration and laminar organization of the cerebral cortex. *Neuron* 22:277–289.

Arnaud, L., Ballif, B. A., and Cooper, J. A. (2003a). Regulation of protein tyrosine kinase signaling by substrate degradation during brain development. *Mol. Cell. Biol.* 23:9293–9302.

Arnaud, L., Ballif, B. A., Forster, E., and Cooper, J. A. (2003b). Fyn tyrosine kinase is a critical regulator of disabled-1 during brain development. *Curr. Biol.* 13:9–17.

Assadi, A. H., Zhang, G., Beffert, U., McNeil, R. S., Renfro, A. L., Niu, S., Quattrocchi, C. C., Antalffy, B. A., Sheldon, M., Armstrong, D. D., Wynshaw-Boris, A., Herz, J., D'Arcangelo, G., and Clark, G. D. (2003). Interaction of reelin signaling and Lis1 in brain development. *Nature Genet.* 35:270–276.

Ballif, B. A., Arnaud, L., and Cooper, J. A. (2003). Tyrosine phosphorylation of disabled-1 is essential for reelin-stimulated activation of Akt and Src family kinases. *Brain Res. Mol. Brain Res.* 117:152–159.

Ballif, B. A., Arnaud, L., Arthur, W. T., Guris, D., Imamoto, A., and Cooper, J. A. (2004). Activation of a Dab1/CrkL/C3G/Rap1 pathway in reelin-stimulated neurons. *Curr. Biol.* 14:606–610.

Beffert, U., Morfini, G., Bock, H. H., Reyna, H., Brady, S. T., and Herz, J. (2002). Reelin-mediated signaling locally regulates protein kinase B/Akt and glycogen synthase kinase 3beta. *J. Biol. Chem.* 277:49958–49964.

Beffert, U., Weeber, E. J., Morfini, G., Ko, J., Brady, S. T., Tsai, L. H., Sweatt, J. D., and Herz, J. (2004). Reelin and cyclin-dependent kinase 5-dependent signals cooperate in regulating neuronal migration and synaptic transmission. *J. Neurosci.* 24:1897–1906.

Beffert, U., Weeber, E. J., Durudas, A., Qiu, S., Masiulis, I., Sweatt, J. D., Li, W. P., Adelmann, G., Frotscher, M., Hammer, R. E., and Herz, J. (2005). Modulation of synaptic plasticity and memory by reelin involves differential splicing of the lipoprotein receptor Apoer2. *Neuron* 47:567–579.

Beffert, U., Durudas, A., Weeber, E. J., Stolt, P. C., Giehl, K. M., Sweatt, J. D., Hammer, R. E., and Herz, J. (2006). Functional dissection of Reelin signaling by site-directed disruption of Disabled-1 adaptor binding to apolipoprotein E receptor 2: distinct roles in development and synaptic plasticity. *J. Neurosci.* 26:2041–2052.

Benhayon, D., Magdaleno, S., and Curran, T. (2003). Binding of purified reelin to ApoER2 and VLDLR mediates tyrosine phosphorylation of disabled-1. *Brain Res. Mol. Brain Res.* 112:33–45.

Bladt, F., Aippersbach, E., Gelkop, S., Strasser, G. A., Nash, P., Tafuri, A., Gertler, F. B., and Pawson, T. (2003). The murine Nck SH2/SH3 adaptors are important for the development of mesoderm-derived embryonic structures and for regulating the cellular actin network. *Mol. Cell. Biol.* 23:4586–4597.

Bock, H. H., and Herz, J. (2003). Reelin activates SRC family tyrosine kinases in neurons. *Curr. Biol.* 13:18–26.

Bock, H. H., Jossin, Y., Liu, P., Forster, E., May, P., Goffinet, A. M., and Herz, J. (2003). Phosphatidylinositol 3-kinase interacts with the adaptor protein Dab1 in response to reelin signaling and is required for normal cortical lamination. *J. Biol. Chem.* 278:38772–38779.

Bock, H. H., Jossin, Y., May, P., Bergner, O., and Herz, J. (2004). Apolipoprotein E receptors are required for reelin-induced proteasomal degradation of the neuronal adaptor protein disabled-1. *J. Biol. Chem.* 279:33471–33479.

Brich, J., Shie, F. S., Howell, B. W., Li, R., Tus, K., Wakeland, E. K., Jin, L. W., Mumby, M., Churchill, G., Herz, J., and Cooper, J. A. (2003). Genetic modulation of tau phosphorylation in the mouse. *J. Neurosci.* 23:187–192.

Calderwood, D. A., Fujioka, Y., de Pereda, J. M., Garcia-Alvarez, B., Nakamoto, T., Margolis, B., McGlade, C. J., Liddington, R. C., and Ginsberg, M. H. (2003). Integrin beta cytoplasmic domain interactions with phosphotyrosine-binding domains: a structural prototype for diversity in integrin signaling. *Proc. Natl. Acad. Sci. USA* 100:2272–2277.

Cariboni, A., Rakic, S., Liapi, A., Maggi, R., Goffinet, A., and Parnavelas, J. G. (2005). Reelin provides an inhibitory signal in the migration of gonadotropin-releasing hormone neurons. *Development* 132:4709–4718.

Caviness, V. S., Jr., and Rakic, P. (1978). Mechanisms of cortical development: a view from mutations in mice. *Annu. Rev. Neurosci.* 1:297–326.

Chae, T., Kwon, Y. T., Bronson, R., Dikkes, P., Li, E., and Tsai, L. H. (1997). Mice lacking p35, a neuronal specific activator of Cdk5, display cortical lamination defects, seizures, and adult lethality. *Neuron* 18:29–42.

Chen, K., Ochalski, P. G., Tran, T. S., Sahir, N., Schubert, M., Pramatarova, A., and Howell, B. W. (2004). Interaction between Dab1 and CrkII is promoted by reelin signaling. *J. Cell Sci.* 117:4527–4536.

Chen, M., She, H., Kim, A., Woodley, D. T., and Li, W. (2000). Nckbeta adapter regulates actin polymerization in NIH 3T3 fibroblasts in response to platelet-derived growth factor bb. *Mol. Cell. Biol.* 20:7867–7880.

Chen, Y., Beffert, U., Ertunc, M., Tang, T. S., Kavalali, E. T., Bezprozvanny, I., and Herz, J. (2005). Reelin modulates NMDA receptor activity in cortical neurons. *J. Neurosci.* 25:8209–8216.

Cowan, C. A., and Henkemeyer, M. (2001). The SH2/SH3 adaptor Grb4 transduces B-ephrin reverse signals. *Nature* 413:174–179.

D'Arcangelo, G., Miao, G. G., Chen, S. C., Soares, H. D., Morgan, J. I., and Curran, T. (1995). A protein related to extracellular matrix proteins deleted in the mouse mutant reeler. *Nature (London)* 374:719–723.

D'Arcangelo, G., Homayouni, R., Keshvara, L., Rice, D. S., Sheldon, M., and Curran, T. (1999). Reelin is a ligand for lipoprotein receptors. *Neuron* 24:471–479.

de Jong, R., van Wijk, A., Heisterkamp, N., and Groffen, J. (1998). C3G is tyrosine-phosphorylated after integrin-mediated cell adhesion in normal but not in Bcr/Abl expressing cells. *Oncogene* 17:2805–2810.

DeLong, G. R., and Sidman, R. L. (1970). Alignment defect of reaggregating cells in cultures of developing brains of reeler mutant mice. *Dev. Biol.* 22:584–600.

Dulabon, L., Olson, E. C., Taglienti, M. G., Eisenhuth, S., McGrath, B., Walsh, C. A., Kreidberg, J. A., and Anton, E. S. (2000). Reelin binds alpha3beta1 integrin and inhibits neuronal migration. *Neuron* 27:33–44.

Eden, S., Rohatgi, R., Podtelejnikov, A. V., Mann, M., and Kirschner, M. W. (2002). Mechanism of regulation of WAVE1-induced actin nucleation by Rac1 and Nck. *Nature* 418:790–793.

Feller, S. M. (2001). Crk family adaptors—signalling complex formation and biological roles. *Oncogene* 20:6348–6371.

Feng, Y., and Walsh, C. A. (2001). Protein-protein interactions, cytoskeletal regulation and neuronal migration. *Nature Rev. Neurosci.* 2:408–416.
Feng, Y., and Walsh, C. A. (2004). Mitotic spindle regulation by Nde1 controls cerebral cortical size. *Neuron* 44:279–293.
Forster, E., Tielsch, A., Saum, B., Weiss, K. H., Johanssen, C., Graus-Porta, D., Muller, U., and Frotscher, M. (2002). Reelin, disabled 1, and beta 1 integrins are required for the formation of the radial glial scaffold in the hippocampus. *Proc. Natl. Acad. Sci. USA* 99:13178–13183.
Frese, S., Schubert, W. D., Findeis, A. C., Marquardt, T., Roske, Y. S., Stradal, T. E., and Heinz, D. W. (2006). The phosphotyrosine peptide binding specificity of Nck1 and Nck2 Src homology 2 domains. *J. Biol. Chem.* 281:18236–18245.
Garrity, P. A., Rao, Y., Salecker, I., McGlade, J., Pawson, T., and Zipursky, S. L. (1996). Drosophila photoreceptor axon guidance and targeting requires the dreadlocks SH2/SH3 adapter protein. *Cell* 85:639–650.
Goffinet, A. M. (1990). Cerebellar phenotype of two alleles of the 'reeler' mutation on similar backgrounds. *Brain Res.* 519:355–357.
Gotthardt, M., Trommsdorff, M., Nevitt, M. F., Shelton, J., Richardson, J. A., Stockinger, W., Nimpf, J., and Herz, J. (2000). Interactions of the low density lipoprotein receptor gene family with cytosolic adaptor and scaffold proteins suggest diverse biological functions in cellular communication and signal transduction. *J. Biol. Chem.* 275:25616–25624.
Grant, S. G., O'Dell, T. J., Karl, K. A., Stein, P. L., Soriano, P., and Kandel, E. R. (1992). Impaired long-term potentiation, spatial learning, and hippocampal development in fyn mutant mice. *Science* 258:1903–1910.
Graus-Porta, D., Blaess, S., Senften, M., Littlewood-Evans, A., Damsky, C., Huang, Z., Orban, P., Klein, R., Schittny, J. C., and Muller, U. (2001). Beta1-class integrins regulate the development of laminae and folia in the cerebral and cerebellar cortex. *Neuron* 31:367–379.
Gupta, A., Tsai, L. H., and Wynshaw-Boris, A. (2002). Life is a journey: a genetic look at neocortical development. *Nature Rev. Genet.* 3:342–355.
Guris, D. L., Fantes, J., Tara, D., Druker, B. J., and Imamoto, A. (2001). Mice lacking the homologue of the human 22q11.2 gene CRKL phenocopy neurocristopathies of DiGeorge syndrome. *Nature Genet.* 27:293–298.
Hack, I., Bancila, M., Loulier, K., Carroll, P., and Cremer, H. (2002). Reelin is a detachment signal in tangential chain-migration during postnatal neurogenesis. *Nature Neurosci.* 5:939–945.
Hammond, V., Howell, B., Godinho, L., and Tan, S. S. (2001). disabled-1 functions cell autonomously during radial migration and cortical layering of pyramidal neurons. *J. Neurosci.* 21:8798–8808.
Hartfuss, E., Forster, E., Bock, H. H., Hack, M. A., Leprince, P., Luque, J. M., Herz, J., Frotscher, M., and Gotz, M. (2003). Reelin signaling directly affects radial glia morphology and biochemical maturation. *Development* 130:4597–4609.
Hasegawa, H., Kiyokawa, E., Tanaka, S., Nagashima, K., Gotoh, N., Shibuya, M., Kurata, T., and Matsuda, M. (1996). DOCK180, a major CRK-binding protein, alters cell morphology upon translocation to the cell membrane. *Mol. Cell. Biol.* 16:1770–1776.
Hemmeryckx, B., Reichert, A., Watanabe, M., Kaartinen, V., de Jong, R., Pattengale, P. K., Groffen, J., and Heisterkamp, N. (2002). BCR/ABL P190 transgenic mice develop leukemia in the absence of Crkl. *Oncogene* 21:3225–3231.
Herrick, T. M., and Cooper, J. A. (2002). A hypomorphic allele of dab1 reveals regional differences in reelin-Dab1 signaling during brain development. *Development* 129:787–796.
Hiesberger, T., Trommsdorff, M., Howell, B. W., Goffinet, A., Mumby, M. C., Cooper, J. A., and Herz, J. (1999). Direct binding of reelin to VLDL receptor and ApoE receptor 2 induces tyrosine phosphorylation of disabled-1 and modulates tau phosphorylation. *Neuron* 24:481–489.
Homayouni, R., Rice, D. S., Sheldon, M., and Curran, T. (1999). Disabled-1 binds to the cytoplasmic domain of amyloid precursor-like protein 1. *J. Neurosci.* 19:7507–7515.
Homayouni, R., Magdaleno, S., Keshvara, L., Rice, D. S., and Curran, T. (2003). Interaction of disabled-1 and the GTPase activating protein Dab2IP in mouse brain. *Brain Res. Mol. Brain Res.* 115:121–129.

Howell, B. W., Gertler, F. B., and Cooper, J. A. (1997a). Mouse disabled (mDab1): a Src binding protein implicated in neuronal development. *EMBO J.* 16:121–132.

Howell, B. W., Hawkes, R., Soriano, P., and Cooper, J. A. (1997b). Neuronal position in the developing brain is regulated by mouse *disabled-1. Nature (London)* 389:733–737.

Howell, B. W., Herrick, T. M., and Cooper, J. A. (1999a). Reelin-induced tyrosine phosphorylation of disabled 1 during neuronal positioning. *Genes Dev.* 13:643–648.

Howell, B. W., Lanier, L. M., Frank, R., Gertler, F. B., and Cooper, J. A. (1999b). The disabled 1 phosphotyrosine-binding domain binds to the internalization signals of transmembrane glycoproteins and to phospholipids. *Mol. Cell. Biol.* 19:5179–5188.

Howell, B. W., Herrick, T. M., Hildebrand, J. D., Zhang, Y., and Cooper, J. A. (2000). Dab1 tyrosine phosphorylation sites relay positional signals during mouse brain development. *Curr. Biol.* 10:877–885.

Huang, Y., Magdaleno, S., Hopkins, R., Slaughter, C., Curran, T., and Keshvara, L. (2004). Tyrosine phosphorylated disabled 1 recruits Crk family adapter proteins. *Biochem. Biophys. Res. Commun.* 318:204–212.

Huang, Y., Shah, V., Liu, T., and Keshvara, L. (2005). Signaling through disabled 1 requires phosphoinositide binding. *Biochem. Biophys. Res. Commun.* 331:1460–1468.

Hunter-Schaedle, K. E. (1997). Radial glial cell development and transformation are disturbed in reeler forebrain. *J. Neurobiol.* 33:459–472.

Ichiba, T., Hashimoto, Y., Nakaya, M., Kuraishi, Y., Tanaka, S., Kurata, T., Mochizuki, N., and Matsuda, M. (1999). Activation of C3G guanine nucleotide exchange factor for Rap1 by phosphorylation of tyrosine 504. *J. Biol. Chem.* 274:14376–14381.

Jones, N., Blasutig, I. M., Eremina, V., Ruston, J. M., Bladt, F., Li, H., Huang, H., Larose, L., Li, S. S., Takano, T., Quaggin, S. E., and Pawson, T. (2006). Nck adaptor proteins link nephrin to the actin cytoskeleton of kidney podocytes. *Nature* 440:818–823.

Jossin, Y. (2004). Neuronal migration and the role of reelin during early development of the cerebral cortex. *Mol. Neurobiol.* 30:225–251.

Jossin, Y., Bar, I., Ignatova, N., Tissir, F., Lambert de Rouvroit, C., and Goffinet, A. M. (2003a). The reelin signaling pathway: some recent developments. *Cereb. Cortex* 13:627–633.

Jossin, Y., Ogawa, M., Metin, C., Tissir, F., and Goffinet, A. M. (2003b). Inhibition of SRC family kinases and non-classical protein kinases C induce a reeler-like malformation of cortical plate development. *J. Neurosci.* 23:9953–9959.

Jossin, Y., Ignatova, N., Hiesberger, T., Herz, J., Lambert de Rouvroit, C., and Goffinet, A. M. (2004). The central fragment of reelin, generated by proteolytic processing in vivo, is critical to its function during cortical plate development. *J. Neurosci.* 24:514–521.

Katyal, S., and Godbout, R. (2004). Alternative splicing modulates disabled-1 (Dab1) function in the developing chick retina. *EMBO J.* 23:1878–1888.

Keshvara, L., Benhayon, D., Magdaleno, S., and Curran, T. (2001). Identification of reelin-induced sites of tyrosyl phosphorylation on disabled 1. *J. Biol. Chem.* 276:16008–16014.

Keshvara, L., Magdaleno, S., Benhayon, D., and Curran, T. (2002). Cyclin-dependent kinase 5 phosphorylates disabled 1 independently of reelin signaling. *J. Neurosci.* 22:4869–4877.

Ko, J., Humbert, S., Bronson, R. T., Takahashi, S., Kulkarni, A. B., Li, E., and Tsai, L. H. (2001). p35 and p39 are essential for cyclin-dependent kinase 5 function during neurodevelopment. *J. Neurosci.* 21:6758–6771.

Kuo, G., Arnaud, L., Kronstad-O'Brien, P., and Cooper, J. A. (2005). Absence of Fyn and Src causes a reeler-like phenotype. *J. Neurosci.* 25:8578–8586.

Kwon, Y. T., and Tsai, L. H. (1998). A novel disruption of cortical development in p35(-/-) mice distinct from reeler. *J. Comp. Neurol.* 395:510–522.

Lambert de Rouvroit, C., and Goffinet, A. M. (2001). Neuronal migration. *Mech. Dev.* 105:47–56.

Li, W., Fan, J., and Woodley, D. T. (2001). Nck/Dock: an adapter between cell surface receptors and the actin cytoskeleton. *Oncogene* 20:6403–6417.

Mayer, H., Duit, S., Hauser, C., Schneider, W. J., and Nimpf, J. (2006). Reconstitution of the reelin signaling pathway in fibroblasts demonstrates that Dab1 phosphorylation is independent of receptor localization in lipid rafts. *Mol. Cell. Biol.* 26:19–27.

Morimura, T., Hattori, M., Ogawa, M., and Mikoshiba, K. (2005). Disabled1 regulates the intracellular trafficking of reelin receptors. *J. Biol. Chem.* 280:16901–16908.

Mukherjee, A., Arnaud, L., and Cooper, J. A. (2003). Lipid-dependent recruitment of NSrc to lipid rafts in the brain. *J. Biol. Chem.* 278:40806–40814.

Niu, S., Renfro, A., Quattrocchi, C. C., Sheldon, M., and D'Arcangelo, G. (2004). Reelin promotes hippocampal dendrite development through the VLDLR/ApoER2-Dab1 pathway. *Neuron* 41:71–84.

Ogawa, M., Miyata, T., Nakajima, K., Yagyu, K., Seike, M., Ikenaka, K., Yamamoto, H., and Mikoshiba, K. (1995). The reeler gene-associated antigen on Cajal-Retzius neurons is a crucial molecule for laminar organization of cortical neurons. *Neuron* 14:899–912.

Ohba, Y., Ikuta, K., Ogura, A., Matsuda, J., Mochizuki, N., Nagashima, K., Kurokawa, K., Mayer, B. J., Maki, K., Miyazaki, J., and Matsuda, M. (2001). Requirement for C3G-dependent Rap1 activation for cell adhesion and embryogenesis. *EMBO J.* 20:3333–3341.

Ohkubo, N., Lee, Y. D., Morishima, A., Terashima, T., Kikkawa, S., Tohyama, M., Sakanaka, M., Tanaka, J., Maeda, N., Vitek, M. P., and Mitsuda, N. (2003). Apolipoprotein E and reelin ligands modulate tau phosphorylation through an apolipoprotein E receptor/disabled-1/glycogen synthase kinase-3beta cascade. *FASEB J.* 17:295–297.

Ohshima, T., Ward, J. M., Huh, C. G., Longenecker, G., Veeranna, Pant, H. C., Brady, R. O., Martin, L. J., and Kulkarni, A. B. (1996). Targeted disruption of the cyclin-dependent kinase 5 gene results in abnormal corticogenesis, neuronal pathology and perinatal death. *Proc. Natl. Acad. Sci. USA* 93:11173–11178.

Ohshima, T., Ogawa, M., Veeranna, Hirasawa, M., Longenecker, G., Ishiguro, K., Pant, H. C., Brady, R. O., Kulkarni, A. B., and Mikoshiba, K. (2001). Synergistic contributions of cyclin-dependent kinase 5/p35 and reelin/Dab1 to the positioning of cortical neurons in the developing mouse brain. *Proc. Natl. Acad. Sci. USA* 98:2764–2769.

Ohshima, T., Ogawa, M., Takeuchi, K., Takahashi, S., Kulkarni, A. B., and Mikoshiba, K. (2002). Cyclin-dependent kinase 5/p35 contributes synergistically with reelin/Dab1 to the positioning of facial branchiomotor and inferior olive neurons in the developing mouse hindbrain. *J. Neurosci.* 22:4036–4044.

Ohshima, T., Suzuki, H., Morimura, T., Ogawa, M., and Mikoshiba, K. (2007). Modulation of reelin signaling by cyclin-dependent kinase 5. *Brain Res.* 1140:84–95.

Olson, E. C., Kim, S., and Walsh, C. A. (2006). Impaired neuronal positioning and dendritogenesis in the neocortex after cell-autonomous Dab1 suppression. *J. Neurosci.* 26:1767–1775.

O'Sullivan, E., Kinnon, C., and Brickell, P. (1999). Wiskott-Aldrich syndrome protein, WASP. *Int. J. Biochem. Cell Biol.* 31:383–387.

Park, T. J., Boyd, K., and Curran, T. (2006). Cardiovascular and craniofacial defects in Crk-null mice. *Mol. Cell. Biol.* 26:6272–6282.

Pawson, T. (1995). Protein modules and signalling networks. *Nature (London)* 373:573–580.

Pinto-Lord, M. C., Evrard, P., and Caviness, V. S., Jr. (1982). Obstructed neuronal migration along radial glial fibers in the neocortex of the reeler mouse: a Golgi-EM analysis. *Brain Res.* 256:379–393.

Ponniah, S., Wang, D. Z., Lim, K. L., and Pallen, C. J. (1999). Targeted disruption of the tyrosine phosphatase PTPalpha leads to constitutive downregulation of the kinases Src and Fyn. *Curr. Biol.* 9:535–538.

Pramatarova, A., Ochalski, P. G., Chen, K., Gropman, A., Myers, S., Min, K. T., and Howell, B. W. (2003). Nck beta interacts with tyrosine-phosphorylated disabled 1 and redistributes in reelin-stimulated neurons. *Mol. Cell. Biol.* 23:7210–7221.

Rakic, S., Davis, C., Molnar, Z., Nikolic, M., and Parnavelas, J. G. (2006). Role of p35/Cdk5 in preplate splitting in the developing cerebral cortex. *Cereb. Cortex* 16(Suppl. 1):i35–45.

Reiner, O. (2000). LIS1. let's interact sometimes… (part 1). *Neuron* 28:633–636.

Reiner, O., Carrozzo, R., Shen, Y., Wehnert, M., Faustinella, F., Dobyns, W. B., Caskey, C. T., and Ledbetter, D. H. (1993). Isolation of a Miller-Dieker lissencephaly gene containing G protein beta-subunit-like repeats. *Nature* 364:717–721.

Rice, D. S., and Curran, T. (2001). Role of the reelin signaling pathway in central nervous system development. *Annu. Rev. Neurosci.* 24:1005–1039.

Rice, D. S., Sheldon, M., D'Arcangelo, G., Nakajima, K., Goldowitz, D., and Curran, T. (1998). Disabled-1 acts downstream of reelin in a signaling pathway that controls laminar organization in the mammalian brain. *Development* 125:3719–3729.

Riddell, D. R., Sun, X. M., Stannard, A. K., Soutar, A. K., and Owen, J. S. (2001). Localization of apolipoprotein E receptor 2 to caveolae in the plasma membrane. *J. Lipid Res.* 42:998–1002.

Rohatgi, R., Nollau, P., Ho, H. Y., Kirschner, M. W., and Mayer, B. J. (2001). Nck and phosphatidylinositol 4,5-bisphosphate synergistically activate actin polymerization through the N-WASP-Arp2/3 pathway. *J. Biol. Chem.* 276:26448–26452.

Rossel, M., Loulier, K., Feuillet, C., Alonso, S., and Carroll, P. (2005). Reelin signaling is necessary for a specific step in the migration of hindbrain efferent neurons. *Development* 132:1175–1185.

Sanada, K., Gupta, A., and Tsai, L. H. (2004). Disabled-1-regulated adhesion of migrating neurons to radial glial fiber contributes to neuronal positioning during early corticogenesis. *Neuron* 42:197–211.

Sasaki, Y., Cheng, C., Uchida, Y., Nakajima, O., Ohshima, T., Yagi, T., Taniguchi, M., Nakayama, T., Kishida, R., Kudo, Y., Ohno, S., Nakamura, F., and Goshima, Y. (2002). Fyn and Cdk5 mediate semaphorin-3A signaling, which is involved in regulation of dendrite orientation in cerebral cortex. *Neuron* 35:907–920.

Schmid, R. S., Shelton, S., Stanco, A., Yokota, Y., Kreidberg, J. A., and Anton, E. S. (2004). alpha3beta1 integrin modulates neuronal migration and placement during early stages of cerebral cortical development. *Development* 131:6023–6031.

Schmid, R. S., Jo, R., Shelton, S., Kreidberg, J. A., and Anton, E. S. (2005). Reelin, integrin and Dab1 interactions during embryonic cerebral cortical development. *Cereb. Cortex* 15:1632–1636.

Schmidt, E. K., Fichelson, S., and Feller, S. M. (2004). PI3 kinase is important for Ras, MEK and Erk activation of Epo-stimulated human erythroid progenitors. *BMC Biol.* 2:7.

Sheen, V. L., Ferland, R. J., Harney, M., Hill, R. S., Neal, J., Banham, A. H., Brown, P., Chenn, A., Corbo, J., Hecht, J., Folkerth, R., and Walsh, C. A. (2006). Impaired proliferation and migration in human Miller-Dieker neural precursors. *Ann. Neurol.* 60:137–144.

Sheldon, M., Rice, D. S., D'Arcangelo, G., Yoneshima, H., Nakajima, K., Mikoshiba, K., Howell, B. W., Cooper, J. A., Goldowitz, D., and Curran, T. (1997). Scrambler and yotari disrupt the *disabled* gene and produce a *reeler*-like phenotype in mice. *Nature (London)* 389:730–733.

Shu, T., Ayala, R., Nguyen, M. D., Xie, Z., Gleeson, J. G., and Tsai, L. H. (2004). Ndel1 operates in a common pathway with LIS1 and cytoplasmic dynein to regulate cortical neuronal positioning. *Neuron* 44:263–277.

Sicheri, F., and Kuriyan, J. (1997). Structures of Src-family tyrosine kinases. *Curr. Opin. Struct. Biol.* 7:777–785.

Simo, S., Pujadas, L., Segura, M. F., La Torre, A., Del Rio, J. A., Urena, J. M., Comella, J. X., and Soriano, E. (2007). Reelin induces the detachment of postnatal subventricular zone cells and the expression of the Egr-1 through Erk1/2 activation. *Cereb. Cortex* 17:294–303.

Sondermann, H., and Kuriyan, J. (2005). C2 can do it, too. *Cell* 121:158–160.

Songyang, Z., Shoelson, S. E., Chaudhuri, M., Gish, G., Pawson, T., Haser, W. G., King, F., Roberts, T., Ratnofsky, S., Lechleider, R. J., Neel, B. G., Birge, R. B., Fajardo, J. E., Chou, M. M., Hanafusa, H., Schaffhausen, B., and Cantley, L. C. (1993). SH2 domains recognize specific phosphopeptide sequences. *Cell* 72:767–778.

Soriano, P., Montgomery, C., Geske, R., and Bradley, A. (1991). Targeted disruption of the c-src proto-oncogene leads to osteopetrosis in mice. *Cell* 64:693–702.

Stolt, P. C., Jeon, H., Song, H. K., Herz, J., Eck, M. J., and Blacklow, S. C. (2003). Origins of peptide selectivity and phosphoinositide binding revealed by structures of disabled-1 PTB domain complexes. *Structure (Camb.)* 11:569–579.

Stolt, P. C., Vardar, D., and Blacklow, S. C. (2004). The dual-function disabled-1 PTB domain exhibits site independence in binding phosphoinositide and peptide ligands. *Biochemistry* 43:10979–10987.

Stolt, P. C., Chen, Y., Liu, P., Bock, H. H., Blacklow, S. C., and Herz, J. (2005). Phosphoinositide binding by the disabled-1 PTB domain is necessary for membrane localization and reelin signal transduction. *J. Biol. Chem.* 280:9671–9677.

Strasser, V., Fasching, D., Hauser, C., Mayer, H., Bock, H. H., Hiesberger, T., Herz, J., Weeber, E. J., Sweatt, J. D., Pramatarova, A., Howell, B., Schneider, W. J., and Nimpf, J. (2004). Receptor clustering is involved in reelin signaling. *Mol. Cell. Biol.* 24:1378–1386.

Su, J., Muranjan, M., and Sap, J. (1999). Receptor protein tyrosine phosphatase alpha activates Src-family kinases and controls integrin-mediated responses in fibroblasts. *Curr. Biol.* 9:505–511.

Suetsugu, S., Tezuka, T., Morimura, T., Hattori, M., Mikoshiba, K., Yamamoto, T., and Takenawa, T. (2004). Regulation of actin cytoskeleton by mDab1 through N-WASP and ubiquitination of mDab1. *Biochem. J.* 384:1–8.

Sweet, H. O., Bronson, R. T., Johnson, K. R., Cook, S. A., and Davisson, M. T. (1996). Scrambler, a new neurological mutation of the mouse with abnormalities of neuronal migration. *Mamm. Genome* 7:798–802.

Tabata, H., and Nakajima, K. (2002). Neurons tend to stop migration and differentiate along the cortical internal plexiform zones in the reelin signal-deficient mice. *J. Neurosci. Res.* 69:723–730.

Takeda, H., Matozaki, T., Takada, T., Noguchi, T., Yamao, T., Tsuda, M., Ochi, F., Fukunaga, K., Inagaki, K., and Kasuga, M. (1999). PI 3-kinase gamma and protein kinase C-zeta mediate RAS-independent activation of MAP kinase by a Gi protein-coupled receptor. *EMBO J.* 18:386–395.

Tanaka, S., Morishita, T., Hashimoto, Y., Hattori, S., Nakamura, S., Shibuya, M., Matuoka, K., Takenawa, T., Kurata, T., Nagashima, K., and Matsuda, M. (1994). C3G, a guanine nucleotide-releasing protein expressed ubiquitously, binds to the Src homology 3 domains of CRK and GRB2/ASH proteins. *Proc. Natl. Acad. Sci. USA* 91:3443–3447.

Terashima, T., Inoue, K., Inoue, Y., Yokoyama, M., and Mikoshiba, K. (1986). Observations on the cerebellum of normal-reeler mutant mouse chimera. *J. Comp. Neurol.* 252:264–278.

Tezuka, T., Umemori, H., Akiyama, T., Nakanishi, S., and Yamamoto, T. (1999). PSD-95 promotes Fyn-mediated tyrosine phosphorylation of the N-methyl-D-aspartate receptor subunit NR2A. *Proc. Natl. Acad. Sci. USA* 96:435–440.

Trommsdorff, M., Gotthardt, M., Hiesberger, T., Shelton, J., Stockinger, W., Nimpf, J., Hammer, R. E., Richardson, J. A., and Herz, J. (1999). Reeler/disabled-like disruption of neuronal migration in knockout mice lacking the VLDL receptor and ApoE receptor 2. *Cell* 97:689–701.

Vanhaesebroeck, B., and Alessi, D. R. (2000). The PI3K-PDK1 connection: more than just a road to PKB. *Biochem. J.* 346(Pt 3):561–576.

Ware, M. L., Fox, J. W., Gonzalez, J. L., Davis, N. M., Lambert de Rouvroit, C. L., Russo, C. J., Chua, S. C., Goffinet, A. M., and Walsh, C. A. (1997). Aberrant splicing of a mouse disabled homolog, mdab1, in the scrambler mouse. *Neuron* 19:239–249.

Weeber, E. J., Beffert, U., Jones, C., Christian, J. M., Forster, E., Sweatt, J. D., and Herz, J. (2002). Reelin and ApoE receptors cooperate to enhance hippocampal synaptic plasticity and learning. *J. Biol. Chem.* 7:7.

Xu, M., Arnaud, L., and Cooper, J. A. (2005). Both the phosphoinositide and receptor binding activities of Dab1 are required for reelin-stimulated Dab1 tyrosine phosphorylation. *Brain Res. Mol. Brain Res.* 139:300–305.

Yang, H., Jensen, P., and Goldowitz, D. (2002). The community effect and Purkinje cell migration in the cerebellar cortex: analysis of scrambler chimeric mice. *J. Neurosci.* 22:464–470.

Yip, J. W., Yip, Y. P., Nakajima, K., and Capriotti, C. (2000). Reelin controls position of autonomic neurons in the spinal cord. *Proc. Natl. Acad. Sci. USA* 97:8612–8616.

Yoneshima, H., Nagata, E., Matsumoto, M., Yamada, M., Nakajima, K., Miyata, T., Ogawa, M., and Mikoshiba, K. (1997). A novel neurological mutant mouse, yotari, which exhibits reeler-like phenotype but expresses CR-50 antigen/reelin. *Neurosci. Res.* 29:217–223.

York, R. D., Yao, H., Dillon, T., Ellig, C. L., Eckert, S. P., McCleskey, E. W., and Stork, P. J. (1998). Rap1 mediates sustained MAP kinase activation induced by nerve growth factor. *Nature* 392:622–626.

Yoshiki, A., and Kusakabe, M. (1998). Cerebellar histogenesis as seen in identified cells of normal-reeler mouse chimeras. *Int. J. Dev. Biol.* 42:695–700.

Yu, X. M., Askalan, R., Keil, G. J., 2nd, and Salter, M. W. (1997). NMDA channel regulation by channel-associated protein tyrosine kinase Src. *Science* 275:674–678.

Yuasa, S., Hattori, K., and Yagi, T. (2004). Defective neocortical development in Fyn-tyrosine-kinase-deficient mice. *Neuroreport* 15:819–822.

Yun, M., Keshvara, L., Park, C. G., Zhang, Y. M., Dickerson, J. B., Zheng, J., Rock, C. O., Curran, T., and Park, H. W. (2003). Crystal structures of the dab homology domains of mouse disabled 1 and 2. *J. Biol. Chem.* 278:36572–36581.

Zhao, S., Chai, X., Forster, E., and Frotscher, M. (2004). Reelin is a positional signal for the lamination of dentate granule cells. *Development* 131:5117–5125.

Zheng, X. M., Wang, Y., and Pallen, C. J. (1992). Cell transformation and activation of pp60c-src by overexpression of a protein tyrosine phosphatase. *Nature* 359:336–339.

Chapter 14
The Relationship of Oxytocin and Reelin in the Brain

George D. Pappas and C. Sue Carter

Contents

1 Introduction

Oxytocin (OT) is a small (nine amino acid) peptide, synthesized primarily in the paraventricular and supraoptic nuclei of the hypothalamus. Classically associated with functions such as birth and lactation, OT can also influence social behavior, the hypothalamus–pituitary–adrenal (HPA) axis, and may have a role in the regulation of neural development.

2 Oxytocin and Oxytocin Receptors

OT receptors (OTRs) consist of several transmembrane receptors of the G protein family, distributed throughout the nervous system, especially in hippocampal and hypothalamic nuclei. OTRs also tend to increase initially in the postnatal period

G. D. Pappas
Department of Psychiatry, Psychiatric Institute, College of Medicine, University of Illinois at Chicago, 1601 West Taylor Street MC912, Chicago, IL 60612
e-mail: gdpappas@uic.edu

C. S. Carter
Department of Psychiatry, Psychiatric Institute, College of Medicine, University of Illinois at Chicago, 1601 West Taylor Street MC912, Chicago, IL 60612

S. H. Fatemi (ed.), *Reelin Glycoprotein: Structure, Biology and Roles in Health and Disease.*

and, at least in rats, may decline in adulthood in areas such as the cortex (Tribollet *et al.*, 1989; Shapiro and Insel, 1990), possibly supporting the hypothesis that the effects of OT may differ between developing and adult animals.

OT has been implicated in cellular proliferation; stem cells treated with OT proliferate and begin to express OTRs (Green *et al.*, 2001). Both OT and reelin are particularly significant during development, and the absence of either may interfere with normal brain development. In addition, reelin is critical to the development of the GABAergic system which modulates the release of OT (Liu *et al.*, 2005).

3 Oxytocin and Reelin Haploinsufficient Reeler Mouse Model

Availability of the reelin haploinsufficient (+/–) reeler mouse (HRM) provides a model for examining the role of reelin in the development of the OT system and especially in the expression of the OTR. In rodents, OT expression increases during the immediate postnatal period. Serum levels of OT are low in certain types of autism (Green *et al.*, 2001). It has been reported that OT infusion may reduce repetitive symptoms in autism spectrum disorder (Hollander *et al.*, 2003).

To date, there has been one published study directly examining possible interactions between OTR and reelin (Liu *et al.*, 2005). In this study, we used immunocytochemistry and *in situ* hybridization in the haploinsufficient (+/–) (HRM) versus wild-type (+/+; WTM) adult mice to quantify OTR abundance in the piriform cortex, the neocortex, the hippocampus, and the retrosplenial cortex. Light microscopy of central nervous system (CNS) sections from the normal WTM brain revealed OTR mRNA and protein to be abundant throughout the cortical regions of the brain, including the neocortex, allocortex, and archicortex. In the +/– HRM, the number of OTR-positive cells in the same cortical areas as those studied in the WTM are 25 to 50% decreased from the wild-type mice (Fig. 14.1, 14.2).

Both reelin and OT have been implicated in autism and in schizophrenia, since marked deficits in OTRs and reelin occur in specific cortical regions (Liu *et al.*, 2005). The neocortex and retrosplenial area of the cortex are implicated in both memory and emotion (Carter, 2003). In addition, reductions in OTR binding were apparent in most areas of the hippocampus (archicortex), including the dentate gyrus of mice. These findings on the distribution of OTR in the cortex are consistent with our earlier findings on the downregulation of reelin in these same brain regions (Pappas *et al.*, 2001, 2003). Preliminary analyses of other regions containing OTRs, including the reticular thalamus and the central and basolateral amygdala, do not show striking differences between the wild-type (+/+) and the reeler (+/–) mice (Liu *et al.*, 2005). In general, cortical and hippocampal deficits of either reelin or OTRs might be expected to be associated with reductions in memory and learning, especially in the cortex of the social environment. Many studies have implicated OT in social memory (see Gimpl and Fahrenholz, 2001). In mice, both maternal experience associated with increases in endogenous OT and treatment of virgin mice with intraventricular injection of OT were associated with improved

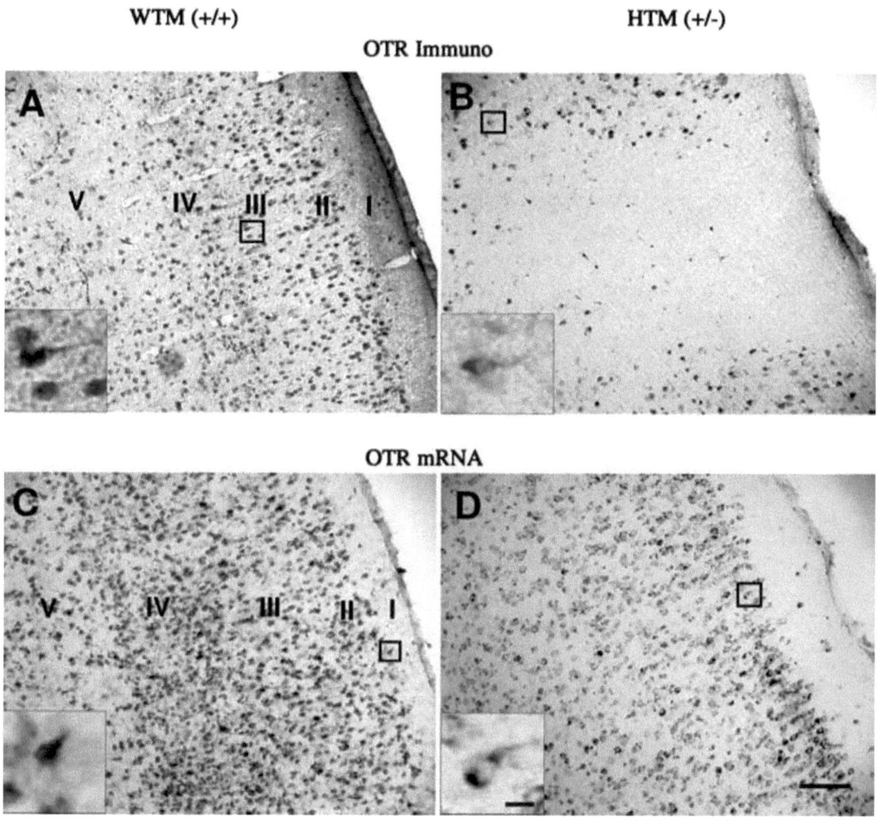

Fig. 14.1 Oxytocin receptor (OTR) **(A, B)** and OTR mRNA **(C, D)** in the motor cortex. Immunolabeling in the wild-type mice (WTM) showed reactivity throughout layers II to V **(A)**, while the haploinsufficient reeler mice (HRM) have lower OTR immuno-reactivity **(B)**. WTM (+/+) show abundant mRNA activity **(C)**, but a lower level of OTR mRNA in the HRM (+/-) **(D)**. Scale bar equals100μm and in inset equals 20μm. (Liu, Pappas, and Carter, 2005; reprinted with permission www.ingentaconnect.com/content/maney/nres)

spatial memory, probably as a result of activity in the hippocampus (Tomizawa *et al.*, 2003).

4 Oxytocin and Autism

Reelin and OT play a role in regulating affect and mood. Downregulation of reelin has been correlated with schizophrenia (Impagnatiello *et al.*, 1998; Fatemi *et al.*, 2000; Costa *et al.*, 2001). It has been proposed that mice haploinsufficient for reelin (+/−) offer an animal model that may also have relevance for the features of autism (Liu *et al.*, 2005; Tueting *et al.*, 2006). Preliminary studies of social behavior in

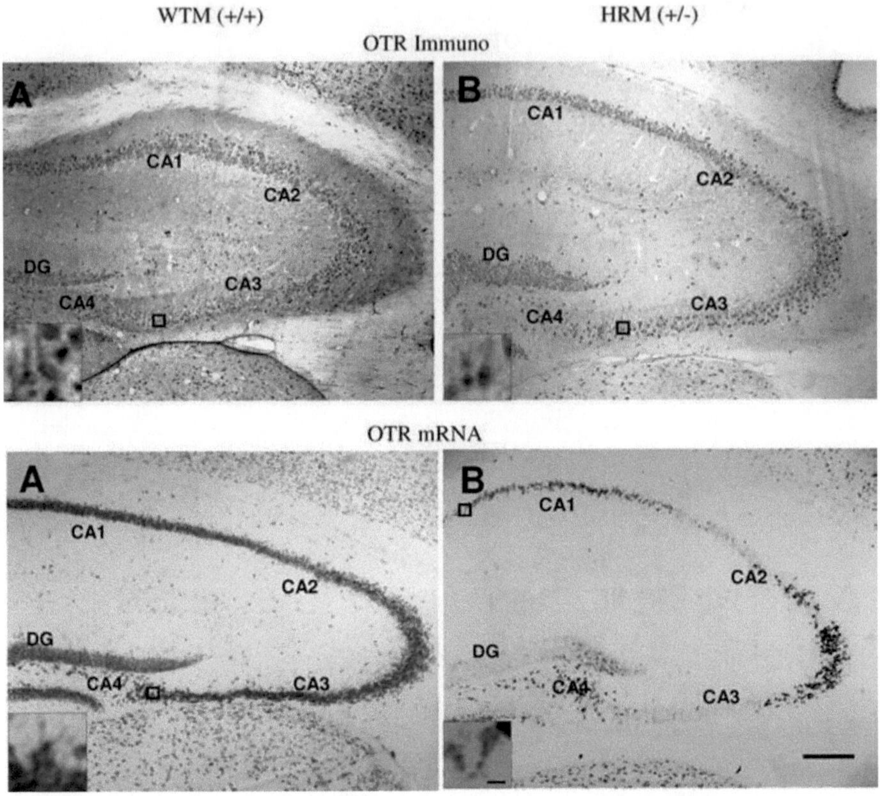

Fig. 14.2 Oxytocin receptor (OTR) **(A, B)** and OTR mRNA **(C, D)** in the hippocampus and dentate gyrus. OTR immunoreactivity is expressed in cells in CA1, 2, 3 and 4 and the dentate gyrus **(A)** in the wild-type mice (WTM), There is a significant change in OTR immuno-reactivity in the haploinsufficient mice (HRM) **(C)**. OTR mRNA also is localized to the cells of CA1, 2, 3 and 4 in the WTM (+/+) **(C)**. However, in the HRM (+/-) reduction are seen in mRNA localization in the hippocampus. All cellular areas of the hippocampus and dentate gyrus show diminished expression of OTR mRNA, with the possible exception of a small portion of CA3 and the hilus of the dentate gyrus **(D)**. Scale bar equals 200μm and in inset equals 10μm. (Liu, Pappas, and Carter, 2005; reprinted with permission www.ingentaconnect.com/content/maney/nres)

(+/−) reeler mice, with levels of reelin that are about half those seen in normal animals, reveal reductions in social interactions and a failure to show social recognition (Doueriri and Guidotti, unpublished data).

We have observed dramatic reductions in OTRs or mRNA for the OTRs (especially in cortical–hippocampal area) in (+/−) reeler mice (Liu *et al.*, 2005). The origins of this difference remain to be studied. However, the promoter region of the reelin gene has CpG islands susceptible to epigenetic methylation (Chen *et al.*, 2002). Hypermethylation of the reelin gene promoter results in a decrease in the transcription of reelin mRNA, due to chromatin and promoter remodeling (Dong *et al.*, 2005). The OTR also contains CpG islands and is susceptible to epigenetic regulation (Kimura *et al.*, 2003). Hypermethylation and gene silencing, induced by developmental events, such as exposure to neuropeptides, differential maternal stimulation (Meaney and Szyf, 2005), or viral infections (Shi *et al.*, 2003), might influence subsequent reelin expression. In rats, the offspring of mothers exposed to high levels of maternal care experienced reduced reactivity to stressors, as well as increased expression of the hippocampal glucocorticoid receptor gene in later life. More recently, these authors have reported that treatment of adult mice with a histone deacetylase inhibitor reversed the effects of low levels of maternal stimulation on reelin (downregulation), while the effects of high levels of early postnatal maternal care (upregulation of reelin) were reversed by methionine, capable of increasing methylation of the gene for reelin (Weaver *et al.*, 2006).

Most studies of reelin-deficient mice have been conducted in males, and thus little is known regarding sex differences in the consequences of reduction in reelin. However, Purkinje cell loss in reelin-deficient mice, compared to the wild-type, was observed only in males. In that study it was reported that "females were spared" (Hadj-Sahraqui *et al.*, 1996). A recent report dealing with Purkinje cell development reports that at ages between P10 and P18, Purkinje cell numbers were decreased in male reelin-deficient mice, while wild-type male and female littermates, and female mice with only one functional gene for reelin (+/rl) displayed normal Purkinje cell numbers. These investigators hypothesize that "reelin haploinsufficiency may be compensated by estrogens" (Assenza *et al.*, 2005). It is also possible that OT alone or in conjunction with estrogen plays a role in the sexually dimorphic consequences of reelin deficiency.

Reelin also regulates GABA. Genes for a subset of the GABA receptors are found on the 15q11-13 loci, another autism susceptibility locus (AUTSL) (Cook *et al.*, 1998), also involved in Prader Willi syndrome, and exhibit an associated deficit in OT. GABA plays a role in many aspects of brain function, especially under conditions when inhibitory processes are required. Functional interactions among OT, reelin, and GABA could be of critical importance to the features of autism.

In addition to OT, another neuropeptide, arginine vasopressin (AVP), has also been implicated in the autism spectrum disorder (ASD) (Insel *et al.*, 1999; Young *et al.*, 2002; Leckman and Herman, 2002; Winslow, 2005; Welch and Ruggiero, 2005). Changes in either OT or AVP or their receptors could be capable of influencing the features of ASD. Relevant to possible mechanisms through which OT or AVP might influence autism, is the fact that the effects of these neuropeptides on brain and behavior are sexually dimorphic, especially during the course of development (Carter, 2003; Bales *et al.*, 2004a,b, 2006; Yamamoto *et al.*, 2004; Bielsky *et al.*, 2005; Thompson *et al.*, 2006). We hypothesize here that AVP, which has a unique role in males, must be present in optimal levels to be protective against ASD. AVP is androgen dependent, and males are more sensitive to AVP, especially during development. Either excess AVP or disruptions in the ASD system could play a role in development of the traits of autism. In contrast, OT, in some cases more abundant in females, normally may be protective.

The capacity of OT or AVP to modify either reelin, the OTR, or AVPRs would have life-long consequences for physiology and behavior. Deficiencies or atypical expression of reelin/GABA, through the primary actions of reelin, or through changes in systems directly dependent on OT or AVP, might confer vulnerability to certain features of autism. Such changes could be manifest as increases in anxiety or low levels of sociality, as well as cognitive deficits resulting from changes in the organization of laminar brain areas.

5 Metabolism, Experience, and Neuropeptides

Research on neuropeptides, such as OT and AVP, tends to focus on the expression of either the peptides themselves or their receptors. However, the functional availability of these compounds depends on their dynamic synthesis and degradation. For example, peptidases and proteases regulate the production and degradation of both OT and AVP (Mitsui *et al.*, 2004; Tsujimoto and Hattori, 2005). These enzymes are regulated by genetics and differ between males and females but can also be affected by diet, salt intake, and stress-related changes in anxiety or trauma across the life span (Maes *et al.*, 1998, 1999, 2001). Of particular interest are prolyl endopeptidases (PEP); these enzymes regulate the metabolism of various peptides, including OT and AVP. Changes in PEPs may have regionally specific effects; for example, inhibition of PEP produces an increase in AVP in the septal nucleus (Miura *et al.*, 1995). Because AVP in the septal area is sexually dimorphic, individual or sex differences in the effects of enzymes such as PEP, especially in this region, offer a possible substrate for individual differences in behavioral reactivity.

Other sources of possible variance in these systems could come from individual differences in lifestyle across the life span. Exercise and fitness have been shown to modulate the autonomic actions of OT, with increased responsivity to OT as a function of increased fitness (Michelini *et al.*, 2003). As another example, diet might have the capacity to influence enzymes that regulate neuropeptides, such as OT and AVP, as well as steroid production (Cameron, 1991). The consequences of diet and exercise, even in early life, might be sexually dimorphic and individually variable.

6 Summary

It is almost impossible to consider the interrelationship of reelin and OT without their role in the development of ASD.

Both reelin and OT play a role in regulating affect and mood. Downregulation of reelin has been strongly correlated with schizophrenia and autism (Fatemi *et al.*, 2005), and it is proposed that reelin haploinsufficient (+/–) reeler mice may serve as a model for neural deficits seen in both schizophrenia and autism (see Liu *et al.*, 2005).

Any theory regarding the causes of ASD might be able to account for the striking male-bias in the occurrence of these disorders. Sex differences in the central regulation and expression of OT and AVP may help in understanding the features of ASD. However, differentiating the possible roles of OT, AVP, and reelin in the features of ASD is not simple.

OT and AVP have some shared functions; for example, both peptides, administered exogenously, can promote positive social interactions and pair bond formation (Cho *et al.*, 1999). In general, centrally active AVP seems to be associated with increased vigilance, anxiety, arousal, and activation, while OT has behavioral and neural effects associated with reduced anxiety, relaxation, growth, and restoration (Carter, 1998; Uvnas-Moberg, 1998). In addition, OT may protect the central and autonomic nervous system against overreactivity or even shutdown, especially in the face of extreme challenges (Porges, 2001).

Exaggerated activity or abnormal activity in systems that rely on AVP, possibly due to increased exposure to androgens (Baron-Cohen, 2002), would be consistent with several features of ASD. Sexually dimorphic effects of OT and AVP, including actions that extend beyond the nervous system to influence metabolic or immune reactions, also might be critical links to uncovering the mechanisms underlying the causes and effects of ASD.

OT is estrogen-dependent and in some cases is higher in females (Carter, 2003). OT can regulate responses to stressors and inflammation and also can be inhibited by stressful experiences. Of potential relevance to ASD is the fact that AVP in the extended amygdale–lateral septal axis of the nervous system is sexually dimorphic (higher in males) (Carter, 2003). In addition, males appear to be more sensitive than females, especially during development, to the actions of AVP. There are several examples in which females appear to be remarkably insensitive to AVP or its absence. Insensitivity to AVP or a lack of dependence on this peptide might be protective in females against the features of ASD. Females might also be protected either directly or indirectly by OT. Experience associated with reductions in fear and an increased sense of safety or trust would be expected to be protective in ASD and related disorders that are characterized by high levels of anxiety (Porges, 2001; Corbett *et al.*, 2006). Sexually dimorphic differences in coping mechanisms, including the willingness to use social interactions to reduce anxiety, could be another mechanism through which males and females might differ in the expression of the features of ASD. Knowledge of natural ways to stimulate the release of endogenous OT or to inhibit "excess" AVP might be protective against the development of the features of ASD, perhaps even remediating the expression of ASD-like behaviors in later life. However, we cannot, at this point, exclude the possibility that disruptions in systems that rely on AVP might also increase the vulnerability to ASD.

References

Assenza, G., Biamonte, F., Cesa, R., Strata, P., and Keller, F. (2005). Interaction between reelin and estrogens on Purkinje cells during development: A model of cerebellar pathology in autism and related disorders. *Soc. Neurosci. Abstr*. 251.2.

Bales, K., Pfeifer, L., and Carter, C. S. (2004a). Sex differences and developmental effects of manipulations of oxytocin on anxiety and alloparenting in prairie voles. *Dev. Psychobiol.* 44:123–131.

Bales, K. L., Kim, A. J., Lewis-Reese, A. D., and Carter, C. S. (2004b). Both oxytocin and vasopressin may influence alloparental care in male prairie voles. *Horm. Behav.* 45:354–361.
Bales, K. L., Kramer, K. M., Lewis-Reese, A. D., and Carter, C. S. (2006). Effects of stress on parental care are sexually dimorphic in prairie voles. *Physiol. Behav.* 87:424–429.
Baron-Cohen, S. (2002). The extreme male brain theory of autism. *Trends Cogn. Sci.* 6:248–254.
Bielsky, I. F., Hu, S. B., and Young, L. J. (2005a). Sexual dimorphism in the vasopressin system: lack of an altered behavioral phenotype in female V1a receptor knockout mice. *Behav. Brain Res.* 164:132–136.
Cameron, J. L. (1991). Metabolic cues for the onset of puberty. *Horm. Res.* 36:97–103.
Carter, C. S. (1998). Neuroendocrine perspectives on social attachment and love. *Psychoneuroendocrinology* 23:779–818.
Carter, C. S. (2003). Developmental consequences of oxytocin. *Physiol. Behav.* 79:383–397.
Chen, Y., Sharma, R., Costa, R. H., Costa, E., and Grayson, D. R. (2002). On the epigenetic regulation of the human reelin promoter. *Nucleic Acids Res.* 3:2930–2939.
Cho, M. M., DeVries, A. C., Williams, J. R., and Carter, C. S. (1999). The effects of oxytocin and vasopressin on partner preferences in male and female prairie voles (Microtus ochrogaster). *Behav. Neurosci.* 113:1071–1080.
Cook, E. H., Jr., Courchesne, R. Y., Cox, M. J., Lord, C., Gonen, D., Cuter, S. J., Lincon, A., Nix, K., Haas, R., Levethanal, B. L., and Courchesne, E. (1998). Linkage-disequilibrium mapping of autistic disorder, with 15q11–13 markers. *Am. J. Hum. Genet.* 62:1077–1083.
Corbett, B. A., Mendoza, S., Abdullah, M., Wegelin, J. A., and Levine, S. (2006). Cortisol circadian rhythms and response to stress in children with autism. *Psychoneuroendocrinology* 31:59–68.
Costa, E., Davis, J., Grayson, D. R., Guidotti, A., Pappas, G. D., and Pesold, C. (2001). Dendritic spine hypoplasticity and downregulation of reelin and GABAergic tone in schizophrenia vulnerability. *Neurobiol. Dis.* 8:723–742.
Dong, E., Agis-Balboa, C., Simonini, M. V., Grayson, D. R., Costa, E., and Guidotti, A. I. (2005). Reelin and glutamic acid decarboxylase$_{67}$ promoter remodeling in an epigenetic methionine-induced mouse model of schizophrenia. *Proc. Natl. Acad. Sci. USA* 102:12578–12583.
Fatemi, S. H., Earle, J. A., and McMenomy, T. (2000). Reduction in reelin immunoreactivity in hippocampus of subjects with schizophrenia, bipolar disorder and major depression. *Mol. Psychiatry* 5:654–663.
Fatemi, S. H., Snow, A. V., Stary, J. M., Araghi-Niknam, M., Reutiman, T. J., Lee, S., Brooks, A. I., and Pearce, D. A. (2005). Reelin signaling is impaired in autism. *Biol. Psychiatry* 57:777–787.
Gimpl, G., and Fahrenholz, F. (2001). The oxytocin receptor system: structure, function, and regulation. *Physiol. Rev.* 81:629–683.
Green, L., Fein, D., Modahl, C., Feinstein, C., Waterhouse, L., and Morris, M. (2001). Oxytocin and autistic disorder: Alterations in peptide forms. *Biol. Psychiatry* 50:609–613.
Hadj-Sahraqui, N., Frederic, F., Delhayi-Bouchaud, N., and Mariani, J. (1996). Gender effect on Purkinje cell loss in the cerebellum of the heterozygous reeler mouse. *J. Neurosci.* 11:45–58.
Hollander, E., Novotny, S., Hanratty, M., Yaffe, R., DeCaria, C. M., Aronwitz, B. R., and Mosovich, S. (2003). Oxytocin infusion reduces repetitive behaviors in adults with autistic and Asperger's disorders. *Neuropsychopharmacology* 28:193–198.
Impagnatiello, F., Guidotti, A. R., Pesold, C., Dwivedi, Y., Caruncho, H., Pisu, M. G., Uzunov, D. P., Smalheiser, N. R., Davis, J. M., Pandey, G. N., Pappas, G. D., Tueting, P., Sharma, R. P., and Costa, E. (1998). A decrease of reelin expression as a putative vulnerability factor in schizophrenia. *Proc. Natl. Acad. Sci. USA* 95:15718–15723.
Insel, T. R., O'Brien, D. J., and Leckman, J. F. (1999). Oxytocin, vasopressin, and autism: Is there a connection? *Biol. Psychiatry* 45:145–157.
Kimura, T., Saji, F., Nishimori, K., Ogita, K., Nakamura, H., Koyama, M., and Murata, Y. (2003). Molecular regulation of the oxytocin receptor in peripheral organs. *J. Mol. Endocrinol.* 30:109–115.

Leckman, J. F., and Herman, A. E. (2002). Maternal behavior and developmental psychopathology. *Biol. Psychiatry* 51:27–43.
Liu, W. S., Pappas, G. D., and Carter, C. S. (2005). Oxytocin receptors are reduced in cortical regions in haploinsufficient reeler (+/–) mice. *Neurol. Res.* 27:339–345.
Maes, M., Goossens, F., Lin, A., De Meester, I., Van Gastel, A., and Scharpe, S. (1998). Effects of psychological stress on serum prolyl endopeptidase and dipeptidyl peptidase IV activity in humans: higher serum prolyl endopeptidase activity is related to stress-induced anxiety. *Psychoneuroendocrinology* 23:485–495.
Maes, M., Lin, A. H., Bonaccorso, S., Goossens, F., Van Gastel, A., Pioli, R., Delmeire, L., and Scharpe, S. (1999). Higher serum prolyl endopeptidase activity in patients with post-traumatic stress disorder. *J. Affect. Disord.* 53:27–34.
Maes, M., Monteleone, P., Bencivenga, R., Goossens, F., Maj, M., van West, D., Bosmans, E., and Scharpe, S. (2001). Lower serum activity of prolyl endopeptidase in anorexia and bulimia nervosa. *Psychoneuroendocrinology* 26:17–26.
Meaney, M. J., and Szyf, M. (2005). Maternal care as a model for experience-dependent chromatin plasticity? *Trends Neurosci.* 28:456–463.
Michelini, L. C., Marcelo, M. C., Amico, J., and Morris, M. (2003). Oxytocinergic regulation of cardiovascular function: studies in oxytocin-deficient mice. *Am. J. Physiol. Heart Circ. Physiol.* 284:H2269–H2276.
Mitsui, T., Nomura, S., Itakura, A., and Mizutaini, S. (2004). Role of aminopeptidases in the blood pressure regulation. *Biol. Pharmacol. Bull.* 27:768–771.
Miura, N., Shibata, S., and Watanabe, S. (1995). Increase in the septal vasopressin content by prolyl endopeptidase inhibitors in rats. *Neurosci. Lett.* 196:128–130.
Pappas, G. D., Kriho, V., and Pesold, C. (2001). Reelin in the extracellular matrix and dendritic spines of the cortex and hippocampus: a comparison between wild-type and heterozygous reeler mice by immunoelectron microscopy. *J. Neurocytol.* 30:413–425.
Pappas, G. D., Kriho, V., Liu, W. S., Tremolizzo, L., Lugli, G., and Larson, J. (2003). Immunocytochemical localization of reelin in the olfactory bulb of the heterozygous reeler mouse: an animal model for schizophrenia. *Neurol. Res.* 25:819–830.
Porges, S. W. (2001). The polyvagal theory: phylogenetic substrates of a social nervous system. *Int. J. Psychophysiol.* 42:123–146.
Shapiro, L. E., and Insel, T. R. (1990). Infant's response to social separation reflects adult differences in affiliative behavior: a comparative developmental study in prairie and montane voles. *Dev. Psychobiol.* 23:375–393.
Shi, L., Fatemi, S. H., Sidwell, R. W., and Patterson, P. H. (2003). Maternal influenza infection causes marked behavioral and pharmacological changes in the offspring. *J. Neurosci.* 23:297–302.
Thompson, R. R., George, K., Walton, J. C., Orr, S. P., and Benson, J. (2006). Sex-specific influences of vasopressin on human social communication. *Proc. Natl. Acad. Sci.. USA* 103:7889–7894.
Tomizawa, K., Iga, N., Lu, Y. F., Moriwaki, A., Matsushita, M., Li, S. T., Miyamoto, O., Itano, T., and Matsui, H. (2003). Oxytocin improves long-lasting spatial memory during motherhood through MAP kinase cascade. *Nature Neurosci.* 6:384–390.
Tribollet, E., Charpak, S., Schmidt, A., Dubois, D. M., and Dreifuss, J. J. (1989). Appearance and transient expression of OT receptors in fetal, infant, peripubertal rat brain studied by autoradiography and electrophysiology. *J. Neurosci.* 9:1764–1773.
Tsujimoto, M., and Hattori, A. (2005). The oxytocinase subfamily of M1 aminopeptidases. *Biochim. Biophys. Acta* 1751:9–18.
Tueting, P., Doueiri, M. S., Guidotti, A., Davis, J. M., and Costa, E. (2006). Reelin downregulation in mice and psychosis endophenotypes. *Neurosci. Biobehav. Rev.* 30:1067–1077.
Uvnas-Moberg, K. (1998). Oxytocin may mediate the benefits of positive social interaction and emotions. *Psychoneuroendocrinology* 23:819–835.
Weaver, I. C., Champagne, F. A., Brown, S. E., Dymov, S., Sharma, S., Meaney, M. J., and Szyf, M. (2006). Reversal of maternal programming of stress responses in adult offspring through

methyl supplementation: altering epigenetic marking later in life. *J. Neurosci.* 25:11045–11054.

Welch, M. G., and Ruggiero, D. A. (2005). Predicted role of secretin and oxytocin in the treatment of behavioral and developmental disorders: implications for autism. *Int. Rev. Neurobiol.* 71:273–314.

Winslow, J. T. (2005). Neuropeptides and non-human primate social deficits associated with pathogenic rearing experience. *Int. J. Dev. Neurosci.* 23:245–251.

Yamamoto, Y., Cushing, B. S., Kramer, K. M., Epperson, P. D., Hoffman, G. E., and Carter, C. S. (2004). Neonatal manipulations of oxytocin alter expression of oxytocin and vasopressin immunoreactive cells in the paraventricular nucleus of the hypothalamus in a gender-specific manner. *Neuroscience* 125:947–955.

Young, L. J., Pitkow, L. J., and Ferguson, J. N. (2002). Neuropeptides and social behavior: animal models relevant to autism. *Mol. Psychiatry* 7:S38–S39.

Chapter 15
Reelin and Thyroid Hormone

Manuel Álvarez-Dolado

Contents

1 Introduction

Thyroid hormone [3,5,3'-triiodothyronine (T3) and thyroxine (T4)] is essential for proper brain development. In humans, the lack of adequate T3 levels during the perinatal period leads to cretinism, a syndrome associated with mental retardation and neurological deficits, such as ataxia, spasticity, and deafness (for review, see Legrand, 1984; Dussault and Ruel, 1987; Braverman and Utiger, 2000; Bernal, 2005a). These alterations are due to misregulation of the gene expression controlled by T3 through its interaction with nuclear receptors, which act as ligand-modulated transcription factors (Muñoz and Bernal, 1997; Forrest and Vennström, 2000; Yen *et al.*, 2006).

In experimental animals, hypothyroidism causes an array of morphological abnormalities in the neonatal brain (Dussault and Ruel, 1987; de Long, 1990). An important alteration is the reduction in myelination, as a consequence of downregulation of the major myelin proteins, and disruption of oligodendrocyte differentiation (Rodríguez-Peña, 1999; Billon *et al.*, 2002). Additional effects of T3 deficiency are a marked delay in neuronal migration, and alterations of neuronal size, packing density, and dendritic morphology (Patel *et al.*, 1976; Legrand, 1984; Berbel *et al.*,

M. Álvarez-Dolado
Cellular Regeneration Laboratory, Centro de Investigación Príncipe Felipe (CIPF), Avenida Autopista del Saler 16-3, Valencia 46013, Spain
e-mail: mdolado@cipf.es

S. H. Fatemi (ed.), *Reelin Glycoprotein: Structure, Biology and Roles in Health and Disease.*

1993, 2001; Lucio *et al.*, 1997). Thus, for instance, migration of granule neurons from the external toward the internal granule layer of the cerebellum is retarded, and defects in the positioning of Purkinje cells are observed (Patel *et al.*, 1976; Lauder, 1979; Legrand, 1984; Clos and Legrand, 1990). Furthermore, in the cerebral cortex, there are significant abnormalities in lamination, reflecting migration defects (Berbel *et al.*, 1993, 2001; Lucio *et al.*, 1997).

In the last two decades, a number of genes with a putative role in these alterations have been identified as regulated by T3. They include those coding for cytoskeletal and extracellular matrix proteins (tau, actin, tenascin-C), neurotropins and their receptors, cell adhesion molecules (N-CAM, L1, TAG-1), transcription factors, and intracellular signaling proteins (RC3, Rhes) (Bernal, 2002, 2005a). Among these genes, *reelin* is of special relevance, given its direct implication and key role in processes, such as cell migration and neuronal positioning.

2 Developmental Regulation of *Reelin* and *Dab1* by T3

Reelin-deficient mice manifest many of the features observed in the hypothyroid brain (Goffinet, 1980; Derer, 1985; Schiffmann *et al.*, 1997; Alcántara *et al.*, 1998). This led A. Muñoz's and E. Soriano's groups to perform a complete developmental study of *reelin* and *dab1* expression in hypothyroid rats (Álvarez-Dolado *et al.*, 1999). Quantification by Northern blot showed that *reelin* expression is decreased by 60% in the hypothyroid cortex at postnatal day 0 (P0). More detailed studies using *in situ* hybridization and immunohistochemistry evidenced a complex regulation of *reelin* expression by T3 (Álvarez-Dolado *et al.*, 1999). In general, the *reelin* regional and laminar expression patterns are not altered during hypothyroidism. However, both the number of labeled neurons, and their intensity of labeling in the hippocampus and layers I and V/VI of the neocortex, are significantly lower than those in control rats, particularly at P0 (Fig. 15.1A,B). At P5, these differences in the expression levels are weaker than at previous stages. The number of reelin-positive neurons in cortical layer I and hippocampus is not affected; in contrast, in layers II–VI it decreases in hypothyroid rats. At later stages, the pattern of expression in the cortex and hippocampus of hypothyroid rats gradually becomes equal to that in controls. Consistent with the mRNA expression pattern, immunolocalization of Reelin protein in hypothyroid brains shows a marked deficiency in the neocortex and hippocampus at E18–P0 (Fig. 15.1C,D). This deficiency tends to disappear with development. Collectively, the data show that *reelin* expression levels are decreased at perinatal stages in the cortex and hippocampus of hypothyroid rats, whereas they appear to reach normal levels at later postnatal stages.

Strikingly, in the cerebellum and olfactory bulb, hypothyroidism modifies the *reelin* expression in a completely different way (Álvarez-Dolado *et al.*, 1999; Manzano *et al.*, 2003). At E18–P0, both *reelin* transcripts and immunoreactivity levels are lower in the cerebellum of hypothyroid rats. In contrast, no remarkable differences are observed in the primordium of the olfactory bulb at these ages.

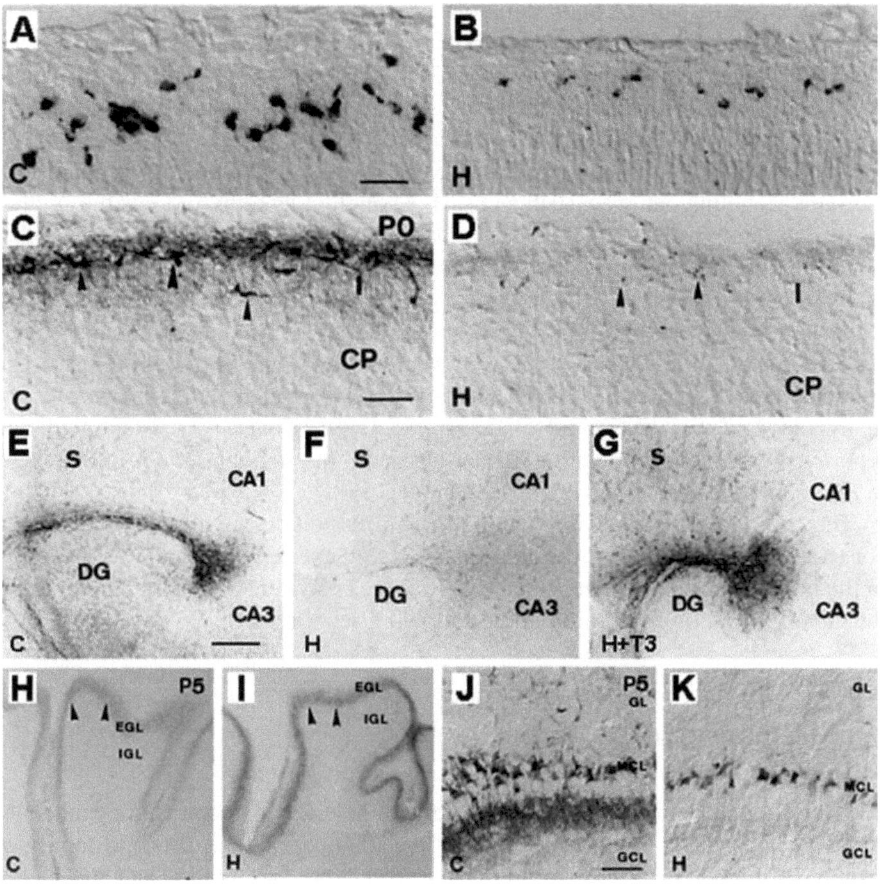

Fig. 15.1 Effects of hypothyroidism on *reelin* RNA and protein expression in the neonatal brain. (**A, B**) Pattern of *reelin* RNA expression in the neocortex of control (**A**) and hypothyroid (**B**) rats at P0. (**C, D**) Photomicrographs showing the distribution of CR50 antibody immunostaining in layer I of control (**C**) and hypothyroid rats (**D**) at P0. Some CR50-positive Cajal-Retzius cells are indicated by arrowheads. Note the decreased staining in hypothyroid animals. Cortical layers are indicated to the right. (**E–G**) Reelin expression detected by CR50 immunostaining in hippocampal organotypic slice cultures. (**E**) Slice from euthyroid rats incubated for 6 days in standard serum. (**F**) Slice from hypothyroid rats incubated for 6 days in thyroid-depleted serum. (**G**) Slices from hypothyroid rats incubated for 6 days in T3/T4-depleted serum supplemented with 500 nM T3. Note that the reduced expression levels in hypothyroid slices are rescued by T3 treatment. (**H–K**) Patterns of Reelin distribution in the cerebellum (**H**, **I**) and olfactory bulb (**J**, **K**) of control (**H**, **J**) and hypothyroid (**I**, **K**) rats at P5. Note the increased Reelin levels in the hypothyroid cerebellum and the opposite in the olfactory bulb. Abbreviations: C, control; CA3, CA1, hippocampal subdivisions CA3 and CA1; CP, cortical plate; DG, dentate gyrus; EGL, external granule cell layer; GCL, granule cell layer; GL, glomerular cell layer; H, hypothyroid; I, cortical layer I; IGL, internal granule cell layer; MCL, mitral cell layer; ML, molecular layer; S, stratum lacunosum-moleculare. Scale bars: **A**, 40 μm (applies to **A–D**); **E**, 200 μm (applies to **E–I**); **J**, 50 μm (applies to **J** and **K**). (Figure modified from Álvarez-Dolado *et al.*, 1999. © *The Journal of Neuroscience*) (*See Color Plates*)

At P5–P15, *reelin* mRNA and immunolabeling are prominent in both the EGL and the IGL of the cerebellum, where, contrarily to previous ages, their levels are clearly elevated in hypothyroid rats (Fig. 15.1H,I). In contrast, in the olfactory bulb, decreased levels of RNA and protein are noticed at P5 in hypothyroid brains (Fig. 15.1J,K), whereas no changes are detected at P15.

Alteration of reelin levels affects the expression of other proteins implicated in the same signaling pathway, such as Dab1. Though in hypothyroid rats the synthesis of *dab1* transcripts is unaltered, the Dab1 immunoreactivity is higher than that of controls in the cortex and hippocampus at E18–P0. Conversely, at P5 the levels of Dab1 are lower in hypothyroid than in control rats. This alteration in Dab1 expression is also observed in the cerebellum. Finally, at later developmental stages (P15, P25), no differences are seen between controls and hypothyroid rats. The results indicate that the levels of Dab1 are inversely correlated with those of Reelin in hypothyroid rats (Álvarez-Dolado *et al.*, 1999). This is consistent with the reported abnormal accumulation of Dab1 protein, but normal *dab1* mRNA levels, in *reeler* mutant mice (Rice *et al.*, 1998).

The complex pattern of *reelin* and *dab1* expression during hypothyroidism is a consequence of the intricate mechanism of action exerted by T3 and its receptors in the brain. I will discuss the complexity of the T3 system and its implications on the control of *reelin* expression in the following sections.

3 Mechanism of T3 Action

T3 controls gene expression by interacting with its nuclear receptors (Oppenheimer and Schwartz, 1997; Forrest and Vennström, 2000; Yen *et al.*, 2006), a family of proteins with several functional domains, especially the ligand- and the DNA-binding domains (Mangelsdorf *et al.*, 1995; Weiss and Ramos, 2004). There are two receptor genes, designated TRα and TRβ, that encode nine protein products generated by alternative splicing and differential promoter usage. Of the nine, only three isoforms (α1, β1, β2) have an intact T3-binding domain and are able to interact with DNA-specific nucleotide sequences (T3REs: thyroid response elements) present in target genes (Lazar, 1993; Muñoz and Bernal, 1997). The physiological role of the other six nonreceptor proteins remains to be solved (Gauthier *et al.*, 2001).

Expression of T3 receptors in the brain starts around E12–14 and is predominantly located in neurons but has also been detected in oligodendrocytes, astrocytes, and microglia (Lima *et al.*, 1998, 2001; Rodríguez-Peña, 1999). TRα1 isoform is widely expressed throughout the brain and accounts for 70–80% of total T3 binding capacity (Ercan-Fang *et al.*, 1996). The TRβ gene is also expressed in the brain, with a more discrete pattern of expression (Forrest *et al.*, 1990). The current view is that the different physiological roles of each receptor depend on their particular patterns of temporal and regional expression (Forrest *et al.*, 1990; Forrest and Vennström, 2000). They can regulate gene transcription through the activation of positive T3REs or repression of negative T3REs. In general, TRα1 receptor acts as a base expression controller whereas TRβ isoforms exert a more specific and timely regulation. There are examples of differential gene regulation in the same cells, specifically exerted through TRα and not TRβ receptor (Manzano *et al.*, 2003).

Their function as upregulators or downregulators of gene expression may also depend on the interaction with other nuclear receptors and transcription cofactors (Nishihara *et al.*, 2004; Moore and Guy, 2005). Their functional interaction as heterodimers with other nuclear receptors, such as retinoid X receptor (RXR) and retinoic acid receptor (RAR), is essential for understanding the mechanism of gene control by T3 (Muñoz and Bernal, 1997; Forrest and Vennström, 2000).

Finally, to add more complexity, T3 levels are not equal in all brain areas. They depend on the activity of deiodinases that convert T4 into T3 (Bianco *et al.*, 2002; Bernal, 2005a), and membrane transporters that control T4/T3 flux through the blood–brain barrier and choroid plexus (Dickson *et al.*, 1987; Abe *et al.*, 2002; Bernal, 2005b). In addition, several studies have indicated posttranscriptional regulatory effects of T3 on mRNA stabilization, processing, and translation, or on posttranslational mechanisms (Aniello *et al.*, 1991).

4 Molecular Mechanism of T3 Control on *reelin* Expression

The precise mechanism of T3 action on *reelin* expression, whether transcriptional or posttranscriptional, remains to be determined. The lack of a strict correlation between the changes in RNA and protein levels supports the idea that *reelin* expression is regulated by T3 at both levels. Location of T3REs in the promoter or intronic sequences of the *reelin* gene remains to be confirmed. This would provide direct evidence of the effects of T3 at the transcriptional level. Nonetheless, the fact that T3 treatment of hypothyroid organotypic brain slices *in vitro* restores the normal *reelin* expression levels strongly suggests a direct effect of this hormone (Fig. 15.1E–G) (Álvarez-Dolado *et al.*, 1999). In addition, *reelin* expression levels are specifically restored in the hypothyroid cerebellum after GC-1 treatment, a specific TRβ agonist (Manzano *et al.*, 2003).

However, other indirect *in vivo* mechanisms cannot be ruled out. For instance, BDNF has been found to negatively regulate the expression of *reelin* (Ringstedt *et al.*, 1998). Previous studies indicated that BDNF expression is diminished at P15 and later ages in the cerebellum of hypothyroid rats (Neveu and Arenas, 1996), although the levels in the cortex and hippocampus remain unchanged (Álvarez-Dolado *et al.*, 1994). Therefore, the increased expression of *reelin* in the hypothyroid cerebellum at P5–P15 may be secondary to the modulation of BDNF levels by T3 in this region.

The differences observed in the effect of the hormone in distinct brain regions, especially in the cerebellum, suggest that T3 may cooperate with locally acting factors, or that hormone action is modulated by region- or cell-specific proteins. Thus, we should consider the physiological levels of T3 in each region, based on the expression and activity of deiodinases and transporters. Interestingly, a strong correlation between *reelin* expression levels and the presence of different deiodinase subtypes in the cerebellum has been reported (Verhoelst *et al.*, 2005). Finally, alterations in the expression pattern of other extracellular matrix proteins and cell adhesion molecules (tenascin-C, L1, TAG-1, N-CAM) during hypothyroidism may also affect *reelin* expression (Iglesias *et al.*, 1996; Álvarez-Dolado *et al.*, 1998, 2000, 2001).

5 Biological Implications of *Reelin* Control by T3

Given the drastic phenotype caused by the lack of Reelin and its important role during brain development, the finding that T3 influences *reelin* expression has been of fundamental importance in better understanding the basis of the alterations that occur in the hypothyroid brain during development.

Reelin and Dab1 are critical for neuronal migration which, in turn, is responsible for lamination and precise cellular localization during CNS development. These processes are severely affected by hypothyroidsm. An abnormal laminar distribution has been reported in the auditory cortex of hypothyroid rats, including an increased number of neurons in layers V/VI, a concomitant decrease in layers II to IV, and the abnormal presence of neurons in the subcortical white matter (Berbel *et al.*, 1993; Lucio *et al.*, 1997). These cytoarchitectonic abnormalities most probably reflect migration defects in the cortex. Also, it has been shown that iodine deficiency causes an impaired maturation of hippocampal radial glial cells, which are involved in neuronal migration (Martínez-Galán *et al.*, 1997). Additionally, hypothyroidism affects the migration of cells from germinative zones toward the olfactory bulb and caudate putamen, as well as the migration of granule neurons from the external toward the internal granule layer of the cerebellum (Patel *et al.*, 1976; Lu and Brown, 1977; Legrand, 1984). As a result, the precise timing to establish appropriate neuronal connections is disrupted, and there is a decrease in the number and density of synaptic contacts. Finally, ectopic localization of Purkinje cells is a typical abnormality found in the hypothyroid cerebellum, which remarkably also occurs to much higher extent in *reeler* mice (Mariani *et al.*, 1977; Legrand, 1984; Miyata *et al.*, 1997). This array of abnormalities is very likely a consequence of the reduction in Reelin content reported in the hypothyroid brain during the perinatal period.

6 Conclusions

The finding that T3 regulates *reelin* expression has been essential in better understanding the basis of the alterations that occur in the hypothyroid brain during development. It explains most of the observed anomalies and has helped to find other molecules regulated by T3 that are implicated in neuronal migration. Further efforts should be devoted to finding other hormones and factors that may be implicated in the mechanisms governing *reelin* expression in the brain.

References

Abe, T., Suzuki, T., Unno, M., Tokui, T., and Ito, S. (2002). Thyroid hormone transporters: recent advances. *Trends Endocrinol. Metab.* 13:215–220.

Alcántara, S., Ruiz, M., D'Arcangelo, G., Ezan, F., de Lecea, L., Curran, T., Sotelo, C., and Soriano, E. (1998). Regional and cellular patterns of reelin mRNA expression in the forebrain of the developing and adult mouse. *J. Neurosci.* 18:7779–7799.

Álvarez-Dolado, M., Iglesias, T., Rodríguez-Peña, A., Bernal, J., and Muñoz, A. (1994). Expression of neurotrophins and the trk family of neurotrophin receptors in normal and hypothyroid rat brain. *Mol. Brain Res.* 27:249–257.

Álvarez-Dolado, M., González-Sancho, J., Bernal, J., and Muñoz, A. (1998). Developmental expression of tenascin-C is altered by hypothyroidism in the rat brain. *Neuroscience* 84:309–322.

Álvarez-Dolado, M., Ruiz, M., Del Río, J. A., Alcántara, S., Burgaya, F., Sheldon, M., Nakajima, K., Bernal, J., Howell, B. W., Curran, T., Soriano, E., and Muñoz, A. (1999). Thyroid hormone regulates reelin and dab1 expression during brain development. *J. Neurosci.* 19:6979–6993.

Álvarez-Dolado, M., Cuadrado, A., Navarro-Yubero, C., Sonderegger, P., Furley, A. J., Bernal, J., and Munoz, A. (2000). Regulation of the L1 cell adhesion molecule by thyroid hormone in the developing brain. *Mol. Cell. Neurosci.* 4:499–514.

Álvarez-Dolado, M., Figueroa, A., Kozlov, S., Sonderegger, P., Furley, A. J., and Munoz, A. (2001). Thyroid hormone regulates TAG-1 expression in the developing rat brain. *Eur. J. Neurosci.* 8:1209–1218.

Aniello, F., Couchie, D., Bridoux, A. M., Gripois, D., and Nunez, J. (1991). Splicing of juvenile and adult *tau* m-RNA variants is regulated by thyroid hormone. *Proc. Natl. Acad. Sci. USA*88:4035–4039.

Berbel, P., Guadaño-Ferraz, A., Martínez, M., Quiles, J., Balboa, R., and Innocenti, G. (1993). Organization of auditory callosal connections in hypothyroid rats. *Eur. J. Neurosci.* 5:1465–1478.

Berbel, P., Ausó, E., García-Velasco, J. V., Molina, M. L., and Camacho, M. (2001). Role of thyroid hormones in the maturation and organisation of rat barrel cortex. *Neurosci.* 107:383–394.

Bernal, J. (2002). Action of thyroid hormone in brain. *J. Endocrinol. Invest.* 25:268–288.

Bernal, J. (2005a). Thyroid hormones and brain development. *Vitam. Horm.* 71: 95–122.

Bernal, J. (2005b). The significance of thyroid hormone transporters in the brain. *Endocrinology* 146:1698–1700.

Bianco, A. C., Salvatore, D., Gereben, B., Berry, M. J., and Larsen, P. R. (2002). Biochemistry, cellular and molecular biology, and physiological roles of the iodothyronine selenodeiodinases. *Endocr. Rev.* 23:38–89.

Billon, N., Jolicoeur, C., Tokumoto, Y., Vennstrom, B., and Raff, M. (2002). Normal timing of oligodendrocyte development depends on thyroid hormone receptor alpha 1 (TRalpha1). *EMBO J.* 21:6452–6460.

Braverman, L. E., and Utiger, R. D. (2000). *Werner and Ingbar's The Thyroid: A Fundamental and Clinical Text,* 8th ed. Lippincott Williams and Wilkins, Philadelphia.

Clos, J., and Legrand, C. H. (1990). An interaction between thyroid hormone and nerve growth factor promotes the development of hippocampus, olfactory bulbs and cerebellum: a comparative biochemical study of normal and hypothyroid rats. *Growth Factor* 3:205–220.

deLong, G. R. (1990). The effect of iodine deficiency on neuromuscular development. *IDD Newsletter* 6:1–12.

Derer, P. (1985). Comparative localization of Cajal-Retzius cells in the neocortex of normal and *reeler* mutant mice fetuses. *Neurosci. Lett.* 54:1–6.

Dickson, P. W., Aldred, A. R., Menting, J. G. T., Marley, P. D., Sawyer, W. H., and Schreiber, G. (1987). Thyroxine transport in choroid plexus. *J. Biol. Chem.* 262:13907–13915.

Dussault, J. H., and Ruel, J. (1987). Thyroid hormones and brain development. *Annu. Rev. Physiol.* 49:321–334.

Ercan-Fang, S., Schwartz, H. L., and Oppenheimer, J. H. (1996). Isoform specific 3,5,3'-triiodothyronine receptor binding capacity and messenger ribonucleic acid content in rat adenohypophysis: effect of thyroidal state and comparison with extrapituitary tissues. *Endocrinology* 137:3228–3233.

Forrest, D., and Vennström, B. (2000). Functions of thyroid hormone receptors in mice. *Thyroid* 1:41–52.

Forrest, D., Sjoberg, M., and Vennstrom, B. (1990). Contrasting developmental and tissue-specific expression of alpha and beta thyroid hormone receptor genes. *EMBO J.* 9:1519–1528.

Gauthier, K., Plateroti, M., Harvey, C. B., Williams, G. R., Weiss, R. E., Refeto,V.S., Willott, J. F., Sundin, V., Roux, J. P., Malaval, L., Hara, M., Samarut, J., and Chassande, O. (2001). Genetic analysis reveals different functions for the products of the thyroid hormone receptor alpha locus. *Mol. Cell. Biol.* 21:4748–4760.

Goffinet, A. M. (1980). The cerebral cortex of the *reeler* embryo (an electron microscopic analysis). *Anat. Embryol.* 159:199–210.

Iglesias, T., Caubín, J., Stunnenberg, H. G., Zaballos, A., Bernal, J., and Muñoz, A. (1996). Thyroid hormone-dependent transcriptional repression of neuronal cell adhesion molecule during brain maturation. *EMBO J.* 15:4307–4316.

Lauder, J. M. (1979). Granule cell migration in the developing rat cerebellum. Influence of neonatal hypo- and hyper-thyroidism. *Dev. Biol.* 70:105–115.

Lazar, M. A. (1993). Thyroid hormone receptors: multiple forms, multiple possibilities. *Endocr. Rev.* 14:184–193.

Legrand, J. (1984). Effects of thyroid hormones on central nervous system. In: Yanai, J. (ed.), *Neurobehavioural Teratology*. Elsevier/North-Holland, Amsterdam, pp. 331–363.

Lima, F. R., Goncalves, N., Gomes, F. C., de Freitas, M. S., and Moura Neto, V. (1998). Thyroid hormone action on astroglial cells from distinct brain regions during development. *Int. J. Dev. Neurosci.* 16:19–27.

Lima, F. R., Gervais, A., Colin, C., Izembart, M., Neto, V. M., and Mallat, M. (2001). Regulation of microglial development: a novel role for thyroid hormone. *J. Neurosci.* 21:2028–2038.

Lu, E., and Brown, W. (1977). The developing caudate nucleus in the euthyroid and hypothyroid rat. *J. Comp. Neurol.* 171:261–284.

Lucio, R. A., García, J. V., Cerezo, J. R., Pacheco, P., Innocenti, G. M., and Berbel, P. (1997). The development of auditory callosal connections in normal and hypothyroid rats. *Cereb. Cortex* 7:303–316.

Mangelsdorf, D. J., Thummel, C., Beato, M., Herrlich, P., Schutz, G., Umesono, K., Blumberg, B., Kastner, P., Mark, M., Chambon, P., and Evans, R. M. (1995). The nuclear receptor superfamily: the second decade. *Cell* 83:835–839.

Manzano, J., Morte, B., Scanlan, T. S., and Bernal, J. (2003). Differential effects of triiodothyronine and the thyroid hormone receptor β-specific agonist GC-1 on thyroid hormone target genes in the brain. *Endocrinology* 144:5480–5487.

Mariani, J., Crepel, F., Mikoshiba, K., Changeux, J. P., and Sotelo, C. (1977). Anatomical, physiological and biochemical studies of the cerebellum from *reeler* mutant mouse. *Philos. Trans. R. Soc. London Ser. B Biol. Sci.* 281:1–28.

Martínez-Galán, J., Pedraza, P., Santacana, M., Escobar del Rey, F., Morreale de Escobar, G., and Ruiz-Marcos, A. (1997). Early effects of iodine deficiency on radial glial cells of the hippocampus of the rat fetus. A model of neurological cretinism. *J. Clin. Invest.* 99:2701–2709.

Miyata, T., Nakajima, K., Aruga, J., Mikoshiba, K., and Ogawa, M. (1997). Regulation of Purkinje cell alignment by reelin as revealed with CR-50 antibody. *J. Neurosci.* 15:3599–3609.

Moore, J. M., and Guy, R. K. (2005). Coregulator interactions with the thyroid hormone receptor. *Mol. Cell. Proteomics* 4:475–482.

Muñoz, A., and Bernal, J. (1997). Biological activities of thyroid hormone receptors. *Eur. J. Endocrinol.* 137:433–445.

Neveu, I., and Arenas, E. (1996). Neurotrophins promote the survival and development of neurons in the cerebellum of hypothyroid rats *in vivo*. *J. Cell Biol.* 133:631–646.

Nishihara, E., O'Malley, B. W., and Xu, J. (2004). Nuclear receptor co-regulators are new players in nervous system development and function. *Mol. Neurobiol.* 3:307–325.

Oppenheimer, J. H., and Schwartz, H. L. (1997). Molecular basis of thyroid hormone-dependent brain development. *Endocr. Rev.* 18:462–475.

Patel, A.J., Rabie, A., Lewis, P., and Balazs, R. (1976). Effects of thyroid deficiency on postnatal cell formation in the rat brain. A biochemical investigation. *Brain Res.* 104:33–48.
Rice, D.S., Sheldon, M., D'Arcangelo, G., Nakajima, K., Goldowitz, D., and Curran, T. (1998). Disabled-1 acts downstream of *Reelin* in a signaling pathway that controls laminar organization in the mammalian brain. *Development* 125:3719–3729.
Ringstedt, T., Linnarsson, S., Wagner, J., Lendahl, U., Kokaia, Z., Arenas, E., Ernfors, P., and Ibañez, C. F. (1998). BDNF regulates reelin expression and Cajal-Retzius cell development in the cerebral cortex. *Neuron* 21:305–315.
Rodríguez-Peña, A. (1999). Oligodendrocyte development and thyroid hormone. *J. Neurobiol.* 40:497–512.
Schiffmann, S. N., Bernier, B., and Goffinet, A. (1997). *reelin* mRNA expression during mouse brain development. *Eur. J. Neurosci.* 9:1055–1071.
Verhoelst, C. H., Roelens, S. A., and Darras, V. M. (2005). Role of spatiotemporal expression of iodothyronine deiodinase proteins in cerebellar cell organization. *Brain Res. Bull.* 67:196–202.
Weiss, R. E., and Ramos, H. E. (2004). Thyroid hormone receptor subtypes and their interaction with steroid receptor coactivators. *Vitam. Horm.* 68:185–207.
Yen, P. M., Ando, S., Feng, X., Liu, Y., Maruvada, P., and Xia, X. (2006). Thyroid hormone action at the cellular, genomic and target gene levels. *Mol. Cell. Endocrinol.* 246:121–127.

Chapter 16
A Tale of Two Genes: Reelin and BDNF

Thomas Ringstedt

Contents

1 Introduction

BDNF is a survival factor for the Cajal-Retzius cells in the marginal zone, which are an important source of Reelin in the neocortex. BDNF is also a negative regulator of Reelin expression in both Cajal-Retzius cells and GABAergic cells in the cortical plate. BDNF and Reelin act in parallel to regulate many processes during neural development and maintenance, including cell migration and neural plasticity. Frequently, BDNF and Reelin have opposite influences on the processes they regulate, suggesting that BDNF-induced downregulation of Reelin is involved. Reelin is an important regulator of neural migration during neocortex formation. BDNF seems to influence this process both directly and indirectly via regulation of Reelin expression. Moreover, epileptic seizures increase BDNF levels while decreasing Reelin levels, and BDNF and Reelin seem to have opposite roles in mediating the effects of the seizures. Mental disorders, in particular schizophrenia, involve alterations in BDNF and Reelin expression. Again, the changes are mainly opposite, and a

T. Ringstedt
Neonatal Unit, Karolinska Institutet, Astrid Lindgren Children's Hospital, Q2:07, SE-171 76, Stockholm, Sweden
e-mail: thomas.ringstedt@ki.se

S. H. Fatemi (ed.), *Reelin Glycoprotein: Structure, Biology and Roles in Health and Disease.*

negative regulation of Reelin by BDNF has been suggested. In contrast, hippocampal LTP is promoted by both Reelin and BDNF signaling. Finally, there is overlap in the epigenetic regulation and signaling pathways of BDNF and Reelin.

2 BDNF and the Neurotrophins

In the 1950s, a diffusible factor that increased innervation of internal organs by promoting neuronal survival was isolated. It was named *nerve growth factor* (NGF) (Cohen and Levi-Montalcini, 1957). A similar, but distinct protein was purified from pig brain in 1982, and was named *brain-derived neurotrophic factor* (BDNF) (Barde *et al.*, 1982). This was the birth of the NGF family of neurotrophic factors, or the neurotrophins. Since then, two more factors have been added to the family in mammals: neurotrophin 3 (NT-3) (Hohn *et al.*, 1990) and neurotrophin 4 (NT-4) (Hallbook *et al.*, 1991). The neurotrophins bind to the Trk family of tyrosine kinase receptors: NGF to TrkA, BDNF and NT-4 to TrkB, and NT-3 to TrkC (but also to a certain extent to TrkA and TrkB). They also bind to the P75 neurotrophin receptor (P75NTR) with equal affinity. The original concept of neurotrophins as target derived survival factors for innervating neurons, still holds. Competition for neurotrophic factors weeds out ill-positioned neurons during the period of naturally occurring cell death. Interestingly, neurotrophin homologues are not found in invertebrates. It is therefore possible that the plasticity inferred by an extrinsic regulation of neuronal survival (as opposed to cell-intrinsic regulation) has coevolved with higher neuronal complexity. While the neurotrophin roles as survival factors in the peripheral nervous system occur during embryonic development, brain neurons mostly seem to develop neurotrophic dependency postnatally, if at all. In addition to promoting survival, neurotrophins are now known to affect neuronal differentiation, maturation, migration, axonal guidance, and plasticity. BDNF in particular stands out as an important regulator of these processes, often paralleling the effects of reelin.

2.1 BDNF Is a Survival Factor for Cajal-Retzius Cells

The Cajal-Retzius (CR) cells are early born cells that are part of the embryonic preplate, marginal zone, and hippocampus. During development, they are the primary producers of Reelin in the neocortex and the hippocampus. It has long been believed that the Reelin produced by CR cells is essential for correct lamination of the neocortex by regulating positioning of migrating neurons (D'Arcangelo *et al.*, 1995; Ogawa *et al.*, 1995; Super *et al.*, 2000), although this has lately been put in doubt (Yoshida *et al.*, 2006). The later fate of the CR cells is unclear. It has been suggested that they die or differentiate into other cell types, but at least a subgroup of this (probably) heterogeneous cell population remains in the adult brain (Meyer *et al.*, 1999; Riedel *et al.*, 2003).

BDNF is only weakly expressed in the mouse brain before birth, but the expression increases rapidly during the first weeks of life (Friedman *et al.*, 1991; Timmusk *et al.*, 1994) and remains high during adulthood. BDNF mutant mice investigated at postnatal day 18 (P18, about their longest survival time) display significantly reduced CR cell numbers in the marginal zone/layer I compared to wild-type littermates (Ringstedt *et al.*, 1998). Thus, BDNF acts as a survival factor for CR cells in the postnatal brain neocortex. However, CR cells in murine hippocampal brain slices exceed their *in vivo* survival time and remain up to 14 days after explantation (the longest experimental period), regardless of whether the slices are derived from wild-type or BDNF mutant embryos, suggesting that BDNF is not a survival factor for hippocampal CR cells (Marty *et al.*, 1996).

2.2 *BDNF Regulates Reelin Expression*

BDNF is also a negative regulator of Reelin expression. In wild-type mice, Reelin expression by the marginal zone CR cells decreases during the first 3 postnatal weeks, inversely correlated with the postnatal rise in BDNF expression. In BDNF mutant mice, CR cell Reelin expression remains constant from birth until P11. At P18 (when the survival effect of BDNF becomes evident), Reelin levels in the remaining CR cells drop to wild-type levels. Conversely, BDNF treatment of dissociated cultures of cortical neurons reduces Reelin levels. In transgenic mice, nestin-driven overexpression of BDNF in the brain (hereafter referred to as nestin-BDNF), reduces CR cell Reelin expression to 50% of wild-type (at embryonic day E18.5) (Ringstedt *et al.*, 1998). Reelin expression by GABAergic cells in the cortical plate is highly reduced or absent in the nestin-BDNF mice. Expression of the downstream effector of Reelin, Dab1, increases (Alcantara *et al.*, 2006), probably as a direct effect of the reduced Reelin level (Howell *et al.*, 1999). BDNF further increases expression of calretinin in neocortical (Alcantara *et al.*, 2006) and hippocampal (Marty *et al.*, 1996) CR cells.

In addition to the reduced Reelin expression, nestin-BDNF mice display polymicrogyria at E18.5 (Ringstedt *et al.*, 1998; Alcantara *et al.*, 2006), and the normal bilayered organization of the neocortical marginal zone, with CR cells close to the pial membranes and GABAergic neurons in the inner part, is disturbed. Beginning at E16, the CR cells are organized in ectopic clusters spaced by empty stretches. At E18.5 the CR cells are enlarged and their axons project abnormally deep into the neural cortex (Alcantara *et al.*, 2006). The CR cell clusters occupy the sulci of the polymicrogyria, while GABAergic neurons are found in the gyri. The laminar distribution of cells in the cortical plate is altered as revealed by BrdU labeling. However, only the migration of late born (E14 to E16) cells is affected. These are present in increased proportions in the interstitial and marginal zones of nestin-BDNF mice. BrdU labeling of early born cells (E11) revealed that unlike the Reeler mouse, the preplate is split in nestin-BDNF mice. Interestingly, positioning of BrdU-labeled cells differs between gyri and sulci, and the sulci contains 33% less BrdU-positive cells. Together with the asymmetric presence of CR and GABAergic cells in sulci and gyri, this hints at the polymicrogyria being shaped by differentiated migration in areas influenced by CR or GABAergic cells, respectively (Alcantara *et al.*, 2006). Interestingly, a conditional knockout of the β1 integrin receptor, which can act as a Reelin receptor (Dulabon *et al.*, 2000), also results in CR cell clustering and polymicrogyria (Graus-Porta *et al.*, 2001; Magdaleno and Curran, 2001). In sum, these studies suggest that Reelin and BDNF interact to regulate cortical plate development.

2.3 *BDNF Regulates Neocortical Cell Migration Independent of Reelin*

The alternate distribution of CR and GABAergic cells in the marginal zone of the nestin-BDNF embryos indicates that tangential migration is affected, since the GABAergic interneurons (Ang *et al.*, 2003), and at least the majority of the CR cells (Yoshida *et al.*, 2006), enter via tangential migration. Tangential migration of interneurons is independent of Reelin signaling (Pla *et al.*, 2006), and Reelin is not essential for CR cell migration, since CR cells are distributed normally along the marginal zone of Reeler mice (mutant for Reelin) (Derer, 1985). Intraventricular injection of BDNF in E13 mice results in altered cell positioning only 2 days later, a period that might be too short for the occurrence of changes in cell migration due to reduced Reelin expression. Also, the CR cells seem unaffected, although Reelin levels have not been investigated (Ohmiya *et al.*, 2002). Application of NT-4 to cortical slice cultures or intraventricular injection in E14 mouse embryos produces a phenotype related to that observed in the nestin-BDNF mice (Brunstrom *et al.*, 1997). Increased numbers of both CR and GABAergic cells enter the marginal zone, probably via tangential migration from the ganglionic eminence and/or cortical hem. NT-4-induced clustering of CR-like cells resembles that observed in the nestin-BDNF mice, but whether NT-4 also affects Reelin expression has not been investigated. In parallel, BDNF does not induce increased cell number in the marginal zone when used at the same dose as NT-4 (20 ng/ml), nor when used at 10-fold higher doses. Only extremely high doses (1 mg/ml) of BDNF are sufficient to increase marginal zone cell number (Brunstrom *et al.*, 1997). NT-4 shares the TrkB receptor with BDNF, but downstream signaling can still proceed differently (Minichiello *et al.*, 1998). However, both BDNF and NT-4 are equally potent in inducing lateral migration of GFP-labeled cells from E14–E16 ganglionic eminence explants into the interstitial and marginal zone of isochronic cortical explants (Polleux *et al.*, 2002). This was demonstrated to be a direct effect, mediated by the PI3-kinase pathway, one of the pathways known to be induced by neurotrophins. Chemotaxic stimulation of embryonic cortical neurons by BDNF or NT-4 has also been demonstrated *in vitro* (Behar *et al.*, 1997). BDNF increases neocortical expression of axon guidance receptors Robo1 and Robo2 (Alcantara *et al.*, 2006), which also are involved in cell migration (Andrews *et al.*, 2006). Both BDNF and NT-4 are expressed at low levels in the embryonic neocortex, although NT-4 expression precedes that of BDNF (Friedman *et al.*, 1991; Timmusk *et al.*, 1993). In conclusion, BDNF regulates cell migration in the neocortex both directly, and indirectly via regulation of the expression of Reelin and other potential mediators of cell migration.

3 Reelin and BDNF Mediate the Effects of Epileptic Seizures

The hippocampal CR cells synthesize and secrete Reelin, which is an important regulator of hippocampal development (Del Rio *et al.*, 1997; Frotscher *et al.*, 2003). However, Reelin expression remains in the hippocampus, even after the disappearance of CR cells (Haas *et al.*, 2000). Granule cell dispersion (GCD), a widening of the dentate gyrus granule cell layer, has been reported after mesial-temporal lobe epilepsy in humans (Houser, 1990). Mimicking epilepsy in rodents by kainic acid-induced seizures results in GCD. A similar phenotype is observed in Reeler mice and in mice mutant for the ApoER2 and VLDLR Reelin receptors (Rakic and Caviness, 1995; D'Arcangelo *et al.*, 1999). Reelin expression is downregulated after seizures, before GCD occurs. GCD is also induced after experimentally induced downregulation of Reelin expression by the blocking antibody CR-50, indicating that GCD is regulated by the altered Reelin levels after epileptic seizures. Thus, in addition to its function during hippocampal development, Reelin seems to have a role in maintaining hippocampal integrity throughout adult life (Heinrich *et al.*, 2006). BDNF, on the other hand, is upregulated after kainic acid-induced seizures. If seizure induction is followed by antisense block of BDNF synthesis, or K252a block of Trk receptors, GCD does not occur (Guilhem *et al.*, 1996). Given that BDNF is a negative regulator of Reelin, it is conceivable that BDNF at least in part is responsible for the downregulation of Reelin expression and induction of GCD after epileptic seizures. There is a strong link between BDNF and epileptic seizures. BDNF protein and mRNA levels are elevated in the temporal lobe of human epileptic brains (Takahashi *et al.*, 1999; Murray *et al.*, 2000). Experimental induction of seizures in rats by lesions (Isackson *et al.*, 1991) or kindling (Ernfors *et al.*, 1991) increases BDNF mRNA levels in many brain regions, including hippocampus and neocortex. Long-term administration of BDNF to rat hippocampus results in spontaneous seizures in 25% of the animals (Scharfman *et al.*, 2002). Conditional knockout of BDNF or its TrkB receptor in neurons results in a mild impairment (BDNF) or complete abolishment (TrkB) of kindling-induced epileptic seizures in mice (He *et al.*, 2004). Thus, TrkB signaling by BDNF and other neurotrophins is part of the epileptogenic process.

4 Reelin and BDNF Promote Neuronal Plasticity

Both BDNF and reelin have been implicated in plasticity, the modulation of synaptic strength, in hippocampus and neocortex. Induction of hippocampal long-term potentiation (LTP), a plasticity event essential for memory formation, is completely blocked in hippocampal slices by the addition of a general antagonist against LDL receptors, including the reelin ApoER2 and VLDL receptors (Bu and Schwartz, 1998). Mice mutant for ApoER2 or VLDLR display memory formation deficits (in contextual fear conditioning) (Weeber *et al.*, 2002). Addition of Reelin results in an immediate enhancement of LTP in wild-type hippocampal slices, but not in slices from ApoER2 or VLDLR mutant mice (Weeber *et al.*, 2002). Reelin-enhanced LTP is mediated through interaction with postsynaptic NMDA receptors: a splice variant of ApoER2 causes phosphorylation of the NMDA receptor subunits NR2A and NR2B in the postsynaptic density of excitatory synapses (Beffert *et al.*, 2005). The expression of this ApoER2 splice variant is triggered by behavioral activity (Beffert *et al.*, 2005). Similarly, BDNF (and other neurotrophins) are induced by neuronal activity (Ernfors *et al.*, 1991), and BDNF expression in the hippocampus parallels the ability to undergo LTP. Addition of BDNF to hippocampal slice cultures promotes LTP induction (Figurov *et al.*, 1996), while LTP is impaired in hippocampal slices from BDNF mutant mice (Korte *et al.*, 1995). Although there is some controversy about the site of BDNF action (Xu *et al.*, 2000), BDNF seems to have a robust postsynaptic effect on LTP (Kovalchuk *et al.*, 2002). Like Reelin, BDNF potentiates the NMDA response to glutamate, but by phosphorylation of the NR1 subunit (Suen *et al.*, 1997; Levine *et al.*, 1998).

In addition to its role in hippocampus, BDNF promotes LTP in the visual cortex (Akaneya *et al.*, 1997; Jiang *et al.*, 2001). Another form of plasticity in the visual cortex of higher mammals is the formation of ocular dominance columns. Axons from the visual system that enter layer IV of the visual cortex have their terminals segregated into eye-specific ocular dominance columns during postnatal development. Addition of exogenous BDNF or NT-4 (Cabelli *et al.*, 1995), or a TrkB antagonist (Cabelli *et al.*, 1997) inhibits this process in cats. Blocking one eye during development (monocular deprivation, MD) alters the size of the ocular dominance columns in favor of the active eye, but only if MD occurs during a certain time-window: the critical period for ocular dominance plasticity. Dark-rearing animals can delay this time-window into adulthood. Transgenic overexpression of BDNF in mouse postnatal neocortex shortens the critical period for ocular dominance plasticity and accelerates maturation of the visual cortex (Huang *et al.*, 1999). Contrary to wild-type mice, dark-rearing the BDNF-overexpressing mice does not delay the plasticity window, indicating that BDNF overexpression can replace the influence of visual experience (Gianfranceschi *et al.*, 2003). In a recent study that used differential display to compare gene expression in the visual neocortex of dark- and light-reared cats and mice, the Reelin signaling pathway gene Dab1 was found to be differently regulated. Dab1 expression was high 5 weeks postnatally in light-reared cats, but low in dark-reared cats. The reverse was true 20 weeks postnatally. Thus, Dab1 expression coincides with the peaks of plasticity in light- and dark-reared animals, respectively, indicating a role for Reelin signaling in visual cortex plasticity (Yang *et al.*, 2006). BDNF expression normally increases after eye opening and during the critical period for ocular dominance plasticity, but this increase in BDNF expression is, like the critical period, delayed in dark-reared animals. Thus, the peaks in BDNF and Dab1 expression can be expected to overlap in the visual cortex. Interestingly, increased Dab1 expression has been shown to correlate with low Reelin levels (Howell *et al.*, 1999). It is therefore possible that the change in Dab1 expression is caused by the negative regulation of Reelin expression by BDNF.

5 Reelin and BDNF Are Involved in Mental Disorders

Altered levels of Reelin and BDNF have been reported in the brains and sera of patients with schizophrenia or autism. Reelin has been reported to be expressed at lower levels in the hippocampus (Impagnatiello *et al.*, 1998; Fatemi *et al.*, 2000; Guidotti *et al.*, 2000; Knable *et al.*, 2004) and temporal and prefrontal neocortex (Impagnatiello *et al.*, 1998; Guidotti *et al.*, 2000) of schizophrenic patients, than in normal controls. BDNF, on the other hand, has been reported to display an increased expression in the hippocampus (Takahashi *et al.*, 2000; Iritani *et al.*, 2003) and neocortex (Takahashi *et al.*, 2000; Durany *et al.*, 2001; Iritani *et al.*, 2003) of schizophrenic patients. Its receptor, TrkB, was reported to be downregulated in the hippocampus and prefrontal cortex (Takahashi *et al.*, 2000). Although a reduced BDNF expression in hippocampus also has been reported (Durany *et al.*, 2001; Knable *et al.*, 2004), the schizophrenic change in expression seems to be quite opposite that described for Reelin. A negative regulation of Reelin by BDNF in schizophrenic patients has, therefore, been suggested (Takahashi *et al.*, 2000). However, the increase in BDNF levels may also be due to a defective secretion, a hypothesis that is supported by observations of decreased serum levels of BDNF in schizophrenic patients (Karege *et al.*, 2002; Toyooka *et al.*, 2002; Iritani *et al.*, 2003). Interestingly, levels of the unprocessed form of Reelin are increased in the blood of schizophrenic patients (Fatemi *et al.*, 2001). That BDNF might be part of the etiology of schizophrenia is supported by the association of a BDNF polymorphism (C270T) with schizophrenia (Szekeres *et al.*, 2003). Associations between Reelin polymorphisms and autism have been described (Persico *et al.*, 2001; Serajee *et al.*, 2006). Reduced levels of Reelin (Fatemi *et al.*, 2005b) have been described in the cerebellum and parts of the neocortex of autistic subjects. BDNF, on the other hand, displays increased levels in the basal forebrain of autistic patients (Perry *et al.*, 2001), and BDNF hyperactivity has also been proposed as a cause of autism (Tsai, 2005). However, the lack of overlap makes a direct regulation of Reelin levels by BDNF unlikely as a cause of the described alterations. Both higher (Miyazaki *et al.*, 2004) and lower (Hashimoto *et al.*, 2006) serum levels of BDNF in autistic patients have been reported.

Reelin levels are also lower in the prefrontal cortex, hippocampus, and cerebellum of subjects with bipolar disorder (Fatemi *et al.*, 2000, 2005a; Guidotti *et al.*, 2000; Knable *et al.*, 2004). No observations of altered BDNF levels in the brains of bipolar patients have been reported, but intracerebral administration of BDNF may have antidepressant effects in animals (Siuciak *et al.*, 1997). There are also associations between BDNF polymorphisms, cognitive ability (Egan *et al.*, 2003), neuroticism

(Sen *et al.*, 2003), and bipolar disorder in human subjects (Neves-Pereira *et al.*, 2002; Sklar *et al.*, 2002).

6 Overlap Between the Epigenetic Regulation and Signal Transduction of BDNF and Reelin

Epigenetic modifications of DNA are self-perpetuating modifications of DNA and histone proteins that affect transcription. The polygenic nature of epigenetic modifications makes this an interesting concept from a psychiatric perspective. Hypermethylation of the Reelin promoter in the brain of schizophrenic patients has been reported (Abdolmaleky *et al.*, 2005). DNA methylation is an epigenetic modification that decreases transcription by interfering with transcription factors. Both the Reelin and the BDNF promoters are cytosine methylated by DNA (cytosine-5) methyltransferase (DNMT), and can undergo acute changes in methylation status (Levenson *et al.*, 2006). Furthermore, activation of the PKC signaling pathway decreases Reelin methylation in the hippocampus (Levenson *et al.*, 2006). The PKC signaling pathway is important for hippocampal plasticity, and can be activated by BDNF/TrkB signaling.

There is a large overlap in the processes regulated by Reelin and BDNF. It is therefore not surprising that they signal in part via convergent pathways. Reelin signaling via the ApoER2 or VLDL receptors proceeds via dimerization and tyrosine phosphorylation of Dab1, which allows it to interact with SH2 domain proteins, among them phosphatidylinositol 3-kinase (PI3K) (Beffert *et al.*, 2002; Bock *et al.*, 2003). BDNF binds to and dimerizes TrkB, thereby inducing tyrosine phosphorylation. Phosphorylated TrkB also interacts with SH2 domain proteins and stimulates PI3K signaling via the SH2 adapter protein Shc (Bibel and Barde, 2000). Furthermore, BDNF can activate cyclin-dependent kinase 5 (CDK5) (Tokuoka *et al.*, 2000; Wang *et al.*, 2006), which acts on the cytoskeleton. CDK5 acts in parallel with Reelin to regulate cell positioning (Ohshima and Mikoshiba, 2002; Beffert *et al.*, 2004), but has also been shown to phosphorylate Dab1 (Keshvara *et al.*, 2002) and modulate Reelin signaling (Ohshima *et al.*, 2007). BDNF and Reelin signaling also converge in regulation of the activity-regulated cytoskeletal-associated protein (Arc). Arc mRNA is present in dendrites where it is locally translated and involved in synaptic stabilization (like during LTP). BDNF promotes translation (Yin *et al.*, 2002) and transcription (Ying *et al.*, 2002) of Arc mRNA, while Reelin via the $\alpha 3\beta 1$ integrin receptor promotes Arc mRNA translation (Dulabon *et al.*, 2000; Dong *et al.*, 2003).

In sum, BDNF and Reelin act in parallel during brain development and maintenance. The reports so far suggest multiple interactions at several different levels, such as through the use of convergent signaling pathways and through the regulation of CR cell survival and Reelin expression by BDNF. Yet further investigation is needed to elucidate the precise nature of the cross-talk between Reelin and BDNF. Their ability to regulate cell positioning and plasticity in the brain makes them likely candidates in the etiology of mental illness.

Acknowledgments I thank Dr. Ola Hermanson and Zachi Horn for valuable comments on the manuscript.

References

Abdolmaleky, H. M., Thiagalingam, S., and Wilcox, M. (2005). Genetics and epigenetics in major psychiatric disorders: dilemmas, achievements, applications, and future scope. *Am. J. Pharmacogenomics* 5:149–160.

Akaneya, Y., Tsumoto, T., Kinoshita, S., and Hatanaka, H. (1997). Brain-derived neurotrophic factor enhances long-term potentiation in rat visual cortex. *J. Neurosci.* 17:6707–6716.

Alcantara, S., Pozas, E., Ibanez, C. F., and Soriano, E. (2006). BDNF-modulated spatial organization of Cajal-Retzius and GABAergic neurons in the marginal zone plays a role in the development of cortical organization. *Cereb. Cortex* 16:487–499.

Andrews, W., Liapi, A., Plachez, C., Camurri, L., Zhang, J., Mori, S., Murakami, F., Parnavelas, J. G., Sundaresan, V., and Richards, L. J. (2006). Robo1 regulates the development of major axon tracts and interneuron migration in the forebrain. *Development* 133:2243–2252.

Ang, E. S., Jr., Haydar, T. F., Gluncic, V., and Rakic, P. (2003). Four-dimensional migratory coordinates of GABAergic interneurons in the developing mouse cortex. *J. Neurosci.* 23:5805–5815.

Barde, Y. A., Edgar, D., and Thoenen, H. (1982). Purification of a new neurotrophic factor from mammalian brain. ***EMBO*** J. 1:549–553.

Beffert, U., Morfini, G., Bock, H. H., Reyna, H., Brady, S. T., and Herz, J. (2002). Reelin-mediated signaling locally regulates protein kinase B/Akt and glycogen synthase kinase 3beta. *J. Biol. Chem.* 277:49958–49964.

Beffert, U., Weeber, E. J., Morfini, G., Ko, J., Brady, S. T., Tsai, L. H., Sweatt, J. D., and Herz, J. (2004). Reelin and cyclin-dependent kinase 5-dependent signals cooperate in regulating neuronal migration and synaptic transmission. *J. Neurosci.* 24:1897–1906.

Beffert, U., Weeber, E. J., Durudas, A., Qiu, S., Masiulis, I., Sweatt, J. D., Li, W. P., Adelmann, G., Frotscher, M., Hammer, R. E., and Herz, J. (2005). Modulation of synaptic plasticity and memory by reelin involves differential splicing of the lipoprotein receptor Apoer2. *Neuron* 47:567–579.

Behar, T. N., Dugich-Djordjevic, M. M., Li, Y. X., Ma, W., Somogyi, R., Wen, X., Brown, E., Scott, C., McKay, R. D., and Barker, J. L. (1997). Neurotrophins stimulate chemotaxis of embryonic cortical neurons. *Eur. J. Neurosci.* 9:2561–2570.

Bibel, M., and Barde, Y. A. (2000). Neurotrophins: key regulators of cell fate and cell shape in the vertebrate nervous system. *Genes Dev.* 14:2919–2937.

Bock, H. H., Jossin, Y., Liu, P., Forster, E., May, P., Goffinet, A. M., and Herz, J. (2003). Phosphatidylinositol 3-kinase interacts with the adaptor protein Dab1 in response to reelin signaling and is required for normal cortical lamination. *J. Biol. Chem.* 278:38772–38779.

Brunstrom, J. E., Gray-Swain, M. R., Osborne, P. A., and Pearlman, A. L. (1997). Neuronal heterotopias in the developing cerebral cortex produced by neurotrophin-4. *Neuron* 18:505–517.

Bu, G., and Schwartz, A. L. (1998). RAP, a novel type of ER chaperone. *Trends Cell Biol.* 8:272–276.

Cabelli, R. J., Hohn, A., and Shatz, C. J. (1995). Inhibition of ocular dominance column formation by infusion of NT-4/5 or BDNF. *Science* 267:1662–1666.

Cabelli, R. J., Shelton, D. L., Segal, R. A., and Shatz, C. J. (1997). Blockade of endogenous ligands of trkB inhibits formation of ocular dominance columns. *Neuron* 19:63–76.

Cohen, S., and Levi-Montalcini, R. (1957). Purification and properties of a nerve growth-promoting factor isolated from mouse sarcoma 180. *Cancer Res.* 17:15–20.

D'Arcangelo, G., Miao, G. G., Chen, S. C., Soares, H. D., Morgan, J. I., and Curran, T. (1995). A protein related to extracellular matrix proteins deleted in the mouse mutant reeler. *Nature* 374:719–723.

D'Arcangelo, G., Homayouni, R., Keshvara, L., Rice, D. S., Sheldon, M., and Curran, T. (1999). Reelin is a ligand for lipoprotein receptors. *Neuron* 24:471–479.

Del Rio, J. A., Heimrich, B., Borrell, V., Forster, E., Drakew, A., Alcantara, S., Nakajima, K., Miyata, T., Ogawa, M., Mikoshiba, K., Derer, P., Frotscher, M., and Soriano, E. (1997). A role for Cajal-Retzius cells and reelin in the development of hippocampal connections. *Nature* 385:70–74.

Derer, P. (1985). Comparative localization of Cajal-Retzius cells in the neocortex of normal and reeler mutant mice fetuses. *Neurosci. Lett.* 54:1–6.

Dong, E., Caruncho, H., Liu, W. S., Smalheiser, N. R., Grayson, D. R., Costa, E., and Guidotti, A. (2003). A reelin–integrin receptor interaction regulates Arc mRNA translation in synaptoneurosomes. *Proc. Natl. Acad. Sci. USA* 100:5479–5484.

Dulabon, L., Olson, E. C., Taglienti, M. G., Eisenhuth, S., McGrath, B., Walsh, C. A., Kreidberg, J. A., and Anton, E. S. (2000). Reelin binds alpha3beta1 integrin and inhibits neuronal migration. *Neuron* 27:33–44.

Durany, N., Michel, T., Zochling, R., Boissl, K. W., Cruz-Sanchez, F. F., Riederer, P., and Thome, J. (2001). Brain-derived neurotrophic factor and neurotrophin 3 in schizophrenic psychoses. *Schizophr. Res.* 52:79–86.

Egan, M. F., Kojima, M., Callicott, J. H., Goldberg, T. E., Kolachana, B. S., Bertolino, A., Zaitsev, E., Gold, B., Goldman, D., Dean, M., Lu, B., and Weinberger, D. R. (2003). The BDNF val66met polymorphism affects activity-dependent secretion of BDNF and human memory and hippocampal function. *Cell* 112:257–269.

Ernfors, P., Bengzon, J., Kokaia, Z., Persson, H., and Lindvall, O. (1991). Increased levels of messenger RNAs for neurotrophic factors in the brain during kindling epileptogenesis. *Neuron* 7:165–176.

Fatemi, S. H., Earle, J. A., and McMenomy, T. (2000). Reduction in reelin immunoreactivity in hippocampus of subjects with schizophrenia, bipolar disorder and major depression. *Mol. Psychiatry* 5:654–663.

Fatemi, S. H., Kroll, J. L., and Stary, J. M. (2001). Altered levels of reelin and its isoforms in schizophrenia and mood disorders. *Neuroreport* 12:3209–3215.

Fatemi, S. H., Stary, J. M., Earle, J. A., Araghi-Niknam, M., and Eagan, E. (2005a). GABAergic dysfunction in schizophrenia and mood disorders as reflected by decreased levels of glutamic acid decarboxylase 65 and 67 kDa and reelin proteins in cerebellum. *Schizophr. Res.* 72:109–122.

Fatemi, S. H., Snow, A. V., Stary, J. M., Araghi-Niknam, M., Reutiman, T. J., Lee, S., Brooks, A. I., and Pearce, D. A. (2005b). Reelin signaling is impaired in autism. *Biol. Psychiatry* 57:777–787.

Figurov, A., Pozzo-Miller, L. D., Olafsson, P., Wang, T., and Lu, B. (1996). Regulation of synaptic responses to high-frequency stimulation and LTP by neurotrophins in the hippocampus. *Nature* 381:706–709.

Friedman, W. J., Olson, L., and Persson, H. (1991). Cells that express brain-derived neurotrophic factor mRNA in the developing postnatal rat brain. *Eur. J. Neurosci.* 3:688–697.

Frotscher, M., Haas, C. A., and Forster, E. (2003). Reelin controls granule cell migration in the dentate gyrus by acting on the radial glial scaffold. *Cereb. Cortex* 13:634–640.

Gianfranceschi, L., Siciliano, R., Walls, J., Morales, B., Kirkwood, A., Huang, Z. J., Tonegawa, S., and Maffei, L. (2003). Visual cortex is rescued from the effects of dark rearing by overexpression of BDNF. *Proc. Natl. Acad. Sci. USA* 100:12486–12491.

Graus-Porta, D., Blaess, S., Senften, M., Littlewood-Evans, A., Damsky, C., Huang, Z., Orban, P., Klein, R., Schittny, J. C., and Muller, U. (2001). Beta1-class integrins regulate the development of laminae and folia in the cerebral and cerebellar cortex. *Neuron* 31:367–379.

Guidotti, A., Auta, J., Davis, J. M., Di-Giorgi-Gerevini, V., Dwivedi, Y., Grayson, D. R., Impagnatiello, F., Pandey, G., Pesold, C., Sharma, R., Uzunov, D., and Costa, E. (2000). Decrease in reelin and glutamic acid decarboxylase67 (GAD67) expression in schizophrenia and bipolar disorder: a postmortem brain study. *Arch. Gen. Psychiatry* 57:1061–1069.

Guilhem, D., Dreyfus, P. A., Makiura, Y., Suzuki, F., and Onteniente, B. (1996). Short increase of BDNF messenger RNA triggers kainic acid-induced neuronal hypertrophy in adult mice. *Neuroscience* 72:923–931.

Haas, C. A., Deller, T., Krsnik, Z., Tielsch, A., Woods, A., and Frotscher, M. (2000). Entorhinal cortex lesion does not alter reelin messenger RNA expression in the dentate gyrus of young and adult rats. *Neuroscience* 97:25–31.

Hallbook, F., Ibanez, C. F., and Persson, H. (1991). Evolutionary studies of the nerve growth factor family reveal a novel member abundantly expressed in Xenopus ovary. *Neuron* 6:845–858.

Hashimoto, K., Iwata, Y., Nakamura, K., Tsujii, M., Tsuchiya, K. J., Sekine, Y., Suzuki, K., Minabe, Y., Takei, N., Iyo, M., and Mori, N. (2006). Reduced serum levels of brain-derived neurotrophic factor in adult male patients with autism. *Prog. Neuropsychopharmacol. Biol. Psychiatry* 30:1529–1531.

He, X. P., Kotloski, R., Nef, S., Luikart, B. W., Parada, L. F., and McNamara, J. O. (2004). Conditional deletion of TrkB but not BDNF prevents epileptogenesis in the kindling model. *Neuron* 43:31–42.

Heinrich, C., Nitta, N., Flubacher, A., Muller, M., Fahrner, A., Kirsch, M., Freiman, T., Suzuki, F., Depaulis, A., Frotscher, M., and Haas, C. A. (2006). Reelin deficiency and displacement of mature neurons, but not neurogenesis, underlie the formation of granule cell dispersion in the epileptic hippocampus. *J. Neurosci.* 26:4701–4713.

Hohn, A., Leibrock, J., Bailey, K., and Barde, Y. A. (1990). Identification and characterization of a novel member of the nerve growth factor/brain-derived neurotrophic factor family. *Nature* 344:339–341.

Houser, C. R. (1990). Granule cell dispersion in the dentate gyrus of humans with temporal lobe epilepsy. *Brain Res.* 535:195–204.

Howell, B. W., Herrick, T. M., and Cooper, J. A. (1999). Reelin-induced tyrosine [corrected] phosphorylation of disabled 1 during neuronal positioning. *Genes Dev.* 13:643–648.

Huang, Z. J., Kirkwood, A., Pizzorusso, T., Porciatti, V., Morales, B., Bear, M. F., Maffei, L., and Tonegawa, S. (1999). BDNF regulates the maturation of inhibition and the critical period of plasticity in mouse visual cortex. *Cell* 98:739–755.

Impagnatiello, F., Guidotti, A. R., Pesold, C., Dwivedi, Y., Caruncho, H., Pisu, M. G., Uzunov, D. P., Smalheiser, N. R., Davis, J. M., Pandey, G. N., Pappas, G. D., Tueting, P., Sharma, R. P., and Costa, E. (1998). A decrease of reelin expression as a putative vulnerability factor in schizophrenia. *Proc. Natl. Acad. Sci. USA* 95:15718–15723.

Iritani, S., Niizato, K., Nawa, H., Ikeda, K., and Emson, P. C. (2003). Immunohistochemical study of brain-derived neurotrophic factor and its receptor, TrkB, in the hippocampal formation of schizophrenic brains. *Prog. Neuropsychopharmacol. Biol. Psychiatry* 27:801–807.

Isackson, P. J., Huntsman, M. M., Murray, K. D., and Gall, C. M. (1991). BDNF mRNA expression is increased in adult rat forebrain after limbic seizures: temporal patterns of induction distinct from NGF. *Neuron* 6:937–948.

Jiang, B., Akaneya, Y., Ohshima, M., Ichisaka, S., Hata, Y., and Tsumoto, T. (2001). Brain-derived neurotrophic factor induces long-lasting potentiation of synaptic transmission in visual cortex in vivo in young rats, but not in the adult. *Eur. J. Neurosci.* 14:1219–1228.

Karege, F., Perret, G., Bondolfi, G., Schwald, M., Bertschy, G., and Aubry, J. M. (2002). Decreased serum brain-derived neurotrophic factor levels in major depressed patients. *Psychiatry Res.* 109:143–148.

Keshvara, L., Magdaleno, S., Benhayon, D., and Curran, T. (2002). Cyclin-dependent kinase 5 phosphorylates disabled 1 independently of reelin signaling. *J. Neurosci.* 22:4869–4877.

Knable, M. B., Barci, B. M., Webster, M. J., Meador-Woodruff, J., and Torrey, E. F. (2004). Molecular abnormalities of the hippocampus in severe psychiatric illness: postmortem findings from the Stanley Neuropathology Consortium. *Mol. Psychiatry* 9:609–620.

Korte, M., Carroll, P., Wolf, E., Brem, G., Thoenen, H., and Bonhoeffer, T. (1995). Hippocampal long-term potentiation is impaired in mice lacking brain-derived neurotrophic factor. *Proc. Natl. Acad. Sci. USA* 92:8856–8860.

Kovalchuk, Y., Hanse, E., Kafitz, K. W., and Konnerth, A. (2002). Postsynaptic induction of BDNF-mediated long-term potentiation. *Science* 295:1729–1734.

Levenson, J. M., Roth, T. L., Lubin, F. D., Miller, C. A., Huang, I. C., Desai, P., Malone, L. M., and Sweatt, J. D. (2006). Evidence that DNA (cytosine-5) methyltransferase regulates synaptic plasticity in the hippocampus. *J. Biol. Chem.* 281:15763–15773.

Levine, E. S., Crozier, R. A., Black, I. B., and Plummer, M. R. (1998). Brain-derived neurotrophic factor modulates hippocampal synaptic transmission by increasing N-methyl-D-aspartic acid receptor activity. *Proc. Natl. Acad. Sci. USA* 95:10235–10239.

Magdaleno, S. M., and Curran, T. (2001). Brain development: integrins and the reelin pathway. *Curr. Biol.* 11:R1032–1035.

Marty, S., Carroll, P., Cellerino, A., Castren, E., Staiger, V., Thoenen, H., and Lindholm, D. (1996). Brain-derived neurotrophic factor promotes the differentiation of various hippocampal nonpyramidal neurons, including Cajal-Retzius cells, in organotypic slice cultures. *J. Neurosci.* 16:675–687.

Meyer, G., Goffinet, A. M., and Fairen, A. (1999). What is a Cajal-Retzius cell? A reassessment of a classical cell type based on recent observations in the developing neocortex. *Cereb. Cortex* 9:765–775.

Minichiello, L., Casagranda, F., Tatche, R. S., Stucky, C. L., Postigo, A., Lewin, G. R., Davies, A. M., and Klein, R. (1998). Point mutation in trkB causes loss of NT4-dependent neurons without major effects on diverse BDNF responses. *Neuron* 21:335–345.

Miyazaki, K., Narita, N., Sakuta, R., Miyahara, T., Naruse, H., Okado, N., and Narita, M. (2004). Serum neurotrophin concentrations in autism and mental retardation: a pilot study. *Brain Dev.* 26:292–295.

Murray, K. D., Isackson, P. J., Eskin, T. A., King, M. A., Montesinos, S. P., Abraham, L. A., and Roper, S. N. (2000). Altered mRNA expression for brain-derived neurotrophic factor and type II calcium/calmodulin-dependent protein kinase in the hippocampus of patients with intractable temporal lobe epilepsy. *J. Comp. Neurol.* 418:411–422.

Neves-Pereira, M., Mundo, E., Muglia, P., King, N., Macciardi, F., and Kennedy, J. L. (2002). The brain-derived neurotrophic factor gene confers susceptibility to bipolar disorder: evidence from a family-based association study. *Am. J. Hum. Genet.* 71:651–655.

Ogawa, M., Miyata, T., Nakajima, K., Yagyu, K., Seike, M., Ikenaka, K., Yamamoto, H., and Mikoshiba, K. (1995). The reeler gene-associated antigen on Cajal-Retzius neurons is a crucial molecule for laminar organization of cortical neurons. *Neuron* 14:899–912.

Ohmiya, M., Shudai, T., Nitta, A., Nomoto, H., Furukawa, Y., and Furukawa, S. (2002). Brain-derived neurotrophic factor alters cell migration of particular progenitors in the developing mouse cerebral cortex. *Neurosci. Lett.* 317:21–24.

Ohshima, T., and Mikoshiba, K. (2002). Reelin signaling and Cdk5 in the control of neuronal positioning. *Mol. Neurobiol.* 26:153–166.

Ohshima, T., Suzuki, H., Morimura, T., Ogawa, M., and Mikoshiba, K. (2007). Modulation of reelin signaling by cyclin-dependent kinase 5. *Brain Res.* 1140:84–95.

Perry, E. K., Lee, M. L., Martin-Ruiz, C. M., Court, J. A., Volsen, S. G., Merrit, J., Folly, E., Iversen, P. E., Bauman, M. L., Perry, R. H., and Wenk, G. L. (2001). Cholinergic activity in autism: abnormalities in the cerebral cortex and basal forebrain. *Am. J. Psychiatry* 158:1058–1066.

Persico, A. M., D'Agruma, L., Maiorano, N., Totaro, A., Militerni, R., Bravaccio, C., Wassink, T. H., Schneider, C., Melmed, R., Trillo, S., Montecchi, F., Palermo, M., Pascucci, T., Puglisi-Allegra, S., Reichelt, K. L., Conciatori, M., Marino, R., Quattrocchi, C. C., Baldi, A., Zelante, L., Gasparini, P., and Keller, F. (2001). Reelin gene alleles and haplotypes as a factor predisposing to autistic disorder. *Mol. Psychiatry* 6:150–159.

Pla, R., Borrell, V., Flames, N., and Marin, O. (2006). Layer acquisition by cortical GABAergic interneurons is independent of reelin signaling. *J. Neurosci.* 26:6924–6934.

Polleux, F., Whitford, K. L., Dijkhuizen, P. A., Vitalis, T., and Ghosh, A. (2002). Control of cortical interneuron migration by neurotrophins and PI3-kinase signaling. *Development* 129:3147–3160.

Rakic, P., and Caviness, V. S., Jr. (1995). Cortical development: view from neurological mutants two decades later. *Neuron* 14:1101–1104.

Riedel, A., Miettinen, R., Stieler, J., Mikkonen, M., Alafuzoff, I., Soininen, H., and Arendt, T. (2003). Reelin-immunoreactive Cajal-Retzius cells: the entorhinal cortex in normal aging and Alzheimer's disease. *Acta Neuropathol. (Berl.)* 106:291–302.

Ringstedt, T., Linnarsson, S., Wagner, J., Lendahl, U., Kokaia, Z., Arenas, E., Ernfors, P., and Ibanez, C. F. (1998). BDNF regulates reelin expression and Cajal-Retzius cell development in the cerebral cortex. *Neuron* 21:305–315.

Scharfman, H. E., Goodman, J. H., Sollas, A. L., and Croll, S. D. (2002). Spontaneous limbic seizures after intrahippocampal infusion of brain-derived neurotrophic factor. *Exp. Neurol.* 174:201–214.

Sen, S., Nesse, R. M., Stoltenberg, S. F., Li, S., Gleiberman, L., Chakravarti, A., Weder, A. B., and Burmeister, M. (2003). A BDNF coding variant is associated with the NEO personality inventory domain neuroticism, a risk factor for depression. *Neuropsychopharmacology* 28:397–401.

Serajee, F. J., Zhong, H., and Mahbubul Huq, A. H. (2006). Association of reelin gene polymorphisms with autism. *Genomics* 87:75–83.

Siuciak, J. A., Lewis, D. R., Wiegand, S. J., and Lindsay, R. M. (1997). Antidepressant-like effect of brain-derived neurotrophic factor (BDNF). *Pharmacol. Biochem. Behav.* 56:131–137.

Sklar, P., Gabriel, S. B., McInnis, M. G., Bennett, P., Lim, Y. M., Tsan, G., Schaffner, S., Kirov, G., Jones, I., Owen, M., Craddock, N., DePaulo, J. R., and Lander, E. S. (2002). Family-based association study of 76 candidate genes in bipolar disorder: BDNF is a potential risk locus. *Mol. Psychiatry* 7:579–593.

Suen, P. C., Wu, K., Levine, E. S., Mount, H. T., Xu, J. L., Lin, S. Y., and Black, I. B. (1997). Brain-derived neurotrophic factor rapidly enhances phosphorylation of the postsynaptic N-methyl-D-aspartate receptor subunit 1. *Proc. Natl. Acad. Sci. USA* 94:8191–8195.

Super, H., Del Rio, J. A., Martinez, A., Perez-Sust, P., and Soriano, E. (2000). Disruption of neuronal migration and radial glia in the developing cerebral cortex following ablation of Cajal-Retzius cells. *Cereb. Cortex* 10:602–613.

Szekeres, G., Juhasz, A., Rimanoczy, A., Keri, S., and Janka, Z. (2003). The C270T polymorphism of the brain-derived neurotrophic factor gene is associated with schizophrenia. *Schizophr. Res.* 65:15–18.

Takahashi, M., Hayashi, S., Kakita, A., Wakabayashi, K., Fukuda, M., Kameyama, S., Tanaka, R., Takahashi, H., and Nawa, H. (1999). Patients with temporal lobe epilepsy show an increase in brain-derived neurotrophic factor protein and its correlation with neuropeptide Y. *Brain Res.* 818:579–582.

Takahashi, M., Shirakawa, O., Toyooka, K., Kitamura, N., Hashimoto, T., Maeda, K., Koizumi, S., Wakabayashi, K., Takahashi, H., Someya, T., and Nawa, H. (2000). Abnormal expression of brain-derived neurotrophic factor and its receptor in the corticolimbic system of schizophrenic patients. *Mol. Psychiatry* 5:293–300.

Timmusk, T., Belluardo, N., Metsis, M., and Persson, H. (1993). Widespread and developmentally regulated expression of neurotrophin-4 mRNA in rat brain and peripheral tissues. *Eur. J. Neurosci.* 5:605–613.

Timmusk, T., Belluardo, N., Persson, H., and Metsis, M. (1994). Developmental regulation of brain-derived neurotrophic factor messenger RNAs transcribed from different promoters in the rat brain. *Neuroscience* 60:287–291.

Tokuoka, H., Saito, T., Yorifuji, H., Wei, F., Kishimoto, T., and Hisanaga, S. (2000). Brain-derived neurotrophic factor-induced phosphorylation of neurofilament-H subunit in primary cultures of embryo rat cortical neurons. *J. Cell Sci.* 113(Pt 6):1059–1068.

Toyooka, K., Asama, K., Watanabe, Y., Muratake, T., Takahashi, M., Someya, T., and Nawa, H. (2002). Decreased levels of brain-derived neurotrophic factor in serum of chronic schizophrenic patients. *Psychiatry Res.* 110:249–257.

Tsai, S. J. (2005). Is autism caused by early hyperactivity of brain-derived neurotrophic factor? *Med. Hypotheses* 65:79–82.

Wang, C. X., Song, J. H., Song, D. K., Yong, V. W., Shuaib, A., and Hao, C. (2006). Cyclin-dependent kinase-5 prevents neuronal apoptosis through ERK-mediated upregulation of Bcl-2. *Cell Death Differ.* 13:1203–1212.

Weeber, E. J., Beffert, U., Jones, C., Christian, J. M., Forster, E., Sweatt, J. D., and Herz, J. (2002). Reelin and ApoE receptors cooperate to enhance hippocampal synaptic plasticity and learning. *J. Biol. Chem.* 277:39944–39952.

Xu, B., Gottschalk, W., Chow, A., Wilson, R. I., Schnell, E., Zang, K., Wang, D., Nicoll, R. A., Lu, B., and Reichardt, L. F. (2000). The role of brain-derived neurotrophic factor receptors in the mature hippocampus: modulation of long-term potentiation through a presynaptic mechanism involving TrkB. *J. Neurosci.* 20:6888–6897.

Yang, C. B., Zheng, Y. T., Kiser, P. J., and Mower, G. D. (2006). Identification of disabled-1 as a candidate gene for critical period neuroplasticity in cat and mouse visual cortex. *Eur. J. Neurosci.* 23:2804–2808.

Yin, Y., Edelman, G. M., and Vanderklish, P. W. (2002). The brain-derived neurotrophic factor enhances synthesis of Arc in synaptoneurosomes. *Proc. Natl. Acad. Sci. USA*99:2368–2373.

Ying, S. W., Futter, M., Rosenblum, K., Webber, M. J., Hunt, S. P., Bliss, T. V., and Bramham, C. R. (2002). Brain-derived neurotrophic factor induces long-term potentiation in intact adult hippocampus: requirement for ERK activation coupled to CREB and upregulation of Arc synthesis. *J. Neurosci.* 22:1532–1540.

Yoshida, M., Assimacopoulos, S., Jones, K. R., and Grove, E. A. (2006). Massive loss of Cajal-Retzius cells does not disrupt neocortical layer order. *Development* 133:537–545.

Chapter 17
Reelin, Liver, and Lymphatics

Brigitte Samama and Nelly Boehm

Contents

1 Introduction

Reelin, as an extracellular glycoprotein involved in neuronal migration and cerebral cortex layering pattern, has received much attention since the discovery of the gene responsible for the disturbed central nervous system development in the reeler mouse, because of its fundamental functions in cerebral development, its modulatory effects on synaptic plasticity in adult rodents, and its potential involvement in psychiatric disorders (for recent reviews and references see: Jossin, 2004; D'Arcangelo, 2005; Fatemi, 2005; Forster *et al.*, 2006; Herz and Chen, 2006). Our knowledge about effects that reelin may have on peripheral organs, however, remains very scarce. Reelin mRNA and protein have been detected during development and adulthood in several peripheral organs. In the present chapter, we will focus on the presence of reelin in liver and lymphatics and discuss some hypotheses about the functional significance of this presence.

B. Samama
Institut d'Histologie, Faculté de Médecine, Université Louis Pasteur, 4 rue Kirschleger, 67085 Strasbourg Cedex, France
e-mail: Brigitte.Samama@medecine.u-strasbg.fr

N. Boehm
Institut d'Histologie, Faculté de Médecine, Université Louis Pasteur, 4 rue Kirschleger, 67085 Strasbourg Cedex, France
email: Nelly.Boehm@medecine.u-strasbg.fr

2 Localization of Reelin in Peripheral Tissues

2.1 Reelin in Liver Cells

Hirotsune *et al.* (1995) observed by RT-PCR the presence of mRNA in adult mouse liver, while DeSilva *et al.* (1997) detected by Northern hybridization the presence of transcripts in fetal human liver. A cellular localization of hepatic reelin was first provided by Ikeda and Terashima (1997) who observed by *in situ* hybridization very strong mRNA expression in livers of fetal mice from embryonic day 9.5 (E9.5) to E16.5 and adults; these authors concluded that reelin was localized in sinusoid endothelial cells. Reelin protein has also been detected in liver; Smalheiser *et al.* (2000) noted that "in well perfused rat liver, little reelin-like immunoreactivity was observed, except for a zone surrounding the sinusoids." By immunocytochemistry using a reelin antibody specific for the N-terminal domain of the protein, generously provided by Dr. Goffinet (De Bergeyck *et al.*, 1998), we observed the presence of reelin-like immunoreactivity in livers of humans (Fig. 17.1A), mice and rats during development (Fig. 17.1B,C), and in adult livers of rats (Fig. 17.1D) (Samama and Boehm, 2005, and unpublished data). Reelin was present in the youngest embryos we studied [i.e., E10.5 in mice, E12.5 in rats, and gestational week 5 (GW5) in humans]. The stained cells lined the sinusoids and appeared as spindle-shaped or stellate cells; their number progressively decreased during development, probably because of the expansion of hepatocyte number; in adult rats, reelin-positive cells were scattered in liver parenchyma (Fig. 17.1D). The cells were present in the space of Disse, between hepatocytes and endothelial cells, and had the same distribution as glial fibrillary acidic protein (GFAP)-positive cells (Buniatian *et al.*, 1997), suggesting that these cells were stellate cells (Ito cells) (Fig. 17.1E); they were identified by immunoelectron microscopy in rat adult liver as stellate cells by their localization between sinusoid endothelial cells and hepatocytes and by their lipid content (Fig. 17.1F,G).

2.2 Reelin in Lymphatics

When analyzing reelin immunoreactivity in vessels, we observed a unique staining in endothelial lining of lymphatic capillaries but not of blood capillaries or vessels (Samama and Boehm, 2005, and unpublished data). Staining was present at E13 in rats in the jugular lymphatic sac (Fig. 17.2A) and in scattered mesenchymal cells in the whole body, and at later fetal stages, in many lymphatics. This staining was also present in lymphatics of adult rats and mice as illustrated in Fig. 17.2B–D, where a strong immunostaining in rat ovarian medulla (Fig. 17.2B) or around Peyer's patches in rat gut (Fig. 17.2C,D) can be seen. In human fetuses, we first observed a clear staining at GW7 in clefts we could identify as lymphatic capillaries. These clefts were especially prominent in skin and lungs; in addition, scattered

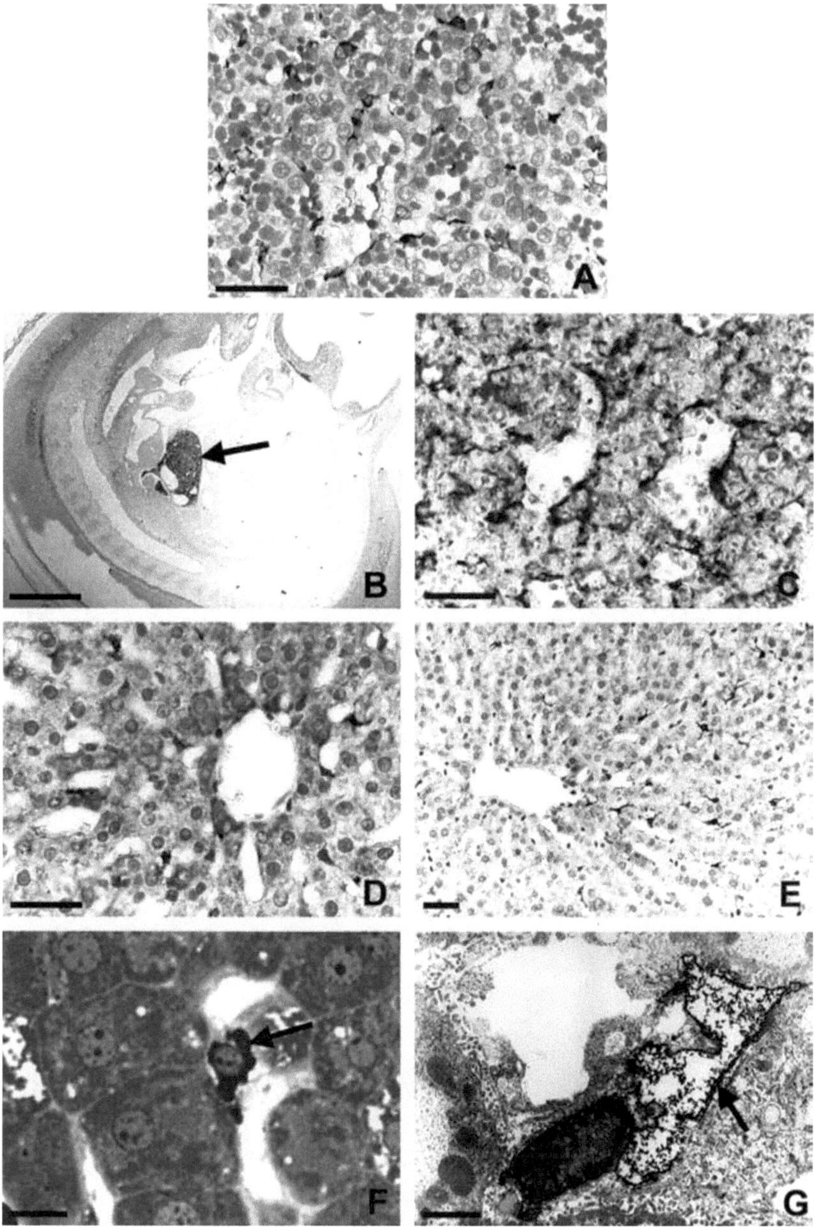

Fig. 17.1 Reelin (**A–D**, **F**, **G**) and GFAP (**E**) expression in human (**A**) and rat (**B–G**) liver. (**A**) Reelin immunostaining in stellate cells of human fetus at GW7. (**B**) Reelin immunostaining in liver of rat fetus at E13 (arrow). (**C**) Reelin immunostaining in stellate cells of rat fetus at E13; **C** is a high magnification of **B**. (**D**) Reelin immunostaining in adult rat stellate cells. (**E**) GFAP immunostaining in adult rat stellate cells. (**F**) Reelin immunostaining in a stellate cell of adult rat observed on a semithin section stained with toluidine blue (arrow). (**G**) Reelin immunostaining in a stellate cell of adult rat: electron microscopic examination; staining is observed in rough endoplasmic reticulum (arrow). Scale bars = 40 μm (**A**, **C–E**), 800 μm (**B**), 10 μm (**F**), and 2 μm (**G**) (*See Color Plates*)

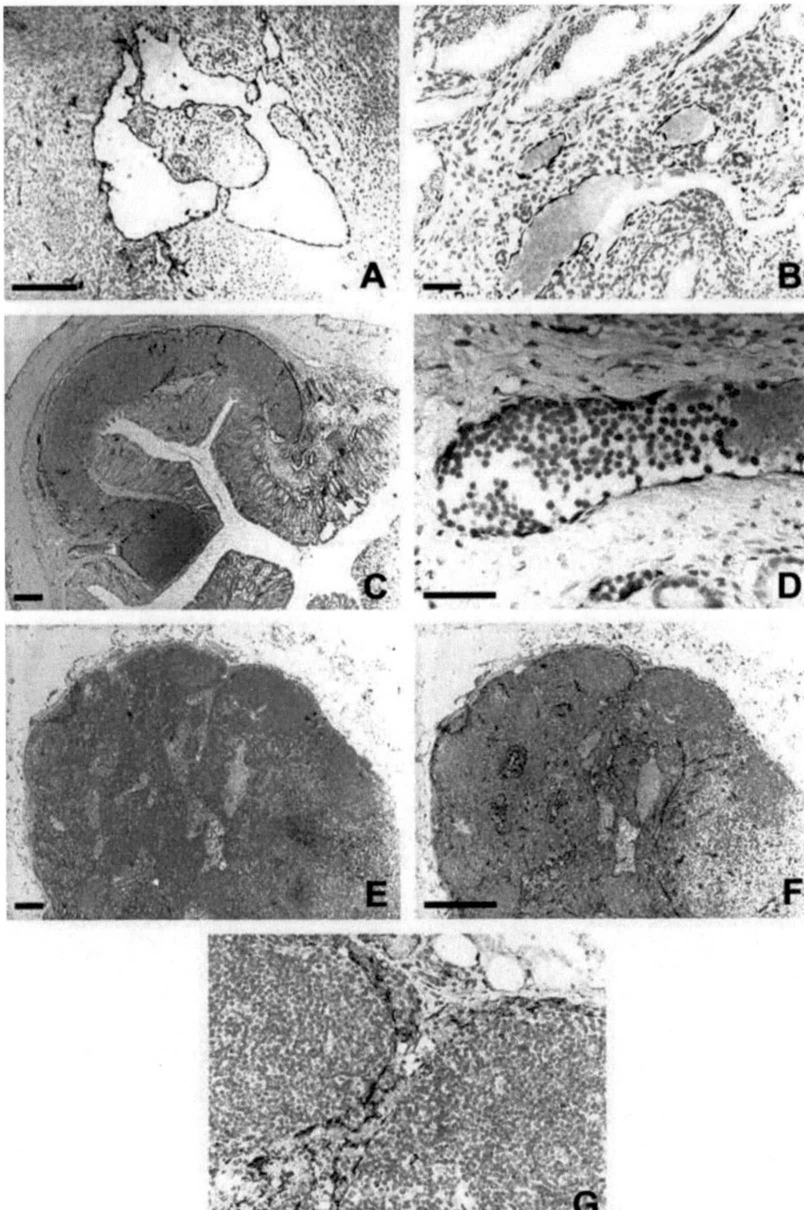

Fig. 17.2 Reelin (**A–E**) and CD31 (**F, G**) expression in rat fetus (**A**), adult rat (**B–D**), and adult human (**E–G**). **(A)** Reelin immunostaining in the jugular lymphatic sac of rat fetus at E13. **(B)** Reelin immunostaining in lymphatics of adult rat ovarian medulla. **(C, D)** Reelin immunostaining of lymphatics around Peyer's patches in adult rat gut; **D** is a high magnification of **C**. **(E)**Absence of reelin immunostaining in adult human lymph node. **(F, G)** CD31 immunostaining in adult human lymph node; **G** is a high magnification of **F**. Scale bars = 150 μm (**A, C, E**) and 40 μm (**B, D, F, G**) (*See Color Plates*)

elongated cells in mesenchyme were stained (Samama and Boehm, 2005). When comparing this staining with that observed with blood vessel markers or whole vasculature markers, reelin-positive cells were clearly present in lymphatic endothelial cells. No reelin immunoreactivity could be detected in lymph nodes of rats and humans as illustrated in Fig. 17.2E–G.

3 Functional Significance of Liver and Lymphatic Reelin

The presence of reelin-immunoreactive cells in peripheral organs raises two questions: Is the synthesized glycoprotein secreted? What may be its function?

3.1 Is Peripheral Reelin a Secreted Glycoprotein?

In the central nervous system during development, reelin is a glycoprotein of the extracellular matrix secreted by several groups of neurons. In the marginal zone of the cerebral cortex, reelin secreted by Cajal-Retzius cells is involved in the correct layering of cortical neurons in an inside-out manner (Curran and D'Arcangelo, 1998; Tissir and Goffinet, 2003; Soriano and Del Rio, 2005). Reelin immunoreactivity, most probably originating locally and not from the plasma, has also been detected in cerebrospinal fluid of humans as two main fragments, reflecting *in vivo* cleavage at two principal processing sites (Ignatova *et al.*, 2004). In peripheral blood, reelin has been detected as full-length protein and as two cleaved fragments in the serum of rats, mice, and humans (Smalheiser *et al.*, 2000; Lugli *et al.*, 2003). No correlation between modifications of cerebrospinal fluid reelin and plasma reelin has been detected in normal human subjects and Alzheimer's disease patients, suggesting that cerebrospinal fluid and peripheral blood reelin have different origins (Botella-Lopez *et al.*, 2006). Reelin, which is detectable in adult rat liver extracts, is also detected in conditioned medium when dissociated liver cells are placed in serum-free medium (Smalheiser *et al.*, 2000). Reelin production *in vitro* is modulated by hormonal influences, since dexamethasone greatly inhibits the accumulation of reelin in both cells and conditioned medium (Smalheiser *et al.*, 2000). All of these data suggest that peripheral reelin may be secreted and participates in the circulating pool of reelin. However, Roberts *et al.* (2005), in a postmortem study of human brains, observed that while the intracellular localization of reelin was quite similar in adult and fetal cortex, extracellular labeling was absent in adults and present in fetuses. After discussion of technical pitfalls, the authors concluded that reelin in the adult cortex was not a secreted protein. In our experiments, we did not observe reelin immunoreactivity in the extracellular matrix of liver or lymphatics. Further studies would be necessary to ascertain the nature of reelin as a secreted or a strictly intracellular glycoprotein. The last hypothesis raises the question of an intracrine effect of reelin as suggested or demonstrated for some peptides (Re and Cook, 2006).

3.2 *Role of Peripheral Reelin?*

3.2.1 Role of Reelin in Liver

Stellate cells (also referred to as Ito cells, fat-storing cells, lipocytes) represent a minor cell population of the liver (Geersts, 2004). Indeed, the adult hepatic lobule consists predominantly of anastomosing plates of hepatocytes limiting sinusoids feeding into the central venules. Stellate cells lie in the space of Disse between the endothelial lining and vascular domain of hepatocytes; this space contains scattered collagen fibrils but lacks basal laminae. The cells extend long cytoplasmic processes around sinusoids and their cytoplasm is characterized by the presence of lipid droplets, which have been demonstrated to store the major part of vitamin A of the whole body. They have important functions in adults, such as vitamin A storage and metabolism, production and remodeling of Disse space extracellular matrix, production of growth factors and cytokines, and regulation of sinusoidal lumen (for review, see Senoo, 2004). These cells present two phenotypes: when quiescent, they exhibit the fat-storing phenotype of the normal liver; when activated, they proliferate and display a myofibroblast-like phenotype, characteristic of stellate cells' response to liver injury (Senoo, 2004; Gressner and Weiskirchen, 2006). The origin of stellate cells remains debated. The liver develops from the anterior endoderm as a hepatic bud growing in the septum transversum, where mesoderm-derived cells promote growth of hepatocytes (Le Douarin, 1975). It is believed, although the exact lineage has not yet been demonstrated, that septum transversum mesenchymal cells give rise to stellate cells (Geersts, 2004). Recently, it has been shown that they do not derive from neural crest (Cassiman *et al.*, 2006). Morphological aspects of liver stellate cell development have been reviewed by Enzan *et al.* (1997). At E10 in mice and rats and at GW5 in humans, when hepatic cords grow into the mesenchyme of the septum transversum, sinusoids are still present and probable progenitors of stellate cells are trapped in the subendothelial space. At E12–14 in mice and rats and GW6–8 in humans, stellate cells are characterized by one or more lipid droplets; this developmental pattern corresponds to our observation of reelin in these cells. One may then suggest that reelin could play a role as a paracrine factor in liver development. However, Ikeda and Terashima (1997), when comparing reeler and wild-type mice, did not notice any difference in liver morphology; it is not excluded that subtle differences may exist, for example, at the ultrastructural level.

Hepatic stellate cells are believed to be a major source of collagen type I production during hepatic fibrosis. In response to liver injury, quiescent cells undergo rapid activation with loss of vitamin A storage and upregulation of α-smooth muscle actin and desmin, Kobold *et al.* (2002) studied reelin expression during activation of stellate cells *in vitro*. They observed that expression was restricted to those cells, was absent in other liver myofibroblasts, and remained stable, while other activation markers were up- or downregulated. During liver injury *in vivo*, they observed an upregulation of reelin in both hepatocytes and stellate cells mainly in the damaged zones, while in the periportal undamaged zone the number of reelin-positive cells remained more or less unchanged. However, at present, no role can be attributed to this upregulation, since reeler mice did not show differences over the complete time course of liver injury as compared to heterozygous and wild-type mice (Kobold *et al.*, 2002).

3.2.2 Role of Reelin in Lymphatics

Concerning lymphangiogenesis, the developmental origin of lymphatic endothelial cells from deep embryonic veins or mesenchymal lymphangioblasts remains to be clearly determined. At present, a dual origin of lymphendothelial cells is mostly accepted. A first type of cells is located in specific segments of the venous system and mostly present in segments where blood and lymph vessels are fused, permanently or transiently. A second type of lymphendothelial cell derives from scattered mesenchymal cells (Wilting and Becker, 2006, and references therein). Interestingly, we observed reelin immunoreactivity in both lymphatic capillaries and scattered elongated cells, suggesting that reelin may be an early marker of lymphatic endothelial cells. The role lymphatic reelin may play in lymphangiogenesis can only be highly speculative. To our knowledge, there is no report about lymphatic malformations in reeler mice. However, Hong *et al.* (2000) reported an autosomal recessive form of lissencephaly (smooth brain) with severe abnormalities of cerebellum, hippocampus, and brainstem, associated with two mutations in the human gene encoding reelin, resulting in low or undetectable levels of reelin protein in blood. It is noteworthy that some patients showed persistent lymphedema neonatally, resulting in accumulation of chylous ascites fluid in one patient who required peritoneal shunting (Hourihane *et al.*, 1993). Lymphedema is caused by insufficient lymph transport, as a result of lymphatic hypoplasia, impaired lymphatic function, or obstruction of lymph flow. That observation points to a possible role of reelin in lymphangiogenesis and lymphatic structure homeostasis.

It is noteworthy that both liver stellate cells and lymphatic endothelial cells are surrounded by a poor extracellular matrix, mostly lacking a continuous basal lamina, although collagen IV is secreted in the space of Disse and in the subendothelial space of lymphatics. Moreover, lymphatic capillaries do not recruit pericytes as do blood capillaries. Although we studied the collecting lymphatic vessels extensively, afferent and efferent lymphatics of lymph nodes were never reelin immunoreactive (unpublished data).

3.3 Reelin Signaling Pathway

Some clues about reelin functions in peripheral organs may come from the localization of the reelin signaling pathway. In the developing cortex, transmission of reelin signal to migrating neurons involves preliminary binding to reelin receptors. Two main types of receptors are known: lipoprotein receptors, i.e., very-low-density lipoprotein receptor (VLDLR) and apolipoprotein E receptor 2 (ApoER2) (Tissir and Goffinet, 2003; May *et al.*, 2005) and α3β1 integrin (Dulabon *et al.*, 2000). In peripheral organs, VLDLRs are abundant in heart, muscle, adipose tissue, and brain and are barely detectable in liver. They are also present in macrophages and endothelial cells of capillaries and arterioles, and in endothelial cells of coronaries but not in aorta or in veins or venules. In human liver, only sinusoidal lining cells but not hepatocytes express this receptor (review in Takahashi *et al.*, 2004, and references therein). ApoER2 transcripts are numerous in postmitotic neurons, testis, and ovary (Kim *et al.*, 1996).

In brain, binding of reelin to ApoER2 and VLDLR induces tyrosine phosphorylation of an adapter protein, disabled-1 (Dab1) by Src family kinases (Jossin *et al.*, 2003; Forster *et al.*, 2006; Stolt and Bock, 2006), resulting in nucleation of multiprotein complexes which modulate cytoskeleton dynamics. Dab1 is highly expressed and tyrosine phosphorylated in developing central nervous system and serves as a substrate for Src family kinases. In the central nervous system, reelin-expressing cells are adjacent to Dab1-immunoreactive cells, which are targets for reelin and are disturbed in their migration in the absence of reelin. In human and mouse cerebral cortex, during development, Dab1, VLDLR, and ApoER2 are expressed during neuronal migration in immature neurons of the cortical plate and may thus be responsive to reelin secreted by Cajal-Retzius cells in the marginal zone (Meyer *et al.*, 2003; Perez-Garcia *et al.*, 2004). Similarly, reelin protein is secreted by neurons adjacent to migrating sympathetic neurons, which, in turn, express Dab1 (Yip *et al.*, 2000, 2003, 2004; Kubasak *et al.*, 2004). Dab1 is mainly expressed in fetal and adult brain; during development, Howell *et al.* (1997) reported Dab1 expression and phosphorylation in some peripheral nerves in the mouse; in cultured cells, only P19 embryonal carcinoma (EC) cells and hematopoietic cell lines expressed Dab1. Smalheiser *et al.* (2000) noted that Dab1 immunoreactivity was present in the posterior lobe of the rat pituitary gland while reelin was present in the intermediate lobe. In human tooth, reelin is expressed in fully differentiated odontoblasts and the reelin signaling pathway is present in the trigeminal ganglion, suggesting that reelin might be involved in the terminal innervation of the dentin–pulp complex (Maurin *et al.*, 2004). Dab1 protein was not detectable in the adult rat liver either by immunocytochemistry or by Western blotting (Smalheiser *et al.*, 2000). We did not observe Dab1 immunoreactivity in livers of fetal and adult mouse and rat (unpublished data). Moreover, Dab1 mutants have no obvious peripheral phenotype.

An alternative hypothesis would be that reelin, secreted in the extracellular matrix, blood, or lymph may use a signaling pathway different from that previously described. It has been shown that reeler mice of both sexes had a reduced number of gonadotropin-releasing hormone (GnRH) neurons in the hypothalamus and that seminiferous tubules were reduced in number and dilated (Cariboni *et al.*, 2005). Reelin expressed along the intracerebral route of these migrating cells has an inhibitory role in guiding these neurons. However, mutant mice lacking reelin receptors or Dab1 have a normal complement of GnRH neurons, showing that the effect of reelin is independent of Dab1.

It is now clear that although several groups observed the presence of reelin in peripheral organs, such as liver and lymphatics and in peripheral blood, the significance of peripheral reelin during development and adulthood needs to be determined.

Acknowledgments This study was supported in part by the Conseil Scientifique de la Faculté de Médecine de Strasbourg and by the Programme Hospitalier de Recherche Clinique 2002, Hôpitaux Universitaires de Strasbourg.

We are grateful to Patricia Bos, Roland Bury, Josiane Meder, and Anne-Laure Burry for their excellent technical assistance.

References

Botella-Lopez, A., Burgaya, F., Gavin, R., Garcia-Ayllon, M.S., Gomez-Tortosa, E., Pena-Casanova, J., Urena, J.M., Del Rio, J.A., Blesa, R., Soriano, E., and Saez-Valero, J. (2006). Reelin expression and glycosylation patterns are altered in Alzheimer's disease. *Proc. Natl. Acad. Sci. USA* 103:5573–5578.

Buniatian, G., Hamprecht, B., and Gebhardt, R. (1996). Glial fibrillary acidic protein as a marker of perisinusoidal stellate cells that can distinguish between the normal and myofibroblast-like phenotypes. *Biol. Cell.* 87:65–73.

Cariboni, A., Rakic, S., Liapi, A., Maggi, R., Goffinet, A., and Parnavelas, J. G. (2005). Reelin provides an inhibitory signal in the migration of gonadotropin-releasing hormone neurons. *Development* 132:4709–4718.

Cassiman, D., Barlow, A., Vander Borght, S., Libbrecht, L., and Pachnis, V. (2006). Hepatic stellate cells do not derive from the neutral crest. *J. Hepatol.* 44:1098–1104.

Curran, T., and D'Arcangelo, G. (1998). Role of reelin in the control of brain development. *Brain Res. Brain Res. Rev.* 26:285–294.

D'Arcangelo, G. (2005). The reeler mouse: anatomy of a mutant. *Int. Rev. Neurobiol.* 71:383–417.

De Bergeyck, V., Naerhuyzen, B., Goffinet, A. M., and Lambert de Rouvroit, C. (1998). A panel of monoclonal antibodies against reelin, the extracellular matrix protein defective in reeler mutant mice. *J. Neurosci. Methods* 82:17–24.

DeSilva, U., D'Arcangelo, G., Braden, V. V., Chen, J., Miao, G. G., Curran, T., and Green, E. D. (1997). The human reelin gene: isolation, sequencing and mapping on chromosome 7. *Genome Res.* 7:157–164.

Dulabon, L., Olson, E. C., Taglienti, M. G., Eisenhuth, S., McGrath, B, Walsh, C. A., Kreidberg, J. A., and Anton, E. S. (2000). Reelin binds alpha3beta1 integrin and inhibits neuronal migration. *Neuron* 27:33–44.

Enzan, H., Himeno, H., Hiroi, M., Kiyoku, H., Saibara, T., and Onishi, S. (1997). Development of hepatic sinusoidal structure with special reference to the Ito cells. *Microsc. Res. Tech.* 39:336–349.

Fatemi, S. H. (2005). Reelin glycoprotein in autism and schizophrenia. *Int. Rev. Neurobiol.* 71:179–187.

Forster, E., Jossin, Y., Zhao, S., Chai, X., Frotscher, M., and Goffinet A. M. (2006). Recent progress in understanding the role of reelin in radial neuronal migration, with specific emphasis on the dentate gyrus. *Eur. J. Neurosci.* 23:901–909.

Geersts, A. (2004). On the origin of stellate cells: mesodermal, endodermal or neuro-ectodermal? *J. Hepatol.* 40:331–334.

Gressner, A. M., and Weiskirchen, R. (2006). Modern pathogenetic concepts of liver fibrosis suggest stellate cells and TGF-beta as major players and therapeutic targets. *J. Cell. Mol. Med.* 10:76–99.

Herz, J., and Chen, Y. (2006). Reelin, lipoprotein receptors and synaptic plasticity. *Nature Rev. Neurosci.* 7:850–859.

Hirotsune, S., Takahara, T., Sasaki, N., Hirose, K., Yoshiki, A., Ohashi, T., Kusakabe, M., Murakami, Y., Muramatsu, M., Watanabe, S., Nakao, K., Katsuki, M., and Hayashizaki, Y. (1995). The reeler gene encodes a protein with an EGF-like motif expressed by pioneer neurons. *Nature Genet.* 10:77–83.

Hong, S. E., Shugart, Y. Y., Huang, D. T., Shahwan, S. A., Grant, P. E., Hourihane, J. O., Martin, N. D., and Walsh, C. A. (2000). Autosomal recessive lissencephaly with cerebellar hypoplasia is associated with human RELN mutations. *Nature Genet.* 26:93–96.

Hourihane, J. O., Bennett, C. P., Chaudhuri, R., Robb, S. A., and Martin, N. D. (1993). A sibship with a neuronal migration defect, cerebellar hypoplasia and congenital lymphedema. *Neuropediatrics* 24:43–46.

Howell, B. W., Gertler, F. B., and Cooper, J. A. (1997). Mouse disabled (mDab1): a Src binding protein implicated in neural development. *EMBO J.* 16:121–132.

Ignatova, N., Sindic, C. J., and Goffinet, A. M. (2004). Characterization of the various forms of the reelin protein in the cerebrospinal fluid of normal subjects and in neurological diseases. *Neurobiol. Dis.* 15:326–330.

Ikeda, Y., and Terashima, T. (1997). Expression of reelin, the gene responsible for the reeler mutation in embryonic development and adulthood in the mouse. *Dev. Dyn.* 210:157–172.

Jossin, Y. (2004). Neuronal migration and the role of reelin during early development of the cerebral cortex. *Mol. Neurobiol.* 30:225–251.

Jossin, Y., Bar, I., Ignatova, N., Tissir, F., Lambert de Rouvroit, C., and Goffinet, A. M. (2003). The reelin signalling pathway: some recent developments. *Cereb. Cortex* 13:627–633.

Kim, D. H., Iijima, H., Goto, K., Ishii, H., Kim, H. J., Suzuki, H., Kondo, H., Saeki, S., and Yamamoto, T. (1996). Human apolipoprotein E receptor 2. A novel lipoprotein receptor of the low density lipoprotein receptor family predominantly expressed in brain. *J. Biol. Chem.* 271:8373–8380.

Kobold, D., Grundmann, A., Piscaglia, F., Eisenbach, C., Neubauer, K., Steffgen, J., Ramadori, G., and Knittel, T. (2002). Expression of reelin in hepatic stellate cells and during hepatic tissue repair: a novel marker for the differentiation of HSC from other liver myofibroblasts. *J. Hepatol.* 36:607–613.

Kubasak, M. D., Brooks, R., Chen, S., Villeda, S. A., and Phelps, P. E. (2004). Developmental distribution of reelin-positive cells and their secreted product in the rodent spinal cord. *J. Comp. Neurol.* 468:165–178.

Le Douarin, N. M. (1975). An experimental analysis of liver development. *Med. Biol.* 53:427–455.

Lugli, G., Krueger, J. M., Davis, J. M., Persico, A. M., and Smalheiser, N. R. (2003). Methodological factors influencing measurement and processing of plasma reelin in humans. *BMC Biochem.* 4:9.

Maurin, J. C., Couble, M. L., Didier-Bazes, M., Brisson, C., Magloire, H., and Bleicher, F. (2004). Expression and localization of reelin in human odontoblasts. *Matrix Biol.* 23:277–285.

May, P., Herz, J., and Bock, H. H. (2005). Molecular mechanisms of lipoprotein receptor signalling. *Cell. Mol. Life Sci.* 62:2325–2338.

Meyer, G., Lambert de Rouvroit, C., Goffinet, A. M., and Wahle, P. (2003). Disabled-1 mRNA and protein expression in developing human cortex. *Eur. J. Neurosci.* 17:517–525.

Perez-Garcia, C. G., Tissir, F., Goffinet, A. M., and Meyer, G. (2004). Reelin receptors in developing laminated brain structures of mouse and human. *Eur. J. Neurosci.* 20:2827–2832.

Re, R. N., and Cook, J. L. (2006). The intracrine hypothesis: an update. *Regul. Pept.* 133:1–9.

Roberts, R. C., Xu, L., Roche, J. K., and Kirkpatrick, B. (2005). Ultrastructural localization of reelin in the cortex in post-mortem human brain. *J. Comp. Neurol.* 482:294–308.

Samama, B., and Boehm, N. (2005). Reelin immunoreactivity in lymphatics and liver during development and adult life. *Anat. Rec. A* 285:595–599.

Senoo, H. (2004). Structure and function of hepatic stellate cells. *Med. Electron Microsc.* 37:3–15.

Smalheiser, N. R., Costa, E., Guidotti, A., Impagnatiello, F., Auta, J., Lacor, P., Kriho, V., and Pappas, G. D. (2000). Expression of reelin in adult mammalian blood, liver, pituitary pars intermedia, and adrenal chromaffin cells. *Proc. Natl. Acad. Sci. USA* 97:1281–1286.

Soriano, E., and Del Rio, J. A. (2005). The cells of Cajal-Retzius: still a mystery one century after. *Neuron* 46:389–394.

Stolt, P. C., and Bock, H. H. (2006). Modulation of lipoprotein receptor functions by intracellular adaptor proteins. *Cell Signal.* 18:1560–1571.

Takahashi, S., Sakai, J., Fujino, T., Hattori, H., Zenimaru, Y., Suzuki, J., Miyamori, I., and Yamamoto, T. T. (2004). The very low-density lipoprotein (VLDL) receptor: characterization and functions as a peripheral lipoprotein receptor. *J. Atheroscler. Thromb.* 11:200–208.

Tissir, F., and Goffinet, A. M. (2003). Reelin and brain development. *Nature Rev. Neurosci.* 4:496–505.

Wilting, J., and Becker, J. (2006). Two endothelial cell lines derived from the somite. *Anat. Embryol. (Berl.)* 211(Suppl. 7):57–63.

Yip, J. W., Yip, Y. P., Nakajima, K., and Capriotti, C. (2000). Reelin controls position of autonomic neurons in the spinal cord. *Proc. Natl. Acad. Sci. USA* 97:8612–8616.

Yip, Y. P., Capriotti, C., and Yip, J. W. (2003). Migratory pathway of sympathetic preganglionic neurons in normal and reeler mutant mice. *J. Comp. Neurol.* 460:94–105.

Yip, Y. P., Capriotti, C., Magdaleno, S., Benhayon, D., Curran, T., Nakajima, K., and Yip, J. W. (2004). Components of the reelin signalling pathway are expressed in the spinal cord. *J. Comp. Neurol.* 470:210–219.

Chapter 18
Reelin and Cajal-Retzius Cells

Jean-Marc Mienville

Contents

1 Introduction

Cajal-Retzius (CR) cells comprise a population of neurons found in the marginal layer of the developing cerebral cortex and hippocampus of amniotes. Their name originates from their codiscovery in the 1890s by Santiago Ramón y Cajal, who was using Golgi staining techniques on brain sections from small mammals such as rabbits (Ramón y Cajal, 1891), and by Gustaf Retzius, who referred to the cortical marginal cells he observed in human fetuses as "Cajal's cells" (Retzius, 1893). Despite this common classification, it should be noted that the morphology of CR cells from primates versus small mammals is not homogeneous. Drawings of primate marginal cells provided by Retzius and other authors (Meyer *et al.*, 1999) indicate rather complex and variable morphologies, and Retzius even initially considered "his" CR cells as glia (König, 1978). By contrast, CR cells present in the rat marginal zone—or layer I of the more mature cortex—display fairly homogeneous aspects, so that their identification is straightforward based on three morphologic criteria: fusiform or ovoid shape; bipolarity, i.e., presence of one axon and one dendrite; and tangential orientation of the latter (Fig. 18.1). Due to their facilitated access in a widely used species, a large body of data have been collected regarding

J. -M. Mienville
Laboratoire de Physiologie Cellulaire et Moléculaire, CNRS – UMR 6548, Université de Nice-Sophia Antipolis, Parc Valrose, 06108 Nice cedex 2, France
e-mail: mienvill@unice.fr

S. H. Fatemi (ed.), *Reelin Glycoprotein: Structure, Biology and Roles in Health and Disease.*

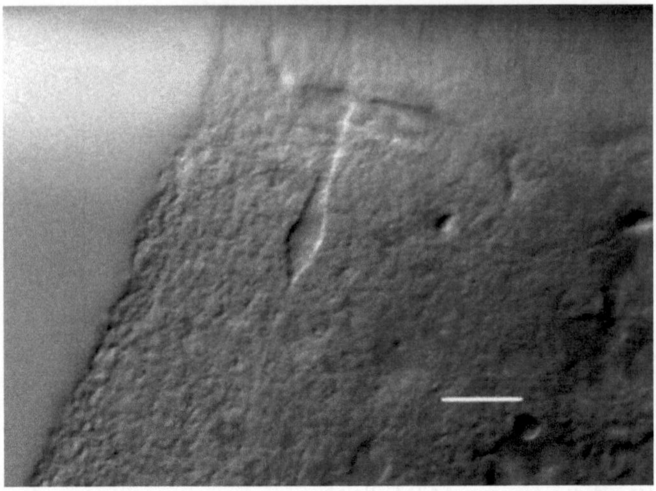

Fig. 18.1 Infrared differential interference contrast image of a Cajal-Retzius cell in a slice of postnatal day 11 rat neocortex. Bar = 20 μm. (Modified from Mienville, 1999, by permission from Oxford Journals)

the physiologic properties of rat CR cells (Mienville, 1999). In 1995, a novel criterion for identifying these neurons emerged with the discovery of reelin, their secreted protein, which is necessary for correct cortical lamination (D'Arcangelo *et al.*, 1995; Ogawa *et al.*, 1995). While reelin is produced by other cells (see below), certainly the combination of morphologic and immunocytochemical criteria now should allow unambiguous identification of CR cells.

Besides aiding with identification, the presence of reelin in CR cells may allow their provisional classification among those cells permitting neuron placement in a developing multilayered structure. This generalization beyond a role in the neocortex proper is necessary, given the presence of CR cells in lower animals devoid of such a structure (i.e., nonmammals). An ontogenic parallel to this observation is provided by the presence of CR cells in the mammalian hippocampus, where reelin is required for proper laminar organization (Nakajima *et al.*, 1997; Fatemi *et al.*, 2000). In the case of the hippocampus, however, the concept of "neuron placement" may be too restrictive, as reelin-producing CR cells also have a role in axonal growth and pathfinding, being required for the layer-specific targeting of developing entorhinal afferents (Del Río *et al.*, 1997). In addition, reelin is not the only signal produced by CR cells. For one thing, it is now generally admitted that they are glutamatergic neurons (Hevner *et al.*, 2003), though CR cell-induced postsynaptic currents have not been demonstrated yet. As for other substances, Clark *et al.* (1997) found that the products of genes whose mutations are responsible for a lissencephalic phenotype are heavily expressed in CR cells (as well as in the ventricular neuroepithelium). As lissencephaly appears to be caused by a developmental migratory defect, this reinforces the organizational role of CR cells, and suggests that some aspects of this role might be independent of reelin production.

Reciprocally, reelin is not exclusively secreted by CR cells. Reelin mRNA has been detected in many zones of the developing CNS, and in several cell types other than CR cells, including mitral cells of the olfactory bulb, granule cells of the cerebellum, and retinal ganglion cells (Schiffmann *et al.*, 1997; Alcántara *et al.*, 1998).

2 Life and Death of CR Cells

CR cells are among the first neurons to be produced during cortical development; they can be detected and are found to express reelin as early as embryonic day 10 in rodents (Ogawa *et al.*, 1995; Alcántara *et al.*, 1998), and at gestational week (GW) 6 in humans (Zecevic *et al.*, 1999). Progenitor commitment to CR cell fate is suppressed by the transcription factor Foxg1, which accounts for the time- and space-limited production of these cells (Hanashima *et al.*, 2004; Muzio and Mallamaci, 2005; Shen *et al.*, 2006). During initial stages of corticogenesis, CR cells, perhaps guided by some autocrine function (Meyer *et al.*, 2003; Perez-Garcia *et al.*, 2004), or more likely by interactions with the meningeal chemokine CXCL12 (Yamazaki *et al.*, 2004; Borrell and Marín, 2006), settle in a phylogenetically and ontogenetically primitive structure called the preplate or primordial plexiform layer (Zecevic *et al.*, 1999). The concept that their site of origin may be the subjacent ventricular zone (Zecevic *et al.*, 1999; Jiménez *et al.*, 2003; Shen *et al.*, 2006) coexists with numerous observations of migratory imports from the ganglionic eminence (Lavdas *et al.*, 1999; Zecevic and Rakic, 2001; Rakic and Zecevic, 2003) and, in fact, from multiple anatomic sites according to time-, origin-, and destination-dependent patterns (Meyer and Wahle, 1999; Meyer *et al.*, 2002a; Rakic and Zecevic, 2003; Takiguchi-Hayashi *et al.*, 2004; Muzio and Mallamaci, 2005; Borrell and Marín, 2006; Yoshida *et al.*, 2006; Cabrera-Socorro *et al.*, 2007; García-Moreno *et al.*, 2007; see Hevner *et al.*, 2003, and Bielle *et al.*, 2005, for a specific focus on mouse CR cells). Upon formation of the cortical plate, CR cells remain in the marginal zone, which is the future layer I of the neocortex.

The survival, function, and subsequent fate of CR cells appear to be determined by a number of extrinsic factors in their environment. For instance, their maintenance during corticogenesis seems to depend on a trophic contribution from meningeal cells, as pharmacologic destruction (Supèr *et al.*, 1997) or genetic alteration (Hartmann *et al.*, 1999) of the latter leads to a major loss of CR cells. Potential candidate factors for meningeal trophic support of CR cells include CXCL12 (also known as stromal cell-derived factor-1; Stumm *et al.*, 2003) and a TGF-β inhibitor (Kim and Pleasure, 2003). Extrinsic monoaminergic afferents may also influence the function and/or fate of CR cells (Naqui *et al.*, 1999; Janušonis *et al.*, 2004). On the pathologic side, expression of reelin by CR cells is decreased by prenatal infection with viruses such as human influenza H1N1 (Fatemi *et al.*, 1999), while exposure to ethanol indirectly disrupts reelin expression in the marginal zone, leading to the formation of brain heterotopias (Mooney *et al.*, 2004).

The normal genesis and differentiation of CR cells also depend on the expression of specific genes such as the transcription factor *Tbr-1* (Lambert de Rouvroit and Goffinet, 1998; Hevner *et al.*, 2001), the *Drosophila* head gap gene orthologs *Emx1* and *Emx2* (Shinozaki *et al.*, 2002; Bishop *et al.*, 2003), *LIS-1*, a gene coding for a microtubule-associated protein (Meyer *et al.*, 2002b), on the integrity of the heat-shock factor 2 → p35/39 → cyclin-dependent kinase 5 pathway (Chang *et al.*, 2006), and on the turning off of the transcription factor COUP-TFI (Studer *et al.*, 2005); in turn, CR cell positioning relies on the correct expression of the integrin *β1* gene (Graus-Porta *et al.*, 2001), while their migration rate is controlled by the transcription factor Pax6 (Stoykova *et al.*, 2003) and by CXCL12 (Borrell and Marín, 2006). As expected, alterations of the above genetic factors lead to abnormalities in cortex development.

Although a matter of occasional debate (Fairén *et al.*, 2002), it is generally admitted that CR cells are not only pioneer but also transient neurons. While some CR-like cells may persist in adult brain (see below), a major disappearance of CR cells occurs between postnatal weeks 2 and 3 in rodents, and around GW 27 in humans. A mechanism of "physiologic excitotoxicity" combining high expression of NMDA receptors and low resting potential (Fig. 18.2) has been proposed to account for this disappearance (Mienville and Pesold, 1999). This view is consistent with the increased survival of hippocampal CR cells in reeler due to decreased glutamatergic entorhinal afferentation (Coulin *et al.*, 2001), and with the massive loss of depolarized cortical CR cells in presenilin-1 (PS1) knockouts (Kilb *et al.*, 2004). Hypotheses evoking programmed cell death also have been proposed. For instance, it has been suggested that p73, a member of the p53 tumor-suppressor family, may be involved in the survival and death of CR cells (Meyer *et al.*, 2002a). Brain-derived neurotrophic factor (BDNF) appears to favor survival of late CR cells, but, interestingly, it is also a downregulating factor of reelin production, and it alters the morphology and organization of CR cells, all events suggesting that BDNF induces a change of function in these cells (Ringstedt *et al.*, 1998; Alcántara *et al.*, 2006). Perhaps relevant to this finding, early human cortical CR cells initially display the same bipolar shape and tangential orientation as those seen in rodents, but subsequently switch (while hippocampal CR cells do not; see Abraham and Meyer, 2003) to the polymorphic aspect originally described by Retzius (Meyer and Goffinet, 1998; Cabrera-Socorro *et al.*, 2007). Such phenotypic changes are compatible with the marked transcriptomic changes observed in developing CR cells (Yamazaki *et al.*, 2004). Around the same early midgestational period, the subpial granular layer supplies a new wave of reelin-expressing cells that resemble CR cells, including those of the "polymorphic" type (Meyer and Wahle, 1999; Rakic and Zecevic, 2003), and that may persist into adulthood (Meyer and Goffinet, 1998). This additional source of reelin, along with the further differentiation of early born CR cells, has been tentatively linked to the dramatically increasing complexity of the primate neocortex and the need to accommodate a protracted neurogenesis (Meyer and Goffinet, 1998; Meyer and Wahle, 1999; Cabrera-Socorro *et al.*, 2007). A similar biphasic scenario has been proposed regarding human hippocampal CR cells (Abraham *et al.*, 2004a). It is noteworthy that CR or CR-like

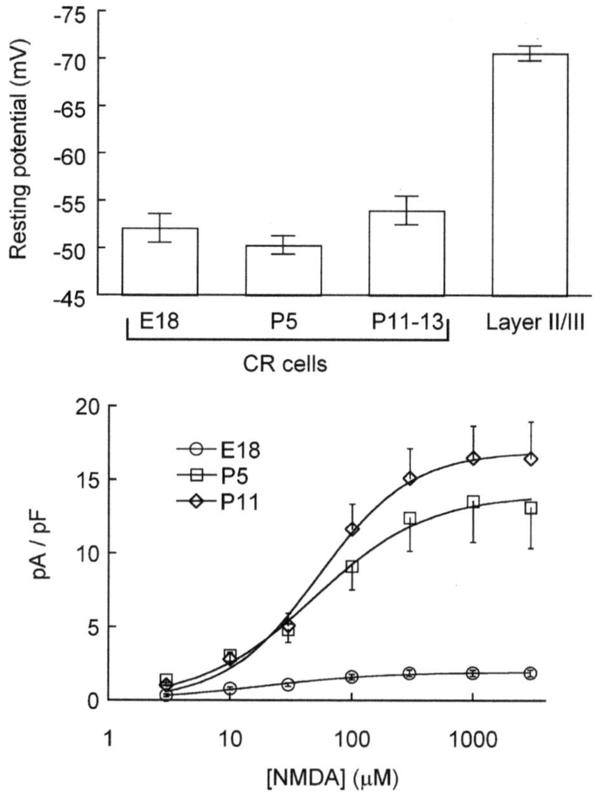

Fig. 18.2 (Upper panel) Low resting potential of rat neocortical Cajal-Retzius cells from embryonic day 18 to postnatal day 13 (mean ± SEM). The mean resting potential of Layer II/III neurons is shown for comparison. (Lower panel) Dose–response curves showing an increase in the density of NMDA receptor-mediated current from embryonic to postnatal stages of development. pA/pF = picoamperes per picofarad. (Modified from Mienville and Pesold, 1999, by permission from the Society of Neuroscience)

cells persist in the adult hippocampus of several mammalian species including mouse (Alcántara *et al.*, 1998), rat (Drakew *et al.*, 1998), macaque (Martínez-Cerdeño *et al.*, 2002), man (Fatemi *et al.*, 2000; Abraham and Meyer, 2003), and pig (Abraham *et al.*, 2004b), somehow consistent with the late survival of CR cells in more primitive brains (Blanton and Kriegstein, 1991).

The observation that CR or other types of reelin-secreting cells are present in the adult brain (Pesold *et al.*, 1998; Zecevic and Rakic, 2001; Abraham *et al.*, 2005) prompts the question of the functional role of reelin in the adult organism. In the cases of the adult entorhinal cortex and postnatal dentate gyrus, owing to the large number of reelin-expressing CR cells therein and the possibility that both structures may be sites of adult neurogenesis, one may hypothesize mere continuity in the same neuron-guiding role throughout life (Riedel *et al.*, 2003; Abraham and Meyer,

2003). On a more general level, an exciting perspective is the putative role of reelin in dendritic spine plasticity (Liu *et al.*, 2001; Abraham *et al.*, 2005; Roberts *et al.*, 2005) and long-term potentiation (Weeber *et al.*, 2002).

3 Reelin-Independent Functions of CR Cells

It is very likely that the most important role of CR cells is related to their capacity to synthesize and secrete reelin. The discovery of this protein and of its key function in cortical lamination was made possible by the availability of the *reeler* mutant mouse (D'Arcangelo *et al.*, 1995; Ogawa *et al.*, 1995). In this animal, several laminated brain structures such as cortex, hippocampus, and cerebellum are disorganized, which results in impaired motor function. It is important to note that CR cells are present in reeler (indeed, they even are found in greater number than in wild-type; see Coulin *et al.*, 2001), various mutations of their *reeler* gene preventing either synthesis or secretion of reelin, with the deleterious consequences just mentioned.

It has now become evident that reelin signaling is not the only developmental function of CR cells. Though it may seem technically difficult to perform selective ablation of CR cells, especially in early embryos, several such attempts indicate that lack (or strong reduction) of CR cells does not yield phenotypes equivalent to that of reeler. For instance, PS1-deficient mice have a dramatically reduced CR cell number, but display cortical dysplasia reminiscent of lissencephaly rather than *reeler* phenotype (Hartmann *et al.*, 1999; but see Hong *et al.*, 2000). *Emx1/2* double mutants incur a complete loss of CR cells and have lamination defects much more severe than those of reeler (Shinozaki *et al.*, 2002). CR cell-depleted $p73^{-/-}$ knockouts fail to develop a hippocampal fissure whereas the latter appears normally in reeler (Meyer *et al.*, 2004). Unexpectedly, in the latter knockouts, lamination of the rostral cortex is not significantly altered despite the ubiquitous loss of CR cells. Similarly, genetic ablation of the cortical hem, a major source of CR cells, does not disrupt lamination of the rostral neocortex (Yoshida *et al.*, 2006). Therefore, the role of reelin and CR cells in rostral areas may have to be redefined.

Nongenetic approaches have used pharmacologic ablation of CR cells versus reelin blockade by CR-50 antibody (or *reeler* mutants) to study the trophic effects of CR cells on axonal growth and pathfinding in the hippocampus. It was found that CR cell ablation prevents the ingrowth of entorhinohippocampal afferents, whereas reelin blockade or *reeler* mutation only affects their collateral branching and innervation density (Del Río *et al.*, 1997). Reelin-independent effects of CR cells can thus be separated; they appear to include both a permissive action on entorhinal afferent ingrowth, possibly via their own projections to the entorhinal cortex (Ceranik *et al.*, 1999; see below), and an inhibitory action on commissural axons via contact inhibition (Borrell *et al.*, 1999). Similar experimental strategies applied to cortical areas indicated that CR cell ablation induces a loss of radial glial processes, the "scaffold" system used by neuroblasts to migrate out of the ventricular zone, and thereby arrests migration. By contrast, the radial glia scaffold persists in reeler, though with a somewhat altered organization (Luque *et al.*, 2003), indicating that the lack of reelin is not the main factor for the loss of radial glia (Supèr *et al.*, 2000). However, the glial scaffold of the dentate gyrus is severely affected in reeler (Frotscher *et al.*, 2003).

The question therefore remains as to the reelin-independent mechanisms used by CR cells to affect developmental processes. As mentioned above, they might secrete soluble signals other than reelin, but our current state of knowledge does not allow expanding on that issue (Derer *et al.*, 2001; Yoshida *et al.*, 2006). Alternatively, the action of CR cells may be of a more physical nature. For instance, it has been proposed that CR cell projections to the entorhinal cortex may serve as a template or guiding scaffold for outgrowing entorhinal afferents (Ceranik *et al.*, 1999). Finally, a promising lead concerns the electrical activity of CR cells. Their participation in network activity of the developing neocortex has been demonstrated both at the single-unit (Soda *et al.*, 2003) and multi-unit recording level (Aguiló *et al.*, 1999). In the latter case, this activity did not reveal any major difference when recorded from reeler mice, confirming a lack of involvement of reelin, and was blocked by neurotransmitter antagonists, suggesting involvement of synaptic contacts. Indeed, spontaneous (Kilb and Luhmann, 2001) and evoked postsynaptic currents (Radnikow *et al.*, 2002) have been demonstrated in CR cells. Altogether, these results suggest that CR cells may function similarly to subplate cells, whose transient synaptic contacts allow establishment of thalamocortical connections (Friauf *et al.*, 1990). In that context, CR cells may serve as a temporary interface necessary for implementing hippocampal (Del Río *et al.*, 1997) as well as cortical layer I inputs. Their developmentally regulated spontaneous firing (Mienville *et al.*, 1999) and possible outputs (Radnikow *et al.*, 2002) in turn may be related to the maturation of the apical dendrites of cortical pyramidal cells (Marín-Padilla, 1998; Mienville, 1999).

4 Reelin Secretion by CR Cells

One may wonder whether this input–output activity of CR cells is also related to triggered reelin secretion. This is very unlikely because reelin secretion follows a constitutive mode, as would be expected for an extracellular matrix protein, and is independent of classical stimulus–secretion coupling processes (Lacor *et al.*, 2000). Studies of the Orleans *reeler* mutation point to the importance of the C-terminal region of reelin for secretion (de Bergeyck *et al.*, 1997), and work on alternatively polyadenylated reelin mRNA has located an important secretory signal to between residues 3328 and 3428 of the protein (Lambert de Rouvroit *et al.*, 1999). From a cellular viewpoint, the secretion of reelin from CR cells is postulated to occur along their axon, from specialized smooth cisterns therein (Derer *et al.*, 2001). However, the potential problem of CR cell axon early myelination (Ramón y Cajal, 1891; Marín-Padilla, 1998) in regard to this type of mechanism needs to be addressed in future studies. Interestingly, Roberts *et al.* (2005) saw reelin labeling in unmyelinated axons of adult and fetal human brains. As with other extracellular matrix proteins, secretion is in equilibrium with production (Lacor *et al.*, 2000; Derer *et al.*, 2001), which still leaves the possibility of a regulation at the transcriptional level. Such a regulation upon CR cells may be operated both upstream, e.g., by DNA topoisomerase IIβ (Lyu and Wang, 2003), and downstream by hormones (Alvarez-Dolado *et al.*, 1999), neurotransmitters (Martínez-Galán *et al.*, 2001), and in general through the electrical input activity mentioned above.

5 CR Cells, Reelin, and Disease

Concomitant with the above-described advances in the understanding of reelin and CR cell physiology, a growing number of neurologic cases are being linked to defects in the reelin pathway. As many chapters herein are devoted to these cases, we limit our account to cases where CR cells specifically appear to be involved. Curiously, several pathologies display CR cell hyperplasia or persistence in cortex and hippocampus from human fetuses and adults. These pathologies include cyclencephaly (Auroux, 1969), polymicrogyria (Eriksson *et al.*, 2001), epilepsy-linked cortical architectural dysplasia (Garbelli *et al.*, 2001), and hippocampal sclerosis linked to temporal lobe epilepsy (TLE; Thom *et al.*, 2002). Regarding TLE, there is disagreement with another study reporting CR cell hypoplasia possibly due to cytotoxic focal ischemia (Haas *et al.*, 2002). Remarkably, CR cell number or morphology appear not to be involved in Alzheimer's disease pathogenesis (Riedel *et al.*, 2003; Miettinen *et al.*, 2005), although the reelin pathway may be involved independently of CR cells. Thus, it may be speculated that fine tuning of CR cell number and reelin levels, both in the embryo and in the adult, is required for proper brain development and for specific neural functions in the adult. In that respect, though not specifically focused on CR cells, it is important to mention the first discovery of human lissencephaly cases diagnosed as reeler mutations (Hong *et al.*, 2000), as it may open the road to more discoveries of neurologic diseases linked to reelin and/or CR cell dysfunction.

6 CR Cells, Reelin, and Evolution

Beyond Cajal's initial (and substantiated) belief that "his special" cells were involved in the brain's ontogenic development, there are now solid reasons to believe that these cells also represent key players in evolutionary drive. Probably the most convincing evidence has emerged recently from a study showing that CR cells of the developing human cortex express *HAR1F*, a novel RNA gene containing a highly evolving "human accelerated region" (Pollard *et al.*, 2006). The fact that this type of noncoding gene is probably only involved in the space-time expression of protein-coding genes is not inconsistent with the phylogenic changes (mirrored by ontogenic changes) observed in CR cells. These changes may be both quantitative, e.g., expressed in CR cell densities and/or degree of reelin expression (Goffinet *et al.*, 1999; Bernier *et al.*, 1999, 2000; Tissir *et al.*, 2003), and qualitative, some peculiar examples being the change of direction of primate CR cell soma and neurites from horizontal to vertical (Retzius, 1893; Zecevic and Rakic, 2001; Cabrera-Socorro *et al.*, 2007), or the exclusive expression in human CR cells of the whole downstream part of the reelin pathway (Perez-Garcia *et al.*, 2004). Similarly, increased proportions of migratory versus local supply of CR cells may correspond to evolutionary steps from reptile to rodent to man (Bielle *et al.*, 2005; Cabrera-Socorro *et al.*, 2007). Consistent with their plasticity, CR cells are evolutionarily conserved, as reflected by their ubiquity over all amniotes and even amphibians. To better understand their developmental roles, it would be essential, whenever possible, to distinguish those related to reelin function from unrelated ones. For instance, it appears that reelin is involved in other, perhaps more fundamental processes than cortical lamination (Kikkawa *et al.*, 2003; Tissir *et al.*, 2003), such as tangential (in addition to radial) migration (Morante-Oria *et al.*, 2003). In fact, the reelin gene is expressed in all vertebrates (Goffinet *et al.*, 1999), and its sequence is highly conserved, consistent with the idea that lability is geared toward space-time expression rather than molecular structure (Bernier *et al.*, 1999). Moreover, this fits a potential regulation by *HAR1F* (Pollard *et al.*, 2006). Thus, for instance, a particular coordination between increased spatialization of CR cells and their reelin secretion timing may have provided a framework for the three-dimensional architecture of the mammalian neocortex (Fairén *et al.*, 2002; Nishikawa *et al.*, 2002; Alcántara *et al.*, 2006).

At the base of their most conspicuous reelin-independent role, the electrical activity of CR cells is probably related to transient priming of maturing cortical wiring. Several aspects of this activity are suggestive of a state of persistent immaturity (Mienville, 1999), which is entirely consistent with high labile potential. To summarize, we may suggest that CR cells were recruited as instructing constituents of the archicortex, where they subserve, in the adult, a still undefined role perhaps linked to synaptic plasticity. Their presumptive high labile potential may have promoted them to an active part of neocortical evolution. The reason for which this "special mission" was coselected with their timed degeneration remains to be discovered.

The past 10 years have seen much progress in our understanding of the detailed mechanisms that govern cortical lamination, wherein CR cells play a pivotal role. As such, these cells and their product(s) should continue inspiring investigators, inasmuch as they stand at a unique crossroads between phylogeny, ontogeny, and disease.

References

Abraham, H., and Meyer, G. (2003). Reelin-expressing neurons in the postnatal and adult human hippocampal formation. *Hippocampus* 13:715–727.

Abraham, H., Perez-Garcia, C. G., and Meyer, G. (2004a). p73 and reelin in Cajal-Retzius cells of the developing human hippocampal formation. *Cereb. Cortex* 14:484–495.

Abraham, H., Toth, Z., and Seress, L. (2004b). A novel population of calretinin-positive neurons comprises reelin-positive Cajal-Retzius cells in the hippocampal formation of the adult domestic pig. *Hippocampus* 14:385–401.

Abraham, H., Tóth, Z., Bari, F., Domoki, F., and Seress, L. (2005). Novel calretinin and reelin expressing neuronal population includes Cajal-Retzius-type cells in the neocortex of adult pigs. *Neuroscience* 136:217–230.

Aguiló, A., Schwartz, T.H., Kumar, V. S., Peterlin, Z. A., Tsiola, A., Soriano, E., and Yuste, R. (1999). Involvement of Cajal-Retzius neurons in spontaneous correlated activity of embryonic and postnatal layer 1 from wild-type and reeler mice. *J. Neurosci.* 19:10856–10868.

Alcántara, S., Ruiz, M., D'Arcangelo, G., Ezan, F., de Lecea, L., Curran, T., Sotelo, C., and Soriano, E. (1998). Regional and cellular patterns of reelin mRNA expression in the forebrain of the developing and adult mouse. *J. Neurosci.* 18:7779–7799.

Alcántara, S., Pozas, E., Ibanez, C. F., and Soriano, E. (2006). BDNF-modulated spatial organization of Cajal-Retzius and GABAergic neurons in the marginal zone plays a role in the development of cortical organization. *Cereb. Cortex* 16:487–499.

Alvarez-Dolado, M., Ruiz, M., Del Río, J. A., Alcántara, S., Burgaya, F., Sheldon, M., Nakajima, K., Bernal, J., Howell, B. W., Curran, T., Soriano, E., and Munoz, A. (1999). Thyroid hormone regulates reelin and dab1 expression during brain development. *J. Neurosci.* 19:6979–6993.

Auroux, M. (1969). Cortex cérébral dans la cyclocéphalie (persistance des cellules de Cajal-Retzius). *Arch. Anat. Histol. Embryol.* 52:43–52.

Bernier, B., Bar, I., Pieau, C., Lambert De Rouvroit, C., and Goffinet, A. M. (1999). Reelin mRNA expression during embryonic brain development in the turtle Emys orbicularis. *J. Comp. Neurol.* 413:463–479.

Bernier, B., Bar, I., D'Arcangelo, G., Curran, T., and Goffinet, A. M. (2000). Reelin mRNA expression during embryonic brain development in the chick. *J. Comp. Neurol.* 422:448–463.

Bielle, F., Griveau, A., Narboux-Neme, N., Vigneau, S., Sigrist, M., Arber, S., Wassef, M., and Pierani, A. (2005). Multiple origins of Cajal-Retzius cells at the borders of the developing pallium. *Nature Neurosci.* 8:1002–1012.

Bishop, K. M., Garel, S., Nakagawa, Y., Rubenstein, J. L., and O'Leary, D. D. (2003). Emx1 and Emx2 cooperate to regulate cortical size, lamination, neuronal differentiation, development of cortical efferents, and thalamocortical pathfinding. *J. Comp. Neurol.* 457:345–360.

Blanton, M. G., and Kriegstein, A. R. (1991). Morphological differentiation of distinct neuronal classes in embryonic turtle cerebral cortex. *J. Comp. Neurol.* 310:558–570.

Borrell, V., and Marín, O. (2006). Meninges control tangential migration of hem-derived Cajal-Retzius cells via CXCL12/CXCR4 signaling. *Nature Neurosci.* 9:1284–1293.

Borrell, V., Ruiz, M., Del Río, J. A., and Soriano, E. (1999). Development of commissural connections in the hippocampus of reeler mice: evidence of an inhibitory influence of Cajal-Retzius cells. *Exp. Neurol.* 156:268–282.

Cabrera-Socorro, A., Hernandez-Acosta, N. C., Gonzalez-Gomez, M., and Meyer, G. (2007). Comparative aspects of p73 and reelin expression in Cajal-Retzius cells and the cortical hem in lizard, mouse and human. *Brain Res.* 1132:59–70.

Ceranik, K., Deng, J., Heimrich, B., Lubke, J., Zhao, S., Forster, E., and Frotscher, M. (1999). Hippocampal Cajal-Retzius cells project to the entorhinal cortex: retrograde tracing and intracellular labelling studies. *Eur. J. Neurosci.* 11:4278–4290.

Chang, Y., Ostling, P., Akerfelt, M., Trouillet, D., Rallu, M., Gitton, Y., El Fatimy, R., Fardeau, V., Le Crom, S., Morange, M., Sistonen, L., and Mezger, V. (2006). Role of heat-shock factor 2 in cerebral cortex formation and as a regulator of p35 expression. *Genes Dev.* 20:836–847.

Clark, G. D., Mizuguchi, M., Antalffy, B., Barnes, J., and Armstrong, D. (1997). Predominant localization of the LIS family of gene products to Cajal-Retzius cells and ventricular neuroepithelium in the developing human cortex. *J. Neuropathol. Exp. Neurol.* 56:1044–1052.

Coulin, C., Drakew, A., Frotscher, M., and Deller, T. (2001). Stereological estimates of total neuron numbers in the hippocampus of adult reeler mutant mice: Evidence for an increased survival of Cajal-Retzius cells. *J. Comp. Neurol.* 439:19–31.

D'Arcangelo, G., Miao, G. G., Chen, S. C., Soares, H. D., Morgan, J. I., and Curran, T. (1995). A protein related to extracellular matrix proteins deleted in the mouse mutant reeler. *Nature* 374:719–723.

de Bergeyck, V., Nakajima, K., Lambert de Rouvroit, C., Naerhuyzen, B., Goffinet, A. M., Miyata, T., Ogawa, M., and Mikoshiba, K. (1997). A truncated reelin protein is produced but not secreted in the 'Orleans' reeler mutation ($Reln^{rl\text{-}Orl}$). *Brain Res. Mol. Brain Res.* 50:85–90.

Del Río, J. A., Heimrich, B., Borrell, V., Förster, E., Drakew, A., Alcántara, S., Nakajima, K., Miyata, T., Ogawa, M., Mikoshiba, K., Derer, P., Frotscher, M., and Soriano, E. (1997). A role for Cajal-Retzius cells and reelin in the development of hippocampal connections. *Nature* 385:70–74.

Derer, P., Derer, M., and Goffinet, A. (2001). Axonal secretion of reelin by Cajal-Retzius cells: evidence from comparison of normal and $Reln^{Orl}$. mutant mice. *J. Comp. Neurol.* 440:136–143.

Drakew, A., Frotscher, M., Deller, T., Ogawa, M., and Heimrich, B. (1998). Developmental distribution of a reeler gene-related antigen in the rat hippocampal formation visualized by CR-50 immunocytochemistry. *Neuroscience* 82:1079–1086.

Eriksson, S. H., Thom, M., Heffernan, J., Lin, W. R., Harding, B. N., Squier, M. V., and Sisodiya, S. M. (2001). Persistent reelin-expressing Cajal-Retzius cells in polymicrogyria. *Brain* 124:1350–1361.

Fairén, A., Morante-Oria, J., and Frassoni, C. (2002). The surface of the developing cerebral cortex: still special cells one century later. *Prog. Brain Res.* 136:281–291.

Fatemi, S. H., Emamian, E. S., Kist, D., Sidwell, R. W., Nakajima, K., Akhter, P., Shier, A., Sheikh, S., and Bailey, K. (1999). Defective corticogenesis and reduction in reelin immunoreactivity in cortex and hippocampus of prenatally infected neonatal mice. *Mol. Psychiatry* 4:145–154.

Fatemi, S. H., Earle, J., and McMenomy, T. (2000). Reduction in reelin immunoreactivity in hippocampus of subjects with schizophrenia, bipolar disorder and major depression. *Mol. Psychiatry* 5:654–663.

Friauf, E., McConnell, S. K., and Shatz, C. J. (1990). Functional synaptic circuits in the subplate during fetal and early postnatal development of cat visual cortex. *J. Neurosci.* 10:2601–2613.

Frotscher, M., Haas, C. A., and Förster, E. (2003). Reelin controls granule cell migration in the dentate gyrus by acting on the radial glial scaffold. *Cereb. Cortex* 13:634–640.

Garbelli, R., Frassoni, C., Ferrario, A., Tassi, L., Bramerio, M., and Spreafico, R. (2001). Cajal-Retzius cell density as marker of type of focal cortical dysplasia. *Neuroreport* 12:2767–2771.

García-Moreno, F., Lopez-Mascaraque, L., and De Carlos, J. A. (2007). Origins and migratory routes of murine Cajal-Retzius cells. *J. Comp. Neurol.* 500:419–432.

Goffinet, A. M., Bar, I., Bernier, B., Trujillo, C., Raynaud, A., and Meyer, G. (1999). Reelin expression during embryonic brain development in lacertilian lizards. *J. Comp. Neurol.* 414:533–550.

Graus-Porta, D., Blaess, S., Senften, M., Littlewood-Evans, A., Damsky, C., Huang, Z., Orban, P., Klein, R., Schittny, J. C., and Muller, U. (2001). Beta1-class integrins regulate the development of laminae and folia in the cerebral and cerebellar cortex. *Neuron* 31:367–379.

Haas, C. A., Dudeck, O., Kirsch, M., Huszka, C., Kann, G., Pollak, S., Zentner, J., and Frotscher, M. (2002). Role for reelin in the development of granule cell dispersion in temporal lobe epilepsy. *J. Neurosci.* 22:5797–5802.

Hanashima, C., Li, S. C., Shen, L., Lai, E., and Fishell, G. (2004). Foxg1 suppresses early cortical cell fate. *Science* 303:56–59.

Hartmann, D., De Strooper, B., and Saftig, P. (1999). Presenilin-1 deficiency leads to loss of Cajal-Retzius neurons and cortical dysplasia similar to human type 2 lissencephaly. *Curr. Biol.* 9:719–727.

Hevner, R. F., Shi, L., Justice, N., Hsueh, Y., Sheng, M., Smiga, S., Bulfone, A., Goffinet, A. M., Campagnoni, A. T., and Rubenstein, J. L. (2001). Tbr1 regulates differentiation of the preplate and layer 6. *Neuron* 29:353–366.

Hevner, R. F., Neogi, T., Englund, C., Daza, R. A., and Fink, A. (2003). Cajal-Retzius cells in the mouse: transcription factors, neurotransmitters, and birthdays suggest a pallial origin. *Brain Res. Dev. Brain Res.* 141:39–53.

Hong, S. E., Shugart, Y. Y., Huang, D. T., Shahwan, S. A., Grant, P. E., Hourihane, J. O., Martin, N. D., and Walsh, C. A. (2000). Autosomal recessive lissencephaly with cerebellar hypoplasia is associated with human RELN mutations. *Nature Genet.* 26:93–96.

Janušonis, S., Gluncic, V., and Rakic, P. (2004). Early serotonergic projections to Cajal-Retzius cells: relevance for cortical development. *J. Neurosci.* 24:1652–1659.

Jiménez, D., Rivera, R., López-Mascaraque, L., and De Carlos, J. A. (2003). Origin of the cortical layer I in rodents. *Dev. Neurosci.* 25:105–115.

Kikkawa, S., Yamamoto, T., Misaki, K., Ikeda, Y., Okado, H., Ogawa, M., Woodhams, P. L., and Terashima, T. (2003). Missplicing resulting from a short deletion in the reelin gene causes reeler-like neuronal disorders in the mutant shaking rat Kawasaki. *J. Comp. Neurol.* 463:303–315.

Kilb, W., and Luhmann, H. J. (2001). Spontaneous GABAergic postsynaptic currents in Cajal-Retzius cells in neonatal rat cerebral cortex. *Eur. J. Neurosci.* 13:1387–1390.

Kilb, W., Hartmann, D., Saftig, P., and Luhmann, H. J. (2004). Altered morphological and electrophysiological properties of Cajal-Retzius cells in cerebral cortex of embryonic presenilin-1 knockout mice. *Eur. J. Neurosci.* 20:2749–2756.

Kim, A. S., and Pleasure, S. J. (2003). Expression of the BMP antagonist Dan during murine forebrain development. *Brain Res. Dev. Brain Res.* 145:159–162.

König, N. (1978). Retzius-Cajal or Cajal-Retzius cells? *Neurosci. Lett.* 9:361–363.

Lacor, P. N., Grayson, D. R., Auta, J., Sugaya, I., Costa, E., and Guidotti, A. (2000). Reelin secretion from glutamatergic neurons in culture is independent from neurotransmitter regulation. *Proc. Natl. Acad. Sci. USA* 97:3556–3561.

Lambert de Rouvroit, C., and Goffinet, A. M. (1998). A new view of early cortical development. *Biochem. Pharmacol.* 56:1403–1409.

Lambert de Rouvroit, C., Bernier, B., Royaux, I., de Bergeyck, V., and Goffinet, A. M. (1999). Evolutionarily conserved, alternative splicing of reelin during brain development. *Exp. Neurol.* 156:229–238.

Lavdas, A. A., Grigoriou, M., Pachnis, V., and Parnavelas, J. G. (1999). The medial ganglionic eminence gives rise to a population of early neurons in the developing cerebral cortex. *J. Neurosci.* 19:7881–7888.

Liu, W. S., Pesold, C., Rodriguez, M. A., Carboni, G., Auta, J., Lacor, P., Larson, J., Condie, B. G., Guidotti, A., and Costa, E. (2001). Down-regulation of dendritic spine and glutamic acid decarboxylase 67 expressions in the reelin haploinsufficient heterozygous reeler mouse. *Proc. Natl. Acad. Sci. USA* 98:3477–3482.

Luque, J. M., Morante-Oria, J., and Fairén, A. (2003). Localization of ApoER2, VLDLR and Dab1 in radial glia: groundwork for a new model of reelin action during cortical development. *Brain Res. Dev. Brain Res.* 140:195–203.

Lyu, Y. L., and Wang, J. C. (2003). Aberrant lamination in the cerebral cortex of mouse embryos lacking DNA topoisomerase IIbeta. *Proc. Natl. Acad. Sci. USA* 100:7123–7128.

Marín-Padilla, M. (1998). Cajal-Retzius cells and the development of the neocortex. *Trends Neurosci.* 21:64–71.

Martínez-Cerdeño, V., Galazo, M. J., Cavada, C., and Clascá, F. (2002). Reelin immunoreactivity in the adult primate brain: intracellular localization in projecting and local circuit neurons of the cerebral cortex, hippocampus and subcortical regions. *Cereb. Cortex* 12:1298–1311.

Martínez-Galán, J. R., Lopez-Bendito, G., Lujan, R., Shigemoto, R., Fairén, A., and Valdeolmillos, M. (2001). Cajal-Retzius cells in early postnatal mouse cortex selectively express functional metabotropic glutamate receptors. *Eur. J. Neurosci.* 13:1147–1154.

Meyer, G., and Goffinet, A. M. (1998). Prenatal development of reelin-immunoreactive neurons in the human neocortex. *J. Comp. Neurol.* 397:29–40.

Meyer, G., and Wahle, P. (1999). The paleocortical ventricle is the origin of reelin-expressing neurons in the marginal zone of the foetal human neocortex. *Eur. J. Neurosci.* 11:3937–3944.

Meyer, G., Goffinet, A. M., and Fairén, A. (1999). What is a Cajal-Retzius cell? A reassessment of a classical cell type based on recent observations in the developing neocortex. *Cereb. Cortex* 9:765–775.
Meyer, G., Perez-Garcia, C. G., Abraham, H., and Caput, D. (2002a). Expression of p73 and reelin in the developing human cortex. *J. Neurosci.* 22:4973–4986.
Meyer, G., Perez-Garcia, C. G., and Gleeson, J. G. (2002b). Selective expression of doublecortin and LIS1 in developing human cortex suggests unique modes of neuronal movement. *Cereb. Cortex* 12:1225–1236.
Meyer, G., ambert de Rouvroit, C., Goffinet, A. M., and Wahle, P. (2003). Disabled-1 mRNA and protein expression in developing human cortex. *Eur. J. Neurosci.* 17:517–525.
Meyer, G., Cabrera Socorro, A., Perez Garcia, C. G., Martinez Millan, L., Walker, N., and Caput, D. (2004). Developmental roles of p73 in Cajal-Retzius cells and cortical patterning. *J. Neurosci.* 24:9878–9887.
Mienville, J. -M. (1999). Cajal-Retzius cell physiology: just in time to bridge the 20th century. *Cereb. Cortex* 9:776–782.
Mienville, J. -M., and Pesold, C. (1999). Low resting potential and postnatal upregulation of NMDA receptors may cause Cajal-Retzius cell death. *J. Neurosci.* 19:1636–1646.
Mienville, J.-M., Maric, I., Maric, D., and Clay, J. R. (1999). Loss of I_A expression and increased excitability in postnatal rat Cajal-Retzius cells. *J. Neurophysiol.* 82:1303–1310.
Miettinen, R., Riedel, A., Kalesnykas, G., Kettunen, H. P., Puolivali, J., Soininen, H., and Arendt, T. (2005). Reelin-immunoreactivity in the hippocampal formation of 9-month-old wildtype mouse: effects of APP/PS1 genotype and ovariectomy. *J. Chem. Neuroanat.* 30:105–118.
Mooney, S. M., Siegenthaler, J. A., and Miller, M. W. (2004). Ethanol induces heterotopias in organotypic cultures of rat cerebral cortex. *Cereb. Cortex* 14:1071–1080.
Morante-Oria, J., Carleton, A., Ortino, B., Kremer, E. J., Fairén, A., and Lledo, P. M. (2003). Subpallial origin of a population of projecting pioneer neurons during corticogenesis. *Proc. Natl. Acad. Sci. USA* 100:12468–12473.
Muzio, L., and Mallamaci, A. (2005). Foxg1 confines Cajal-Retzius neuronogenesis and hippocampal morphogenesis to the dorsomedial pallium. *J. Neurosci.* 25:4435–4441.
Nakajima, K., Mikoshiba, K., Miyata, T., Kudo, C., and Ogawa, M. (1997). Disruption of hippocampal development in vivo by CR-50 mAb against reelin. *Proc. Natl. Acad. Sci. USA* 94:8196–8201.
Naqui, S. Z., Harris, B. S., Thomaidou, D., and Parnavelas, J. G. (1999). The noradrenergic system influences the fate of Cajal-Retzius cells in the developing cerebral cortex. *Brain Res. Dev. Brain Res.* 113:75–82.
Nishikawa, S., Goto, S., Hamasaki, T., Yamada, K., and Ushio, Y. (2002). Involvement of reelin and Cajal-Retzius cells in the developmental formation of vertical columnar structures in the cerebral cortex: evidence from the study of mouse presubicular cortex. *Cereb. Cortex* 12:1024–1030.
Ogawa, M., Miyata, T., Nakajima, K., Yagyu, K., Seike, M., Ikenaka, K., Yamamoto, H., and Mikoshiba, K. (1995). The *reeler* gene-associated antigen on Cajal–Retzius neurons is a crucial molecule for laminar organization of cortical neurons. *Neuron* 14:899–912.
Perez-Garcia, C. G., Tissir, F., Goffinet, A. M., and Meyer, G. (2004). Reelin receptors in developing laminated brain structures of mouse and human. *Eur. J. Neurosci.* 20:2827–2832.
Pesold, C., Impagnatiello, F., Pisu, M. G., Uzunov, D. P., Costa, E., Guidotti, A., and Caruncho, H. J. (1998). Reelin is preferentially expressed in neurons synthesizing gamma-aminobutyric acid in cortex and hippocampus of adult rats. *Proc. Natl. Acad. Sci. USA* 95:3221–3226.
Pollard, K. S., Salama, S. R., Lambert, N., Lambot, M. A., Coppens, S., Pedersen, J. S., Katzman, S., King, B., Onodera, C., Siepel, A., Kern, A. D., Dehay, C., Igel, H., Ares, M., Jr., Vanderhaeghen, P., and Haussler, D. (2006). An RNA gene expressed during cortical development evolved rapidly in humans. *Nature* 443:167–172.
Radnikow, G., Feldmeyer, D., and Lübke, J. (2002). Axonal projection, input and output synapses, and synaptic physiology of Cajal-Retzius cells in the developing rat neocortex. *J. Neurosci.* 22:6908–6919.

Rakic, S., and Zecevic, N. (2003). Emerging complexity of layer I in human cerebral cortex. *Cereb. Cortex* 13:1072–1083.
Ramón y Cajal, S. (1891). Sur la structure de l'écorce cérébrale de quelques mammifères. *Cellule* 7:125–176.
Retzius, G. (1893). Die Cajal'schen Zellen der Grosshirnrinde beim Menschen und bei Säugethieren. *Biol. Untersuch. Neue Folge* 5:1–15.
Riedel, A., Miettinen, R., Stieler, J., Mikkonen, M., Alafuzoff, I., Soininen, H., and Arendt, T. (2003). Reelin-immunoreactive Cajal-Retzius cells: the entorhinal cortex in normal aging and Alzheimer's disease. *Acta Neuropathol. (Berl.)* 106:291–302.
Ringstedt, T., Linnarsson, S., Wagner, J., Lendahl, U., Kokaia, Z., Arenas, E., Ernfors, P., and Ibanez, C.F. (1998). BDNF regulates reelin expression and Cajal-Retzius cell development in the cerebral cortex. *Neuron* 21:305–315.
Roberts, R. C., Xu, L., Roche, J. K., and Kirkpatrick, B. (2005). Ultrastructural localization of reelin in the cortex in post-mortem human brain. *J. Comp. Neurol.* 482:294–308.
Schiffmann, S. N., Bernier, B., and Goffinet, A. M. (1997). Reelin mRNA expression during mouse brain development. *Eur. J. Neurosci.* 9:1055–1071.
Shen, Q., Wang, Y., Dimos, J. T., Fasano, C. A., Phoenix, T. N., Lemischka, I. R., Ivanova, N. B., Stifani, S., Morrisey, E. E., and Temple, S. (2006). The timing of cortical neurogenesis is encoded within lineages of individual progenitor cells. *Nature Neurosci.* 9:743–751.
Shinozaki, K., Miyagi, T., Yoshida, M., Miyata, T., Ogawa, M., Aizawa, S., and Suda, Y. (2002). Absence of Cajal-Retzius cells and subplate neurons associated with defects of tangential cell migration from ganglionic eminence in Emx1/2 double mutant cerebral cortex. *Development* 129:3479–3492.
Soda, T., Nakashima, R., Watanabe, D., Nakajima, K., Pastan, I., and Nakanishi, S. (2003). Segregation and coactivation of developing neocortical layer 1 neurons. *J. Neurosci.* 23:6272–6279.
Stoykova, A., Hatano, O., Gruss, P., and Götz, M. (2003). Increase in reelin-positive cells in the marginal zone of Pax6 mutant mouse cortex. *Cereb. Cortex* 13:560–571.
Studer, M., Filosa, A., and Rubenstein, J. L. (2005). The nuclear receptor COUP-TFI represses differentiation of Cajal-Retzius cells. *Brain Res. Bull.* 66:394–401.
Stumm, R. K., Zhou, C., Ara, T., Lazarini, F., Dubois-Dalcq, M., Nagasawa, T., Hollt, V., and Schulz, S. (2003). CXCR4 regulates interneuron migration in the developing neocortex. *J. Neurosci.* 23:5123–5130.
Supèr, H., Martínez, A., and Soriano, E. (1997). Degeneration of Cajal-Retzius cells in the developing cerebral cortex of the mouse after ablation of meningeal cells by 6-hydroxydopamine. *Brain Res. Dev. Brain Res.* 98:15–20.
Supèr, H., Del Río, J.A., Martinez, A., Perez-Sust, P., and Soriano, E. (2000). Disruption of neuronal migration and radial glia in the developing cerebral cortex following ablation of Cajal-Retzius cells. *Cereb. Cortex* 10:602–613.
Takiguchi-Hayashi, K., Sekiguchi, M., Ashigaki, S., Takamatsu, M., Hasegawa, H., Suzuki-Migishima, R., Yokoyama, M., Nakanishi, S., and Tanabe, Y. (2004). Generation of reelin-positive marginal zone cells from the caudomedial wall of telencephalic vesicles. *J. Neurosci.* 24:2286–2295.
Thom, M., Sisodiya, S. M., Beckett, A., Martinian, L., Lin, W. R., Harkness, W., Mitchell, T. N., Craig, J., Duncan, J., and Scaravilli, F. (2002). Cytoarchitectural abnormalities in hippocampal sclerosis. *J. Neuropathol. Exp. Neurol.* 61:510–519.
Tissir, F., Lambert de Rouvroit, C., Sire, J. Y., Meyer, G., and Goffinet, A. M. (2003). Reelin expression during embryonic brain development in *Crocodylus niloticus*. *J. Comp. Neurol.* 457:250–262.
Weeber, E. J., Beffert, U., Jones, C., Christian, J. M., Forster, E., Sweatt, J. D., and Herz, J. (2002). Reelin and ApoE receptors cooperate to enhance hippocampal synaptic plasticity and learning. *J. Biol. Chem.* 277:39944–39952.

Yamazaki, H., Sekiguchi, M., Takamatsu, M., Tanabe, Y., and Nakanishi, S. (2004). Distinct ontogenic and regional expressions of newly identified Cajal-Retzius cell-specific genes during neocorticogenesis. *Proc. Natl. Acad. Sci. USA* 101:14509–14514.
Yoshida, M., Assimacopoulos, S., Jones, K. R., and Grove, E. A. (2006). Massive loss of Cajal-Retzius cells does not disrupt neocortical layer order. *Development* 133:537–545.
Zecevic, N., and Rakic, P. (2001). Development of layer I neurons in the primate cerebral cortex. *J. Neurosci.* 21:5607–5619.
Zecevic, N., Milosevic, A., Rakic, S., and Marín-Padilla, M. (1999). Early development and composition of the human primordial plexiform layer: an immunohistochemical study. *J. Comp. Neurol.* 412:241–254.

Chapter 19
Reelin and Odontogenesis

Françoise Bleicher, Henry Magloire, Marie-Lise Couble, and Jean-Christophe Maurin

Contents

1 Introduction

Human tooth is made of three different hard tissues: enamel—recovering the crown and corresponding to the most mineralized tissue found in the body; cementum—deposited on the root surface; and dentine—underlying the enamel and cementum and forming the bulk of the tooth. Dentine covers the pulp, which lies in the center

F. Bleicher
University of Lyon, Lyon, F-69000; University of Lyon 1, Lyon, F-69100; CNRS UMR5242, IGFL, LYON, F-69007; INRA, LIMR1288, Lyon, F-69007; INSERM ERI16, Faculté d'Odontologie, Lyon F-69008 France, email: bleicher@sante.univ-lyon1.fr

H. Magloire
University of Lyon, Lyon, F-69000; University of Lyon 1, Lyon, F-69100; CNRS UMR5242, IGFL, LYON, F-69007; INRA, LIMR1288, Lyon, F-69007; INSERM ERI16, Faculté d'Odontologie, Lyon F-69008 France

M. -L. Couble
University of Lyon, Lyon, F-69000; University of Lyon 1, Lyon, F-69100; CNRS UMR5242, IGFL, LYON, F-69007; INRA, LIMR1288, Lyon, F-69007; INSERM ERI16, Faculté d'Odontologie, Lyon F-69008 France

J. -C. Maurin
INSERM ERM 0203, IFR 53, Interfaces Biomatériaux/Tissus Hôtes, Faculté de Chirurgie-Dentaire de Reims, F-51100 Reims France

S. H. Fatemi (ed.), *Reelin Glycoprotein: Structure, Biology and Roles in Health and Disease.*

of the tooth. Pulp, the vital mesenchymal tissue, contains: (1) a specialized layer of cells at its periphery (the odontoblast layer) that is responsible for the dentine matrix synthesis and (2) blood vessels and nerves. During development, tooth pulp acquires a profuse nociceptive and sympathetic innervation from the trigeminal ganglion (TG) and the superior cervical ganglion, respectively.

The main symptom of tooth hypersensitivity, or more precisely, dentine hypersensitivity, is a sharp sudden pain of short duration in response to thermal stimuli, such as intake of cold or hot foods, but may also arise from tactile stimuli, e.g., the use of a toothbrush. Dentine hypersensitivity may arise as a result of loss of enamel by attrition, abrasion, or erosion and/or root surface denudation with exposure of the underlying dentine. Up to 30% of adults experience this situation at some time during their lifetime. And now, an increasing number of adolescents are concerned due to excessive dietary acids, such as citrus juices and carbonated drinks.

The mechanism underlying dentine sensitivity has been widely discussed over the years, and several hypotheses have been proposed to explain this process (Dowell and Addy, 1983). The "hydrodynamic theory," developed in the 1960s, postulates that rapid shifts, in either direction, of the fluids within the dentinal tubules, following a stimulus application, result in activation of a nerve receptor sensitive to pressure, which leads to the transmission of the stimuli in the pulp/inner dentine region of the tooth (Brannstrom, 1962). However, trigeminal sensory nerve endings penetrate no more than 0.1–0.2 mm in coronal dentine tubules. This fact does not fully support the view that intradentinal axons could be directly elicited by stimuli applied to the teeth. Recently, an array of evidence has shown that odontoblasts, rather than nerve fibers, may operate as sensor cells. These neural crest-derived cells play a central role during the formation of dentine. They extend a long cell process in the dentinal tubules and their body is enclosed in a dense network of sensory pulpal axons.

Two kinds of mechanosensitive K^+ channels (KCa and TREK-1) have been identified in human odontoblasts (Allard *et al.*, 2000; Magloire *et al.*, 2003). Moreover, human odontoblasts *in vitro* produce voltage-gated tetrodotoxin (TTX)-sensitive Na^+ currents in response to depolarization under voltage clamp conditions and are able to generate action potentials (Allard *et al.*, 2006). These findings indicate that odontoblasts may be able to convert pain-evoking fluid displacement within dentinal tubules into electrical signals, strengthening their possible role as tooth sensor cells that initiate tooth pain transmission. These results raise the question of how the firing of odontoblasts is transmitted to the neighboring nerve cells. In fact, the way odontoblasts and nerve cells may communicate remains unclear.

Recently, we have shown that reelin, a large extracellular matrix glycoprotein elaborated by odontoblasts, could promote adhesion between nerve endings and cells (Maurin *et al.*, 2004). This close association suggested that odontoblasts and nerve endings may directly interact, although no synaptic structures or any junction could be detected between them (Ibuki *et al.*, 1996).

This chapter will review the development of dental innervation, the signaling molecules involved in this process, and the putative role of reelin in odontogenesis.

2 Establishment of Tooth Innervation

2.1 The Embryonic Stages of Tooth Development

Several studies of the localization of nerve fibers in human and murine teeth have shown that dental axon growth and patterning take place in a spatiotemporally controlled manner and are tightly linked with advancing tooth morphogenesis (Pearson, 1977; Mohamed and Atkinson, 1983; Tsuzuki and Kitamura, 1991; Fristad *et al.*, 1994; Hildebrand *et al.*, 1995; Luukko, 1997; Kettunen *et al.*, 2005). Tooth development is generally divided into three main stages, defined as initiation, morphogenesis, and cell differentiation. Trigeminal axons are present in the maxillary and mandibular processes before any structural signs of tooth formation are detectable. Then, a plexus of nerve fibers is seen in the mesenchyme, beneath the thickened oral epithelium, during the formation of dental lamina, which is the first histological sign of tooth formation. As the dental epithelium thickens and the underlying mesenchyme undergoes condensation, axon sprouts grow toward the mesenchyme and continue to the epithelium as lingual and buccal branches. During the next stages of morphogenesis, at the cap and early bell stages, local tooth primordial axons form a plexus at the base of the primitive dental papilla and come into contact with the dental follicle. Sensory axons first enter the dental pulp in the late bell stage, at the onset of enamel formation (Fig. 19.1A–D; for review, see Hildebrand *et al.*, 1995).

2.2 The Postnatal Stages of Tooth Development

Newly erupted human teeth, with an incomplete root formation, have a sparse innervation. They are less sensitive to electrical or thermal stimulation. A rapid development of sensory pulpal axons arises during root formation after tooth eruption, leading to the formation of the subodontoblastic plexus of Raschkow. The sensory nerve endings originating from the plexus of Raschkow increase in the odontoblastic layer, coiling around the cell bodies and processes of odontoblasts in the dentinal tubule (Fig. 19.1E–H). Dentinal innervation begins at the tips of the pulp horn, and the pioneer axons enter the dentine, just before tooth eruption.

Trigeminal nerve endings are distributed within a gradient, with the greatest innervation at the tip of the pulp horn where the sensitivity is also the greatest. A less dense innervation is observed as distance to the pulp horn tip increases, and root dentine is only slightly innervated. However, a greater number of nerve fibers are observed in human root dentine than in the rodent one. This situation could explain the high sensitivity of exposed human root dentine during thermal stimulation and carious process.

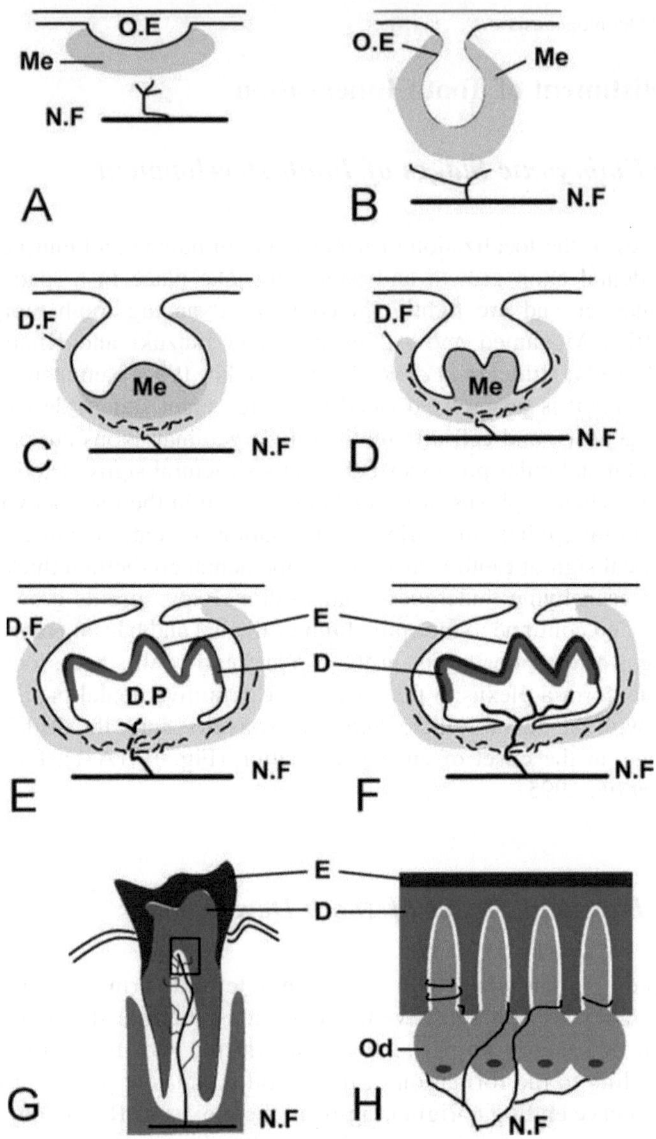

Fig. 19.1 Schematic representation of dental innervation during tooth development from embryonic stages (**A—D**) to postnatal stages (**E—H**). (**A**) Epithelial thickening stage. A plexus of nerve fibers is observed in the mesenchyme beneath the thickened oral epithelium. (**B**) Bud stage. The oral epithelium thickens and the mesenchyme undergoes a condensation. Axon sprouts grow toward the mesenchyme and continue to the epithelium as lingual and buccal branches. (**C**) Cap stage. Local axons form a plexus at the base of the primitive dental papilla and come into contact with the dental follicle. (**D**) Early bell stage. The number of axons increases in the dental follicle. (**E**) Late bell stage. At the onset of amelogenesis and dentinogenesis, the first sensory axons enter the dental papilla. (**F**) During early root formation, the number of pulpal axons increases. (**G**) During tooth eruption and with the advancing root formation, a rapid development of sensory pulpal axons leads to the formation of the subodontoblastic plexus of Raschkow. (**H**) Enlarged schematic representation of the dentin pulp complex innervation. The sensory nerve endings originating from the plexus of Raschkow coil around the cell bodies and processes of odontoblasts in the dentinal tubules. D, dentin; D.F, dental follicle; D.P, dental papilla; E, enamel; Me, mesenchyme; N.F, nerve fiber; Od, odontoblasts; O.E, oral epithelium (*See Color Plates*)

Anterograde axonal transport labeling methods in animal molars have shown that sensory nerves innervating dentine can extend up to 0.1–0.2 mm into the dentinal tubules at the tip of the crown (more than 50% of the dentinal tubule can be innervated in this area) (Byers *et al.*, 2003). They usually penetrate shorter distances in other coronal regions. Some of the nerve fibers appear to form adhesive contact with the odontoblasts inside the dentinal tubules. An extracellular space is retained between these two cells, but the junctions are not of the gap or synaptic form (Ibuki *et al.*, 1996). This space—about 20 nm wide—could be used for the delivery of signaling substances in molecular or ionic forms.

In human teeth, several years elapse between eruption and the root apex closure corresponding to a slow neural maturation. Then, by the time occlusion has been established, most of the dentinal innervation at the tip of the cusp has formed. (for review, see Hildebrand *et al.*, 1995).

3 Molecular Signaling and Pulp Innervation

As described above, the pulpal nerve development is closely coordinated with tooth development. It seems that growth and establishment of nerve terminals are controlled by local molecular signals. The specificity of dental sensory innervation and the highly plastic formed system imply that signals involved in the regulation of axon growth must be selective in terms of sensory modalities. The regulation of the development of tooth sensory nerve supply at a molecular level is beginning to be elucidated. There is accumulating evidence that the molecules involved in the development of the peripheral nervous system are also involved in dental axon guidance (for a review, see Luukko *et al.*, 2005). Among them, neurotrophic factors [nerve growth factor (NGF), brain-derived neurotrophic factor (BDNF), and glial cell-line-derived neurotrophic factor (GDNF)], axon guidance molecules (semaphorins, netrins, and ephrins), and extracellular matrix proteins (laminin, fibronectin, and tenascin) have been shown to play a role in the establishment of the mesenchymal dental axon pathway (Nosrat *et al.*, 1997, 1998; Loes *et al.*, 2002; Mitsiadis *et al.*, 2003; Lillesaar and Fried, 2004; Fried *et al.*, 2005; Kettunen *et al.*, 2005; Maurin *et al.*, 2005).

NGF, GDNF, and netrin-3, which exert positive effects on axon growth, are specifically expressed in the mesenchymal dental axon pathway when pioneer dental axons are navigating toward the early bud stage tooth germ (Luukko *et al.*, 2005). Subsequently, *Ngf* becomes expressed together with *Gdnf* in the dental follicle target field (Luukko *et al.*, 1997; Nosrat *et al.*, 1998). Interestingly, semaphorin 3a, a secreted chemorepellant, is present in the developing tooth in sites that are devoid of nerve fibers, and its pattern of expression is complementary to the one of NGF and GDNF. Together, these molecules draw a corridor in the jaw mesenchyme for the trigeminal axons toward the tooth target area.

Postnatally, laminin-8 (Lm-411) and tenascin-C are expressed by tooth pulp fibroblasts and can therefore participate in the dental mesenchyme innervation (Sahlberg *et al.*, 2001; Fried *et al.*, 2005). Laminin-8 has been shown to promote neurite outgrowth from trigeminal ganglion sensory neurons. Both NGF and BDNF genes are upregulated in the odontoblastic target area of the dental papilla consisting of dentine, predentine, odontoblasts, and the underlying mesenchymal subodontoblastic area before axon ingrowth to the dental papilla (Luukko *et al.*, 1997), though they appear not to initiate axon growth to the dental papilla.

NGF, however, seems to be essential for the development of the pulpal innervation. Indeed, treatment of newborn rats with anti-NGF antibodies reduces the amount of sensory nerve fibers in dental papilla (Qian and Naftel, 1996). Moreover, NGF knockout mice are devoid of pulpal innervation (Byers and Närhi, 1999).

During root formation, NGF, GDNF, and BDNF mRNAs are mainly detected in the dental pulp, the odontoblast layer, and the subodontoblast zone. Their pattern of expression correlates with onset of dentine innervation. Moreover, their receptors, TrkA, GFR-a1, and TrkB, respectively, are expressed by trigeminal ganglion neurons (Nosrat *et al.*, 1997). We have also shown that semaphorin-7A (Sema 7A) is specifically expressed at this stage by odontoblasts and is closely correlated with the odontoblast/dentine innervation process (Maurin *et al.*, 2005).

All of these molecules can then promote the final step of dentinal innervation by guiding pulpal developing axons into this final target field.

4 Pulpal Nerve Fiber Plasticity: Putative Role of Odontoblasts

Neuronal pulpal changes occur throughout the life of the tooth, and remodeling of nerve distribution occurs alongside, with the shift from the primary to the permanent dentition, and with aging and dental injury. During pulpal inflammatory processes (Byers and Närhi, 1999), sensory nerve fibers react to dentine injury by sprouting extensively of their terminal branches in the adjacent surviving pulp. In addition, extent and duration of the response (localized in time and space) depend on the severity and the nature of the injury, as well as the survival of odontoblasts. During reactive dentinogenesis following minor dentine injuries, original odontoblasts are not altered, and the sprouting of sensory nerves is related to the increase of dentinal sensitivity noted in human teeth after cavity drilling (Anderson *et al.*, 1967). There is evidence that neurotrophic factors might be involved in this process, and the transient increase in the level of NGF originating from adjacent pulp cells surrounding the injured pulp has been strongly suggested (Byers *et al.*, 1992). In contrast, a deep dentin cavity or a small pulp exposure causes destruction of odontoblasts, and the absence or reduction in the sensory innervation in the underlying pulp which is probably related to the laying down of reparative dentine by odontoblast-like cells. Thus, the regeneration of pulpal axons seems to be under the control of factors originating from odontoblasts. Laminin, an adhesive molecule bathing the odontoblast layer and, particularly, the α 1 and 2 subunits (Salmivirta *et al.*, 1997), has also been suspected of guiding the regenerating axons (Fried *et al.*, 1992). In this context, the role of extracellular matrix molecules, such as tenascin and reelin, elaborated by odontoblasts, in the growth and orientation of nerve endings during tooth injury remains an open question.

5 Expression of Reelin and Its Receptor During Odontogenesis

Recently, reelin was identified in a subtractive cDNA library from human odontoblasts (Buchaille *et al.*, 2000). This finding raised the question of the role of this molecule, described first as an axon guidance molecule in the central neural system, during tooth development, and particularly during the terminal innervation of the dentine pulp complex.

5.1 *Reelin: A Role in Tooth Morphogenesis?*

To date, very few studies have mentioned the reelin expression during tooth development and developing tooth germ. Ikeda and Terashima (1997), analyzing reelin mRNA localization during mouse development, reported some transient or permanent expressions in different peripheral organs, i.e., liver, kidney, testis, ovary. They

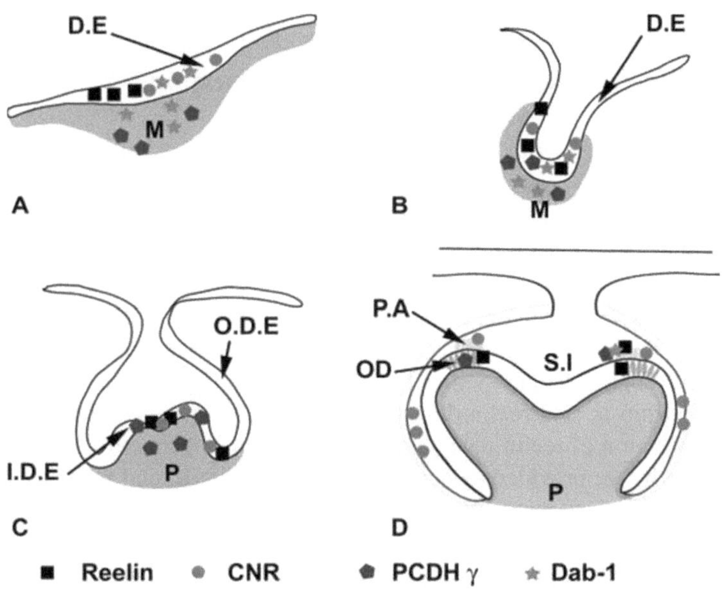

Fig. 19.2 Schematic representation of reelin gene expression and its receptors during successive stages of odontogenesis. Reelin is first detected in the oral epithelium from the initiation stage through the early bell stage. Then, reelin expression shifts in differentiating odontoblasts at the late bell stage. Dab1 is mainly expressed in both oral epithelium and dental mesenchyme during the initiation stages (epithelial thickening and bud stages). CNRs are present in the epithelium through the tooth development whereas PCDH-γ is expressed in both epithelial and mesenchymal compartments. D.E, dental epithelium; I.D.E, inner dental epithelium; M, mesenchyme, OD, odontoblasts; O.D.E, outer dental epithelium; P, dental papilla; P.A, preameloblasts; S.I, stratum intermedium (*See Color Plates*)

observed a positive signal in the tooth at the cap stage. Another study reports the expression patterns of reelin and the new protocadherin families CNRs and PCDH-γ during mouse odontogenesis (Heymann *et al.*, 2001).

During the thickening of the oral epithelium (corresponding to the initiation stage), expression of reelin transcripts is restricted to the epithelial layer, whereas reelin receptors —cadherin-related neuronal receptor (CNR) and protocadherin γ (PCDH-γ) —are detected in the mesenchyme underlying the dental epithelium (Heymann *et al.*, 2001). Dab1 (disabled-1), the reelin cytoplasmic adapter, first similarly distributed in both epithelium and mesenchyme, decreases as a function of the increasing level of differentiation of ameloblasts and odontoblasts. During this last step of tooth development (bell stage), PCDH-γ is also identified in differentiating ameloblasts (Fig. 19.2). Taken together, these findings show that a correlation could exist between expression of reelin and its receptors during odontogenesis, thus giving putative roles in morphogenesis and differentiation processes.

Given the role of reelin in architectonic brain development, however, its expression in mature human odontoblasts (Maurin *et al.*, 2004), and the unique nerve–odontoblast relationship, it is tempting to speculate on a relevant physiological role for reelin in the plasticity of dental pulp axons (adhesion/recognition) in adult teeth.

5.2 Reelin: A Role in Dental Pain Transmission?

In situ hybridization and immunohistochemistry performed on human pulp tissues show a restricted expression of reelin in the odontoblast layer at the gene and protein level (Fig.19.3) (Maurin *et al.*, 2004). *In vitro*, using cultured odontoblasts differentiated from pulp cells (Couble *et al.*, 2000), reelin is identified as large patches in the microenvironment of the odontoblast membrane. In addition, the coculture of trigeminal axons with odontoblasts (Maurin *et al.*, 2004) successfully mimics the *in vivo* situation demonstrating that a single neurite could be associated with an odontoblast cell through a varicosity (bead nerve terminal) (Fig. 19.3). Interestingly, this close relationship is also reelin-dependent. These findings raise the question of a possible expression of reelin receptors by neighboring trigeminal ganglion afferent axons. Recently, we provided evidence that VLDLR, CNR, and Dab1 (assumed as reelin receptors and cytoplasmic adapter) were expressed in rat trigeminal ganglion, suggesting that reelin could constitute a critical functional link between odontoblast cell membrane and nerve terminals (Maurin *et al.*, 2004). This close association between cells could be assumed as the earliest step of tooth pain transmission. Indeed, based on the spatial situation of odontoblasts, nerve endings, and fluid movements in dentinal tubules, nociceptive responses may result from an increase in intradentinal pressure, which, in turn, might activate nerve endings. Thus, reelin could have a pivotal role in the extension and branching of pulp axons in target areas as in other tissues, such as the retina, the hippocampus, or the olfactory system (Borrell *et al.*, 1999; Rice and Curran, 2000; Teillon *et al.*, 2003), thus participating in events underlying the sensory transduction in teeth. A similar role has recently

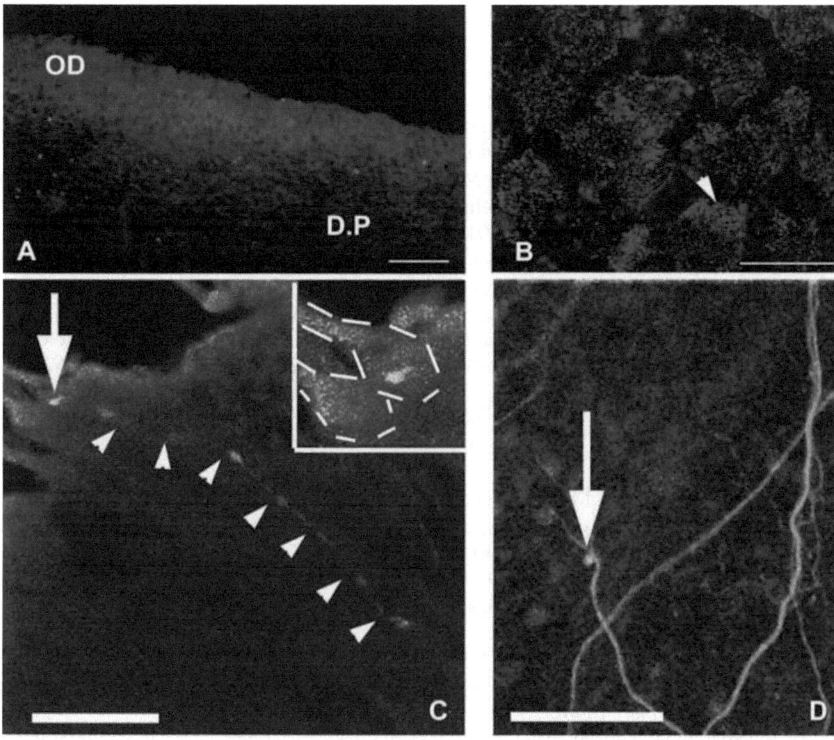

Fig. 19.3 Expression of reelin in human odontoblasts. (**A**) An immunolabeling of reelin performed with anti-reelin antibody 142 shows a signal in the odontoblast layer (OD). No staining is observed in dental pulp cells (D.P) (bar is 100 μm). (**B**) Immunofluorescence labeling with the same antibody, and without permeabilization of the cells, appears as reelin-positive patches localized around the cultured odontoblast cell membrane (arrowhead) (bar is 100 μm). (**C**) A double immunostaining with the monoclonal anti-reelin antibody and a polyclonal anti-neurofilament H on a human dental pulp section was analyzed by confocal microscopy. The nerve fiber course in the pulp can be followed (arrowheads). A yellow patch observed in a nerve varicosity, indicates a colocalization between nerve fiber and reelin close to the odontoblast membrane (arrow and insert) (bar is 20 μm). (**D**) Coculture of human odontoblasts and rat trigeminal ganglion shows the same colocalization (yellow) of reelin (red) and the varicosity (green) in the odontoblast cell layer (bar is 20 μm). [Modified from Maurin *et al.* (2004). Expression and localization of reelin in human odontoblasts. *Matrix Biol.* 23:277–285, with permission from Elsevier] (*See Color Plates*)

been demonstrated using the mutant *reeler* mice showing that reelin signaling is essential for the development of central circuits underlying nociception (Villeda *et al.*, 2006). Accordingly, the putative role of reelin in tooth pain transmission is strengthened by the recent evidence for excitable properties of odontoblasts, concentration of mechanosensitive channels (Allard *et al.*, 2000; Magloire *et al.*, 2003) in the borderline between cell extension, and cell bodies and clustering of key molecules at the site of odontoblast–nerve contact. Finally, taken together, these findings strongly suggest that odontoblasts may operate as sensor cells (Allard *et al.*, 2006).

References

Allard, B., Couble, M. L., Magloire, H., and Bleicher, F. (2000). Characterization and gene expression of high conductance calcium-activated potassium channels displaying mechanosensitivity in human odontoblasts. *J. Biol. Chem.* 275:25556–25561.

Allard, B., Magloire, H., Couble, M. L., Maurin, J. C., and Bleicher, F. (2006). Voltage-gated sodium channels confer excitability to human odontoblasts: possible role in tooth pain transmission. *J. Biol. Chem.* 281:29002–29010.

Anderson, D. J., Matthews, B., and Shelton, L. E. (1967). Variations in the sensitivity of osmotic stimulation of human dentin. *Arch. Oral Biol.* 12:43–47.

Borrell, V., Del Rio, J. A., Alcantara, S., Derer, M., Martinez, A., D'Arcangelo, G., Nakajima, K., Mikoshiba, K., Derer, P., Curran, T., and Soriano, E. (1999). Reelin regulates the development and synaptogenesis of the layer-specific entorhino-hippocampal connections. *J. Neurosci.* 19:1345–1358.

Brannstrom, M. (1962). A hydrodynamic mechanism in the transmission of pain producing stimuli through the dentine. In: Anderson, D. J. (ed.), *Sensory Mechanisms in Dentine*. Pergamon Press, Oxford, pp. 73–79.

Buchaille, R., Couble, M. L., Magloire, H., and Bleicher, F. (2000). A substractive PCR-based cDNA library from human odontoblast cells: identification of novel genes expressed in tooth forming cells. *Matrix Biol.* 19: 421–430.

Byers, M. R., and Närhi, M. V. O. (1999). Dental injury models: experimental tools for understanding neuroinflammatory interactions and polymodal nociceptor functions. *Crit. Rev. Oral Biol. Med.* 10:4–39.

Byers, M. R., Kvinnsland, I., and Bothwell, M. (1992). Analysis of low affinity nerve growth factor receptor during pulpal healing and regeneration of myelinated and unmyelinated axons in replanted teeth. *J. Comp. Neurol.* 326:470–484.

Byers, M. R., Suzuki, H., and Maeda, T. (2003). Dental neuroplasticity, neuro-pulpal interactions, and nerve regeneration. *Microsc. Res. Tech.* 60:503–515.

Couble, M. L., Farges, J. C., Bleicher, F., Perrat-Mabillon, B., Boudeulle, M., and Magloire, H. (2000). Odontoblast differentiation of human dental pulp cells in explant cultures. *Calcif. Tissue Int.* 66:129–138.

Dowell, P., and Addy, M. (1983). Dentine hypersensitivity: a review: Aetiology, symptoms and theories of pain production. *J. Clin. Periodontol.* 10:341–350.

Fried, K., Risling, M., Edwall, L., and Olgart, L. (1992). Immunoelectron microscopic localization of laminin and collagen type IV in normal and denervated tooth pulp of the cat. *Cell Tissue Res.* 270:157–164.

Fried, K., Sime, W., Lillesaar, C., Virtanen, I., Tryggvasson, K., and Patarroyo, M. (2005). Laminins 2 (alpha2beta1gamma1, Lm-211) and 8 (alpha4beta1gamma1, Lm-411) are synthesized and secreted by tooth pulp fibroblasts and differentially promote neurite outgrowth from trigeminal ganglion sensory neurons. *Exp. Cell Res.* 307:329–341.

Fristad, I., Heyeraas, K. J., and Kvinnsland, I. (1994). Nerve fibres and cells immunoreactive to neurochemical markers in developing rat molars and supporting tissues. *Arch. Oral Biol.* 39:633–646.

Heymann, R., Kallenbach, S., Alonso, S., Carroll, P., and Mitsiadis, T. A. (2001). Dynamic expression patterns of the new protocadherin families CNRs and PCDH-gamma during mouse odontogenesis: comparison with reelin expression. *Mech. Dev.* 106: 181–184.

Hildebrand, C., Fried, K., Tuisku, F., and Johansson, C. S. (1995). Teeth and tooth nerves. *Prog. Neurobiol.* 45:165–222.

Ibuki, T., Kido, M. A., Kiyoshima, T., Terada, Y., and Tanaka, T. (1996). An ultrastructural study of the relationship between sensory trigeminal nerves and odontoblasts in rat dentin/pulp as demonstrated by the anterograde transport of wheat germ agglutinin-horseradish peroxidase (WGA-HRP). *J. Dent. Res.* 75:1963–1970.

Ikeda, Y., and Terashima, T. (1997). Expression of reelin, the gene responsible for the reeler mutation, in embryonic development and adulthood in the mouse. *Dev. Dyn.* 210:157–172.

Kettunen, P., Loes, S., Furmanek, T., Fjeld, K., Kvinnsland, I. H., Behar, O., Yagi, T., Fujisawa, H., Vainio, S., Taniguchi, M., and Luukko, K. (2005). Coordination of trigeminal axon navigation and patterning with tooth organ formation: epithelial-mesenchymal interactions, and epithelial Wnt4 and Tgfbeta1 regulate semaphorin 3a expression in the dental mesenchyme. *Development* 132:323–334.

Lillesaar, C., and Fried, K. (2004). Neurites from trigeminal ganglion explants grown *in vitro* are repelled or attracted by tooth-related tissues depending on developmental stage. *Neuroscience* 125:149–161.

Loes, S., Luukko, K., Kvinnsland, I.H., Salminen, M., and Kettunen, P. (2003). Developmentally regulated expression of Netrin-1 and -3 in the embryonic mouse molar tooth germ. *Dev. Dyn.* 227:573–577.

Luukko, K. (1997). Immunohistochemical localization of nerve fibres during development of embryonic rat molar using peripherin and protein gene product 9.5 antibodies. *Arch. Oral Biol.* 42:189–195.

Luukko, K., Arumae, U., Karavanov, A., Moshnyakov, M., Sainio, K., Sariola, M., Saarma, M., and Thesleff, I. (1997). Neurotrophin mRNA expression in the developing tooth suggests multiple roles in innervation and organogenesis. *Dev. Dyn.* 210:117–129.

Luukko, K., Kvinnsland, I. H., and Kettunen, P. (2005). Tissue interactions in the regulation of axon pathfinding during tooth morphogenesis. *Dev. Dyn.* 234:482–488.

Magloire, H., Lesage, F., Couble, M. L., Lazdunski, M., and Bleicher, F. (2003). Expression and localization of TREK-1 K^+ channels in human odontoblasts. *J. Dent. Res.* 82:542–545.

Maurin, J. C., Couble, M. L., Didier-Bazes, M., Brisson, C., Magloire, H., and Bleicher, F. (2004). Expression and localization of reelin in human odontoblasts. *Matrix Biol.* 23:277–285.

Maurin, J. C., Delorme, G., Machuca-Gayet, I., Couble, M. L., Magloire, H., Jurdic, P., and Bleicher, F. (2005). Odontoblast expression of semaphorin 7A during innervation of human dentin. *Matrix Biol.* 24:232–238.

Mitsiadis, T. A., Couble, P., Dicou, E., Rudkin, B. B., and Magloire, H. (2003). Patterns of nerve growth factor (NGF), proNGF, and p75 NGF receptor expression in the rat incisor: comparison with expression in the molar. *Differentiation* 54:161–175.

Mohamed, S. S., and Atkinson, M. E. (1983). A histological study of the innervation of developing mouse teeth. *J. Anat.* 136:735–749.

Nosrat, C. A., Fried, K., Lindskog, S., and Olson, L. (1997). Cellular expression of neurotrophin mRNAs during tooth development. *Cell Tissue Res.* 290:569–580.

Nosrat, C. A., Fried, K., Ebendal, T., and Olson, L. (1998). NGF, BDNF, NT3, NT4 and GDNF in tooth development. *Eur. J. Oral Sci.* 106(Suppl. 1):94–99.

Pearson, A. A. (1977). The early innervation of the developing deciduous teeth. *J. Anat.* 123: 563–577.

Qian, X. B., and Naftel, J. P. (1996). Effects of neonatal exposure to anti-nerve growth factor on the number and size distribution of trigeminal neurones projecting to the molar dental pulp in rats. *Arch. Oral Biol.* 41:359–367.

Rice, D. S., and Curran, T. (2000). Disabled-1 is expressed in type AII amacrine cells in the mouse retina. *J. Comp. Neurol.* 424:327–338.

Sahlberg, C., Aukhil, I., and Thesleff, I. (2001). Tenascin-C in developing mouse teeth: expression of splice variants and stimulation by TGFbeta and FGF. *Eur. J. Oral Sci.* 109:114–124.

Salmivirta, K., Sorokin, L. M., and Ekblom, P. (1997). Differential expression of laminin alpha chains during murine tooth development. *Dev. Dyn.* 210:206–215.

Shimizu, A., Nakakura-Ohshima, K., Noda, T., Maeda, T., and Ohshima, H. (2000). Responses of immunocompetent cells in the dental pulp to replantation during the regeneration process in rat molars. *Cell Tissue Res.* 302:221–233.

Teillon, S. M., Yiu, G., and Walsh, C. A. (2003). Reelin is expressed in the accessory olfactory system, but is not a guidance cue for vomeronasal axons. *Brain Res. Dev. Brain Res.* 140:303–307.

Tsuzuki, H., and Kitamura, H. (1991). Immunohistochemical analysis of pulpal innervation in developing rat molars. *Arch. Oral Biol.* 36:139–146.

Villeda, S. A., Akopians, A. L., Babayan, A. H., Basbaum, A. I., and Phelps, P. E. (2006). Absence of reelin results in altered nociception and aberrant neuronal positioning in the dorsal spinal cord. *Neuroscience* 139:1385–1396.

Chapter 20
Homozygous and Heterozygous Reeler Mouse Mutants

Patricia Tueting, Graziano Pinna, and Erminio Costa

Contents

P. Tueting
Psychiatric Institute/Department of Psychiatry, University of Illinois at Chicago,
1601 West Taylor Street MC912, Chicago, IL 60612
e-mail: ptueting@psych.uic.edu

G. Pinna
Psychiatric Institute/Department of Psychiatry, University of Illinois at Chicago,
1601 West Taylor Street MC912, Chicago, IL 60612

E. Costa
Psychiatric Institute/Department of Psychiatry, University of Illinois at Chicago,
1601 West Taylor Street MC912, Chicago, IL 60612

S. H. Fatemi (ed.), *Reelin Glycoprotein: Structure, Biology and Roles in Health and Disease.*

1 Introduction

Reelin is a large-molecular-weight protein secreted during early embryonic development into the extracellular matrix by Cajal-Retzius cells, a temporary population of GABAergic neurons located in the uppermost layer of the developing cortex (marginal zone). In the adult, reelin is exclusively located in GABAergic interneurons of the upper layers of the cortex, and following secretion into the extracellular matrix, surrounds dendritic spines of pyramidal cells located in every layer (Costa *et al.*, 2002). Glutamic acid decarboxylase 67 (GAD67)-positive interneurons of Layer I also exclusively contain DNA methyltransferase 1 (DNMT1) (Ruzicka *et al.*, 2007).

The RELN gene, located on human chromosome 7q22, encodes a 3460-amino-acid protein that is 94% identical to the mouse protein. The promoter region of RELN is rich in CpG islands and, therefore, prone to DNA methylation which results in transcription downregulation of reelin (Chen *et al.*, 2002). Significantly, GABAergic neurons of postmortem brains of schizophrenia and bipolar disorder patients with psychosis are characterized by both reelin downregulation and DNMT1 upregulation (Veldic *et al.*, 2004, 2005). These findings suggest the possibility that reelin promoter methylation by DNMT1 is involved in the dendritic spine hypoplasticity and GABAergic tone downregulation operative in schizophrenia (Costa *et al.*, 2001).

2 The *Reeler* Mouse and Neurodevelopmental Disorders

The homozygous *reeler* mouse lacks reelin and has been instrumental in deciphering reelin's role in embryonic neurodevelopment. There are several strains of *reeler*; all have abnormalities in the coding region of the reelin gene resulting in various degrees of reelin expression downregulation, but there are discrete strain differences in loss of reelin function depending on which region of the gene is affected (D'Arcangelo, 2006).

Studies of *reeler* mouse have shown that reelin expression is critical for the normal inside-out layering of the cerebral cortex; earlier born neurons normally stop their radial migration in deeper layers of the cortex, and later born neurons navigate around them into upper layers. In *reeler* mutant mice, cortical layering is reversed (Caviness, 1982). During development, GABAergic neurons from the ganglionic eminence migrate tangentially to the cortex, but this migration does not appear to be directly affected by lack of reelin (Hevner *et al.*, 2004). The subsequent radial migration of these interneurons toward an appropriate cortical layer (Yabut *et al.*, 2007) occurs in a manner that preserves interneuron/pyramidal neuron laminar position even though cortical layers are inverted (Manent *et al.*, 2006; Pla *et al.*, 2006). Areas of the brain besides the cortex are abnormal in reelin deficient mouse, among them the hippocampus (Caviness and Sidman, 1973), cerebellum (Goffinet *et al.*, 1984), olfactory bulb (Wyss *et al.*, 1980), striatum (Marrone *et al.*, 2006), and the rostral telencephalic cholinergic system (Sigala *et al.*, 2007). The loss of cellular organization in the cerebellum is responsible for the reeling gait after which the mutant is named. Not surprisingly, given their substantial neurobiological abnormalities, a large number of behavioral dysfunctions in addition to a reeling gait have been recognized in *reeler* mouse (Salinger *et al.*, 2003; Laviola *et al.*, 2006; Marrone *et al.*, 2006).

The homozygous *reeler* mouse has been suggested as a model for understanding abnormal neuronal migration and even for human lissencephaly (D'Arcangelo, 2006). A human equivalent of the null *reeler* mutant apparently does exist, but afflicted individuals are extremely rare (seven individuals have been described to date). They have severe lissencephaly and cerebellar atrophy, probably preventing adult survival (Hong *et al.*, 2000; Chang *et al.*, 2007). It is possible that the null mutant genotype is more prevalent in human populations, but the fetus is usually not viable.

However, if given special care, *reeler* mice can survive into adulthood and may even be fertile, serving as a practical model for studying reelin function. Human stem cells, injected into the lateral ventricles of *reeler* mice, fail to migrate or display typical differentiation patterns, while stem cells injected into wild-type mice (WT) migrate and differentiate normally (Kim *et al.*, 2002). Won *et al.* (2006) have reported that *reeler* mice have a defect in adult neurogenesis, resulting in larger brain infarcts than WT following middle cerebral artery occlusion and larger excitotoxic lesions after intracerebral injection of *N*-methyl-D-aspartate (NMDA). Also, several other mutant or knockout mice that serve as models for severe neurological and psychiatric disorders involve mutations that include components of the reelin signaling pathway, or proteins that are operative in the reelin signaling pathway (recently reviewed by D'Arcangelo, 2006). Appropriate examples are mice with mutations of DAB1 (the adapter protein for reelin), mice heterozygous for Lis1, and mice with mutations of the receptors for reelin [apolipoprotein ER2 receptor (ApoER2) and very-low-density lipoprotein receptor (VLDLR), or integrin $\alpha 3$] (D'Arcangelo, 2006). These mutants may have a phenotype similar to *reeler* and have helped to provide valuable insight into the reelin signaling pathway and the consequences of its disruption. Significantly, reelin is overexpressed in cortex and CSF in several neurodegenerative diseases, including Alzheimer's disease (Botella-Lopez *et al.*, 2006). The mechanism is unknown, but Herz and Chen (2006) have proposed a role for reelin lipoprotein receptors in Alzheimer's disease.

In a potentially significant new development with implications for future understanding of reelin function, Pollard *et al.* (2006) have reported that a noncoding RNA gene (HAR1F) is coexpressed with reelin in Cajal-Retzius neurons of human cortex from 7 to 9 gestational weeks which is a critical period for cortical neuron migration. Presumably, a similar situation is operative in mice.

3 The Heterozygous Reeler Mouse (HRM) as a Model for Nondegenerative Psychiatric Disorders

Disorders such as schizophrenia have long puzzled neuroanatomists because of the failure to find evidence of obvious abnormalities or degeneration and cell loss commensurate with the extreme behavioral manifestations. Therefore, interest in the role of reelin in psychosis escalated when reelin was found to be 50% downregulated in psychotic postmortem brain (Impagnatiello *et al.*, 1998; Guidotti *et al.*, 2000). Downregulation of reelin in schizophrenia and in bipolar disorder has been replicated in other laboratories and brain cohorts and is considered the most consistent finding in schizophrenia postmortem brain tissue (Torrey *et al.*, 2005). Moreover, the decrease in reelin expression has been extended to other similar nondegenerative disorders of neurobiological function, including autism (Fatemi *et al.*, 2001, 2002, 2005), suggesting a link among these psychiatric disorders.

A small number of heterozygous individuals with mutations of the RELN locus have been described in the families of the seven homozygous probands studied to date (Hong *et al.*, 2000; Chang *et al.*, 2007). These heterozygous relatives appear to express various psychiatric pathologies but have no specific psychiatric diagnosis in common. It is possible that they could have other gene mutations in addition to the reelin gene mutation, or that the expression of reelin in GABAergic neurons might vary in individuals due to the embryonic environment, epigenetic factors, or to compensatory mechanisms for the reelin deficit. Supporting a possible epigenetic role in reelin and GABAergic tone downregulation is the finding that DNMT1 is upregulated in GABAergic neurons of psychotic postmortem brain (Veldic *et al.*, 2004, 2005; Abdolmaleky *et al.*, 2005; Ruzicka *et al.*, 2007).

While the null mutant *reeler* mouse is not suitable as a stand-alone model of psychosis because of the mutant's extreme structural brain impairment and associated behavioral abnormalities, knowledge about the role of reelin in the neurodevelopment of *reeler* mouse is important for the HRM model. The schizophrenia literature strongly suggests the importance of neurodevelopment in setting the stage for schizophrenia vulnerability (Rehn and Rees, 2005). Environmental insults, such as drugs, malnutrition, hypoxia, and viral infection, are hypothesized to interact with specific stages of embryonic development and with the presence of schizophrenia-related genes, including reelin. A complex model in which nonspecific genetic factors that increase susceptibility to developmental abnormalities interact with specific genetic factors and epigenetic and compensatory events has been proposed to explain the etiology of schizophrenia (Avila *et al.*, 2003).

3.1 Importance of Reelin in Neurogenesis and Synaptic Plasticity

The pleiotropic nature of the RELN gene accounts for its essential role in normal neurogenesis and synaptic function in the adult cortex and hippocampus. Following reelin secretion into the extracellular matrix by GABAergic interneurons located in the upper cortical layers, reelin surrounds spines of glutamatergic pyramidal cell dendrites, where there is evidence that it plays a role in modulating pyramidal neuron dendritic spine structures and morphology in excitatory synapses, which include the NMDA receptor implicated in various aspects of memory function (Costa *et al.*, 2001; Dong *et al.*, 2003). Importantly, dendritic spine density in prefrontal cortical pyramidal neurons is decreased in schizophrenia (Glantz and Lewis, 2000; Hill *et al.*, 2006).

In support of reelin's role in synaptic function, Qiu *et al.* (2006; Qiu and Weeber, 2007) reported that HRM exhibits profound impairment in hippocampal CA1 excitatory postsynaptic potentials, paired pulse facilitation ratio, long-term potentiation and depression, as well as a reduction in spontaneous inhibitory postsynaptic currents. Qiu *et al.* (2006) also provide evidence that these synaptic impairments in hippocampal plasticity and functional inhibition at excitatory synapses may underlie behavioral deficits in hippocampal associative function, including the deficits in contextual fear conditioning and prepulse inhibition of startle (PPI) in HRM.

3.2 Reelin Downregulation and GABAergic Tone Deficit in Psychosis

Reelin downregulation in psychosis has been linked to GABAergic tone deficit (Costa *et al.*, 2001; Carboni *et al.*, 2004). Glutamic acid decarboxylases 65, 67 (GAD65, 67) catalyze the synthesis of the inhibitory neurotransmitter γ-aminobutyric acid (GABA) from glutamate. Downregulation of the GAD67 isoform is a critical aspect of the HRM model of psychosis (Liu *et al.,* 2001) because, along with reelin, GAD67 mRNA and protein are also downregulated in postmortem cortex and hippocampus of psychotic patients (Impagnatiello *et al.*, 1998; Guidotti *et al.*, 2000; Akbarian and Huang, 2006). GAD65 appears not to be downregulated in psychotic postmortem brain, but this may depend on diagnosis, area of the brain studied, or other factors (Fatemi *et al.*, 2002). Importantly, the localization of the two GAD isoforms may differ in different subpopulations of GABAergic neurons (Volk *et al.*, 2000) and in different brain areas. Thus, there is increasing evidence that the neurobiological abnormalities and behavioral dysfunctions observed in schizophrenia are related to an epigenetically induced and coordinated downregulation of GAD67 in specific types of GABAergic interneurons (Grayson *et al.*, 2005, 2006; Lewis *et al.*, 2005), which is likely to be related to the decrease in reelin and to the increase in DNMT1 expressions in these same neurons (Veldic *et al.*, 2004, 2005; Ruzicka *et al.*, 2007).

GAD67 downregulation in HRM is only about 30% (Liu *et al.*, 2001), i.e., the downregulation of GAD67 expression in HRM is less severe than in psychotic patients where the GAD67 downregulation is approximately 50% (Guidotti *et al.*, 2000). Interestingly, GAD67 heterozygous mice do not reveal downregulation of reelin expression (Liu *et al.*, 2001; Carboni *et al.,* 2004). Together, these findings suggest that a number of factors, in addition to reelin and GAD67, may be contributing to the onset of psychosis.

3.3 Controversy over the Usefulness of the HRM Model

HRM was an obvious animal model for evaluating the role of a 50% downregulation of reelin in psychosis, and it is now apparent that a complete model of reelin downregulation in psychosis needs to include methylation of the reelin gene promoter in GABAergic neurons (Tremolizzo *et al.*, 2002). Unlike the homozygous *reeler,* which shows obvious structural and behavioral impairment, the HRM behavioral and biological phenotype appeared similar to WT at first. Only a closer examination revealed neurobiological and behavioral abnormalities reminiscent of abnormalities found in psychosis. We have recently reviewed the HRM model (Tueting *et al.*, 2006), and consequently, will only highlight and update this evidence here.

Neurobiological findings in HRM (previously summarized in Table 1 of Tueting *et al.*, 2006, p. 1067) are similar to abnormalities found in postmortem brain of psychotic patients and include a decrease in neuropil, especially in frontal cortex, together with increased pyramidal cell density, an increased number of GABAergic cells in underlying white matter, decreased GABA turnover, and indirect evidence for a dysregulation of glutamate, dopamine, and related neuroendocrine functions. Importantly, the decreased spine density on dendrites of pyramidal cells located in Layer III of frontal cortex has been reported both in HRM (Liu *et al.*, 2001) and in psychotic patients (Glantz and Lewis, 2000; Hill *et al.*, 2006). Since publication of our review (Tueting *et al.*, 2006), a report on rostral regions of the basal forebrain and medial cortex indicates significant redistribution of cholinergic neuron innervation in HRM (Sigala *et al.*, 2007). There is also a recent report on Purkinje cell loss in male HRM occurring in an earlier age (Biamonte *et al.*, 2006) than originally reported (Hadj-Sahraoui *et al.*, 1996). Another new report indicates a decrease in the density of μ-opioid receptors in the midbrain of HRM (Ognibene *et al.*, 2007a).

However, it is the HRM behavior, not the substantial neurobiological findings summarized briefly above, that has sparked a controversy over the usefulness of the HRM as a psychosis model. A complete absence of behavioral deficits in HRM on standard behavioral test batteries has been vigorously defended (Salinger *et al.*, 2003; Podhorna and Didriksen, 2004). Tests that were negative in the Salinger *et al.* (2003) and Podhorna and Didriksen (2004) studies included assessments of sensory function, social behavior, anxiety level, spatial working memory, fear conditioning, and PPI. The absence of a behavioral deficit in HRM, despite extreme neurobiological deficit, is explained by the ability of the developing nervous system to compensate for the consequences of the decrease in reelin (Salinger *et al.*, 2003).

On the other hand, the evidence suggesting behavioral abnormalities in HRM compared to WT is substantial and has been recently reviewed (see Table 2 in Tueting *et al.*, 2006, p. 1069) and includes deficits in PPI, social interaction and social recognition, contextual fear conditioning, olfactory discrimination learning, and radial arm maze performance (especially following MK801 administration). In addition, there are new reports of behavioral differences between HRM and WT that have been published since our review. These findings include HRM deficits in executive function (Brigman *et al.*, 2006) and in learning (Krueger *et al.*, 2006), as well as differences between the genotypes in anxiety, risk assessment, motor impulsivity, morphine-induced analgesia (Ognibene *et al.*, 2007a), and in subsonic vocalization and locomotor response during an amphetamine challenge (Laviola *et al.*, 2006). Despite the mounting evidence for the existence of subtle behavioral differences between the two genotypes, the usefulness of the HRM model continues to be discounted (e.g., Patterson, 2006).

There are several possible explanations for failure to find behavioral deficits in HRM in certain circumstances, many of which we have reviewed previously (Tueting *et al.*, 2006). Here we will expose this work to further scrutiny and revisit the controversy focusing on prepulse inhibition of startle (PPI).

3.4 PPI Deficit in HRM

The fact that the PPI deficit observed in psychotic patients and their close relatives who fail to express a psychiatric disorder is especially intriguing, because PPI is considered to be an endophenotype marker for psychosis that can be studied comparatively in humans and animals. Another advantage of PPI is that much is known about the underlying brain circuitry, anatomy, pharmacology, and genetics with respect to psychosis (Koch, 1999; Geyer *et al.*, 2001; Swerdlow *et al.*, 2001; Hauser *et al.*, 2005).

PPI deficit in HRM was originally reported by Tueting *et al.* (1999) and recently replicated by Qiu *et al.* (2006). However, both Salinger *et al.* (2003) and Podhorna and Didriksen (2004) failed to find a significant PPI deficit in HRM compared to WT. Failure to find a deficit in HRM for an accepted endophenotype for psychosis vulnerability is surprising and requires explanation. The issues that need further scrutiny appear to be failure to measure reelin and variation in breeding procedures, as well as variation in specific parameters of the behavioral experiments and in the social environment, which we will explore in greater depth here.

Methods. Mice used in our experiments to be described were progeny of WT (paternal) and HRM (maternal) pairings in our colony of B6C3F, Edinburgh reelin mutation, originally obtained from Jackson Laboratories. The colony has been maintained continuously in our temperature- and light-controlled vivarium for 20+ generations. After weaning at 21 days, mice are normally housed five of the same sex in a plastic cage with random assignment of genotypes to a cage. Mice were genotyped (Tueting *et al.*, 1999) and behaviorally tested (10:00 and 16:00 hr) with the investigator blind to genotype. They were 4–7 months old, older than in previous studies, and perhaps more vulnerable to environmental adverse events and sensitive to epigenetic influences. Importantly, quantitative reelin mRNA levels were measured using RT-PCR including internal standards. Our results expand on our earlier finding of a PPI deficit in HRM using a broader range of prepulse intervals and suggest the possibility that the extent of reelin and GABAergic deficiencies in neural circuits underlying PPI can be compromised by factors, such as experiment duration and social isolation, to such an extent that normal adaptive neural responses underlying PPI may be altered.

3.4.1 PPI Deficit in HRM Is Present over an Extended Range of Silent Interval Prepulse–Startle Delays

That variation in experimental parameters can affect PPI is well documented. Prepulse intensity and duration, startle intensity and duration, prepulse interval, the interstimulus interval, stimulus modality, the nature and number of the different trial types used, and the noise and lighting background all affect PPI. The 100-ms silent prepulse interval is the most common interval used in human and animal PPI evaluation relevant to translational research in psychopharmacology (Geyer *et al.*, 2001). In the current study, we used a wider range of prepulse intervals. The SR-Lab Startle Response System (San Diego Instruments) was programmed so that a trial consisted of a 115-dB startle pulse 30 ms in duration (startle trial) or of the same startle pulse preceded by an 85-dB prepulse 20 ms in duration (prepulse trial). The *Unfilled Interval* program consisted of a random sequence of startle pulse [only] trials and five types of prepulse trials in which the silent interval between prepulse and startle pulse was varied from 40 to 420 ms. The *Filled Interval* program was the same as the *Unfilled Interval* program, except that the prepulse remained on during the prepulse interval up to 20 ms before presentation of the startle pulse. Ratios reflecting the amount of prepulse-induced inhibition of the startle reflex were calculated by subtracting the mean peak amplitude for prepulse trials from the mean peak amplitude for startle pulse [only] trials and dividing by mean peak amplitude for startle pulse trials to normalize, and finally multiplying by 100 to calculate percent inhibition.

In Fig. 20.1A, HRM is compared to WT for the two conditions. When the prepulse was kept on during the prepulse interval (*Filled Interval*), HRM and WT showed equally high levels of prepulse inhibition at all prepulse intervals, and PPI was not systematically related to length of interval. For the *Unfilled Interval* condition, HRM show decreased inhibition compared to WT, and the extent of inhibition decreased as the prepulse interval increased for both genotypes. These findings confirm and extend our earlier finding of PPI deficit in HRM which was based on the 100-ms prepulse silent interval and the same startle intensity (Tueting *et al.*, 1999).

3.4.2 The PPI Deficit in HRM Is Greater in Later Trials of the Session

PPI is generally considered to be a stable and reliable measurement between and within sessions when the same parameters are used. The means shown in Fig. 20.1A were calculated on the basis of all 72 trials in the session. The session of 72 trials was designed in three equivalent 24-trial blocks so that changes in PPI during the session could be detected. For this analysis, the data shown in Fig. 20.1A were collapsed across all five prepulse intervals. Fig. 20.1B shows that the PPI deficit in HRM was significantly larger in later trials of the session for the *unfilled* condition. For the *filled* interval condition, PPI decreased during the session equally for WT and HRM. Startle response amplitude for startle [only] trials decreased (habituated) during the session. However, neither startle response amplitude nor startle habituation differed as a function of genotype (WT versus HRM) or condition (*filled* versus *unfilled* interval). Thus, it is possible that the PPI deficits in HRM compared to WT would fail to be significant in short sessions.

What could explain the differential PPI changes in HRM and WT over the session for the *Unfilled Interval* condition? Since startle response habituation and PPI for the *Filled Interval* condition were similar for HRM and WT over trials, the increased PPI deficit in HRM later in the session for the *Unfilled Interval* condition is probably not related to muscle fatigue, but rather may be related to a functional failure in interconnected neural networks of frontal cortex, hippocampus, amygdala, and nucleus accumbens that underlie inhibition of the startle reflex (Swerdlow *et al.*, 2001). The differential PPI changes in HRM and WT over the

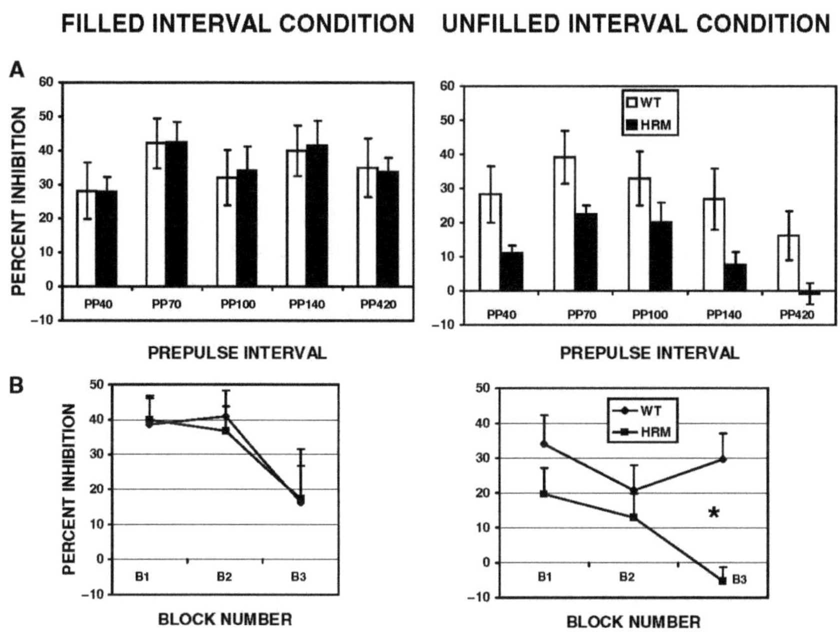

Fig. 20.1 (**A**) HRM show a PPI deficit when the interval between prepulse and startle pulse is silent (*unfilled* interval condition). No deficit is present for the *filled* interval condition not requiring a memory trace of the prepulse. The data were collapsed across intervals and a two-way repeated measures ANOVA performed on the mean PPI difference between the *filled* and *unfilled* conditions. There was a significant effect of genotype (p=0.042) and prepulse interval (p<0.001) [8 male WT, 6 male HRM group-housed mice; 72 trials]. (**B**) PPI in the same experiment was collapsed across the five prepulse intervals and the means shown separately for each of the three consecutive equivalent blocks of 24 trials each. There was no significant difference in PPI between WT and HRM for the *filled* condition for any trial block. An interaction between Genotype and Block was significant for the *unfilled* condition (F=3.550, p=0.045), and post-hoc comparisons (Neuman Keuls) revealed a significant difference between WT and HRM for Block 3. Two-way repeated measures ANOVA (1 factor repetition)

session could be associated with GABAergic tone downregulation and excitatory synapse deficits in HRM with consequent failure of glutamatergic, dopaminergic, acetylcholinergic, or other components of PPI neural circuitry to express the normal compensation for the loss of inhibition.

3.4.3 GABAergic Positive Allosteric Modulators (Agonists) Correct PPI Deficit in HRM in Later Trials of the Session

We initially reported that young male HRM show anxiety on the elevated plus-maze (Tueting *et al.*, 1999), which could be interpreted to reflect a downregulation of GABAergic tone in HRM. To test whether the PPI deficit in HRM is related to GABAergic downregulation, we studied whether imidazenil, a specific positive allosteric modulator of the action of GABA at $GABA_A$ receptors (Costa *et al.*, 2002), is able to correct the PPI deficit operative in the later trials. Fig. 20.2 shows a replication of the PPI deficit in later trials of the session in a sample of female HRM, confirming the finding in males (Fig. 20.1B). (The deficit occurs earlier than in the male study, but the PPI paradigm was also different.) The deficit in the later trials was corrected by subcutaneous injection of as little as 0.3 and 0.6 mg/kg of imidazenil, injected 20 min before the PPI session. Neither mean startle response nor startle habituation was affected by genotype or by imidazenil.

Increased PPI deficit in HRM in later trials of the session could be related to a nonspecific action (stress?) included in the PPI testing procedure, the effects of which accumulate over the session duration in the presence of a downregulation of GABAergic tone, and a related failure of other neuronal systems due to a decrease of synaptic efficacy (Carboni *et al.*, 2004). These systems normally can compensate for some loss of inhibition. This interpretation is supported by the imidazenil-mediated correction of the PPI deficit in HRM in the later trials of the session, since this drug is a positive allosteric modulator of the action of GABA at specific $GABA_A$ receptors, including $\alpha5$ subunits, and is devoid of action on receptors containing $\alpha1$ subunits (Guidotti *et al.*, 2005). Interesting in this regard is the fact that the $\alpha5$ subunit of the $GABA_A$ receptor in mutant mice has been related to PPI deficit (Hauser *et al.*, 2005). In addition, there is a recent report of decreased social interaction in HRM that can be normalized by imidazenil treatment, a finding associated with an upregulation of the $\alpha5$ subunit of the $GABA_A$ receptor by the benzodiazepine (Doueiri *et al.*, 2006).

3.5 PPI in WT is More Sensitive than the HRM Genotype to the Effects of Social Isolation

The two controversial reports that concluded that WT and HRM do not differ in PPI involved a combined sample of group-housed and socially isolated mice (Salinger *et al.*, 2003) or socially isolated mice only (Podhorna and Didriksen, 2004). Fig. 20.3

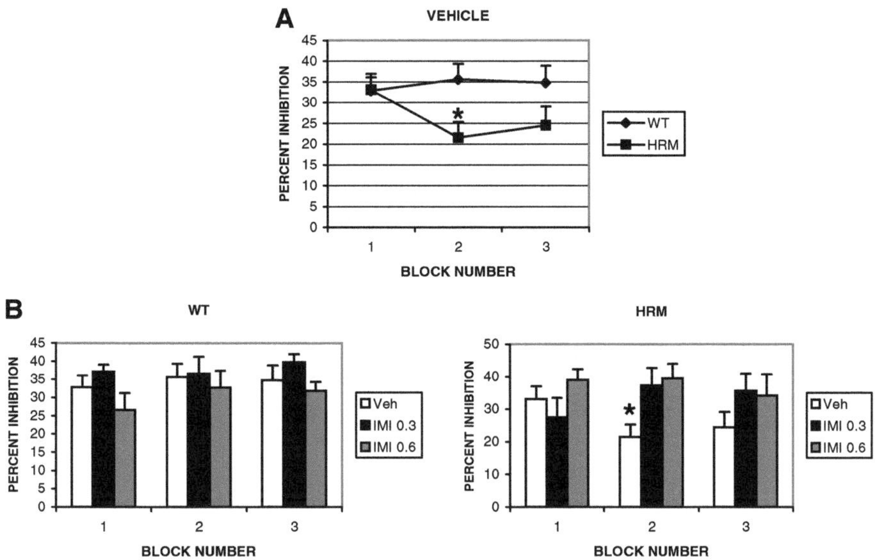

Fig. 20.2 (**A**) There is a significant difference between HRM and WT in Block 2 but not in Block 1. (**B**) Imidazenil (0.3 and 0.6 mg/kg, s.c.) corrects the deficit in PPI present in HRM for the vehicle condition in later trial blocks of the session. Five female WT, five female HRM, 5 months old. (An *unfilled* interval trial sequence consisting of startle only trials was randomly presented with 85- and 80-dB prepulse–startle trials with the prepulse–startle delay constant at 100 ms) $^*p<0.05$

shows that PPI was significantly decreased in HRM compared to WT when mice were group housed, replicating in older mice our earlier findings obtained with young mice (Tueting *et al.*, 1999). After 2 weeks of isolation, however, WT and HRM no longer differed in PPI, due to the fact that percent inhibition (averaged across prepulse intervals) declined for WT following isolation (32 ± 4% in group housed but 15 ± 6% following 2 weeks of isolation, $t=2.689$, $p=0.031$) and failed to change in HRM or even increased slightly (16.4 ± 3.7% in group housed compared to 16.9 ± 4.2% in isolated).

A decrease in PPI in WT following isolation is consistent with other reports showing that PPI decreases in isolated compared to socially housed rodents (Powell and Geyer, 2002; Sakaue *et al.*, 2003). There were no significant differences in mean startle amplitude over blocks as a function of isolation or genotype. Thus, it appears that social isolation in adulthood can obscure the PPI deficit in HRM compared to WT, which could explain the negative findings published by Salinger *et al.* (2003) and Podhorna and Didriksen (2004).

Following 4 weeks of isolation, the reelin expression was measured in the frontal cortex as previously described (Tueting *et al.*, 1999; Tremolizzo *et al.*, 2002), and its expression was 59% lower in HRM than in WT. In addition, reelin expression varied considerably among individuals both within the WT (80–250 pmole/0.5 μg RNA) and within the HRM group (40–100 pmole/0.5 μg mRNA). Presumably, this

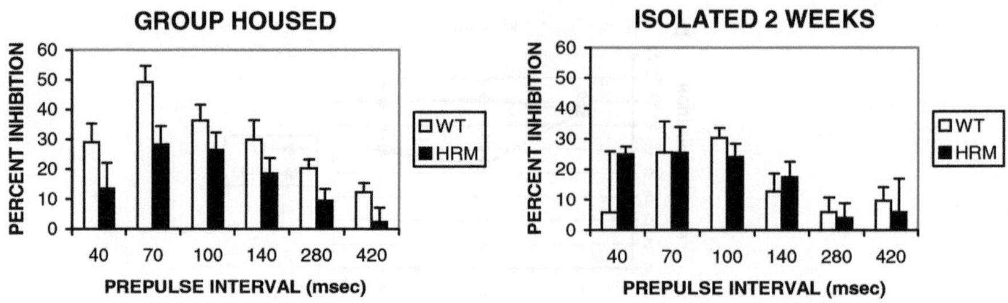

Fig. 20.3 Group-housed male HRM show a deficit in PPI (F=7.509, p=0.017, two-way repeated measures ANOVA with interval as the repeated measure). After 2 weeks of isolation, the difference between WT and HRM was no longer present due to a significant decrease in PPI in WT (t_7=3.071, p=0.018) and no decrease in PPI in HRM. Eight male WT, six male HRM. (Interval trial types collapsed for statistical analysis)

Table 20.1 DNMT1 expression (ratio of DNMT1 to NSE mRNA) is increased in frontal cortex of mice isolated for 3 weeks

	Mean	S.D.
Group housed	1.477	0.289
Socially isolated	2.556*	0.190

* t_7 = 4.321, p =0.003.

increased variance among individual mice is due to epigenetic variance since the coding gene dosage is 100% for WT and 50% for HRM. Moreover, there was a positive correlation between reelin mRNA and PPI when PPI was measured post-isolation and close to sacrifice in both genotypes (WT= +0.25, HRM = +0.854, $p<0.05$). This observation confirms an association between higher levels of reelin and greater prepulse inhibition (implicit in the group housed WT vs HRM difference in PPI) even in mice with the same genetic code for reelin.

The effect of social isolation on HRM and WT was assessed by Salinger *et al.* (2003), but their conclusion that social isolation was without effect was based on higher order statistical interactions that were complex, as both male and female mice and the *reeler* genotype were included in the sample together with WT and HRM. There are known to be substantial sex differences in neurobiological and behavioral responses to social isolation (Pinna *et al.*, 2003, 2004).

We have recently noted a significant increase in DNMT1 in the frontal cortex of isolated Swiss male mice as shown in Table 20.1. An isolation-induced increase in DNMT1 in WT is consistent with methylation of the reelin promoter and could be related to our failure to find a difference in PPI between WT and HRM following isolation. A next step would be to compare DNMT1 expression in HRM and WT in group-housed versus socially isolated mice.

The differential effects of social isolation in HRM and WT mice and the increase in DNMT expression in WT following isolation are findings that are consistent with a complex interaction between genetic and epigenetic factors influencing reelin expression. Such a complex interaction has recently been reported by Laviola *et al.* (2006) in a study involving a comparison between the locomotor activity of *reeler*, WT, and HRM following amphetamine challenge and by Ognibene *et al.* (2007b) in a study of the consequences of maternal separation on infant *reeler*, WT, and HRM.

4 Discussion and Conclusions

4.1 Interaction of Genetic and Epigenetic Influences on Reelin Downregulation

Studies of HRM have increased our understanding of the neurobiological and behavioral consequences of reelin downregulation. However, a new version of a model to study reelin downregulation in psychosis must also consider the extent of methylation of the RELN and GAD67 promoters, in addition to the genetic coding of these genes (Grayson *et al.*, 2005, 2006). In human postmortem psychotic brain, there is evidence that downregulation of reelin is related to hypermethylation of the reelin promoter, as DNMT1 mRNA is overexpressed in cortical GABAergic neurons in which reelin and GAD67 are downregulated (Veldic *et al.*, 2004, 2005; Grayson *et al.*, 2006; Ruzicka *et al.*, 2007). Grayson *et al.* (2005) isolated, bisulfite treatment amplified, and sequenced genomic DNA from the cortices of schizophrenia patients and nonpsychiatric subjects and found increased methylation within the CpG islands of the reelin promoter at positions −134 and −139 (base pairs). These two positions overlap with functionally defined *cis*-acting elements which bind repressor factors which leads to a compromised RELN promoter function. A coordinated hypermethylation of the RELN and GAD67 promoters is likely to be operative (Kundakovic *et al.*, 2007).

Additional evidence for epigenetic influences on reelin and GABAergic downregulation is provided by the fact that methionine exacerbates psychotic symptoms and, when given to WT mice, leads to downregulation of reelin and GAD67 (Tremolizzo *et al.*, 2002) and to an increased recruitment of methyl-binding domain proteins expressed by RELN and GAD67 promoters (Dong *et al.*, 2005). Moreover, HRM and methionine treated WT mice share similar behavioral consequences, presumably due to GABAergic downregulation, including similar PPI deficits and deficits in social interaction and recognition (Tremolizzo *et al.*, 2002, 2005).

Now that it appears that DNMT1 expression is greater in socially isolated than group-housed WT mice, the obvious next step is to compare HRM and WT, with respect to epigenetic changes induced by isolation. These future studies might involve measurement of the expression of DNMT1, reelin, GAD67, and methyl-binding proteins in group-housed and socially isolated HRM and WT. Ideally, such future studies would incorporate measurements of the extent of promoter methylation.

4.2 *Evaluation of Age and Sex in Future Studies of Reelin Downregulation in Psychosis*

Since, in human psychosis vulnerability, age and sex are known to be important variables, these two variables also need to be considered in an animal model to evaluate their role in reelin downregulation in psychosis. Purkinje cell loss has been reported in male, but not female HRM, and is influenced by estrogen (Biamonte *et al.*, 2006) and age (Hadj-Sahraoui *et al.*, 1996). In addition, reelin deficiency in HRM exacerbates neuronal death resulting from ischemic injury (Won *et al.*, 2006), which often increases with age. Our unpublished data suggest that PPI from 1 to 6 months of age is significantly positively correlated with age in WT mice, but not in HRM. Thus, interpreting the relationship between reelin downregulation and behavior will require systematic age-related studies of behavior in both sexes, along with quantitative measurements of reelin and GAD67 expression in the two different genotypes.

4.3 *The Need to Account for Compensatory Mechanisms for Reelin and GABAergic Downregulation*

The fact that multiple transmitter systems (e.g., dopaminergic, serotonergic, cholinergic) and endocrine function are altered both in psychotic patients (e.g., Lewis and Lieberman, 2000) and in HRM (Tueting *et al.*, 2006), suggests a need to explore the mechanisms and the time course of possible compensatory events in those systems that are invoked to balance the reduction of reelin and GABAergic function. Comparing HRM and WT treated with methionine versus saline is a possible strategy that could be explored in greater depth along with direct *ad hoc* measurements of probable indicators of compensations that are operative in other neurochemical pathways. In our studies of deficits in HRM and methionine-treated WT mice (Tremolizzo *et al.*, 2002), we detected similar deficits in PPI and in social behavior whether reelin was reduced by genetic code or by methionine, which is interesting, given that compensatory and epigenetic events for compromised reelin and GABA functions have been presumably occurring since conception in HRM but only for a short time in methionine-treated mice.

4.4 *Experimental Design Issues for the Future*

In evaluating the HRM model for schizophrenia, it is important to consider that conclusions are based on difference scores between WT and HRM. Our failure to find a PPI deficit in HRM following isolation could be explained by the downregulation of RELN and GAD67 genes, and that of other genes mediated by methylation that could differ in extent or in consequence in WT and HRM. Thus, laboratory differences could reflect sample differences in WT only, in HRM only, or in both WT and HRM. This issue becomes crucial when one hypothesizes that the difference score reflects reelin downregulation as a consequence of genetic coding plus epigenetic and compensatory factors.

Given that epigenetic events may not have the same consequences in all subjects because of baseline differences in the current state of the epigenome, the exploration of within-subject treatment strategies may be useful. Within-subject studies are more common in the human psychosis literature, due to the advantage of having each subject serve as their own control, as patients vary considerably from one to another even within the same diagnostic group. Different experimental controls and potential confounds apply in within-study designs than in between-subject designs. For example, ideally our study of pre- and post-social isolation could be replicated using a separate control group of mice not subjected to isolation.

In conclusion, there is mounting evidence that HRM and WT mice differ in behavior and underlying neurobiology in ways consistent with continued use of HRM as a model for psychosis and related disorders. In addition, the interaction between DNA coding sequences and epigenetic modification in GABAergic neurons that underlie reelin and GAD67 downregulation can be best understood by combining DNA coding and epigenetic profiling (Petronis, 2004; Schumacher *et al.*, 2006). Further study of HRM is warranted because reference points for the genetic code are needed in order to interpret the contribution of epigenetic or compensatory influences at each neurodevelopmental stage (Tueting *et al.*, 2006).

Acknowledgment The authors wish to thank Alessandro Guidotti for a critical reading of the manuscript.

References

Abdolmaleky, H. M., Cheng, K. H., Russo, A., Smith, C. L., Faraone, S. V., Wilcox, M., Shafa, R., Glott, S. J., Nguyen, G., Ponte, J. F., Thiagalingam, S., and Tsuan, M. T. (2005). Hypermethylation of the reelin (RELN) promoter in the brain of schizophrenia patients: a preliminary report. *Am. J. Med. Genet. B Neuropsychiatr. Genet.* 134(1):60–66.

Akbarian, S., and Huang, H. S. (2006). Molecular and cellular mechanisms of altered GAD1/GAD67 expression in schizophrenia and related disorders. *Brain Res. Rev.* 52(2):293–304.

Avila, M. T., Sherr, J., Valentine, L. E., Blaxton, T. A., and Thaker, G. K. (2003). Neurodevelopmental interactions conferring risk for schizophrenia: a study of dermatoglyphic markers in patients and relatives. *Schizophr. Bull.* 29(3):595–605.

Biamonte, F., Assenza, G., Marino, R., Caruso, D., Crotti, S., Melcangi, R. C., Cesa, R., Strata, P, and Keller, F. (2006). Interaction between estrogens and reelin in Purkinje cell development. *Neurosci. Abstr. Online* #322.18.

Botella-Lopez, A., Burgay, F., Gavin, R., Garcia-Ayllon, M. S., Gomez-Tortosa, G., Pena-Casanova, J., Urena, J. M., Del Rio, J. A., Blesa, R., Soriano, E., and Saez-Valero, S. (2006). Reelin expression and glycosylation patterns are altered in Alzheimer's disease. *Proc. Natl. Acad. Sci. USA* 103(14):5573–5578.

Brigman, J., Padukiewicz, K. E., Sutherland, M. L., and Rothblat, L. A. (2006). Executive functions in the heterozygous reeler mouse model of schizophrenia. *Behav. Neurosci.* 120(4):984–988.

Carboni, G., Tueting, P., Tremolizzo, L., Sugaya, I., Davis, J., Costa, E., and Guidotti, A. (2004). Enhanced dizocilpine efficacy in heterozygous reeler mice relates to GABA turnover down-regulation. *Neuropsychopharmacology* 46:1070–1081.

Caviness, V. S., Jr. (1982). Neocortical histogenesis in normal and reeler mice; a developmental study based upon [3H] autoradiography. *Brain Res.* 256:293–302.

Caviness, V.S., and Sidman, R. L. (1973). Retrohippocampal, hippocampal and related structures of the forebrain in the *reeler* mutant mouse. *J. Comp. Neurol.* 147:235–254.

Chang, B. S., Duzcan, F., Kim, S., Cinbis, M., Aggarwal, A., Apse, K. A., Ozdel, O., Atmaca, M., Zencir, S., Bagci, H., and Walsh, C. A. (2007). The role of RELN in lissencephaly and neuropsychiatric disease. *Am. J. Med. Genet. B Neuropsychiatr. Genet.* 144(1):58–63.

Chen, Y., Sharma, R. P., Costa, R. H., Costa, E., and Grayson, D. R. (2002). On the epigenetic regulation of the human reelin promoter. *Nucleic Acids Res.* 30(13):2930–2939.

Costa, E., Davis, J., Grayson, D. R., Guidotti, A., Pappas, G. D., and Pesold, C. (2001). Dendritic spine hypoplasticity and downregulation of reelin and GABAergic tone in schizophrenia vulnerability. *Neurobiol. Dis.* 8:723–742.

Costa, E., Davis, J., Pesold, C., Tueting, P., and Guidotti, A. (2002). The heterozygote reeler mouse as a model for the development of a new generation of antipsychotics. *Curr. Opin. Pharmacol.* 2(1):56–62.

D'Arcangelo, G. (2006). Reelin mouse mutants as models of cortical development disorders. *Epilepsy Behav.* 8(1):81–90.

Dong, E., Caruncho, H., Liu, W. S., Smalheiser, N. R., Grayson, D. R., Costa, E., and Guidotti, A. (2003). A reelin-integrin receptor interaction regulates Arc mRNA translation in synaptoneurosomes. *Proc. Natl. Acad. Sci. USA* 100(9):5479–5484.

Dong, E., Agis-Balboa, R. C., Simonini, M. V., Grayson, D. R., Costa, E., and Guidotti, A. (2005). Reelin and glutamic acid decarboxylase 67 promoter remodeling in an epigenetic methionine-induced mouse model of schizophrenia. *Proc. Natl. Acad. Sci. USA* 102(35):12578–12583.

Doueiri, M., Hanbauer, I., Zhubi, A., Pinna, G., Guidotti, A., Davis, J., and Costa, E. (2006). Pharmacological modification of cognitive and social behavior dysfunction in heterozygous reeler mice (HRM). *Soc. Neurosci. Abstr. Online.*

Fatemi, S. H., Stary, J. M., Halt, A. R., and Realmuto, G. R. (2001). Dysregulation of Reelin and Bcl-2 proteins in autistic cerebellum. *J. Autism Dev. Disord.* 31(6):529–535.

Fatemi, S. H., Halt, A. R., Stary, J. M., Kanodia, R., Schulz, S. C., and Realmuto, G. R. (2002). Glutamic acid decarboxylase 65 and 67 kDa proteins are reduced in autistic parietal and cerebellar cortices. *Biol. Psychiatry* 52:805–810.

Fatemi, S. H., Snow, A. V., Stary, J. M., Araghi-Niknam, M., Reutiman, T. J., Lee, S., Brooks, A. I., and Pearce D. A. (2005). Reelin signaling is impaired in autism. *Biol. Psychiatry* 57(7):777–787.

Geyer, M. A., Krebs-Thompson, K., and Braff, D. L. (2001). Pharmacological studies of prepulse inhibition models of sensorimotor deficits in schizophrenia: a decade in review. *Psychopharmacology* 156(2–3):117–154.

Glantz, L. A., and Lewis, D. A. (2000). Decreased dendritic spine density on prefrontal cortical pyramidal neurons in schizophrenia. *Arch.Gen. Psychiatry* 57:65–73.

Goffinet, A. M., So, K. F., Yamamoto, M., Edwards, M., and Caviness, V. S. (1984). Architectonic and hodological organization of the cerebellum in *reeler* mutant mice. *Brain Res.* 318:263–276.

Grayson, D. R., Jia, X., Chen, Y., Sharma, R. P., Mitchell, C. P., Guidotti, A., and Costa, E. (2005). Reelin promoter hypermethylation in schizophrenia. *Proc. Natl. Acad. Sci. USA* 102:9341–9346.

Grayson, D. R., Chen, Y., Costa, E., Dong, E., Guidotti, A., Kundakovic, M., and Sharma, R. P. (2006). The human reelin gene: transcription factors (+), repressors (–) and the methylation switch (+/–) in schizophrenia. *Pharmacol. Ther.* 111(1):272–286.

Guidotti, A., Auta, J., Davis, J. M., Di-Giorgi-Gerevini, V., Dwivedi, Y., Grayson, D. R., Impagnatiello, F., Pandey, G., Pesold, C., Sharma, R., Uzunov, D., and Costa, E. (2000). Decrease in reelin and glutamic acid decarboxylase 67 (GAD 67) expression in schizophrenia and bipolar disorder: a postmortem brain study. *Arch. Gen. Psychiatry* 57(11):1061–1069.

Guidotti, A., Auta, J., Davis, J. M., Dong, E., Grayson, D. R., Veldic, M., Zhang, X., and Costa, E. (2005). GABAergic dysfunction in schizophrenia: new treatment strategies on the horizon. *Psychopharmacology (Berlin)* 180(2):191–205.

Hadj-Sahraoui, N., Frederic, F., Delhayi-Bouchaud, N., and Mariani, J. (1996). Gender effect on Purkinje cell loss in the cerebellum of the heterozygous reeler mouse. *J. Neurogenet.* 11:45–58.

Hauser, J., Rudolph, U., Keist, R., Mohler, H., Feldon, J., and Yee, B. K. (2005). Hippocampal $\alpha 5$ subunit-containing $GABA_A$ receptors modulate the expression of prepulse inhibition. *Mol. Psychiatry* 10(2):201–207.

Herz, J., and Chen, Y. (2006). Reelin, lipoprotein receptors and synaptic plasticity. *Nature Rev. Neurosci.* 7:850–859.

Hevner, R. F., Daza, R. A., Englund, C., Kohtz, J., and Fink, A. (2004). Postnatal shifts of interneuron position in the neocortex of normal and *reeler* mice; evidence for inward radial migration. *Neuroscience* 124:605–618.

Hill, J. J., Hashimoto, T., and Lewis, D. A. (2006). Molecular mechanisms contributing to dendritic spine alterations in the prefrontal cortex of subjects with schizophrenia. *Mol. Psychiatry* 11(6):557–566.

Hong, S. E., Shugart, Y. Y., Huang, D. T., Shahwan, S. A., Grant, P. E., Hourihane, J. O., Martin, N. D., and Walsh, C. A. (2000). Autosomal recessive lissencephaly with cerebellar hypoplasia is associated with human RELN mutation. *Nature Genet.* 26:93–96.

Impagnatiello, F., Guidotti, A., Pesold, C., Dwivedi, Y., Caruncho, H., Pisu, M. G., Smalheiser, N. R., Davis, J. M., Pandey, G. N., Pappas, G. D., Tueting, P., and Costa, E. (1998). A decrease of reelin expression as a putative vulnerability factor in schizophrenia. *Proc. Natl. Acad. Sci. USA* 95:15718–15723.

Kim, H. M., Qu, T., Kriho, V., Lacor, P., Smalheiser, N., Pappas, G. D., Guidotti, A., Costa, E., and Sugaya, K. (2002). Reelin function in neural stem cell biology *Proc. Natl. Acad. Sci. USA* 99(6):4020–4025.

Koch, M. (1999). The neurobiology of startle. *Prog. Neurobiol.* 59:107–128.

Krueger, D. D., Howell, J. L., Hebert, B. F., Olausson, P., Taylor, J. R., and Nairn, A. C. (2006). Assessment of cognitive function in the heterozygous reeler mouse. *Psychopharmacology (Berlin)* 189(1):85–104.

Kundakovic, M., Chen, Y., Costa, E., and Grayson, D. R. (2007). DNA methyltransferase inhibitors coordinately induce expression of the human reelin and GAD67 genes. *Mol. Pharmacol.* 71(3):644–653.

Laviola, G., Adriani, W., Gaudino, C., Marino, R., and Keller, F. (2006). Paradoxical effects of prenatal acetylcholinesterase blockade on neuro-behavioral development and drug-induced stereotypies in reeler mutant mice. *Psychopharmacology (Berl)* 187(3):331–344.

Lewis, D. A., and Lieberman, J. A. (2000). Catching up on schizophrenia: natural history and neurobiology. *Neuron* 28:325–334.

Lewis, D. A., Hashimoto, T, and Volk, D. W. (2005). Cortical inhibitory neurons and schizophrenia. *Nature Rev. Neurosci.* 6:312–324.

Liu, W. S., Pesold, C., Rodriguez, M. A., Carboni, G., Auta, J., Lacor, P., Larson, J., Condie, B. G., Guidotti, A., and Costa, E. (2001). Down-regulation of dendritic spine and glutamic acid decarboxylase 67 expressions in the reelin haploinsufficient heterozygous reeler mouse. *Proc. Natl. Acad. Sci. USA* 98(6):3477–3482.

Manent, J. B., Jorquera, I., Ben-Ari, Y., Aniksztein, L., and Represa, A. (2006). Glutamate acting on AMPA but not NMDA receptors modulates the migration of hippocampal interneurons. *J. Neurosci.* 26(22):5901–5909.

Marrone, M. C., Marinelli, S., Biamonte, F., Keller, F., Sgobio, C. A., Ammassari-Teule, M., Bernardi, G., and Mercuri, N. B. (2006). Altered cortico-striatal synaptic plasticity and related behavioural impairments in reeler mice. *Eur. J. Neurosci.* 24:2061–2070.

Ognibene, E., Adriani, W., Granstrem, O., Pieretti, S., and Laviola, G. (2007a). Impulsivity-anxiety-related behavior and profiles of morphine-induced analgesia in heterozygous reeler mice. *Brain Res.* 1131(1):173–180.

Ognibene, E., Adriani, W., Macri, S., and Laviola, G. (2007b). Neurobehavioural disorders in the infant reeler mouse model: interaction of genetic vulnerability and consequences of maternal separation. *Behav. Brain Res.* 177:142–149.

Patterson, P. H. (2006). Modeling features of autism in animals. In: Moldin, S. O., and Rubenstein, J. L. R. (eds.), *Understanding Autism: From Basic Neuroscience to Treatment.* CRC Taylor & Francis, New York, pp. 277–301.

Petronis, A. (2004). The origin of schizophrenia: genetic thesis, epigenetic antithesis, and resolving synthesis. *Biol. Psychiatry* 55(10):965–970.

Pinna, G., Dong, E., Matsumoto, K., Costa, E., and Guidotti, A. (2003). In socially isolated mice, the reversal of allopregnanolone down-regulation mediates the anti-aggressive action of fluoxetine. *Proc. Natl. Acad. Sci. USA* 100(4):2035–2040.

Pinna, G., Agis-Balboa, R. C., Doueiri, M. S., Guidotti, A., and Costa, E. (2004). Brain neurosteroids in gender-related aggression induced by social isolation. *Crit. Rev. Neurobiol.* 16(1–2):75–82.

Pla, R., Borrell, V., Flames, N., and Marin, O. (2006). Layer acquisition by cortical GABAergic interneurons is independent of reelin signaling. *J. Neurosci.* 26(26):6924–6934.

Podhorna, J., and Didriksen, M. (2004). The heterozygous reeler mouse: behavioural phenotype. *Behav. Brain Res.* 153:43–54.

Pollard, K. S., Salama, S. R., Lambert, N., Lambot, M. A., Coppens, S., Pedersen, J. S., Katzman, S., King, B., Onodera, C., Siepel, A., Kern, A. D., Dehay, C., Igel, H., Ares, M., Jr., Vanderhaeghen, P., and Haussler, D. (2006). An RNA gene expressed during cortical development evolved rapidly in humans. *Nature* 443(7108):167–172.

Powell, S. B., and Geyer, M. A. (2002). Developmental markers of psychiatric disorders as identified by sensorimotor gating. *Neurotox. Res.* 4(5–6):489–502.

Qiu, S., and Weeber, E. J. (2007). Reelin signaling facilitates maturation of CA1 glutamatergic synapses. *J. Neurophysiol.* 97(3):2312–2321.

Qiu, S., Korwek, K. M., Pratt-Davis, A. R., Peters, M., Bergman, M. Y., and Weeber, E. J. (2006). Cognitive disruption and altered hippocampus synaptic function in reelin haploinsufficient mice. *Neurobiol. Learn. Mem.* 85:228–242.

Rehn, A. E., and Rees, S. M. (2005). Investigating the neurodevelopmental hypothesis of schizophrenia. *Clin. Exp. Pharm. Phys.* 32:687–696.

Ruzicka, W. B., Zhubi, A., Veldic, M., Grayson, D. R., Costa, E., and Guidotti, A. (2007). Selective epigenetic alteration of Layer 1 GABAergic neurons isolated from prefrontal cortex of schizophrenia patients using laser-assisted microdissection. *Mol. Psychiatry* 12(4):385–397.

Sakaue, M., Ago, Y., Baba, A., and Matsuda, T. (2003). The 5-HT1A receptor agonist MKC-242 reverses isolation rearing-induced deficits of prepulse inhibition in mice. *Psychopharmacology (Berlin)* 170(1):73–79.

Salinger, W. L., Ladrow, P. L., and Wheeler, C. B. (2003). Behavioural phenotype of the reeler mutant mouse: effects of Reln gene dosage and social isolation. *Behav. Neurosci.* 117:1257–1274.

Schumacher, A., Kapranov, P., Kaminsky, Z., Flanagan, J., Assadzadah, A., Yau, P., Virtanen, C., Winegarden, N., Checng, J., Gingeras, T., and Petronis, A. (2006). Microarray-based DNA methylation profiling technology and applications. *Nucleic Acids Res.* 34(2):528–542.

Sigala, S., Zoli, M., Palozzolo, F., Faccoli, S., Zanard, A., Mercuri, N. B., and Spano, P. (2007). Selective disarrangement of the rostral telencephalic cholinergic system in heterozygous reeler mice. *Neuroscience* 144(3):834–844.

Swerdlow, N. R., Geyer, M. A., and Braff, D. L. (2001). Neural circuit regulation of prepulse inhibition of startle in the rat: current knowledge and future challenges. *Psychopharmacology (Berl)* 156:194–215.

Torrey, E. F., Barci, B. M., Webster, M. J., Bartko, J. J., Meador-Woodruff, J. H., and Knable, M. B. (2005). Neurochemical markers for schizophrenia, bipolar disorder, and major depression in postmortem brains. *Biol. Psychiatry* 57:252–260.

Tremolizzo, L., Carboni, G., Ruzicka, W. B., Mitchell, C. P., Sugaya, I., Tueting, P., Sharma, R., Grayson, D. R., Costa, E., and Guidotti, A. (2002). An epigenetic mouse model for molecular and behavioral neuropathologies related to schizophrenia vulnerability. *Proc. Natl. Acad. Sci. USA* 99:17095–17100.

Tremolizzo, L., Doueiri, M. S., Dong, E., Grayson, D. R., Davis, D., Pinna, G., Tueting, P., Rodriguez-Menendez, V., Costa, E., and Guidotti, A. (2005). Valproate corrects the schizophrenia-like epigenetic behavioral modifications induced by methionine in mice. *Biol. Psychiatry* 57:500–509.

Tueting, P., Costa, E., Dwivedi, Y., Guidotti, A., Impagnatiello, F., Manev, R., and Pesold, C. (1999). The phenotypic characteristics of heterozygous reeler mouse. N*euroReport*10(6):1329–1334.

Tueting, P., Doueiri, M. S., Guidotti, A., Davis, J. M., and Costa, E. (2006). Reelin down-regulation in mice and psychosis endophenotypes. *Neurosci. Biobehav. Rev*. 30:1065–1077.

Veldic, M., Caruncho, H. J., Liu, W. S., Davis, J. K., Satta, R., Grayson, D. R., Guidotti, A., and Costa, E. (2004). DNA-methyltransferase 1 mRNA is selectively overexpressed in telencephalic GABAergic interneurons of schizophrenia brains. *Proc. Natl. Acad. Sci. USA* 101(1):348–353.

Veldic, M., Guidotti, A., Maloku, E., Davis, J. M., and Costa, E. (2005). In psychosis, cortical interneurons overexpress DNA-methyltransferase 1. *Proc. Natl. Acad. Sci. U.S.A.* 102:2152–2157.

Volk, D. W., Austin, M. C., Pierri, J. N., Sampson, A. R., and Lewis, D. A. (2000). Decreased glutamic acid decarboxylase67 messenger RNA expression in a subset of prefrontal cortical gamma-aminobutyric acid neurons in subjects with schizophrenia. *Arch. Gen. Psychiatry* 57(3):237–245.

Won, S. J., Kim, S. E., Xie, L., Wang, Y., Mao, X. O., Jin, K., and Greenberg, D. A. (2006). Reelin-deficient mice show impaired neurogenesis and increased stroke size. *Exp. Neurol.* 198:250–259.

Wyss, J. M., Stanfield, B. B., and Cowan, W. M. (1980). Structural abnormalities in the olfactory bulb of the Reeler mouse. *Brain Res.* 188:566–571.

Yabut, O., Renfro, A., Niu, S., Swann, J. W., Marin, O., and D'Arcangelo, G. (2007). Abnormal laminar position and dendrite development of interneurons in the reeler forebrain. *Brain Res.* 1140C:75–83.

Chapter 21
Reelin and Lissencephaly

Elena Parrini and Renzo Guerrini

Contents

1 Introduction

The development of the human cerebral cortex is a dynamic process that can be divided into partially overlapping stages occurring during several gestational weeks (Barkovich *et al.*, 2005). Migration of postmitotic neurons from the ventricular zone to form the cortical plate comprises one of the most critical stages in brain development. When migration is complete, the cortex is a six-layered structure, with each layer comprising different types of neurons that form discrete connections within the CNS and perform distinct functions (O'Rourke *et al.*, 1992). When neurons reach their destination, they stop migrating and order themselves into specific "architectonic" patterns in brain development (Fig. 21.1A). Understanding this complex process has progressed based on studies of human malformations and mouse models with deficient neuronal migration, particularly the malformation known as lissencephaly (LIS).

The term *lissencephaly*, derived from the Greek words *lissos* meaning smooth and *enkephalos* meaning brain, is a neuronal migration disorder characterized by absent (agyria) or decreased (pachygyria) convolutions, producing a smooth cere-

E. Parrini
Pediatric Neurology Unit and Laboratories, Children's Hospital A. Meyer-University of Florence, viale Pieraccini 24, Florence, Italy
e-mail: e.parrini@meyer.it

R. Guerrini
Pediatric Neurology Unit and Laboratories, Children's Hospital A. Meyer-University of Florence, viale Pieraccini 24, Florence, Italy
e-mail: r.guerrini@meyer.it

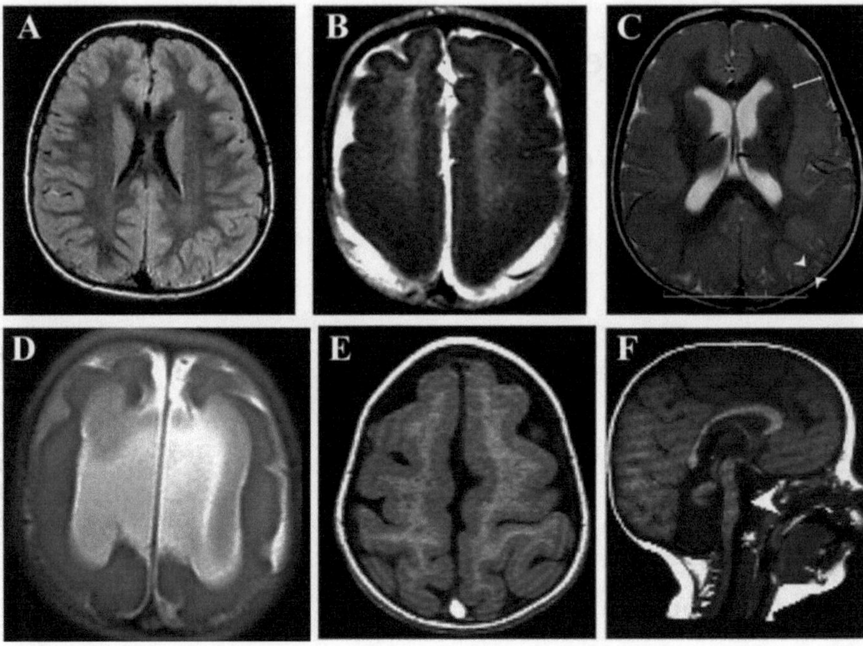

Fig. 21.1 (**A**) Brain MRI scan; axial section of a normal brain. (**B**) Axial section: the cortex in the posterior brain is completely smooth, in the frontal lobes the gyral pattern is simplified and the cortex is thickened. This 4-year-old boy has infantile spasms and a deletion involving the *LIS1* gene. (**C**) Brain MRI scan; axial section: lissencephaly in a girl with *DCX* mutation. There is a typical anterior > posterior malformation pattern, cortical thickness is around 2 cm in the frontal lobes (single arrow) and around 4 mm in the posterior brain (doublearrowheads). (**D**) Axial section: 1-year-old boy with X-linked lissencephaly, with corpus callosum agenesis and ambiguous genitalia due to mutation of the *ARX* gene. Note absence of the corpus callosum with ventriculomegaly and lissencephaly. (**E, F**) Brain MRI scans of two patients from a family with LCH type b and a mutation in the *RELN* gene. (**E**) Axial section: the cortex is thickened and the gyral pattern is simplified. (**F**) Sagittal section: the cerebellum is severely reduced in size, with hypoplasia of the inferior vermis and of the hemispheres. The pons (arrowhead) is reduced in size. [Reprinted by permission from Macmillan Publishers Ltd. (Hong *et al.*, *Nature Genet.* 2000; 26:93–96), copyright (2000)]

bral surface (Barkovich *et al.*, 2005). The cytoarchitecture consists of four primitive layers including an outer marginal layer, which contains Cajal-Retzius neurons (layer I), a superficial cellular layer, which contains numerous large and disorganized pyramidal neurons (layer II) corresponding to the true cortex, a variable cell-sparse layer (layer III), and a deep cellular layer composed of medium and small neurons, which extends more than half the width of the mantle (layer IV) (Kato and Dobyns, 2003). The white matter, which is severely reduced in volume, occasionally contains individual neurons or collection of neurons forming heterotopia.

Mechanisms by which cell migration into the cortical plate stops at the appropriate location have been elucidated through the characterization of the *Reeler* mutant mouse (Caviness and Sidman, 1973). In this animal model, the cortical pattern is opposite with respect to the normal inside-to-outside development of the cerebral cortex. This observation suggests *Reln* to be required for the normal inside-to-outside positioning of cells as they migrate from the ventricular zone; a first component of a signaling pathway guiding cells to the correct location in the cerebral cortex.

2 Lissencephaly Categories

Several different LIS types have been recognized. The most common type, known as classical (or type 1) LIS, features a very thick cortex (10–20 mm rather than the normal 4 mm) and no other major brain malformations. This type of LIS is caused by mutations of the *LIS1* gene (Reiner *et al.*, 1993) and of the *DCX* (or *XLIS*) gene (des Portes *et al.,* 1998; Gleeson *et al.,* 1998). *LIS1* mutations result in more severe LIS in the posterior brain regions (posterior>anterior gradient) (Fig. 21.1B), whereas *DCX* mutations result in more severe LIS in the anterior brain regions (anterior>posterior gradient) (Fig. 21.1C) (Pilz *et al.*, 1998; Dobyns *et al.*, 1999). The interaction of both DCX and LIS1 with microtubules may explain the striking similarities between the lissencephalic phenotypes produced by mutations in these two genes.

Classical LIS is rare with a prevalence of 11.7 per million births. All patients have early developmental delay, early diffuse hypotonia, later spastic quadriplegia and opisthotonus, and eventual severe or profound mental retardation. Rarely, patients with pachygyria may have moderate mental and motor impairment. Some children with LIS have lived more than 20 years, although anecdotal experience suggests that the life span is less than 10 years in most patients. Seizures occur in over 90% of children, with onset before 6 months in about 75%. About 80% of children have infantile spasms, although EEG does not show typical hypsarrhythmia. Later, most children have mixed seizure disorders. As most clinical and neurophysiological studies on children with LIS were conducted before genetic distinction between *DCX* and *LIS1* was made, it is unknown whether these two forms have distinctive electroclinical patterns.

The *LIS1* and *DCX* genes do not account for all known cases of classical LIS, and additional LIS syndromes have been described (Walsh, 1999). Miller-Dieker syndrome (MDS) is caused by a contiguous gene deletion. Classical LIS is accompanied by distinct dysmorphic facial features, including prominent forehead, flattened ear helices, short nose, and anteverted nares. Deletions of 17p13.3, including the *LIS1* gene, are found in almost 100% of patients (Dobyns *et al.*, 1993). Deletion of two additional genes, *CRK* and *14-3-3e*, telomeric to *LIS1,* may contribute to the most severe LIS grade and dysmorphic features observed in MDS. X-linked LIS with corpus callosum agenesis and ambiguous genitalia (XLAG) features LIS with posterior-to-anterior gradient and only moderate increase of the cortical thickness (6–7 mm), absent corpus callosum, and ambiguous genitalia with micropenis and cryptorchidism (Bonneau *et al.*, 2002; Kato *et al.*, 2004) (Fig. 21.1D). Mutations of the X-linked *ARX* gene were identified in individuals with XLAG and in some female relatives (Kitamura *et al.*, 2002). The mutations of the *ARX* gene in XLAG patients are predominantly premature terminations.

3 Lissencephaly with Cerebellar Hypoplasia (LCH)

Malformations in the LIS spectrum can be associated with significant cerebellar underdevelopment and have recently been referred to as lissencephaly with cerebellar hypoplasia (LCH).

Six different subtypes of LCH have been described in patients with LCH with heterogeneous clinical presentations (Ross *et al.*, 2001). Phenotypic features included small head circumference, cortical malformation ranging from agyria to simplification of the gyral pattern, and from near-normal cortical thickness to marked thickening of the cortical gray matter. Cerebellar manifestations range from midline hypoplasia to diffuse volume reduction and disturbed foliation. In the LCHb subgroup the cerebral cortex is pachygyric with mild cortical thickening (4–10 mm). An anterior predominant gradient with fewer, broader gyri in the frontal cortex has been reported. Despite the moderate thickening of cerebral cortex, the hippocampal formation could not be clearly identified. The presumptive hippocampus was straightened and suggested to have marked disorganization of the CA regions and dentate gyrus with marked reduction in the anterior–posterior extent of the parahippocampal cortex. The entire cerebellum was severely hypoplastic with absent folia.

An autosomal recessive form of LCH type b associated with severe abnormalities of the cerebellum, hippocampus, and brainstem was described in two consanguineous pedigrees (Hong *et al.*, 2000). In these patients, the cortex was thickened, and the gyral pattern was simplified (Fig. 21.1E). Both of these abnormalities were more severe frontally and temporally, so that the thickness and gyral pattern of the occipital and parietal cortex were relatively normal. The hippocampus appeared flattened, lacking definable upper and lower blades. The subcortical white matter was decreased in amount but consistently normal in its signal characteristics. The corpus callosum was thin over its entire rostrocaudal extent. The lateral ventricles were enlarged, and the cerebellum was severely reduced in size, with hypoplasia of the inferior vermis and cerebellar hemispheres, devoid of any detectable folds or normal architecture (Fig. 21.1F). The pons was reduced in size in superior–inferior and anteroposterior extent. Affected individuals presented dysmorphic facial features, including bitemporal hollowing, sloping of forehead, widely set eyes, and prominent nasal bridge. Affected children in one family had congenital lymphedema, hypotonia, severe developmental delay, and generalized seizures that were controlled by drugs. Severe hypotonia, developmental delay, and seizures were also reported in the other pedigree. Affected children in both families carried mutations in the *RELN* gene (7q22.1), leading to disrupted splicing of *RELN* cDNA. Western blotting revealed low or undetectable amounts of RELN protein in the serum.

4 Reelin *(RELN)* and Lissencephaly

RELN encodes a large (388 kDa) extracellular matrix protein that acts on migrating cortical neurons by binding to the very-low-density lipoprotein receptor (VLDLR), the apolipoprotein E receptor 2, $\alpha 3\beta 1$ integrin, and cadherin-related receptors

(CNRs) (D'Arcangelo *et al.*, 1995). Mutations of the mouse homologues of *RELN* cause brain defects in mice that resemble LCH (Lambert de Rouvroit and Goffinet, 1998). In mice, *Reln* mutations cause cerebellar hypoplasia, abnormal cortical neuronal migration in the cerebrum, and abnormal axonal connectivity (Lambert de Rouvroit and Goffinet, 1998; Gonzalez *et al.*, 1997). Neurons in affected mice fail to reach their correct location in the developing brain, disrupting the organization of the cerebellar and cerebral cortices and other laminated regions. In this animal model, the cortical layering appears inverted (Caviness, 1976; Ogawa *et al.*, 1995). Thus, Reln is thought to control cell–cell interactions critical for cell positioning in the brain.

RELN mutation analysis is indicated in patients with LCH and an autosomal recessive pattern of inheritance. However, no clear indication on the range of severity of the malformation can be clearly defined at present, due to the paucity of reported cases with proven *RELN* gene defect.

As *RELN* is encoded by 65 exons, covering more than 400 kb of genomic DNA and 12 kb of coding cDNA (DeSilva *et al.*, 1997) the genetic test should be performed in *RELN* cDNA using RT-PCR amplification of RNA from the patients. Rare patients with autosomal recessive LCH, severe epilepsy, mental retardation, and a chromosomal rearrangement causing the disruption of *RELN,* with absence of encoded protein, have been identified (Zaki *et al.*, 2007; Chang *et al.*, 2007). For this reason, cytogenetic analysis, including high-resolution karyotype, FISH, or array-CGH analysis, should be performed as a complement or alternative to *RELN* mutation analysis in patients with LCH.

References

Barkovich, A. J., Kuzniecky, R.I., Jackson, G. D., Guerrini, R., and Dobyns, W. B. (2005). A developmental and genetic classification for malformations of cortical development. *Neurology* 65(12):1873–1887.

Bonneau, D., Toutain, A., Laquerriere, A., Marret, S., Saugier-Veber, P., Barthez, M. A., Radi, S., Biran-Mucignat, V., Rodriguez, D., and Gelot, A. (2002). X-linked lissencephaly with absent corpus callosum and ambiguous genitalia (XLAG): clinical, magnetic resonance imaging, and neuropathological findings. *Ann. Neurol.* 51:340–349.

Caviness, V. S., Jr. (1976). Patterns of cell and fiber distribution in the neocortex of the reeler mutant mouse. *J. Comp. Neurol.* 170:435–447.

Caviness, V. S., Jr., and Sidman, R. L. (1973). Time of origin of corresponding cell classes in the cerebral cortex of normal and reeler mutant mice: an autoradiographic analysis. *J. Comp. Neurol.* 148:141–151.

Chang, B. S., Duzcan, F., Kim, S., Cinbis, M., Aggarwal, A., Apse, K. A., Ozdel, O., Atmaca, M., Zencir, S., Bagci, H., and Walsh, C. A. (2007). The role of RELN in lissencephaly and neuropsychiatric disease. *Am. J. Med. Genet. B Neuropsychiatr. Genet.* 144(1):58–63.

D'Arcangelo, G., Miao, G. G., Chen, S. C., Soares, H. D., Morgan, J. I., and Curran, T. (1995). A protein related to extracellular matrix proteins deleted in the mouse mutant reeler. *Nature* 374:719–723.

DeSilva, U., D'Arcangelo, G., Braden, V. V., Chen, J., Miao, G. G., Curran, T., and Green, E. D. (1997). The human reelin gene: isolation, sequencing, and mapping on chromosome 7. *Genome Res.* 7:157–164.

des Portes, V., Pinard, J. M., Billuart, P., Vinet, M. C., Koulakoff, A., Carrie, A., Gelot, A., Dupuis, E., Motte, J., Berwald-Netter, Y., Catala, M., Kahn, A., Beldjord, C., and Chelly, J. (1998). A novel CNS gene required for neuronal migration and involved in X-linked subcortical laminar heterotopia and lissencephaly syndrome. *Cell* 92:51–61.

Dobyns, W. B., Reiner, O., Carrozzo, R., and Ledbetter, D. H. (1993). Lissencephaly. A human brain malformation associated with deletion of the LIS1 gene located at chromosome 17p13. *J. Am. Med. Assoc.* 270:2838–2842.

Dobyns, W. B., Truwit, C. L., Ross, M. E., Matsumoto, N., Pilz, D. T., Ledbetter, D. H., Gleeson, J. G., Walsh, C. A., and Barkovich, A. J. (1999). Differences in the gyral pattern distinguish chromosome 17-linked and X-linked lissencephaly. *Neurology* 53:270–277.

Gleeson, J. G., Allen, K. M., Fox, J. W., Lamperti, E. D., Berkovic, S., Scheffer, I., Cooper, E. C., Dobyns, W. B., Minnerath, S. R., Ross, M. E., and Walsh, C. A. (1998). doublecortin, a brain-specific gene mutated in human X-linked lissencephaly and double cortex syndrome, encodes a putative signaling protein. *Cell* 92:63–72.

Gonzalez, J. L., Russo, C. J., Goldowitz, D., Sweet, H. O., Davisson, M. T., and Walsh, C. A. (1997). Birthdate and cell marker analysis of scrambler: a novel mutation affecting cortical development with a reeler-like phenotype. *J. Neurosci.* 17:9204–9211.

Hong, S. E., Shugart, Y. Y., Huang, D. T., Shahwan, S. A., Grant, P. E., Hourihane, J. O., Martin, N. D., and Walsh, C. A. (2000). Autosomal recessive lissencephaly with cerebellar hypoplasia is associated with human RELN mutations. *Nat. Genet.* 26(1):93–96.

Kato, M., and Dobyns, W. B. (2003). Lissencephaly and the molecular basis of neuronal migration. *Hum. Mol. Genet.* 12:89–96.

Kato, M., Das, S., Petras, K., Kitamura, K., Morohashi, K., Abuelo, D. N., Barr, M., Bonneau, D., Brady, A. F., Carpenter, N. J., Cipero, K. L., Frisone, F., Fukuda, T., Guerrini, R., Iida, E., Itoh, M., Lewanda, A. F., Nanba, Y., Oka, A., Proud, V. K., Saugier-Veber, P., Schelley, S. L., Selicorni, A., Shaner, R., Silengo, M., Stewart, F., Sugiyama, N., Toyama, J., Toutain, A., Vargas, A. L., Yanazawa, M., Zackai, E. H., and Dobyns, W. B. (2004). Mutations of ARX are associated with striking pleiotropy and consistent genotype-phenotype correlation. *Hum Mutat.* 23(2):147–159.

Kitamura, K., Yanazawa, M., Sugiyama, N., Miura, H., Iizuka-Kogo, A., Kusaka, M., Omichi, K., Suzuki, R., Kato-Fukui, Y., Kamiirisa, K., Matsuo, M., Kamijo, S., Kasahara, M., Yoshioka, H., Ogata, T., Fukuda, T., Kondo, I., Kato, M., Dobyns, W. B., Yokoyama, M., and Morohashi, K. (2002). Mutation of ARX causes abnormal development of forebrain and testes in mice and X-linked lissencephaly with abnormal genitalia in humans. *Nature Genet.* 32:359–369.

Lambert de Rouvroit, C., and Goffinet, A. M. (1998). The reeler mouse as a model of brain development. *Adv. Anat. Embryol. Cell Biol.* 150:1–106.

Ogawa, M., Miyata, T., Nakajima, K., Yagyu, K., Seike, M., Ikenaka, K., Yamamoto, H., and Mikoshiba, K. (1995). The reeler gene-associated antigen on Cajal–Retzius neurons is a crucial molecule for laminar organization of cortical neurons. *Neuron* 14:899–912.

O'Rourke, N. A., Dailey, M. E., Smith, S. J., and McConnell, S. K. (1992). Diverse migratory pathways in the developing cerebral cortex. *Science* 258:299–302.

Pilz, D. T., Matsumoto, N., Minnerath, S., Mills, P., Gleeson, J. G., Allen, K. M., Walsh, C. A., Barkovich, A. J., Dobyns, W. B., Ledbetter, D. H., and Ross, M. E. (1998). LIS1 and XLIS (DCX) mutations cause most classical lissencephaly, but different patterns of malformation. *Hum. Mol. Genet.* 7:2029–2037.

Reiner, O., Carrozzo, R., Shen, Y., Wehnert, M., Faustinella, F., Dobyns, W. B., Caskey, C. T., and Ledbetter, D. H. (1993). Isolation of a Miller-Dieker lissencephaly gene containing G protein β-subunit-like repeats. *Nature* 364:717–721.

Ross, M. E., Swanson, K., and Dobyns, W. B. (2001). Lissencephaly with cerebellar hypoplasia (LCH): a heterogeneous group of cortical malformations. *Neuropediatrics* 32(5):256–263.

Walsh, C. A. (1999). Genetic malformations of the human cerebral cortex. *Neuron* 23:19–29.

Zaki, M., Shehab, M., El-Aleem, A. A., Abdel-Salam, G., Koeller, H. B., Ilkin, Y., Ross, M. E., Dobyns, W. B., and Gleeson, J. G. (2007). Identification of a novel recessive RELN mutation using a homozygous balanced reciprocal translocation. *Am. J. Med. Genet. A* 143:939–944.

Chapter 22
The Role of Reelin in Etiology and Treatment of Psychiatric Disorders

S. Hossein Fatemi, Teri J. Reutiman, and Timothy D. Folsom

Contents

1 Introduction

There are many brain proteins that participate in the early growth and development of the mammalian central nervous system. Reelin is a glycoprotein that helps guide brain development in an orderly fashion. Changes in the level of this protein or its receptors or downstream proteins may cause abnormal corticogenesis and alter

S.H. Fatemi
Departments of Psychiatry, Pharmacology, and Neuroscience, University of Minnesota Medical School, 420 Delaware Street SE, MMC 392, Minneapolis, MN 55455
e-mail: fatem002@umn.edu

T.J. Reutiman
Department of Psychiatry, University of Minnesota Medical School, 420 Delaware Street SE, MMC 392, Minneapolis, MN 55455

T.D. Folsom
Department of Psychiatry, University of Minnesota Medical School, 420 Delaware Street SE, MMC 392, Minneapolis, MN 55455

S. H. Fatemi (ed.), *Reelin Glycoprotein: Structure, Biology and Roles in Health and Disease.*

synaptic plasticity. These changes have also been observed in a number of neuropsychiatric disorders. We will discuss more about this protein and its possible involvement in various neuropsychiatric disorders.

2 Structure of Reelin

The Reelin gene (Reln) is localized to chromosome 7 in man (DeSilva *et al.*, 1997), and its protein product has a relative molecular mass of 388 kDa (Ogawa *et al.*, 1995; D'Arcangelo *et al.*, 1995). On SDS-PAGE, Reelin appears as several protein bands, ranging from 410 to 330, 180 kDa, and several smaller fragments (Smalheiser *et al.*, 2000; Fatemi *et al.*, 2002; Lugli *et al.*, 2003; Ignatova *et al.*, 2004). Reelin is a secreted extracellular matrix protein containing 3461 amino acids (DeBergeyck *et al.*, 1998). Reelin contains a signal peptide followed by an N-terminal sequence and a hinge region upstream from eight Reelin repeats of 350–390 amino acids (DeBergeyck *et al.*, 1998). Each Reelin repeat is composed of two subrepeats separated by an EGF motif (DeBergeyck *et al.*, 1998). The Reelin protein ends with a highly basic C-terminus composed of 33 amino acids (DeBergeyck *et al.*, 1998). An epitope known as the CR-50 is localized near the N-terminus (D'Arcangelo *et al.*, 1997) and is composed of amino acids 230–346 of Reelin glycoprotein (Utsunomiya-Tate *et al.*, 2000). This epitope is essential for Reelin–Reelin electrostatic interactions that produce a soluble string-like homopolymer, composed of up to 40 or more regularly-repeated monomers, which form *in vivo* (Utsunomiya-Tate *et al.*, 2000). Mutated Reelin, which lacks a CR-50 epitope, fails to form homopolymers, and is, thereby, unable to transduce the Reelin signa1 (Utsunomiya-Tate *et al.*, 2000). Reelin binds several proteins as likely receptors including apolipoprotein E receptor 2 (ApoER2), very-low-density liproprotein receptor (VLDLR), and $\alpha 3\beta 1$ integrin protein (D'Arcangelo *et al.*, 1999; Hiesberger *et al.*, 1999; Dulabon *et al.*, 2000). Reelin binding to ApoER2 and VLDLR receptors induces clustering of the latter receptors, causing dimerization/oligomerization of the adapter protein, disabled-1 (Dab1), on the cytosolic aspect of the plasma membrane (Strasser *et al.*, 2004) with eventual tyrosine phosphorylation of Dab1 adapter protein (Cooper and Howell, 1999), facilitating the transduction of signaling pathway from the Reelin-producing cells [GABAergic neurons (Pesold *et al.*, 1990) or Cajal-Retzius cells of layer I (Fatemi *et al.*, 1999)] to downstream receptor sites on cortical pyramidal cells (Rodriguez *et al.*, 2000). *In vivo*, Reelin is processed by cleavage at two locations, i.e., between repeats 2 and 3 and repeats 6 and 7 (de Rouvroit *et al.*, 1999), resulting in three final fragments (Jossin *et al.*, 2004). The central Reelin fragment is composed of repeats 3–6 and is necessary and sufficient for receptor binding to ApoER2 and VLDLR proteins, causing Dab1 phosphorylation in neuronal cultures (Jossin *et al.*, 2004) and able to rescue the reeler phenotype in embryonic brain cultures. Furthermore, Reelin also activates serine-threonine kinases (p35/Cdk5) and Src-tyrosine kinase family (Fyn-kinase), also leading to phosphorylation of Dab1 (Keshvara *et al.*, 2001; Beffert *et al.*, 2002; Arnaud *et al.*, 2003 a,b). Phosphorylated Dab1 can become the substrate for various kinases, leading to a number of important events, such as synaptic and dendritic spine plasticity (Rodriguez *et al.*, 2000), neurotransmission (Keshvara *et al.*, 2001; Beffert *et al.*, 2002; Arnaud *et al.*, 2003a,b), and inhibition of the level of glycogen synthase-kinase 3β (GSK3β), leading to modulation of pathways of cell survival and growth (Beffert *et al.*, 2002) (Fig. 22.1). Additionally, phosphorylated Dab1 is a substrate for

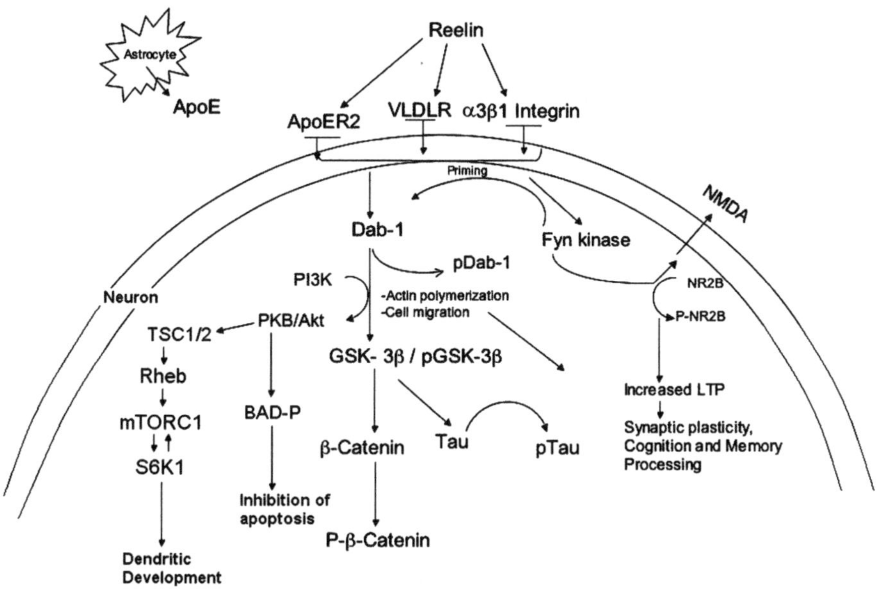

Fig. 22.1 The Reelin signaling system and cognition. Extracellular Reelin glycoprotein is secreted by Cajal-Retzius cells and certain cortical and hippocampal GABAergic cells and cerebellar granule cells. Reelin can bind its receptors ApoER2, VLDLR, and α3β1 integrin directly, initiating the signaling system in the effector cells, i.e., cortical pyramidal cells. Reelin induction of the cascade leads to clustering of the receptors causing dimerization/oligomerization of Dab1 protein and activation of Src-tyrosine kinase family/Fyn-kinase leading to tyrosine phosphorylation of Dab1 protein in a positive-feedback loop. Interaction between Dab1, N-WASP, and ARP 2/3 complex, causes formation of microspikes or filopodia which are important in processes of cell migration and synaptic plasticity. Finally, phosphorylation of a subpopulation of Dab1 molecules causes degradation of Dab1 via ubiquitination, resulting in termination of Reelin signaling cascade. Downstream effector proteins involved in Reelin signaling path include phosphatidylinositol- 3-kinase (PI3K) and protein kinase B (PKB/Akt), which further impact on three other important molecules, glycogen synthase kinase (GSK3β), β-catenin, and tau. Activation of Akt causes phosphorylation of BAD at serine 136 which leads to inhibition of apoptosis. Activation of PI3K and Akt following Dab1 phosphorylation leads to activation of mTor-S6K1 pathway which results in dendritic development. The latter proteins can modulate pathways, affecting cell proliferation, apoptosis, and neurodegeneration, respectively. Finally, Reelin has a direct effect on enhancement of long-term potentiation (LTP), via direct involvement of its receptors VLDLR and ApoER2. Alternately, tyrosine phosphorylation of NR2B subunit of NMDA receptor by Fyn kinase is essential for induction of LTP and modulation of synaptic plasticity, potentially converging on Reelin's role in cognition and memory processing (Fatemi, 2005; Jossin and Goffinet, 2007; Ohkubo *et al.*, 2007). [Modified from Fatemi, S. H. (2005). Reelin glycoprotein in autism and schizophrenia. *Int. Rev. Neurobiol.* 71:179–187]

polyubiquitination-dependent degradation leading to degradation of a subpopulation of Dab1 molecules, via the proteosome pathway (Arnaud *et al.*, 2003a). Dab1 degradation may be an important factor in fine-tuning the Reelin signal and response to it in the CNS (Arnaud *et al.*, 2003a).

Recent work by Suetsugu and co-workers (2004) explains the mechanisms through which Reelin stimulation of Dab1 affects migration of cells. Following induction of Reelin signaling system, Dab1 activates N-WASP [a neuronal type of Wiskott-Aldrich syndrome protein capable of inducing long actin microspikes (Miki *et al.*, 1998)] and stimulates actin polymerization through the Arp 2/3 complex [actin-related proteins 2 and 3, which are essential for initiation of actin assembly (Welch *et al.*, 1997)], causing formation of microspikes or filopodia. Phosphorylation of Dab1 upon Reelin stimulation and via Fyn-Src kinase mediation, causes ubiquitination of Dab1 in a Cbl-dependent manner [Casitas B lymphoma protein, a ubiquitin ligase (Arnaud *et al.*, 2003a,b; Duan *et al.*, 2004)], leading to inhibition of filopodium induction (Figure 22.1) and eventual arrest in cell migration. This mechanism may also underlie abnormal cell migration during brain development observed in the reeler mouse (Ogawa *et al.*, 1995; Arnaud *et al.*, 2003a,b) (vide infra).

Very recently, Jossin and Goffinet (2007) showed that Reelin activates the mTor (mammalian target of rapamycin)-S6K1 (S6 kinase 1) pathway following phosphorylation of Dab1 and activation of PI3K and Akt (PKB, protein kinase B). Moreover, it was proposed that PI3K helps in radial migration of cortical neurons through the intermediate zone independent of Reelin and Akt (Jossin and Goffinet, 2007). In an additional study, Ohkubo *et al.* (2007) showed that Reelin binding of ApoE receptor and activation of the PI3K/Akt pathway causes phosphorylation of Bcl2/Bcl-x associated death promoter (BAD) which helps to protect cells from apoptosis and promotes survival of mature neurons in the brain (Ohkubo *et al.*, 2007).

3 Reelin Mutant Mice

Mutation of the gene for Reelin, as seen in homozygous reeler mutant mice (Goffinet, 1979, 1984), leads to development of ataxia and a reeling gait in the affected mice. Additionally, absence of the Reln gene during embryogenesis leads to development of a brain with multiple histologic defects including a reversal of the normal layering of the brain (Falconer, 1951; Goffinet, 1984, 1992), abnormal positioning of the neurons, and aberrant orientation of cell bodies and nerve fibers (Falconer, 1951; Goffinet, 1984, 1992). The reeler cerebellum is hypoplastic (Magdaleno *et al.*, 2002) and the Purkinje cell number is reduced (Hadj-Sahraoui *et al.*, 1996). Mutations involving ApoER2, VLDLR, and $\alpha 3\beta 1$ integrin receptors result in defective cortical lamination and abnormal neuronal migration (Trommsdorff *et al.*, 1999; Dulabon *et al.*, 2000). Additionally, mice that lack either Reelin or both VLDLR and ApoER2 receptors, exhibit hyperphosphorylation of the tau protein, resulting in dysregulation of neuronal microtubule function (Hiesberger *et al.*, 1999). Several other reeler-like phenotypes have also been described which produce various neurologic phenotypes similar to the reeler homozygous mutant (for a detailed discussion see Fatemi, 2001). More interestingly, several experimental paradigms and haploinsufficiency in Reln gene in mice also cause decreases in Reelin production with resultant cortical and behavioral abnormalities (Fatemi *et al.*, 1999; Tueting *et al.*, 1999; Fatemi, 2001; Janusonis *et al.*, 2004). In the heterozygous reeler mutation, there is a 50% reduction in Reelin protein and mRNA, decrease in dendritic spine density in frontal cortex, neuropil hypoplasticity, decreased GAD67 expression, and decreased GABA turnover (Carboni *et al.*, 2004). Additionally, the heterozygous reeler mutant mice exhibit decreased prepulse inhibition (Tueting *et al.*, 1999, 2005), a phenomenon observed in schizophrenia and autism (McAlonan *et al.*, 2002; Meincke *et al.*, 2004). Prenatal human influenza viral infection in midterm pregnant mice leads to abnormal corticogenesis (Fatemi *et al.*, 1999), decrease in brain Reelin protein content (Fatemi *et al.*, 1999), and reduced prepulse inhibition (Shi *et al.*, 2003). Finally, exposure of rat pups to 5-methoxytryptamine leads to reductions in brain and blood Reelin levels and abnormal corticogenesis (Janusonis *et al.*, 2004).

4 Reelin's Presence in All Vertebrates

Reelin protein is present in all vertebrates and conserved through evolution (Tissir and Goffinet, 2003). Additionally, the wide distribution of Reelin in the adult lamprey brain is consistent with the existence of different roles for this protein not related to CNS development in the vertebrates (Perez-Costas *et al.*, 2004). For example, Reelin expression in brains of male European starlings was widely distributed in the forebrain including in areas incorporating new neurons in adulthood such as in and around the song control nucleus (Absil *et al.*, 2003). Reelin expression is highly sensitive to testosterone, decreasing markedly in response to exogenous administration of this hormone (Absil *et al.*, 2003). Thus, here, Reelin expression in the brain varies seasonally and could therefore provide a signal that could modulate the seasonal effects in the incorporation of new neurons in the song control system (Absil *et al.*, 2003). In mammals, including rodents, Reelin production begins as early as day 9.5 in the embryonic mouse brain (Ogawa *et al.*, 1995; Ikeda and Terashima, 1997). The cells synthesizing Reelin are Cajal-Retzius cells which act as pathfinding neurons that help in early laminar organization of the cortex (Ogawa *et al.*, 1995). In the adult mammalian brain, Reelin is localized to layer I cortical Cajal-Retzius cells, cortical GABAergic interneurons in layers II–IV (Impagnatiello *et al.*, 1998), cerebellar granule cells (Lacor *et al.*, 2000), and hippocampal interneurons (Fatemi *et al.*, 2000). Presence of Reelin-positive cells in the adult hippocampus indicates that Reelin function is not restricted to the embryonic period but may continue throughout adult life (Abraham and Meyer, 2003).

A recent report demonstrates coexpression of Reelin and Dab1 in Cajal-Retzius cells during cortical development, and in cortical pyramidal cells in the adult CNS (Deguchi *et al.*, 2003).

It is now clearly established that Reelin protein serves a dual purpose in mammalian brain: Embryologically, it guides neurons and radial glial cells to their correct positions in the developing brain (Forster *et al.*, 2002; Luque *et al.*, 2003). After the fetal phase of brain development, levels of Reelin begin to decrease, reaching a plateau by late childhood and remaining constant thereafter in mice (M. Araghi-Niknam and S.H. Fatemi, unpublished data). Moreover, Reelin is largely replaced by Reelin-expressing GABAergic interneurons that are dispersed throughout the mammalian neocortex (Impagnatiello *et al.*, 1998) and hippocampus (Fatemi *et al.*, 2000; Abraham and Meyer, 2003). Levels of the Reelin receptors ApoER2, VLDLR, and α3β1 integrin and the adapter protein Dab1, which are all essential to the Reelin signaling system, remain expressed in adult brain (Abraham and Meyer, 2003).

5 Reelin and Its Receptors

Previous work by Rodriguez *et al.* (2000) showed an association between Reelin and its receptor α3β1 integrin with synaptic structures, raising the possibility of a potential role in neurotransmission. Recent reports by J. Herz's laboratory (Weeber *et al.*, 2002; Herz and Chen, 2006) show that Reelin has a direct effect on enhancement of long-term potentiation (LTP) in hippocampus which is abolished when hippocampus slice cultures are used from VLDLR and ApoER2 knockout mice lacking the receptors for Reelin. These investigators further report that Reelin and ApoE receptors cooperate to enhance hippocampal synaptic plasticity and learning (Weeber *et al.*, 2002). Moreover, mice that lack the Reelin receptors ApoER2 or VLDLR have pronounced defects in memory formation and LTP (Weeber *et al.*, 2002). More recent work by Beffert *et al.* (2005, 2006) has further demonstrated the importance of ApoER2 on LTP. An amino acid sequence encoded by an exon on the intracellular domain of ApoER2 is required for Reelin-induced tyrosine phosphorylation of NMDA receptor subunits (Beffert *et al.*, 2005). Mice lacking this sequence performed poorly in learning and memory tasks (Beffert *et al.*, 2005). Beffert *et al.* (2006) further demonstrated that mice that have ApoER2 lacking a sequence required for interaction with Dab1 similarly have abnormalities in LTP and behavior which are different still from knockout mice (Beffert *et al.*, 2006).

A recent report by Barr *et al.* (2007) showed the importance of VLDLR and ApoER2 in regulating sensorimotor gating in mice. VLDLR knockout mice revealed deficits in a crossmodal PPI task involving the presentation of acoustic and tactile stimuli while ApoER2 heterozygous and knockout mice showed significant increased crossmodal PPI (Barr *et al.*, 2007).

6 Reelin's Role in Psychiatric Disorders

Several studies now implicate the pathological involvement of Reelin gene or its protein product in six neuropsychiatric disorders, namely, schizophrenia (Impagnatiello *et al.*, 1998; Fatemi *et al.*, 2000; Guidotti *et al.*, 2000; Costa *et al.*, 2003a; Eastwood *et al.*, 2003; Eastwood and Harrison, 2003; Abdolmaleky *et al.*, 2004; Knable *et al.*, 2004; Fatemi *et al.*, 2005a; Wedenoja *et al.*, 2007), autism (Persico *et al.*, 2001; Zhang *et al.*, 2002; Fatemi *et al.*, 2002, 2004, 2005b; Lugli *et al.*, 2003), bipolar disorder (Impagnatiello *et al.*, 1998; Fatemi *et al.*, 2000; Guidotti *et al.*, 2000; Knable *et al.*, 2004), major depression (Fatemi *et al.*, 2001; Knable *et al.*, 2004), lissencephaly (Hong *et al.*, 2000; Miyata *et al.*, 2003), and Alzheimer's disease (Saez-Valero *et al.*, 2003; Botella-Lopez *et al.*, 2006).

7 Reelin in Schizophrenia, Bipolar Disorder, and Major Depression

Neuroanatomical studies have shown Reelin abnormalities throughout the brain in patients with schizophrenia, bipolar disorder, and major depression. Reelin expression has consistently been shown to be decreased in all three disorders. Impagnatiello *et al.* (1998) used Northern and Western blotting and immunocytochemistry to show reductions in Reelin mRNA and protein in cerebellar, hippocampal, and frontal cortices of patients with schizophrenia and psychotic bipolar disorder. These authors suggested that Reelin might be a vulnerability factor in development of psychosis (Impagnatiello *et al.*, 1998). Later, Guidotti *et al.* (2000) confirmed and extended these observations in postmortem frontal cortex of additional subjects with schizophrenia and psychotic bipolar disorder. Reduction in Reelin was associated with a concurrent decrease in glutamic acid decarboxylase 67-kDa (GAD67) protein in the same postmortem brains (Guidotti *et al.*, 2000). A later immunocytochemical report (Fatemi *et al.*, 2000) showed significant reductions in Reelin immunoreactivity in the hippocampus of patients with schizophrenia, nonpsychotic bipolar disorder, and major depression, suggesting that Reelin deficiency may not be limited to subjects with psychosis alone (Fatemi *et al.*, 2000). Knable *et al.* (2004) analyzed molecular abnormalities of the hippocampus in severe psychiatric illness and reconfirmed that GABAergic marker Reelin was decreased in schizophrenia, bipolar disorder, and depression, attesting to reported GABAergic dysfunction in all three disorders. Fatemi *et al.* (2005a) subsequently demonstrated significant reductions in Reelin, as well as GAD65 and 67-kDa proteins, in cerebella of subjects with schizophrenia, bipolar disorder, and major depression, providing further evidence that GABAergic dysfunction is apparent in these disorders (Fatemi *et al.*, 2005a). The cerebellar decreases in GAD65 and 67-kDa proteins have been replicated in brains of subjects with schizophrenia by the laboratory of N. Perrone-Bizzozero (Bullock *et al.*, 2006), asserting the biological importance of both enzymes in the pathology of psychiatric disorders. Eastwood *et al.* (2003), also showed a trend for reduction in Reelin mRNA in cerebella of subjects with schizophrenia. Interestingly, these reductions in Reelin mRNA correlated negatively with expression of semaphorin 3A mRNA (Eastwood *et al.*, 2003). The authors suggested that these findings were consistent with an early neurodevelopmental origin for schizophrenia, and that the reciprocal changes in Reelin and semaphorin 3A may be indicative of a mechanism that affects the balance between inhibitory and trophic factors regulating synaptogenesis (Eastwood *et al.*, 2003). Eastwood and Harrison extended their work to superior temporal cortex and discovered reductions in Reelin mRNA in interstitial white matter neurons (cells representing the adult remnants of the cortical subplate) in brains of subjects with schizophrenia, supporting the contention that the origins of schizophrenia may be neurodevelopmental (Eastwood and Harrison, 2003).

Recent evidence indicates that hypermethylation of the Reelin gene promoter may be responsible for decreased expression of Reelin in brains of subjects with schizophrenia (Costa *et al.*, 2003a; Abdolmaleky *et al.*, 2004). Costa and co-workers have posited the opinion that alterations in chromatin remodeling related to a selective upregulation of DNA-5-cytosine methyltransferase (DNMT) expression in GABAergic neurons of schizophrenic prefrontal cortex may induce a hypermethylation of Reelin and GAD67 promoter CpG islands, which subsequently downregulate their expression (Costa *et al.*, 2003a). These authors suggest that targeting this deficit with inhibitors of histone deacetylases (HDAC) may reduce the DNMT upregulation via covalent modification of nucleosomal histone tails, potentially upregulating Reelin expression in schizophrenic brain (Costa *et al.*, 2003a,b; Abdolmaleky *et al.*, 2004). Indeed, Veldic *et al.* (2004) have recently shown that mRNA for DNA-methyltransferase 1, which catalyzes the methylation of promoter CpG islands, is increased in cortical GABAergic interneurons but not in pyramidal neurons of schizophrenic brains. More recently, Dong *et al.* (2007) demonstrated in mice that following treatment with L-methionine, which increases RELN and GAD67 promoter cytosine-5-hypermethylation, treatment with HDACs valproate and MS-275 dramatically accelerated RELN and GAD67 promoter demethylation (Dong *et al.*, 2007). This suggests the possibility that RELN reactivation may take place due to antipsychotic-induced demethylation (Dong *et al.*, 2007).

Despite these biochemical findings, two recent reports failed to find any association between Reln gene polymorphisms and schizophrenia (Akahane *et al.*, 2002; Chen *et al.*, 2002). Akahane *et al.* examined the polymorphic CGG repeat in the 5´ untranslated region of the Reln gene in 150 schizophrenic and 150 controls matched for age, sex, and ethnicity and found no evidence for any significant association of schizophrenia with polymorphisms for Reln or VLDLR genes (Akahane *et al.*, 2002). By the same token, Chen *et al.* studied a single nucleotide polymorphism at the 5′ promoter region of the human Reln gene in 279 Han Chinese schizophrenic patients and 255 controls and could not demonstrate any significant associations in the Reln gene polymorphisms and schizophrenia (Chen *et al.*, 2002).

The lack of conclusive results associating RELN, among other genes, with schizophrenia has led to the study of quantifiable traits and endophenotypes (Gottesman and Gould, 2003). A study by Wedenoja *et al.*, while demonstrating a lack of association between RELN and a clinical diagnosis of schizophrenia in a Finnish population sample, found that RELN was associated with a number of cognitive traits known to be affected in schizophrenia (Wedenoja *et al.*, 2007). Specifically, they found associations between a RELN intragenic microsatellite marker (RELNSAT6) and attention and working memory, verbal learning and memory, and executive function (Wedenoja *et al.,* 2007). These findings are consistent with RELN's role in synaptic plasticity (Costa *et al.*, 2002; Fatemi, 2005), which is crucial for cognitive abilities (Garlick, 2002).

8 Reelin in Autism

8.1 Brain Abnormalities

In a series of postmortem studies, Fatemi *et al.* (2001, 2004, 2005b) demonstrated reductions in Reelin protein in several brain sites in autism. Brain levels of Reelin 410 kDa were reduced significantly in frontal (Area 9) and cerebellar areas and nonsignificantly in parietal (Area 40) cortex of autistic subjects versus controls. There was also a trend for reduction in Reelin 410 kDa in autistic children, indicating that the reduced Reelin levels were present from childhood (Fatemi *et al.*, 2004). Brain levels of Reelin 330 kDa were reduced significantly in Area 9 and nonsignificantly in Area 40 and cerebellum (Fatemi *et al.*, 2005b). Brain levels of Reelin 180 kDa were reduced significantly in cerebellum and Area 9 and nonsignificantly in Area 40 (Fatemi *et al.*, 2001, 2005b). These results are significant as pathologic findings of the brain are prevalent throughout the brain of subjects with autism including the frontal, parietal, and cerebellar cortices (Palmen *et al.*, 2004).

8.2 Blood Abnormalities

Smalheiser *et al.* (2000) identified three Reelin bands in rat mouse and human blood but these bands were absent in blood from homozygous reeler (*rl/rl*) mice while heterozygous reeler (*rl*/+) mice expressed half as much as wildtype (Smalheiser *et al.*, 2000). Measurement of blood Reelin levels in subjects with autism showed reductions in 410- and 330-kDa species in the subjects with autism versus matched controls (Fatemi *et al.*, 2002). Janusonis *et al.* (2004) found that brain and blood levels of Reelin were decreased in newborn pups (postnatal day 0) born to mice that had been treated with 5-methoxytryptamine. This result suggests that disruption of the serotonergic system, which is known to be altered in autism (Chugani, 2002), produces altered Reelin expression (Janusonis *et al.*, 2004). Finally, Lugli *et al.* (2003) confirmed Fatemi's work, showing significant reductions in 330-kDa plasma protein in a selected group of autistic subjects.

8.3 Genetic Polymorphisms

These biochemical data are bolstered by two association studies showing significant linkage between Reln gene polymorphisms and autism (Persico *et al.*, 2001;

Zhang *et al.*, 2002). Recently, Persico *et al.* (2001) described a significant association between autism and Reln gene variants using case–control and family-based designs. They showed a significant association between autistic disorder and the length of a polymorphic GGC repeat located immediately 5' of the Reln gene ATG initiation codon. A further link to autism was also established for specific haplotypes defined by single-base substitutions located in a splice junction of exon 6 and within the coding sequence of exon 50 (Persico *et al.*, 2001). These investigators also showed preferential transmission of "long" triplet repeat alleles (i.e., >11 repeats) to autistic patients and correlated this phenomenon with decreases in blood Reelin 330-kDa levels in the autistic offspring (Lugli *et al.*, 2003). These authors concluded that transmission of "long" alleles from either parent significantly enhanced the overall probability of a child being affected by autism (Persico *et al.*, 2001; Lugli *et al.*, 2003).

In a subsequent report, Zhang *et al.* (2002) did not observe any evidence for expansion or instability of transmission of GGC repeats in the autistic subjects but were able to confirm, using a family-based association test, that larger alleles were transmitted higher than expected in the affected children indirectly supporting the work of Persico *et al.* (2001). In contrast, four reports failed to detect any genetic linkage between Reln gene polymorphisms and autism (Krebs *et al.*, 2002; Bonora *et al.*, 2003; Devlin *et al.*, 2004; Li *et al.*, 2004). Krebs *et al.* (2002) performed a transmission disequilibrium test analysis of the 5' UTR polymorphism in 167 families, including 218 affected subjects, and could not show any association between this GGC polymorphism of the Reln gene and autism in a population of mixed European descent. Bonora *et al.* (2003), using a positional candidate gene approach, found novel missense variants in the Reln gene with low frequency but could not support a major role for Reln in autism in IMGSAC and German singleton families. Devlin *et al.* (2004) used a large independent family-based sample from the NIH Collaborative Programs of Excellence in Autism (CPEA) Network and could not find any significant association between Reln gene alleles and autism. Finally, Li *et al.* (2004) also could not find any evidence for an association between WNT2 and Reln polymorphisms and autism. However, these authors (Li *et al.*, 2004) felt that "association studies of DNA variations are often ineffective in addressing functional alteration of gene products at the level of gene expression" and suggested additional biochemical studies of brain and blood products to further assess the involvement of the Reln gene in autism. Persico *et al.* have demonstrated using *in vitro* studies that the long triplet GGC repeats blunted Reln gene expression (Persico *et al.*, 2006). The authors suggest that this may account for the decreased Reelin expression in subjects with autism (Persico *et al.*, 2006). Finally, Serajee *et al.*, in a study of 34 Reln single nucleotide polymorphisms (SNP) in a sample of Caucasian families, found an association of autism with a C/T SNP in intron 59 of the Reln gene (Serajee *et al.*, 2005).

Despite the controversial nature of genetic association studies, Rakic and co-workers (Janusonis *et al.*, 2004) have developed a potential animal model for autism which links prenatal serotonergic abnormalities to reduced brain and blood Reelin levels and abnormal brain development, indicating the relevance of biochemical/neuroanatomic studies pertaining to the Reelin signaling system in autism.

Other behavioral and biochemical data also show that reductions in levels of Reelin in brain or blood, following postnatal hypoxia (Curristin *et al.*, 2002), prenatal viral infection in midgestation (Fatemi *et al.*, 1999; Shi *et al.*, 2003), immunological challenge with Poly I:C (a viral mimic) (Meyer *et al.*, 2007) and in heterozygous reeler mutants (Tueting *et al.*, 1999) cause abnormalities in behavior such as decrease in prepulse inhibition (PPI), increase in anxiety, and decrease in memory formation. Additionally, mutations in the RELN gene have been associated with significant learning disability, hypoplastic cerebellum, ataxia, and cognitive decline in man and mouse (Goffinet, 1992).

9 Reelin in Lissencephaly

Reelin mutations have also been discovered in a variant of lissencephaly, whereby the affected individuals have very low or undetectable levels of Reelin in their sera (Hong *et al.*, 2000; Chang *et al.*, 2007). Both Chang *et al.* (2007) and Hong *et al.* (2000) reported that the affected children exhibited congenital lymphedema and hypotonia with brain showing moderate lissencephaly and profound cerebellar hypoplasia. Miyata *et al.* (2003), in a study of a 7-day-old neonate born at 38 gestational weeks with lissencephaly with cerebellar hypoplasia, found altered Reelin expression. Assadi *et al.* (2003) developed compound mutant mice, with disruptions in the Reln gene and PAFAH1B1 (encoding LIS1), which exhibited a higher incidence of hydrocephalus and enhanced cortical and hippocampal layering defects, implicating involvement of both genes in normal brain development.

10 Reelin in Alzheimer's Disease

Finally, Saez-Valero *et al.* (2003) measured Reelin 180-kDa levels in CSF of 13 healthy controls, 14 fronto-temporal dementia, and 20 Alzheimer's disease patients. They reported significant increases in CSF 180-kDa Reelin species in both dementias versus controls, suggesting the involvement of Reelin in neurodegenerative disorders (Saez-Valero *et al.*, 2003). More recently, Botella-Lopez *et al.* (2006) found a significant increase in CSF 180-kDa Reelin, and a significant increase in 180-kDa Reelin and total Reelin in frontal cortex in subjects with Alzheimer's disease (Botella-Lopez *et al.*, 2006). Additionally, they found an increase in Reelin/GAPDH mRNA in frontal cortex (Botella-Lopez *et al.*, 2006). In contrast, Ignatova *et al.* (2004) measured CSF Reelin in adults and children and found no correlation with age or neurologic disease (Alzheimer's dementia, multiple sclerosis). However, the latter investigators used a scoring technique which was semiquantitative and had a smaller *N* for each patient population (Ignatova *et al.*, 2004).

11 Effects of Psychotropic Medications on Reelin Expression in Rat Brain

Our laboratory has investigated whether chronic administration of psychotropic medications (clozapine, fluoxetine, haloperidol, lithium, olanzapine, and valproic acid) used in the treatment of psychiatric disorders (schizophrenia, major depression, bipolar disorder, etc.) alters mRNA and protein levels for Reelin in rat frontal cortex (FC). FC of drug-treated rats (21 days of intraperitoneal injections) versus saline-treated controls were subjected to SDS-PAGE and Western blotting. Additionally, rat FC mRNAs were also subjected to qRT-PCR. Levels of Reelin were significantly altered in several drug-treated rat FC groups versus controls. These data suggest that changes are due to the psychotropic medications, and that the changes in Reelin expression may help explain the efficacy of these drugs.

11.1 Results

We measured protein levels for Reelin using SDS-PAGE and Western blotting and mRNAs by qRT-PCR for each of the six drug-treatment groups. Reelin molecules appeared on SDS-PAGE as multimeric bands ranging from ~410 to ~330 to ~180 kDa. All values were normalized against β-actin. There were no significant differences in levels of β-actin in the drug-treated brains versus controls.

11.1.1 Clozapine

In clozapine-treated rat FC, Reelin protein showed significant downregulation of the 410- and 180-kDa isoforms ($p=0.0024$ and 0.0099, respectively) while the 330-kDa isoform showed a nonsignificant downregulation. Reln mRNA was significantly upregulated ($p=0.0006$) versus controls (Fig. 22.2).

11.1.2 Fluoxetine

Fluoxetine-treated rat FC showed nonsignificant downregulation of all three isoforms of Reelin protein. Reln mRNA was significantly upregulated ($p=0.0003$) versus controls (Fig. 22.3).

11.1.3 Haloperidol

Reelin protein showed nonsignificant downregulation of the 410- and 330-kDa isoforms while the 180-kDa isoform was significantly downregulated ($p=0.044$) in

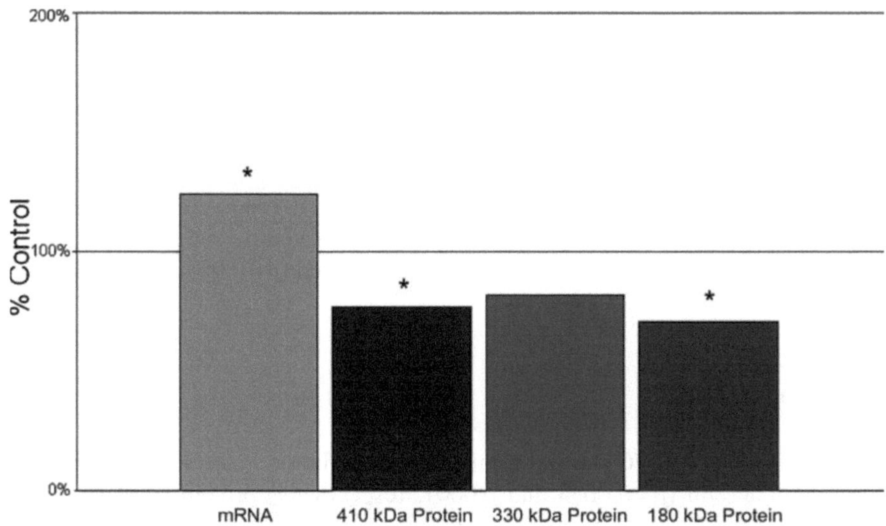

Fig. 22.2 The impact of clozapine on rat brain levels of Reelin. In clozapine-treated rat FC, Reelin protein showed significant downregulation of the 410- and 180-kDa isoforms while Reln mRNA was significantly upregulated versus controls (*See Color Plates*)

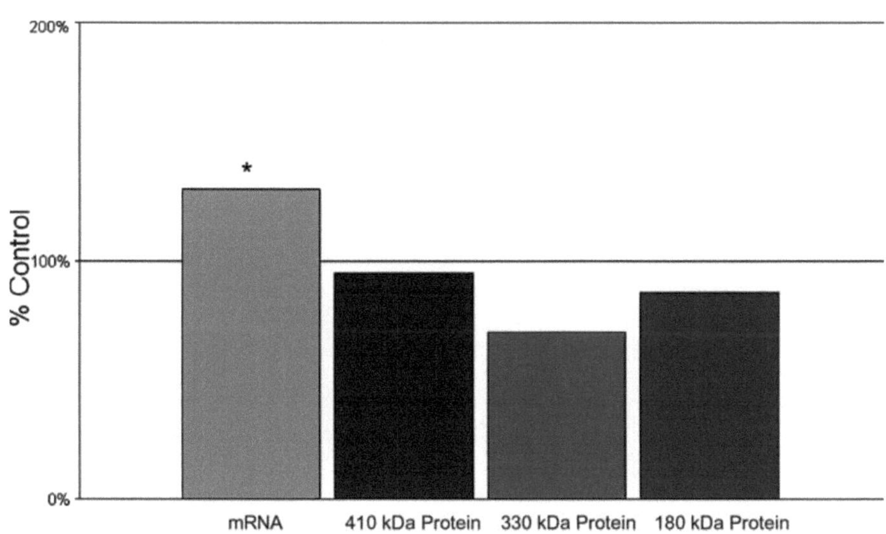

Fig. 22.3 The impact of fluoxetine on rat brain levels of Reelin. Reln mRNA was significantly upregulated in fluoxetine-treated rat FC versus controls (*See Color Plates*)

haloperidol-treated rat FC. Reln mRNA level was significantly downregulated in haloperidol versus control rat FC ($p \leq 0.0001$) (Fig. 22.4).

11.1.4 Lithium

Reelin showed nonsignificant downregulation of the 410- and 330-kDa isoforms in lithium-treated FC, while the 180-kDa isoform was significantly downregulated ($p = 0.0066$). Reln mRNA was significantly upregulated ($p = 0.0047$) in lithium versus control rat FC (Fig. 22.5).

11.1.5 Olanzapine

Olanzapine-treated rat FC showed significant upregulation of the 410- and 180-kDa isoforms of Reelin ($p = 0.0033$ and 0.0001, respectively) proteins (Fatemi *et al.*, 2006), while the 330-kDa protein isoform was nonsignificantly upregulated. Reln mRNA was significantly upregulated ($p = 0.0259$) (Fig. 22.6).

11.1.6 Valproic Acid

The 410- and 330-kDa isoforms of Reelin showed nonsignificant upregulation, while the 180-kDa isoform showed nonsignificant downregulation in VPA-treated

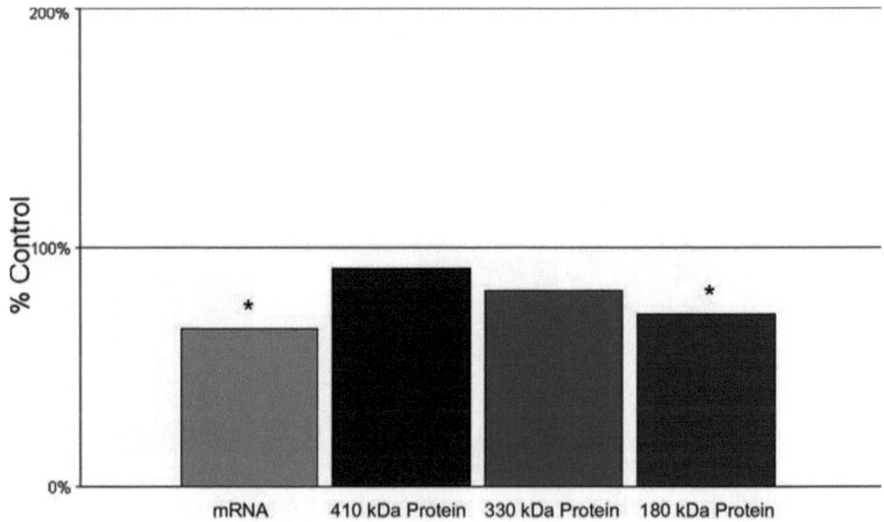

Fig. 22.4 The impact of haloperidol on rat brain levels of Reelin. Reelin protein showed the 180-kDa isoform was significantly downregulated as was Reln mRNA level in haloperidol versus control rat FC (*See Color Plates*)

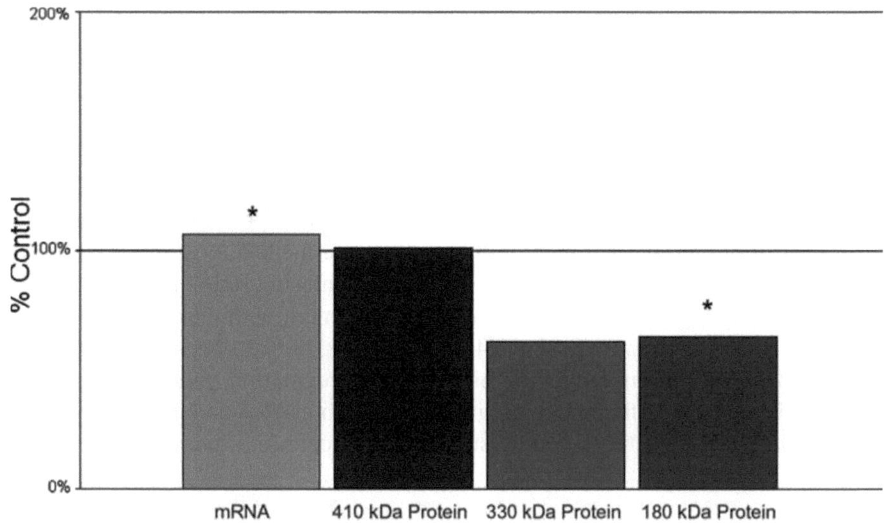

Fig. 22.5 The impact of lithium on rat brain levels of Reelin. The 180-kDa isoform of Reelin was significantly downregulated following chronic treatment with lithium. In contrast, Reln mRNA was significantly upregulated in lithium versus control rat FC (*See Color Plates*)

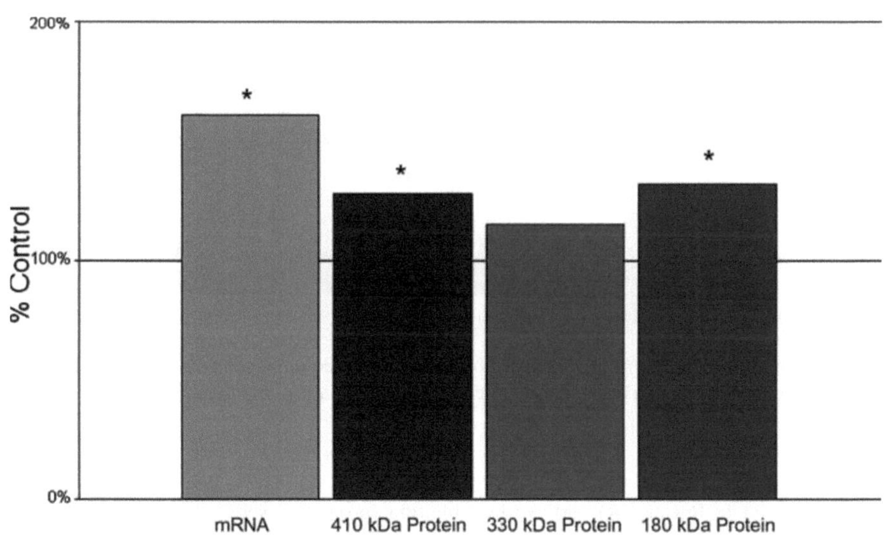

Fig. 22.6 The impact of olanzapine on rat brain levels of Reelin. Olanzapine-treated rat FC showed significant upregulation of the 410- and 180-kDa isoforms of Reelin. Reln mRNA was also significantly upregulated (*See Color Plates*)

versus control rat FC. Reln mRNA was significantly downregulated ($p<0.0001$) (Fig. 22.7).

11.1.7 Summary

Table 22.1 summarizes the changes in Reelin mRNA and protein expression in rat FC as a result of chronic treatment with psychotropic medications. Importantly, all of the drugs tested altered Reln mRNA with each increasing Reln mRNA except for haloperidol and valproic acid. In contrast, only treatment with olanzapine led to an increase in Reelin protein for both the 410- and 180-kDa isoforms, while clozapine, haloperidol, and lithium led to significant downregulation. Additionally, only haloperidol caused significant downregulation in both mRNA and protein levels for Reelin. Taken together, these results suggest that Reelin is a target of commonly

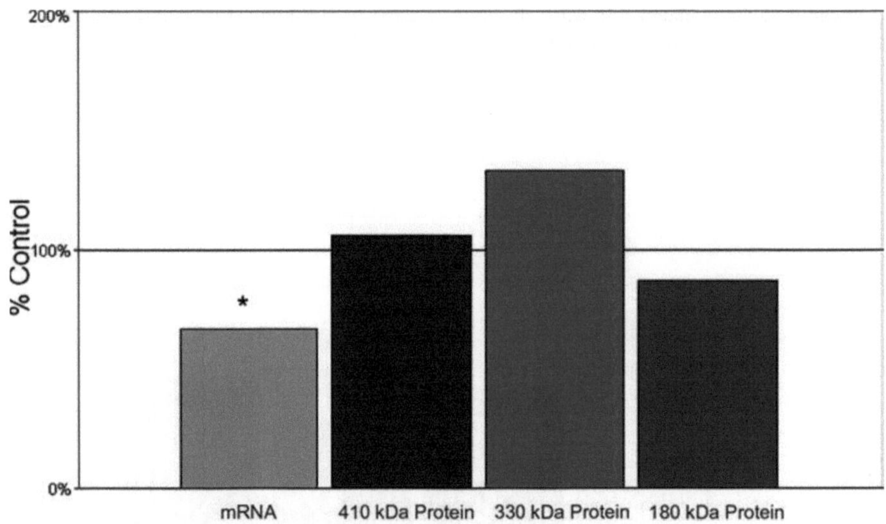

Fig. 22.7 The impact of valproic acid on rat brain levels of Reelin. Reln mRNA was significantly downregulated in rat FC as a result of treatment with VPA (*See Color Plates*)

Table 22.1 qRT-PCR (mRNA) and Western blotting (protein) results on drug-treated versus control rat FC

		Clozapine	Fluoxetine	Haloperidol	Lithium	Olanzapine	VPA
Reelin	Protein	↓*	nc	↓*	↓*	↑*	nc
	mRNA	↑*	↑*	↓*	↑*	↑*	↓*

* $p<0.05$; nc, no change.

used psychotropic drugs and altered expression of Reelin may partly explain their efficacies.

12 Conclusions

Altered Reelin expression is associated with a number of psychiatric disorders. Reelin expression is increased in cerebrospinal fluid (Saez-Valero *et al.*, 2003) and in frontal cortex (Botella-Lopez *et al.*, 2006) in subjects with Alzheimer's disease. More commonly, however, Reelin expression is decreased in various psychiatric disorders. Reelin deficiency has been observed in schizophrenia, bipolar disorder, major depression, autism, and lissencephaly. In brains from subjects with schizophrenia, reduced Reelin expression has been observed in cerebellum (Impagnatiello *et al.*, 1998; Eastwood *et al.*, 2003; Fatemi *et al.*, 2005a), frontal cortex (Impagnatiello *et al.*, 1998; Guidotti *et al.*, 2000), hippocampus (Impagnatiello *et al.*, 1998; Fatemi *et al.*, 2000), and superior temporal cortex (Eastwood and Harrison, 2003). Decreased Reelin expression in bipolar disorder has been observed in hippocampus (Impagnatiello *et al.*, 1998; Fatemi *et al.*, 2000; Knable *et al.*, 2004), cerebellum (Impagnatiello *et al.*, 1998; Fatemi *et al.*, 2005a), and frontal cortex (Impagnatiello *et al.*, 1998; Guidotti *et al.*, 2000). Subjects with major depression display reduced Reelin in cerebellum (Fatemi *et al.*, 2005a) and hippocampus (Knable *et al.*, 2004; Fatemi *et al.*, 2005a). In autism, decreased Reelin was observed in frontal cortex and cerebellum (Fatemi *et al.*, 2005). Decreased Reelin has also been observed in blood (Persico *et al.*, 2001; Fatemi *et al.*, 2002) with Persico *et al.* (2001) correlating this decrease in blood Reelin with the preferential transmission of "long" triplet repeat alleles of Reln (Lugli *et al.*, 2003). Reduced Reelin expression has been observed in brain (Miyata *et al.*, 2003) and blood (Hong *et al.*, 2000) in lissencephaly.

Various mechanisms may be operational in these neuropsychiatric disorders where Reelin production may be affected selectively by various mutations or selective hypermethylation of the Reelin gene promoter (Costa *et al.*, 2003a; Abdolmaleky *et al.*, 2004), causing either profound (schizophrenia, autism, lissencephaly) or moderate (bipolar disorder, major depression) cognitive deficits (Impagnatiello *et al.*, 1998; Fatemi *et al.*, 2000; Guidotti *et al.*, 2000; Hong *et al.*, 2000) associated with their respective Reelin levels. The overall picture emerging from these reports suggests that Reelin deficiency may be associated not only with vulnerability to developing psychosis, but also to the development of cognitive dysfunction as clinical symptoms often observed in various neuropsychiatric disorders, such as bipolar disorder (Impagnatiello *et al.*, 1998; Guidotti *et al.*, 2000), major depression (Fatemi *et al.*, 2000), autism (Fatemi *et al.*, 2002, 2005a), and lissencephaly (Hong *et al.*, 2000). This hypothesis is also supported by animal studies (Rodriguez *et al.*, 2000) linking Reelin–integrin interactions with synaptic plasticity. Association of ApoER2 and LDL receptor family with Reelin protein may also link certain neurodegenerative disorders, such as Alzheimer's dementia (Bothwell and Giniger, 2000;

Helbecque and Amouyel, 2000) with dysregulation of the Reelin signaling system.

In summary, Reelin glycoprotein acts as a protease both during embryogenesis and in the adult brain. Absence of Reelin during development leads to abnormal corticogenesis, Purkinje cell loss, and ataxia. Reductions in levels of Reelin during adult life may cause cognitive deficits, as seen in autism, schizophrenia, bipolar disorder, and lissencephaly. Moreover, Reelin is involved in a signaling pathway that underlies memory formation, LTP, and synaptic plasticity. Thus, Reelin's role in early growth and development of the central nervous system makes it an important candidate gene in investigating psychiatric disorders which are associated with gross morphological changes in brain and/or cognitive deficits. Reelin may also have other undefined roles in health and disease because of its presence in diverse areas of the body. Future biochemical, genetic, and neuroanatomic studies will surely expand our knowledge about this important protein and determine its involvement in various neurodevelopmental disorders.

Acknowledgments Parts of this chapter were adapted from Dr. Fatemi's previous publications on Reelin (such as Fatemi, 2001, 2002, 2005). The work of Dr. Fatemi has been supported by the National Institutes of Health (#1R01 HD046589-01A2 and 1R01HD052074-01A2), Stanley Medical Research Institute, March of Dimes, The Jonty Foundation, and the Kunin Fund of St. Paul Foundation. I am grateful for secretarial assistance by Ms. Laurie Iversen and Danielle Johansson. I would like to thank the Harvard Brain Tissue Research Center, University of Miami Brain Endowment Bank, and University of Maryland Brain Bank for providing tissues for my research.

References

Abdolmaleky, H. M., Smith, C. L., Farone, S. V., Shafa, R., Stone, W., Glatt, S. J., and Tsuang, M. T. (2004). Methylomics in psychiatry: modulation of gene–environment interactions may be through DNA methylation. *Am. J. Med. Genet.* 127B:51–59.

Abraham, H., and Meyer, G. (2003). Reelin-expressing neurons in the postnatal and adult human hippocampal formation. *Hippocampus* 13:715–727.

Absil, P., Pinxten, R., Balthazart, J., and Eens, M. (2003). Effects of testosterone on reelin expression in the brain of male European starlings. *Cell Tissue Res.* 312:81–98.

Akahane, A., Kunugi, H., Tanaka, H., and Nanko, S. (2002). Association analysis of polymorphic CGG repeat in 5' UTR of the reelin and VLDLR genes with schizophrenia. *Schizophr. Res.* 58(1):37–41.

Arnaud, L., Ballif, B. A., and Cooper, J. A. (2003a). Regulation of protein tyrosine kinase signalling by substrate degradation during brain development. *Mol. Cell Biol.* 23:9293–9302.

Arnaud, L., Ballif, B. A., Forster, E., and Cooper, J. A. (2003b). Fyn tyrosine kinase is a critical regulator of disabled-1 during brain development. *Curr. Biol.* 13:9–17.

Assadi, A. H., Zhang, G., Beffert, U., McNeil, R. S., Renfro, A. L., Niu, S., Quattrocchi, C. C., Antalffy, B. A., Sheldon, M., Armstrong, D. D., Wynshaw-Boris, A., Herz, J., D'Arcangelo, G., and Clark, G. D. (2003). Interaction of reelin signaling and Lis1 in brain development. *Nature Genet.* 35:270–276.

Barr, A. M., Fish, K. N., and Markou, A. (2007). The reelin receptors VLDLR and ApoER2 regulate sensorimotor gating in mice. *Neuropharmacology* 52:1114–1123.

Beffert, U., Morfini, G., Bock, H. H., Reyna, H., Brady, S. T., and Herz, J. (2002). Reelin-mediated signaling locally regulates protein kinase B/Akt and glycogen synthase kinase 3β. *J. Biol. Chem.* 51:49958–49964.

Beffert, U., Weeber, E. J., Durudas, A., Qiu, S., Masiulis, I., Sweatt, J. D., Li, W. P., Adelmann, G., Frotscher, M., Hammer, R. E., and Herz, J. (2005). Modulation of synaptic plasticity and memory by reelin involves differential splicing of the lipoprotein receptor Apoer2. *Neuron* 47:567–579.

Beffert, U., Durudas, A., Weeber, E. J., Stolt, P. C., Giehl, K. M., Sweatt, J. D., Hammer, R. E., and Herz, J. (2006). Functional dissection of reelin signaling by site-directed disruption of disabled-1 adaptor binding to apolipoprotein E receptor 2: distinct roles in development and synaptic plasticity. *J. Neurosci.* 26:2041–2052.

Bonora, E., Beyer, K. S., Lamb, J. A., Parr, J. R., Klauck, S. M., Benner, A., Paolucci, M., Abbott, A., Ragoussis, I., Poustka, A., Bailey, A. J., and Monaco, A. P., and the International Molecular Genetic Study of Autism Consortium (IMGSAC). (2003). Analysis of reelin as a candidate gene for autism. *Mol. Psychiatry* 10:885–892.

Botella-Lopez, A., Burgaya, F., Gavin, R., Garcia-Ayllon, M. S., Gomez-Tortosa, E., Pena-Casanova, J., Urena, J. M., Del Rio, J. A., Blesa, R., Soriano, E., and Saez-Valero, J. (2006). Reelin expression and glycosylation patterns are altered in Alzheimer's disease. *Proc. Natl. Acad. Sci. USA* 103:5573–5578.

Bothwell, M., and Giniger, E. (2000). Alzheimer's disease: neurodevelopment converges with neurodegeneration. *Cell* 102:271–273.

Bullock, W. M., Paz, R. D., Roberts, R. C., Andreasen, N. C., and Perrone-Bizzozero, N. I. (2006). Gene expression alterations in the cerebellum of patients with schizophrenia revealed by DNA microarray analysis. *J. Neurochem.* 96(suppl 1):34.

Carboni, G., Tueting, P., Tremolizzo, L., Sugaya, I., Davis, J., Costa, E., and Guidotti, A. (2004). Enhanced dizocilpine efficacy in heterozygous reeler mice relates to GABA turnover down-regulation. *Neuropharmacology* 46:1070–1081.

Chang, B. S., Duzcan, F., Kim, S., Cinbis, M., Aggarwal, A., Apse, K. A., Ozdel, O., Atmaca, M., Zencir, S., Bagci, H., and Walsh, C. A. (2007). The role of RELN in lissencephaly and neuropsychiatric disease. *Am. J. Med. Genet. B Neuropsychiatr. Genet.* 144:58–63.

Chen, M. I., Chen, S. Y., Huang, C. H., and Chen, C. H. (2002). Identification of a single nucleotide polymorphism at the 5' promoter region of human reelin gene and association study with schizophrenia. *Mol. Psychiatry* 7:447–448.

Chugani, D. C. (2002). Role of altered brain serotonin mechanisms in autism. *Mol. Psychiatry* 7(Suppl. 2): S16–S17.

Cooper, J., and Howell, B.W. (1999). Lipoprotein receptors: signaling function in the brain? *Cell* 97:671–674.

Costa, E., Chen, Y., Davis, J., Dong, E., Noh, J. S., Tremolizzo, L., Veldic, M., Grayson, D. R., and Guidotti, A. (2002). Reelin and schizophrenia: a disease at the interface of the genome and the epigenome. *Mol. Interv.* 2:47–57.

Costa, E., Grayson, D. R., and Guidotti, A. (2003a). Epigenetic downregulation of GABAergic function in schizophrenia; potential for pharmacological intervention? *Mol. Interv.* 3:220–229.

Costa, E., Grayson, D. R., Mitchell, C. P., Tremolizzo, L., Veldic, M., and Guidotti, A. (2003b). GABAergic cortical neuron chromatin as a putative target to treat schizophrenia vulnerability. *Crit. Rev. Neurobiol.* 15:121–142.

Curristin, S. M., Cao, A., Stewart, W. B., Zhang, H., Madri, J. A., Morrow, J. S., and Ment, L. R. (2002). Disrupted synaptic development in the hypoxic newborn brain. *Proc. Natl. Acad. Sci. USA* 99:16729–16734.

D'Arcangelo, G., Miao, G. G., Chon, S. C., Soares, H. D., Morgan, J. I., and Curran, T. (1995). A protein related to extracellular matrix proteins detected in the mouse mutant reeler. *Nature* 374:719–723.

D'Arcangelo, G., Nakajima, K., Miyata, T., Ogawa, M., Mikoshiba, K., and Curran, T. (1997). Reelin is a secreted glycoprotein recognized by the CR-50 monoclonal antibody. *J. Neurosci.* 17:23–31.

D'Arcangelo, G., Homayouni, R., Keshvara, L., Rice, D. S., Sheldon, M., and Curran, T. (1999). Reelin is a ligand for lipoprotein receptors. *Neuron* 24:471–479.

DeBergeyck, V., Naerhuyzen, B., Goffinet, A. M., and Lambert de Rouvroit, C. (1998). A panel of monoclonal antibodies against reelin, the extracellular matrix protein defective in reeler mutant mice. *J. Neurosci. Methods* 82:17–24.

Deguchi, K., Inoue, K., Avila, W. E., Lopez-Terrada, D., Antalffy, B. A., Quattrocchi, C. C., Sheldon, M., Mikoshiba, K., D'Arcangelo, G., and Armstrong, D. L. (2003). Reelin and disabled-1 expression in developing and mature human cortical neurons. *J. Neuropathol. Exp. Neurol.* 62:676–684.

de Rouvroit, C. L., deBergeyck, V., Cortvrindt, C., Bar, I., Eeckhout, Y., and Goffinet, A. M. (1999). Reelin, the extracellular matrix protein deficient in reeler mutant mice, is processed by a metalloproteinase. *Exp. Neurol.* 156:214–217.

DeSilva, U., D'Arcangelo, G., Braden, V. Y., Chen, J., Miso, G. G., Curran, T., and Green, E. D. (1997). The human reelin gene: isolation, sequencing, and mapping on chromosome 7. *Genome Res.*7:157–164.

Devlin, B., Bennett, P., Dawson, G., Figlewicz, D. A., Grigorenko, E. L., McMahon, W., Minshew, N., Pauls, D., Smith, M., Spence, M. A., Rodier, P. M., Stodgell, C., and Schellenberg, G. D.; CPEA Genetics Network. (2004). Alleles of a reelin CGG repeat do not convey liability to autism in a sample from the CPEA network. *Am. J. Med. Genet.* 126B:46–50.

Dong, E., Costa, E., Grayson, D. R., and Guidotti, A. (2007). Histone hyperacetylation induces demethylation of reelin and 67-kDa glutamic acid decarboxylase promoters. *Proc. Natl. Acad. Sci. USA* 104:4676–4681.

Duan, L., Reddi, A. L., Ghosh, A., Dimri, M., and Band, H. (2004). The Cbl family and other ubiquitin ligases: destructive forces in control of antigen receptor signalling. *Immunity* 21:7–17.

Dulabon, L., Olson, E. C., Taglienti, M. G., Eisenhuth, S., McGrath, B., Walsh, C. A., Kreidberg, J. A., and Anton, E. S. (2000). Reelin binds alpha 3 beta 1 integrin and inhibits neuronal migration. *Neuron* 27:33–44.

Eastwood, S. L., and Harrison, P. J. (2003). Interstitial white matter neurons express less reelin and are abnormally distributed in schizophrenia: towards an integration of molecular and morphologic aspects of the neurodevelopmental hypothesis. *Mol. Psychiatry* 8:821–831.

Eastwood, S. L., Law, A. J., Everall, I. P., and Harrison, P. J. (2003). The axonal chemorepellant semaphorin 3A is increased in the cerebellum in schizophrenia and may contribute to its synaptic pathology. *Mol. Psychiatry* 8:148–155.

Falconer, D. S. (1951). Two new mutants, "Trembler" and "Reeler", with neurological actions in the house mouse. *J. Genet.* 50:192–201.

Fatemi, S. H. (2001). Reelin mutations in mouse and man: from reeler mouse to schizophrenia, mood disorders, autism and lissencephaly. *Mol. Psychiatry* 6:129–133.

Fatemi, S. H. (2002). The role of reelin in pathology of autism. *Mol. Psychiatry* 7: 919–920.

Fatemi, S. H. (2005). Reelin glycoprotein in autism and schizophrenia. *Int. Rev. Neurobiol.* 71:179–187.

Fatemi, S. H., Emamian, E. S., Kist, D., Sidwell, R. W., Nakajima, K., Akhter, P., Shier, A., Sheikh, S., and Bailey, K. (1999). Defective corticogenesis and reduction in reelin immunoreactivity in cortex and hippocampus of prenatally infected neonatal mice. *Mol. Psychiatry* 4:145–154.

Fatemi, S. H., Earle, J. A., and McMenomy, T. (2000). Reduction in reelin immunoreactivity in hippocampus of subjects with schizophrenia, bipolar disorder and major depression. *Mol. Psychiatry* 5:654–663.

Fatemi, S. H., Stary, J. M., Hart, A. R., and Realmuto, G. R. (2001). Dysregulation of reelin and Bcl-2 proteins in autistic cerebellum. *J. Autism Dev. Disord.* 31:529–535.

Fatemi, S. H., Stary, J. M., and Egan, E. A. (2002). Reduced blood levels of reelin as a vulnerability factor in pathophysiology of autistic disorder. *Cell. Mol. Neurobiol.* 22:139–152.

Fatemi, S. H., Snow, A. V., and Stary, J. M. (2004). Reelin glycoprotein is reduced in cerebellum and areas 9 & 40 of autistic brains. *Biol. Psychiatry* 55:806.

Fatemi, S. H., Stary, J. M., Araghi-Niknam, M., and Egan, E. (2005a). GABAergic dysfunction in schizophrenia and mood disorders as reflected by decreased levels of reelin and GAD 65 & 67 kDa proteins in cerebellum. *Schizophr. Res.* 72:109–122.

Fatemi, S. H., Snow, A. V., Stary, J. M., Araghi-Niknam, M., Reutiman, T. J., Lee, S., Brooks, A. I., and Pearce, D. A. (2005b). Reelin signaling is impaired in autism. *Biol. Psychiatry* 57:777–787.

Fatemi, S. H., Reutiman, T. J., Folsom, T. D., Bell, C., Nos, L., Fried, P., Pearce, D. A., Singh, S., Siderovski, D. P., Willard, F. S., and Fukuda, M. (2006). Chronic olanzapine treatment causes differential expression of genes in frontal cortex of rats as revealed by DNA microarray technique. *Neuropsychopharm.* 31(9):1888–1899.

Forster, E., Tielsch, A., Saum, B., Weiss, K. H., Johanssen, C., Graus-Porta, D., Muller, U., and Frotscher, M. (2002). Reelin, disabled 1, and beta 1 integrins are required for the radial glial scaffold in the hippocampus. *Proc. Natl. Acad. Sci. USA* 99:13178–13183.

Garlick, D. (2002). Understanding the nature of the general factor of intelligence: the role of individual differences in neural plasticity as an explanatory mechanism. *Psychol. Rev.* 109:116–136.

Goffinet, A. M. (1979). An early developmental defect in the cerebral cortex of the Reelor mouse. *Anat. Embryol.* 157:205–218.

Goffinet, A. M. (1984). Events governing organization of postmigratory neurons: studies on brain development in normal and reeler mice. *Brain Res.* 319:261–296.

Goffinet, A. M. (1992). The reeler gene: a clue to brain development and evolution. *Int. J. Dev. Biol.* 36:101–107.

Gottesman, I. I., and Gould, T. D. (2003). The endophenotype concept in psychiatry: etymology and strategic intentions. *Am. J. Psychiatry* 160:636–645.

Guidotti, A. R., Auta, J., Davis, J., Di-Giorgi-Gerevini, V., Dwivedi, Y., Grayson, D. R., Impagnatiello, F., Pandey, G., Pesold, C., Sharma, R., Uzunov, D., and Costa, E. (2000). Decrease in reelin and glutamic acid decarboxylase 67 (GAD 67) expression in schizophrenia and bipolar disorder: a postmortem brain study. *Arch. Gen. Psychiatry* 57:1061–1069.

Hadj-Sahraoui, N., Frederic, F., Delhaye-Bouchaud, N., and Mariani, J. (1996). Gender effect on Purkinje cell loss in the cerebellum of the heterozygous reeler mouse. *J. Neurogenet.* 11:45–58.

Helbecque, N., and Amouyel, P. (2000). Very low density lipoprotein in Alzheimer diseases. *Microsc. Res.Tech.* 50:273–277.

Herz, J., and Chen, Y. (2006). Reelin, lipoprotein receptors and synaptic plasticity. *Nature Rev. Neurosci.* 7:850–859.

Hiesberger, T., Trommsdorff, M., Howell, B. W., Goffinet, A., Mumby, M. C., Cooper, J. A., and Herz, J. (1999). Direct biding of reelin to VLDL receptor and ApoE receptor 2 induces tyrosine phosphorylation of disabled-1 and modulates tau phosphorylation. *Neuron* 24:481–489.

Hong, S. E., Shugart, Y. Y., Huang, D. T., Shahwan, S. A., Grant, P. E., Hourihane, J. O., Martin, N. D., and Walsh, C. A. (2000). Autosomal recessive lissencephaly with cerebellar hypoplasia is associated with human RELN mutations. *Nature Genet.* 26:93–96.

Ignatova, N., Sindic, C. J. M., and Goffinet, A. M. (2004). Characterization of the various forms of the reelin protein in the cerebrospinal fluid of normal subjects and in neurological diseases. *Neurobiol. Dis.* 15:326–330.

Ikeda, Y., and Terashima, T. (1997). Expression of reelin, the gene responsible for the reeler mutation, in embryonic development and adulthood in the mouse. *Dev. Dyn.* 210:157–172.

Impagnatiello, F., Guidotti, A. R., Pesold, C., Dwivedi, Y., Caruncho, H., Pisu, M. G., Uzuniv, D. P., Smalheiser, N. R., Davis, J. M., Pandey, G. N., Pappas, G. D., Tueting, P., Sharma, R. P., and Costa, E. (1998). A decrease of reelin expression as a putative vulnerability factor in schizophrenia. *Proc. Natl. Acad. Sci. USA* 95:15718–15723.

Janusonis, S., Gluncic, V., and Rakic, P. (2004). Early serotonergic projections to Cajal-Retzius cells: relevance for cortical development. *J. Neurosci.* 24:1652–1659.

Jossin, Y., and Goffinet, A. M. (2007). Reelin signals through PI3K and Akt to control cortical development and through mTor to regulate dendritic growth. *Mol. Cell. Biol.* 27:7113–7124.

Jossin, Y., Ignatova, N., Hiesberger, T., Herz, J., Lambert de Rouvroit, C., and Goffinet, A. M. (2004). The central fragment of reelin, generated by proteolytic processing *in vivo*, is critical to its function during cortical plate development. *J. Neurosci.* 24:514–521.

Keshvara, L., Benhayon, D., Magdaleno, S., and Curran, T. (2001). Identification of reelin-induced sites of tyrosyl phosphorylation on disabled-1. *J. Biol. Chem.* 276:16008–16014.

Knable, M. B., Barci, B. M., Webster, M. J., Meador-Woodruff, J., and Torrey, E. F. (2004). Molecular abnormalities of the hippocampus in severe psychiatric illness: postmortem findings from the Stanley Neuropathology Consortium. *Mol. Psychiatry* 9:609–620.

Krebs, M. O., Betancur, C., Leroy, S., Bourdel, M. C., Gillberg, C., and Leboyer, M; Paris Autism Research International Sibpair (PARIS) study. (2002). Absence of association between a polymorphic GGC repeat in the 5' untranslated region of the reelin gene and autism. *Mol. Psychiatry* 7:801–804.

Lacor, P., Grayson, D. R., Auto, J., Sugaya, I., Costa, E., and Guidotti, A. (2000). Reelin secretion from glutamatergic neurons in culture is independent from neurotransmitter regulation. *Proc. Natl. Acad. Sci. USA* 97:3556–3561.

Li, J., Nguyen, L., Gleason, C., Lotspeich, L., Spiker, D., Risch, N., and Myers, R. M. (2004). Lack of evidence for an association between WNT2 and RELN polymorphisms and autism. *Am. J. Med. Genet.* 126B:51–57.

Lugli, G., Krueger, J. M., Davis, J. M., Persico, A. M., Keller, F., and Smalheiser, N. R. (2003). Methodological factors influencing measurement and processing of plasma reelin in humans. *BMC Biochem.* 4:9.

Luque, J. M., Morante-Oria, J., and Fairen, A. (2003). Localization of ApoER2, VLDLR and Dab-1 in radial glia: groundwork for a new model of reelin action during cortical development. *Dev. Brain Res.* 140:195–203.

Magdaleno, S., Keshvara, L., and Curran, T. (2002). Rescue of ataxia and preplate splitting by ectopic expression of reelin in reeler mice. *Neuron* 33:573–586.

McAlonan, G. M., Daly, E., Kumari, V., Critchley, H. D., van Amelsvoort, T., Suckling, J., Simmons, A., Sigmundsson, T., Greenwood, K., Russell, A., Schmitz, N., Happe, F., Howlin, P., and Murphy, D. G. (2002). Brain anatomy and sensorimotor gating in Asperger's syndrome. *Brain* 125:1594–1606.

Meincke, U., Light, G. A., Geyer, M. A., Braff, D. L., and Gouzoulis-Mayfrank, E. (2004). Sensitization and habituation of the acoustic startle reflex in patients with schizophrenia. *Psychiatry Res.* 126:51–61.

Meyer, U., Nyffeler, M., Yee, B. K., Knuesel, I., and Feldon, J. (2007). Adult brain and behavioral pathological markers of preimmune challenge during early/middle and late fetal development in mice. *Brain Behav. Immun.* in press.

Miki, H., Sasaki, T., Takai, Y., and Takenawa, T. (1998). Induction of filopodium formation by a WASP-related actin-depolymerizing protein N-WASP. *Nature* 391:93–96.

Miyata, H., Chute, D. J., Fink, J., Villablanca, P., and Vinters, H. V. (2003). Lissencephaly with agenesis of corpus callosum and rudimentary dysplastic cerebellum: a subtype of lissencephaly with cerebellar hypoplasia. *Acta Neuropathol. (Berl.)* 107:69–81.

Ogawa, M., Miyata, T., Nakajima, K., Yoguy, K., Seiko, M., Ikenaka, K., Yamamoto, H., and Mikoshiba, K. (1995). The reeler gene-associated antigen on Cajal–Retzius neurons is a crucial molecule for laminar organization of cortical neurons. *Neuron* 14:899–912.

Ohkubo, N., Vitek, M. P., Morishima, A., Suzuki, Y., Miki, T., Maeda, N., and Mitsuda, N. (2007). Reelin signals survival through Src-family kinases that inactivate BAD activity. *J. Neurochem.* 103(2):820–830.

Palmen, S. J., van Engeland, H., Hof, P. R., and Schmitz, C. (2004). Neuropathological findings in autism. *Brain* 127:2572–2583.

Perez-Costas, E., Melendez-Ferro, M., Perez-Garcia, C. G., Caruncho, H. J., and Rodicio, M. C. (2004). Reelin immunoreactivity in the larval sea lamprey brain. *J. Chem. Neuroanat.* 23:211–221.

Persico, A., D'Agruma, L., Maiorano, N., Totaro, A., Militerni, R., Bravaccio, C., Wassink, T. H., Schneider, C., Melmed, R., Trillo, S., Montecchi, F., Palermo, M., Pascucci, T., Puglisi-Allegra, S., Reichelt, K. L., Conciatori, M., Marino, R., Quattrocchi, C. C., Baldi, A., Zelante, L., Gasparini, P., and Keller, F. (2001). Collaborative linkage study of autism: reelin gene alleles and haplotypes as a factor predisposing to autistic disorder. *Mol. Psychiatry* 6:150–159.

Persico, A., Levitt, P., and Pimenta, A. F. (2006). Polymorphic GGC repeat differentially regulates human reelin gene expression levels. *J. Neural Transm.* 113:1373–1382.

Pesold, C., Impagnatiello, F., Pisu, M. G., Uzunov, D. P., Costa, E., Guidotti, A., and Caruncho, H. J. (1990). Reelin is preferentially expressed in neurons synthesizing gamma-aminobutyric acid in cortex and hippocampus of adult rats. *Proc. Natl. Acad. Sci. USA* 95:3221–3226.

Rodriguez, M. A., Pesold, C., Liu, W. S., Kriho, V., Guidotti, A., Pappas, G. D., and Costa, E. (2000). Colocalization of integrin receptors and reelin in dendritic spine post-synaptic densities of adult non-human primate cortex. *Proc. Natl. Acad. Sci. USA* 97:3550–3555.

Saez-Valero, J., Costell, M., Sjogren, M., Andreasen, N., Blennow, K., and Luque, J.M. (2003). Altered levels of cerebrospinal fluid reelin in frontotemporal dementia and Alzheimer's disease. *J. Neurosci. Res.* 72:132–136.

Serajee, F. A., Zhong, H., and Mahbubul Huq, A. H. M. (2005). Association of reelin gene polymorphisms with autism. *Genomics* 87:75–83.

Shi, L., Fatemi, S. H., Sidewell, R. W., and Patterson, P. H. (2003). Maternal influenza infection causes marked behavioral and pharmacological changes in the offspring. *J. Neurosci.* 23:297–302.

Smalheiser, N. R., Costa, E., Guidotti, A., Impagnatiello, F., Auta, J., Lacor, P., Kriho, V., and Pappas, G. D. (2000). Expression of reelin in adult mammalian blood, liver, pituitary pars intermedia, and adrenal chromaffin cells. *Proc. Natl. Acad. Sci. USA* 97:1281–1286.

Strasser, V., Fasching, D., Hauser, C., Mayer, H., Bock, H. H., Hiesberger, T., Herz, J., Weeber, E. J., Sweatt, J. D., Pramatarova, A., Howell, B., Schneider, W. J., and Nimpf, J. (2004). Receptor clustering is involved in reelin signaling. *Mol. Cell. Biol.* 24:1378–1386.

Suetsugu, S., Tezuka, T., Morimura, T., Hattori, M., Mikoshiba, K., Yamamoto, T., and Takenawa, T. (2004). Regulation of actin cytoskeleton by mDab1 through N-WASP and ubiquitination of mDab1. *Biochem. J.* 384:1–8.

Tissir, F., and Goffinet, A. M. (2003). Reelin and brain development. *Nature Rev. Neurosci.* 4:496–505.

Trommsdorff, M., Gotthardt, M., Hiesberger, T., Shelton, J., Stodkinger, W., Nimpf, J., Hammer, R. E., Richardson, J. A., and Herz, J. (1999). Reeler/disabled-like disruption of neuronal migration in knockout mice lacking the VLDL receptor and ApoE receptor 2. *Cell* 97:689–701.

Tueting, P., Costa, E., Dwivedi, Y., Guidotti, A., Impagnatiello, F., Manev, R., and Pesold, C. (1999). The phenotypic characteristics of heterozygous reeler mouse. *NeuroReport* 10:1329–1334.

Tueting, P., Doueiri, M.-S., Davis, J. M., and Guidotti, A. (2005). Prepulse inhibition of startle reflects compromised GABAergic neurotransmission in reeler heterozygous mice. Program No. 936.5 2005 Abstract Viewer and Itinerary Planner. Society for Neuroscience, Washington, DC. Online.

Utsunomiya-Tate, N., Kubo, K. I., Tate, S. C., Kainosho, M., Katayama, E., Nakajima, K., and Mikoshiba K. (2000). Reelin molecules assemble together to form a large protein complex, which is inhibited by the function-blocking CR-50 antibody. *Proc. Natl. Acad. Sci. USA* 97:9729–9734.

Veldic, M., Caruncho, H. J., Liu, W. S., Davis, J., Satta, R., Grayson, D. R., Guidotti, A., and Costa, E. (2004). DNA-methyltransferase 1 mRNA is selectively overexpressed in telencephalic GABAergic interneurons of schizophrenia brains. *Proc. Natl. Acad. Sci. USA* 101:348–353.

Wedenoja, J., Loukola, A., Tuulio-Henricksson, A., Pauino, T., Ekelund, J., Silander, K., Varilo, T., Hekkila, K., Suvisaari, J., Partonen, T., Lonnqvist, J., and Peltonen, L. (2007). Replication of linkage on chromosome 7q22 and association of the regional reelin gene with working memory in schizophrenia families. *Mol. Psychiatry* in press.

Weeber, E. J., Beffert, U., Jones, C., Christian, J. M., Forster, E., Sweatt, J. D., and Herz, J. (2002). Reelin and ApoE receptors cooperate to enhance hippocampal synaptic plasticity and learning. *J. Biol. Chem.* 277:39944–39952.

Welch, M. D., Iwamatsu, A., and Mitchison, T. J. (1997). Actin polymerization is induced by Arp 2/3 protein complex at the surface of Listeria monocytogenes. *Nature* 385:265–269.

Zhang, H., Liu, X., Zhang, C., Muno, E., Macciardi, F., Grayson, D. R., Guidotti, A. R., and Holden, J. J. (2002). Reelin gene alleles and susceptibility for autism spectrum disorders. *Mol. Psychiatry* 7:1012–1017.

Chapter 23
Reelin Downregulation as a Prospective Treatment Target for GABAergic Dysfunction in Schizophrenia

Erminio Costa, Ying Chen, Erbo Dong, Dennis R. Grayson, Alessandro Guidotti, and Marin Veldic

Contents

E. Costa
Psychiatric Institute, Department of Psychiatry, University of Illinois at Chicago, 1601 West Taylor Street MC912, Chicago, IL 60612
e-mail: costa@psych.uic.edu

Y. Chen
Psychiatric Institute, Department of Psychiatry, University of Illinois at Chicago, 1601 West Taylor Street MC912,Chicago, IL 60612

E. Dong
Psychiatric Institute, Department of Psychiatry, University of Illinois at Chicago, 1601 West Taylor Street MC912, Chicago, IL 60612

D.R. Grayson
Psychiatric Institute, Department of Psychiatry, University of Illinois at Chicago, 1601 West Taylor Street MC912, Chicago, IL 60612

A. Guidotti
Psychiatric Institute, Department of Psychiatry, University of Illinois at Chicago, 1601 West Taylor Street MC912, Chicago, IL 60612

M. Veldic
Psychiatric Institute, Department of Psychiatry, University of Illinois at Chicago, 1601 West Taylor Street MC912, Chicago, IL 60612

S. H. Fatemi (ed.), *Reelin Glycoprotein: Structure, Biology and Roles in Health and Disease.*

1 Reelin is a Protein Synthesized Almost Exclusively in GABAergic Neurons During Embryonic and Adult Life

The proper functioning of the mammalian cortex depends on the formation of neuronal networks, including principal projection neurons and interneurons that use glutamate and GABA as transmitters, respectively. In the adult brain, cortical interneurons have been implicated in the regulation of the synaptogenesis and neuronal wiring operative in cortical network formation. These neurons are aspiny, express local projecting axons, and their staining with the Golgi method reveals a soma volume smaller than most cortical neurons. They store and synthesize the neurotransmitter GABA and also frequently synthesize and secrete reelin. In embryonic cortex, reelin is synthesized and secreted by the Cajal-Retzius cells, guides neuronal migration and positioning of pyramidal neurons (D'Arcangelo *et al.*, 1995). However, postnatally during CNS development and maturation, this protein is synthesized and secreted from GABAergic interneurons and harmonizes the functional plastic interaction of neuronal axons, dendrites, and their spines (Costa *et al.*, 2001; Niu *et al.*, 2004). Reelin secreted in the extracellular matrix contributes to the modulation of neuronal excitability, firing frequencies, and the morphological properties of the telencephalic neuronal networks regulating their coordinated activity (Liu *et al.*, 2001; Costa *et al.*, 2001; Weeber *et al.*, 2002; Qiu *et al.*, 2007).

Recent studies suggest that during conscious states, reelin binding to cortical dendritic spines participates in the modulation of synaptic plasticity and memory processes by: (a) consolidating long-term potentiation (LTP) expression in hippocampal slices, (b) harmonizing protein synthesis locally in dendrites and their spines, and (c) regulating maturation and number of dendritic spines.

Several lines of evidence show that in the brain of schizophrenia (SZ) patients, reelin and GAD67 expressions are severely decreased (Akbarian *et al.*, 1995; Impagnatiello *et al.*, 1998; Fatemi *et al.*, 2000; Guidotti *et al.*, 2000, 2005; Benes and Beretta, 2001; Costa *et al.*, 2001; Eastwood and Harrison, 2003; Veldic *et al.*,

2004, 2005, 2007; Woo *et al.*, 2004; Lewis *et al.*, 2005; Ruzicka *et al.*, 2007). Hence, if cognitive function were to be related to cortical dendritic spine density and maturation and should LTP consolidation be modulated by reelin release, it is possible to entertain the hypothesis that in the presence of a reduced amount of reelin release, as may occur in SZ, dendritic spine maturation may be delayed and distorted, resulting in a decrease in the number of spines associated with this cognitive deficit.

The hypothesis, that in SZ patients these cognitive deficits may be related to a disruption of the inhibitory GABAergic synaptic strength in specific corticolimbic circuits due to an insufficient expression of reelin, is addressed in this chapter.

2 Corticolimbic GABAergic Neurons Secrete Reelin, Which is Important in Modulating CNS Neuronal Plasticity

In the mammalian neocortex and hippocampus, reelin is synthesized almost exclusively by GABAergic neurons (Alcantara *et al.*, 1998; Pesold *et al.*, 1998, 1999; Rodriguez *et al.*, 2002), which secrete this protein in the proximity of dendritic spines (Rodriguez *et al.*, 2000; Costa *et al.*, 2001). In the human cortex, reelin mRNA is expressed in GABAergic neurons of every cortical region or layer studied (Table 23.1); however, the percentage of GAD-positive neurons expressing reelin differs in different layers. For example, approximately 100% of GAD65/67-positive neurons express reelin mRNA in the upper cortical layers, whereas in layers V and VI of different cortical areas, only 50 to 30% of the GAD65/67-positive neurons express reelin mRNA (Table 23.1). The neuronal expression of reelin studied with light microscopic immunoreactivity in rodent, human, and nonhuman primate neocortices reveals that reelin-like immunopositive neurons are not present in all cortical layers but in several areas are confined to layers I and II. In addition, in rodents

Table 23.1 Reelin and glutamic acid decarboxylase 65 (GAD65) expression in Brodmann's area 9 GABAergic neurons of nonpsychiatric subjects (NPS), schizophrenia (SZP), and bipolar disorder patients (BDP)

		Layer I	Layer II	Layer III–IV	Layer V	Layer VI
NPS	Reelin	25 ± 0.85	38 ± 1.6	25 ± 0.78	16 ± 0.22	10.8 ± 0.30
	GAD65	26 ± 0.64	42 ± 0.89	35 ± 1.10	21 ± 0.25	21 ± 0.30
SZP	Reelin	16 ± 0.62*	26 ± 1.69*	19 ± 0.76*	14 ± 0.24	9.7 ± 0.30
	GAD65	27 ± 0.65	44 ± 1.23	36 ± 1.0	23 ± 0.30	23 ± 0.22
BDP	Reelin	18 ± 1.7*	26 ± 1.7*	21 ± 0.99	15 ± 0.25	10 ± 0.37
	GAD65	26 ± 0.77	44 ± 1.1	35 ± 0.86	21 ± 0.33	21 ± 0.37

Counts of reelin and GAD65 mRNA-positive neurons in six layers of BA9 in NPS ($n = 27$), SZP ($n = 20$), and BDP ($n = 14$). Differences were calculated by ANOVA and p values were compared by Bonferroni t-test. *Denotes statistically significant differences ($p \leq 0.013$) when SZP or BDP are compared to NPS. Specimens were obtained from Harvard Brain Tissue Resource Center (Belmont, MA).

and primates, a broad band of diffuse extracellular reelin-like immunoreactivity is detectable in cortical layers I, II, and III (Pesold *et al.*, 1998; Guidotti *et al.*, 2000; Rodriguez *et al.*, 2002). These findings suggest that the reelin storage capacity in cortical GABAergic interneurons may vary and that probably, similar to cerebellar granule cells, reelin may be secreted from cortical GABAergic interneurons into the extracellular space by a "constitutive mechanism" (Lacor *et al.*, 2000).

One can consider that reelin may be continuously secreted into extracellular spaces and may undergo rapid metabolic processing by the action of extracellular peptidases. This alternative is supported by the observation that cerebrospinal fluid (CSF) contains a significant amount of reelin processing products (Ignatova *et al.*, 2004). The presence of reelin in the extracellular space of upper cortical layers expressing a high density of pyramidal neuron apical dendrites may have functional significance and likely suggests a putative role for extracellular reelin in the maturation of newly formed dendritic spines. In fact, reelin secreted into the extracellular space adheres to dendritic spine postsynaptic densities (Rodriguez *et al.*, 2000; Costa *et al.*, 2001), and the number of dendritic spines is reduced in the PFC of SZ patients (Glantz and Lewis, 2000; Rosoklija *et al.*, 2000), as well as in the frontal cortex and hippocampus of the heterozygous reeler mouse (HRM) (Liu *et al.*, 2001).

3 Dendritic Postsynaptic Densities and Spines Are Principal Targets of Extracellular Matrix Reelin

We investigated the possible contribution of extracellular reelin to synaptic plasticity in HRM. This mouse model expresses only 50% of the reelin expressed by the wild-type mouse (WTM) (Tueting *et al.*, 1999; Liu *et al.*, 2001). The HRM exhibits: (a) increased density of cortical neuronal packing, (b) a decreased cortical thickness due to neuropil hypoplasia, (c) a marked decrease of dendritic spine density on basal and apical dendritic branches of FC pyramidal neurons, and (d) a decrease in dendritic spines expressed on the basal dendritic branches of CA1 pyramidal neurons of the hippocampus (Liu *et al.*, 2001).

To establish whether a defect in GAD67 expression, similar to that observed in SZ, is also operative in neuropil hypoplasia, we studied the heterozygous GAD67 mouse. This mouse expresses about 50% of GAD67 mRNA levels and a similar decrease in GABA biosynthesis in the FC (Liu *et al.*, 2001; Carboni *et al.*, 2004). At the same time, it expresses normal amounts of reelin and fails to show neuropil hypoplasia or a dendritic spine expression downregulation. These findings, coupled with the immunoelectron-microscopic observation that reelin colocalizes with integrin receptors expressed by dendritic postsynaptic densities (Rodriguez *et al.*, 2000; Liu *et al.*, 2001; Dong *et al.*, 2003), suggest that reelin may be a regulatory factor operative in the expression density of cortical dendritic spines. This plastic function is of particular interest, because the brain neurohistochemical phenotypic traits and behavioral deficits exhibited by HRM are similar to those found in post-mortem brains of psychotic patients.

4 Electrophysiological and Behavioral Action of Reelin

In electrophysiological experiments, the addition of reelin enhances LTP induced by tetanus or high-frequency electrical stimulation in mouse hippocampus CA1 region (Weeber *et al.*, 2002). In recent studies, reelin was shown (via ApoE2 and VLDL receptors) to potentiate the NMDA receptor-mediated current intensity in CA1 pyramidal neurons as a result of tyrosine phosphorylation of

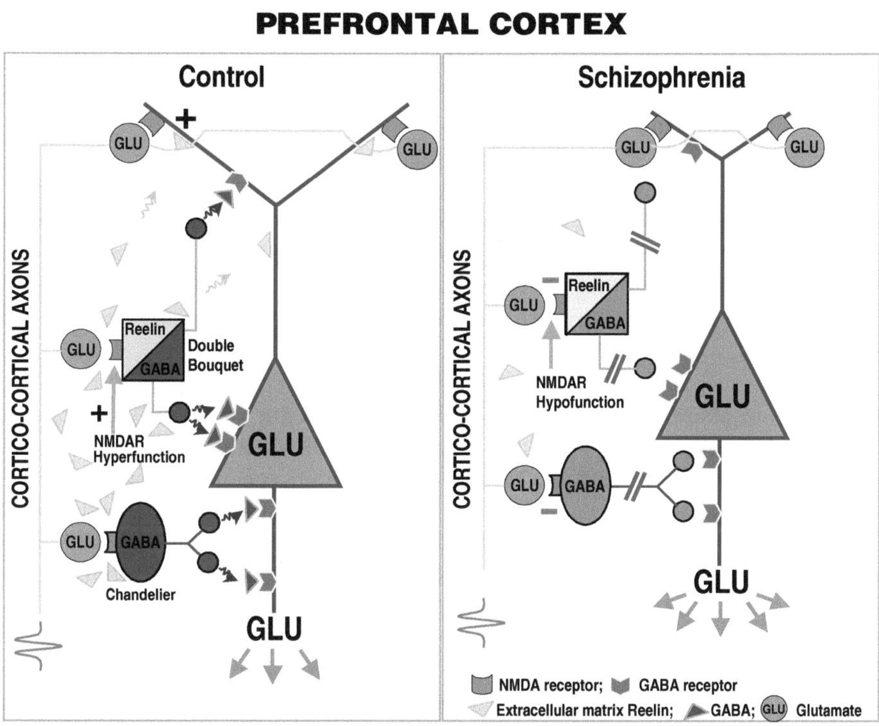

Fig. 23.1 (Control) Reelin expressed in double bouquet or horizontal cells in the upper prefrontal cortex layers is secreted by a constitutive mechanism in the extracellular matrix space and: (a) binds to the apical dendritic branches of pyramidal neurons inducing spine formation by facilitating dendritic resident mRNA translation or (b) binds to dendrites or cell bodies of GABAergic interneurons (double bouquet or chandelier cells), facilitating the action of glutamate at NMDA receptors located on GABAergic interneurons and thereby increasing the release of GABA on apical dendrites, cell bodies, and axon initial segments of pyramidal neurons.

(Schizophrenia) Reelin and GAD67 expression and reelin and GABA release are downregulated. The reelin deficit causes: (a) decreased dendritic spine density on the apical dendrites of pyramidal neurons and (b) hypofunction of NMDA receptors located on double bouquet or chandelier cells, eliciting a further decrease of GABA released on the apical dendrites, cell bodies, or axon initial segments of pyramidal neurons. The deficit of GABAergic neurotransmission results in an increased output of glutamate from the axon terminal of pyramidal neurons (*See Color Plates*)

NMDA receptor subunits NR2A and NR2B (Beffert *et al.*, 2005; Chen *et al.*, 2005). Therefore, reduced expression of reelin in the HRM may result in decreased function of NMDA receptors located on dendrites or cell bodies of GABAergic interneurons innervated by glutamatergic nerve terminals (Fig. 23.1). In addition, the network-driven spontaneous inhibitory postsynaptic currents recorded in CA1 pyramidal neurons of HRM appear to be reduced (Qiu *et al.*, 2006). Thus, the GABA-releasing mechanisms may be impaired in HRM due to: (a) a decreased excitatory drive on GABAergic interneuron function and (b) the reduction of GAD67 expression (Liu *et al.*, 2001; Carboni *et al.*, 2004). Since reelin is localized in the proximity of dendritic postsynaptic densities in cortical and hippocampal regions (Costa *et al.*, 2001; Pappas *et al.*, 2001), it seems plausible to infer that reelin, by influencing the function of NMDA receptors located on GABAergic interneurons, may selectively facilitate the release of GABA, providing a perisynaptic modulatory milieu for excitatory or inhibitory synapse formation and spine maturation.

Behavioral studies of HRM reveal that reelin haploinsufficiency does not result in a novel specific behavioral pattern but rather causes specific modifications characterized by: (a) a deficit of olfactory discrimination learning (Larson *et al.*, 2003), (b) deficits in associative hippocampal learning in contextual fear conditioning tests (Qiu *et al.*, 2006), and (c) enhanced susceptibility to the cognitive impairments induced by dizocilpine detected in the eight-arm radial maze (Carboni *et al.*, 2004).

The enhanced dizocilpine susceptibility in HRM does not appear to be due to differences in pharmacokinetic characteristics because the levels of dizocilpine in brain cortices of HRM and WTM were virtually equal. We also failed to detect differences between HRM and WTM in glutamate brain content and in the rate of [^{13}C]glucose incorporation into glutamate brain pools. In contrast, we found that the conversion index of glutamate into GABA (an indirect estimation of GABA turnover) is decreased in the cortex, hippocampus, and striatum of HRM compared to WTM (Carboni *et al.*, 2004). Qiu *et al.* (2006) reported that downregulation of telencephalic GABAergic transmission may explain the increased susceptibility of HRM to the amnestic action of dizocilpine. Results from our and other laboratories are consistent with the hypothesis that the increased susceptibility of HRM to the amnestic, locomotor, and stereotypic behaviors elicited by dizocilpine may depend on a downregulation of telencephalic GABAergic inhibitory tone. This decrease determines an increase of the intermittent population firing of pyramidal neurons which facilitates increased thalamocortical, corticothalamic, corticocortical, corticostriatal, and corticomesolimbic excitatory transmission. In addition, our findings also point out that in the HRM model, similar to SZ patients, there is a decrease of telencephalic dendritic spine density (Fig. 23.2). Presumably, this decrease in spine density expression is elicited by a combination of GABAergic hypofunction and the downregulation of reelin secretion from GABAergic neurons where reelin is selectively synthesized.

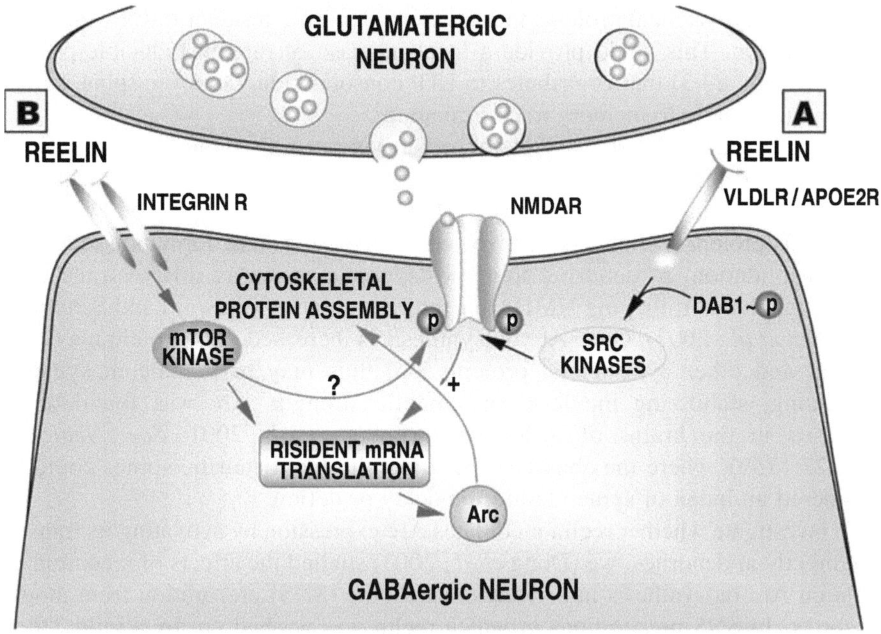

Fig. 23.2 Putative role of reelin in synaptic plasticity. Reelin is depicted binding to a dendritic postsynaptic density of a cortical GABAergic interneuron. Either (A) to VLDL or ApoE2 receptors (VLDLR or APOE2R) or (B) to integrin receptors (INTEGRINR). (A) Reelin modulates NMDA receptor (NMDAR) activity through SRC kinase-mediated tyrosine phosphorylation of the NMDAR intracellular sites (Weeber *et al.*, 2002; Herz and Chen, 2006). (B) Reelin modulates Arc expression and cytoskeletal protein assembly through activation of mTOR kinase (Dong *et al.*, 2003) (*See Color Plates*)

5 Reelin Signaling Pathways and the Translation of Dendritic mRNAs

In adult CNS, the molecular mechanisms whereby extracellular matrix reelin, by adhering to cortical dendritic postsynaptic densities, modulates their plasticity, for instance by either changing the density of glutamatergic receptor expression at synapses, facilitating AMPA receptor subunit insertion (Qiu *et al.*, 2007), or increasing the number of dendritic spines (Costa *et al.*, 2001), are not completely understood.

Recent studies suggest that reelin secreted in the extracellular matrix by GABAergic neurons acts as an important indirect modulator of synaptic plasticity in the adult mammalian brain. In fact, reelin binding at synaptic ApoE2 and VLDL and/or integrin receptors (Weeber *et al.*, 2002; Dong *et al.*, 2003) could stabilize dendritic postsynaptic density expression by providing a molecular scaffold for the assembly of cytoskeletal proteins that facilitate dendritic resident mRNA transport and translation. This could provide a local increase of rapidly inducible protein synthesis (Fig. 23.3) that contributes to LTP consolidation, dendritic spine formation, and ultimately to memory trace formation.

Arc (activity-regulated cytoskeletal protein) is a rapidly inducible cytoskeletal protein whose biosynthesis is encoded by dendrite resident mRNAs located in apical dendrites in spatial proximity to dendritic spines (Steward and Schuman, 2001). This protein is known to be involved in increasing spine formation following LTP consolidation. At dendritic spine postsynaptic sites, Arc mRNA translation is rapidly induced following NMDA receptor stimulation (Steward and Schuman, 2001; Yin *et al.*, 2002). Once Arc biosynthesis is increased, this protein may bind to actin and other cytoskeletal proteins and thus may participate in synaptic remodeling, stabilizing the level of synaptic strength. Arc was found to be decreased in the brains of reeler mice (Lacor *et al.*, 2001, *Soc. Neurosci. Abstr.*27:1759), where the abundance of filopodia-like dendritic spines could be considered an index of spine maturation delay or deficit.

To investigate whether reelin modulates Arc expression by activating its translation directly at dendrites, we (Dong *et al.*, 2003) studied the effects of recombinant reelin on Arc biosynthesis in a synaptoneurosome (SNS) preparation from mouse neocortex. In SNS preparations in which reelin was washed out by a mild Triton X-100 treatment, the application of full-length recombinant mouse reelin results in the displacement of [^{125}I]echistatin binding to integrin receptors with a K_i of 22 pM. On the other hand, echistatin (50–100 nM) completely antagonizes and abates reelin binding to the SNS. The addition of reelin to reelin-free SNS enhances the

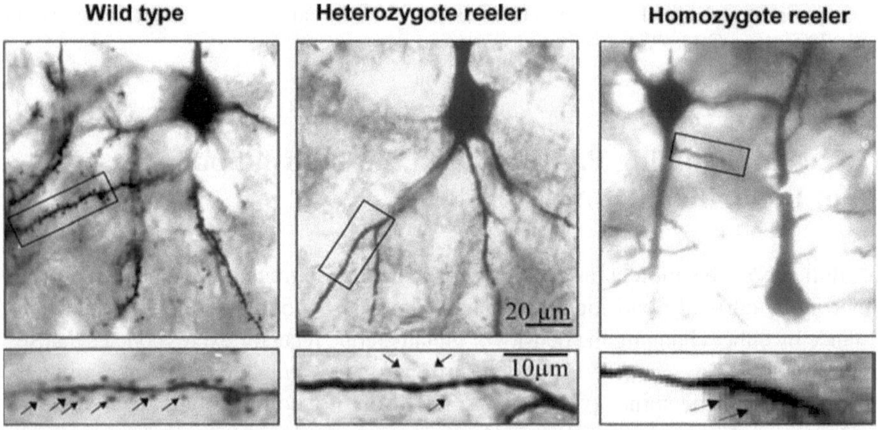

Fig. 23.3 The reeler mouse shows that reelin regulates pyramidal neuron dendritic spine expression. Photomicrographs showing Golgi-impregnated basilar dendritic spines of layer III frontal cortex pyramidal neurons of wild-type (left), heterozygous reeler (middle), and homozygous reeler (right) mice. Note the almost complete absence of spines on dendrites obtained from the heterozygous and homozygous reeler mice. In the homozygous reeler mouse, the laminar structure is disrupted and the pyramidal neuron orientation is altered. (Modified from Liu *et al.*, 2001)

incorporation of [^{35}S]methionine into Arc in a concentration-dependent manner. We also detected that this incorporation is virtually abolished by 50 to 100 nM of rapamycin, a blocker of the mammalian target of rapamycin kinase (mTOR). These data suggest that, in addition to activating Src kinases via ApoE2 and VLDL receptors (Herz and Chen, 2006), reelin may bind to integrin receptors activating mTOR kinases, thereby facilitating dendritic spine resident mRNA translation of Arc and other dendritic mRNAs (Fig. 23.2). Upon reelin addition and activation of mTOR kinase, Arc and other dendritic spine-resident mRNAs very likely acquire a polyadenylation tail that allows them to associate with locally resident polyribosomes to initiate their translation (Richter and Lorenz, 2002).

These findings raise the possibility that reelin binding to integrin receptors at specific spine synapses (i.e., between GABAergic axon terminals and dendritic spines of glutamatergic pyramidal neurons) is a pivotal event which promotes highly selective synaptic modifications during the LTP plasticity associated with memory trace formation.

6 Reelin Deficiency Plays an Important Role in the Pathophysiology of Schizophrenia

Currently, there is compelling evidence that a GABAergic deficit occurs in corticolimbic regions of patients with either SZ or bipolar disorder; this defect may involve a marked (approximately 50%) decreased expression of GAD67 (one of the two brain isoenzymes that synthesize GABA) (Akbarian *et al.*, 1995; Benes and Beretta, 2001; Woo *et al.*, 2004; Guidotti *et al.*, 2005; Lewis *et al.*, 2005) and reelin (Impagnatiello *et al.*, 1998; Fatemi *et al.*, 2000; Guidotti *et al.*, 2000; Eastwood and Harrison, 2003; Veldic *et al.*, 2004, 2005, 2007; Ruzicka *et al.*, 2007) (Table 23.1). As mentioned, this large-molecular-weight protein, which is synthesized and secreted into the extracellular matrix from GABAergic interneurons and which adheres to dendrites of pyramidal neurons or to somata of GABAergic interneurons, appears to modulate the plasticity of dendritic postsynaptic densities via an activation of a protein kinase that phosphorylates the intracellular consensi of NMDA receptor subunits, thereby increasing their affinity for glutamate (Fig. 23.2) (for reviews see Costa *et al.*, 2001, and Herz and Chen, 2006).

Recent clinical observations and postmortem brain studies suggest that the NMDA receptor density on the somata of cortical or hippocampal GABAergic interneurons is decreased in SZ patients (Krystal *et al.*, 1999; Woo *et al.*, 2004). The decrease of NMDA receptors in GABAergic neurons, together with the decrease of reelin present in the extracellular space, may further reduce the glutamate-mediated release of GABA and the consequent decrease of inhibitory input at postsynaptic sites located on apical dendrites, somata, or initial axon segments of pyramidal neurons in SZ patients (Fig. 23.1).

In this scenario, the downregulation of reelin expression is probably responsible for decreases in dendritic spine density, neuropil hypoplasia, and desynchronization associated with the aberrant cortical intermittent population firing that underlie cognitive dysfunctions in SZ (Selemon and Goldman-Rakic, 1999; Glantz and Lewis, 2000; Rosoklija *et al.*, 2000; Black *et al.*, 2004; Spencer *et al.*, 2004).

7 Molecular Basis for Reelin Downregulation in SZ Psychosis

Central to the pursuit of new approaches in SZ treatment is the identification of a molecular basis for the pathophysiological processes that underlie cognitive dysfunction.

The data suggest that in SZ, a defect of reelin secreted from GABAergic neurons may represent a potential pathophysiological target for treatment. Hence, an understanding of whether alterations of genetic or epigenetic mechanisms are responsible for reelin downregulation may offer reelin biosynthesis inhibition as a novel and promising target for antipsychotic drug development.

7.1 Analysis of the Reelin Promoter in Genomic DNA

We mapped the exon/intron structure of the human reelin gene to various BAC sequences present in the human database. The exon/intron structure of *Reln* is remarkably conserved phylogenetically from mouse (Royaux *et al.*, 1997) to human (Chen *et al.*, 2002). The human reelin gene (RELN) maps to chromosome 7q22 (DeSilva *et al.*, 1997) and spans several BAC clones which have been sequenced. We subcloned a 4.2-kb *Eco*RI fragment that contains the entire first exon, 255 bp of the first intron, and some 3.7 kb of 5' flanking DNA. Various unique restriction sites in this clone were used to subclone portions of the region upstream of the ATG start codon into the luciferase reporter construct, pGL-3 basic.

Sequences surrounding the transcriptional start site and first exon form a CpG island, much like that reported for the murine gene (Royaux *et al.*, 1997). Numerous transcription factor search programs identified multiple sequence motifs for different DNA-binding proteins, including CREB, multiple Sp1, and Pax6 sites. We have performed transient transfections and have been able to show that there is an upstream enhancer which contains recognition sites for Tbr1, Sp1, and Pax6. The Sp1 site appears to be critical to the retinoic acid induction of the gene in NT2 cells (Chen *et al.*, 2007).

7.2 Analysis of the 5' UTR

During our initial assessment of potential polymorphic regions in reelin cDNA, we investigated the CGG repeat that is within the 5' untranslated region. This repeat is unusual because of its proximity to the initiation codon and its potential for formation of hairpin structures that may affect transcription/translation. By far, the most common alleles present in the population studied contained 8 and 10 repeats. Of particular interest is that there is no correlation between any of the repeat-specific alleles and disease phenotype. This is similar to a report by another group who found comparable results (Huang and Chen, 2006). Using transient transfection assays, it has been shown that the amounts of reelin-derived reporter activity negatively correlate with the length of the polymorphic repeat (Persico *et al.*, 2006). This suggests that longer repeats may prove detrimental to stable mRNA expression.

7.3 Dnmt1 Overexpression and Reelin Promoter Hypermethylation in Telencephalic GABAergic Neurons of SZ Patients

In spite of the persistent downregulation of reelin mRNA and protein in schizophrenia, evidence of this gene linkage to this disorder is weak. That is, certain reelin alleles are indicated as risk factors only when considered in combination with one of several additional genes that have been implicated (Hall *et al.*, 2007). Very probably, the key factor associated with reelin downregulation is the overexpression of Dnmt1 in telencephalic GABAergic neurons of SZ patients (Veldic *et al.*, 2005, 2007; Ruzicka *et al.*, 2007). Dnmt1 is one member of the DNA methyltransferases, also including Dnmt3a, 3b, and 3L, which methylates genomic DNA in neurons and other cell types (Goll and Bestor, 2005). We have recently reported that in the human cortex, Dnmt1 is expressed at much higher levels than Dnmt3a or 3b in GABAergic interneurons, but its expression is virtually absent in glutamatergic pyramidal neurons. Moreover, Dnmt1 expression is clearly upregulated in cortical GABAergic neurons of layers I and II and in basal ganglion GABAergic medium spiny neurons of SZ brains (Veldic *et al.*, 2005, 2007; Ruzicka *et al.*, 2007). This increase in Dnmt1 correlates with a reproducible decrease in reelin and GAD67 mRNA expression, probably due to promoter hypermethylation (Fig. 23.4).

Consistent with the increased expression of Dnmt1 and the corresponding decrease in reelin and GAD67 expression in cortical GABAergic neurons of SZ patients, we (Grayson *et al.*, 2005) and others (Abdolmaleky *et al.*, 2005) have shown that in SZ patients, portions of the reelin promoter are hypermethylated. We propose that the reduced expression of reelin and also that of GAD67 mRNAs results in a subsequent decrease in interneuron inhibitory tone, which described in the context of SZ appears to be linked to a disruption of pyramidal neuron firing rates (Guidotti *et al.*, 2005; Levenson and Sweatt, 2005; Lewis *et al.*, 2005).

Two reports in the literature show that the reelin promoter is hypermethylated in patients with SZ (Abdolmaleky *et al.*, 2005; Grayson *et al.*, 2005). In the first, the

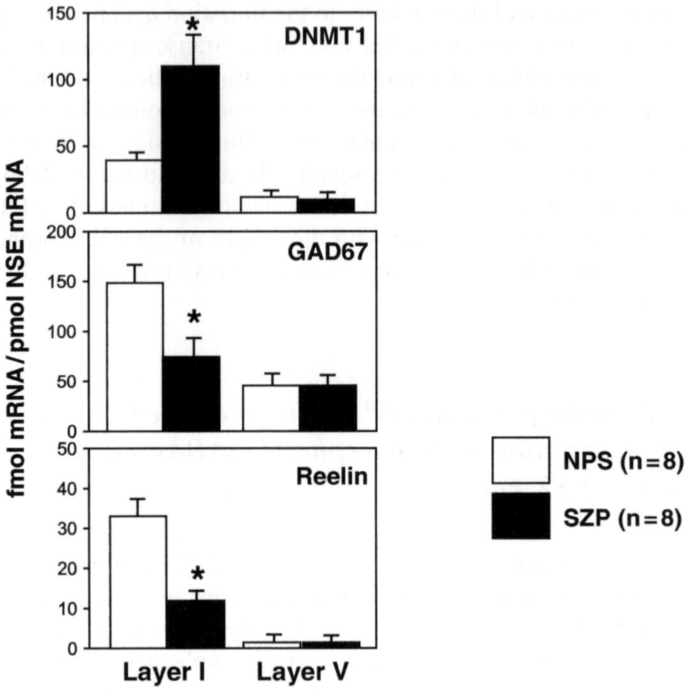

Fig. 23.4 Dnmt1 mRNA is overexpressed and GAD67 and reelin mRNAs are downregulated in prefrontal cortex layer I but not in layer V GABAergic interneurons of SZ patients. Dnmt1, GAD67, reelin, and NSE (neuron-specific enolase) mRNAs were extracted from tissue sections microdissected from layers I and V of Brodmann's area 9 slices and were quantified with nested competitive RT-PCR with internal standards. Data represent the mean ± SE of eight subjects per group. Asterisks denote $p < 0.05$ when SZ patients are compared to nonpsychiatric (NPS) subjects. (From Ruzicka *et al.*, 2007)

authors used Brodmann's areas 9 and 10 from the Harvard Brain Tissue Resource Center to extract genomic DNA. Bisulfite analysis and methylation-specific PCR were used to evaluate the extent of methylation. The authors report an increased amount of methylation in 73% of SZ patients compared with 24% in the control group. The more heavily methylated region was located close to the putative CREB binding site positioned around −420 to −398 relative to the RNA start site (Abdolmaleky *et al.*, 2005).

In a second report of reelin promoter hypermethylation, we (Grayson *et al.*, 2005) examined two different brain cohorts: the Stanley Foundation Neuropathology Consortium and the Harvard Brain Tissue Resource Center. Genomic DNA from occipital cortices was obtained from the first source, while DNA from the PFC was used in the second collection. Bisulfite analysis of genomic DNA followed by nested PCR amplification and sequencing of individual clones was used to identify methylated bases. Interestingly, the analysis of these two patient collections also showed differences in the methylation patterns among SZ subjects and nonpsychiatric subjects. Genomic DNA from SZ brain was more heavily methylated at two positions, −139 and −134, relative to the RNA start site (Fig. 23.5). While the background methylation patterns were different in both brain collections, the results appeared consistent in these two groups. The two more heavily methylated sites reside within a Pax6 binding site that has recently been shown to be relevant for regulating reelin expression in neural progenitor cells (NT2 cells; Chen *et al.*, 2002, 2007).

In addition to the finding that these sites were hypermethylated in SZ patients, it was also established that double-stranded oligos containing the methylated bases

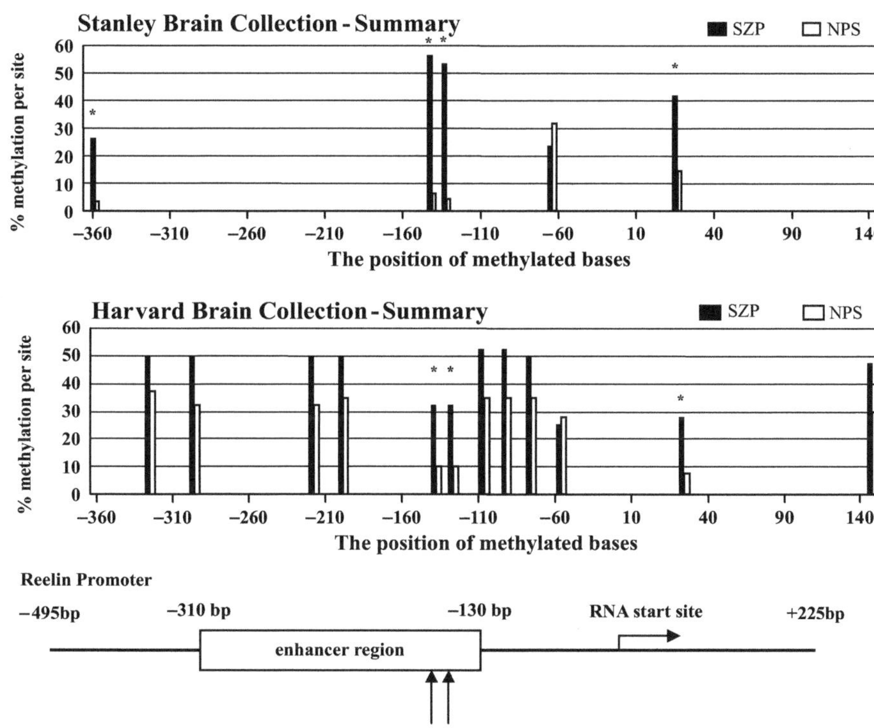

Fig. 23.5 Reelin promoter methylation profile. Genomic DNA was isolated from either occipital cortices (upper) or prefrontal cortices (lower) for bisulfite analysis. (Upper) Summary of data obtained from the Stanley Foundation patients; (lower) data obtained from the Harvard Brain Collection. Following bisulfite conversion, individual DNA strands were amplified, subcloned, and sequenced. The bars show locations of methylated bases. The methylation profiles were different in the two brain collections, which may be due to the brain regions available for this study. A linear representation of the human reelin promoter is shown at the bottom of the figure. Asterisks represent bases consistently methylated in both collections. [Originally published in Grayson *et al.* (2005) and reprinted with permission from the National Academy of Sciences]

bound proteins in nuclear extracts with higher affinity than the nonmethylated oligos. That is, the proteins present in extracts from precursors that do not express the gene showed ~1.7-fold higher affinity for the methylated site compared to the nonmethylated site. This suggests that methyl domain-binding proteins are likely present in NT2 cell nuclear extracts, and that these proteins bind to the methylated bases with a higher affinity. This was confirmed using gel shift competition assays which showed that binding of nuclear proteins was higher at methylated than nonmethylated sites (Grayson *et al.*, 2005).

7.4 Point Mutations

To characterize the function of these two sites, we altered each site independently and together in a manner such that only one (m −141, m −136) or both (m −136/−141) base pairs were altered. The remainder of the −514 promoter was left intact and the constructs were transiently introduced into NT2 cells (Fig. 23.6). The data indicate that the −141 bp mutation had little effect on promoter activity, while the m −136 mutant was only half as active as the parent construct, suggesting that this single base pair substitution within the Pax6 binding site is sufficient to disrupt promoter transcription. Although it remains plausible that methylation of this base acts to inactivate the putative Pax6 binding site, it seems more likely that methyl CpG binding proteins, such as MeCP2, bind to the site to repress activity (Grayson *et al.*, 2005). While these experiments are interesting, they do not clarify how Dnmt1 or

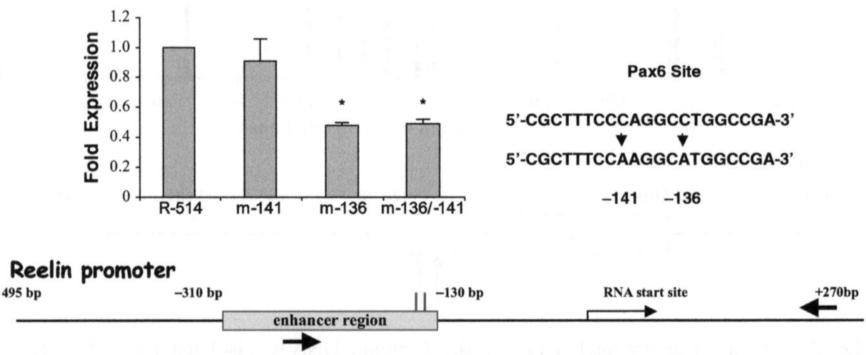

Fig. 23.6 Reelin promoter point mutations. We designed site-directed mutants within the Pax6 binding site that had previously been shown to be more heavily methylated in patients with SZ (Grayson *et al.*, 2005). These corresponded to the double (−141/−136), and single promoter mutants (m −141) and (m −136). These minimal mutants were introduced into NT2 cells using transient transfection assays and reporter activity was measured 36hr later. NT2 cells transfected with the single mutant (m −136) and double mutant construct (m −136/−141) exhibited 50% of the activity of the −514 promoter. **p*, 0.05 expressed as a percent of the SV40 promoter and compared with the reelin −514 promoter for statistical purposes (one-way ANOVA followed by Fisher LSD Method) (*See Color Plates*)

other Dnmts act to induce promoter hypermethylation at subsets of promoters operative in the regulation of GABAergic neurons. Answers to these questions will provide clues to the study of mechanisms that may underlie the etiology of SZ and will provide us with additional targets for drug interventions.

8 Reelin Deficiency in an Epigenetic Reeler Mouse Model Relevant to SZ

8.1 Heuristic Value of the Epigenetic Model

The significance of epigenetic downregulation of reelin and GAD67 expression as pathological entities in SZ morbidity would be strengthened by confirming in animal models that reminiscent of SZ, there is a cause–effect relationship between reelin and GAD67 promoter hypermethylation, reelin and GAD67 expression downregulation, and onset of reelin and behavioral dysfunctions. An epigenetic animal model of SZ could be useful not only to clarify the complex mechanisms of gene expression regulation but also to study their pharmacology to reverse the expression of epigenetic mechanisms operative in SZ. Although the HRM model and several other animal models relevant to SZ have been proposed and studied (for reviews, see Gray, 1998; Lipska and Weinberger, 2000; Moser *et al.*, 2000; Costa *et al.*, 2001; Kilts, 2001; Marcotte *et al.*, 2001; Andres, 2002; Murcia *et al.*, 2005; Tueting *et al.*, 2006), they are all "traditional models" that have been built on genetic but not epigenetic profiling. A new epigenetic model was proposed by Tremolizzo *et al.* (2002), who administered large doses of L-methionine to mice for 15 days to induce reelin and GAD67 promoter CpG-island promoter hypermethylation of genes, such as RELN and GAD67. The rationale for this epigenetic mouse model of SZ is based on several reports of the exacerbation of psychotic symptoms elicited by a 2-week treatment of SZ patients with high daily doses (20–40 g) of L-methionine (Wyatt *et al.*, 1971).

8.2 Protracted MET Treatment Induces Reelin and GAD67 Promoter Hypermethylation

The epigenetic mouse model established by administering large dose regimens of L-methionine (MET, 0.25–1 g/kg twice a day for 3 to 15 days) results in: (a) increased FC content of the methyl donor SAM, (b) reelin promoter hypermethylation, and (c) downregulation of the expression of reelin and GAD67 genes by an extent (~50%) that is comparable to the reelin and GAD67 expression deficit measured in the PFC of SZ patients (Guidotti *et al.*, 2000; Tremolizzo *et al.*, 2002).

The time required for MET treatment to induce a maximum hypermethylation of reelin and GAD67 promoters has been studied using bisulfite DNA promoter sequencing and MSP (methylation-specific PCR) (Tremolizzo *et al.*, 2002, 2005; Dong *et al.*, 2005, 2007). The CpG dinucleotide hypermethylation of RELN and GAD67, but not GAD65 or NSE promoters, increases in a time-dependent manner and reaches its maximum levels within 6–7 days of MET treatment.

Based on bisulfite DNA sequencing data, the analysis of single CpG dinucleotide methylation in the reelin promoter region of mice (from −340 and +160 bp) indicates that after MET treatment, hypermethylation is primarily restricted to CpGs in a region just upstream of the transcriptional start site (Fig. 23.7).

The mechanisms by which MET induces reelin promoter hypermethylation may depend on an increase of brain SAM content that alters high-order chromatin remodeling in GABAergic neurons by: (a) inducing nucleosomal histone (H) tail hypermethylation (i.e., MET—5.2 mmol/kg s.c., twice daily for 15 consecutive days—more than doubled the FC content of dimethyl lysine(K9-H3); and (b) recruiting multifunctional repressor complexes comprising histone methyl transferases (HMTs), histone deacetylases (HDACs), and Dnmts (Burgers *et al.*, 2002; Jenuwein, 2002; Johnstone, 2002; Dong *et al.*, 2005). Further, the recruitment of Dnmt1 methylates reelin and GAD67 promoter CpG dinucleotides.

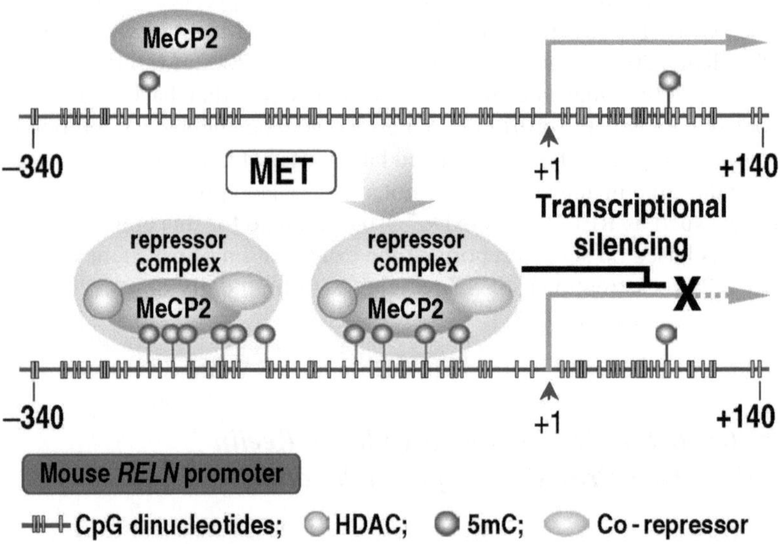

Fig. 23.7 Proposed mechanisms by which mouse *RELN* promoter hypermethylation and recruitment of chromatin remodeling complexes (MeCP2, HDACs, and co-repressors) regulate reelin gene expression. The mouse reelin (*RELN*) promoter region depicted here follows that reported by Tremolizzo *et al.* (2002) and includes the repressor protein complex. Vertical bars represent CpG dinucleotides present in this region. Pink dots denote 5mC present in the sequence. Note the increase of 5mC in MET (methionine)-treated mice. MeCP2 recruits co-repressor complexes including HDACs and induces a state of gene repression (*See Color Plates*)

We have also demonstrated that hypermethylated reelin and GAD67 promoters recruit methyl CpG binding proteins (i.e., MECP2 protein) (Dong *et al.*, 2005), which likely contribute to the transition from a transcriptionally active chromatin (euchromatin) to a transcriptionally repressed chromatin (heterochromatin). The proposed mechanism is shown in Fig. 23.7.

8.3 MET-Induced Epigenetic Mouse Model of SZ to Evaluate Prospective Drugs Capable of Increasing Reelin and GAD67 Expression by Inducing DNA-Demethylation

Mice treated with 0.75 g/kg of MET for 15 days, in addition to a decrease of FC reelin and GAD67 that mimics the decrease of reelin and GAD67 observed in the PFC of SZ patients in its intensity, appear to mimic specific phenotypic aspects of SZ. These include a decreased number of spines in apical dendrites of cortical layer II and III pyramidal neurons, and behavioral deficits such as prepulse inhibition of startle (PPI), social interaction, and cognitive abnormalities (Tremolizzo *et al.*, 2005). Thus, we have inferred that the MET-induced epigenetic mouse model can be used to study whether drugs that reduce MET-induced behavioral abnormalities related to SZ also reduce reelin and GAD67 promoter hypermethylation and reelin and GAD67 mRNA/protein downregulation.

A logical strategy for the treatment of the reelin/GABAergic dysfunction in SZ would be to normalize the SZ-related increase of Dnmt1 expression in corticolimbic GABAergic neurons by reducing the hypermethylation of reelin and GAD67 promoters with the use of inhibitors of Dnmt1 catalytic activity. However, the most potent Dnmt1 inhibitors (5-aza-cytidine, zebularine) available today fail to readily cross the blood–brain barrier when administered systemically, and are only active in the S-phase of the cell cycle (Brueckner and Lyko, 2004). Hence, a new approach to the treatment of SZ may result from a search for drugs that display direct Dnmt1 catalysis inhibitory activity in nondividing differentiated neurons or drugs that inhibit reelin and GAD67 promoter hypermethylation indirectly by inducing DNA-demethylase activity.

In the epigenetic field, a prevailing concept about the steady-state levels of DNA methylation has been that DNA methylation in somatic cells is almost exclusively maintained by the activities of Dnmts. However, accumulating evidence suggests that the induction of active DNA-demethylases may play an important role in regulating the level of methylated cytosines at various functional stages of differentiation in mammalian cells (Szyf, 2005).

There is also evidence that active demethylation of specific genes is not restricted to differentiating cells but may also take place in somatic postdifferentiated cells, including terminally differentiated neurons. For example, reelin and GAD67 promoters are hypermethylated in the cortex and hippocampus of MET-treated mice, but these promoters can undergo rapid demethylation by the administration of the HDAC inhibitors VPA or MS-275 (Dong *et al.*, 2007), presumably via an induction of DNA-demethylases. Since VPA is gaining clinical importance as a coadjuvant of antipsychotic efficacy in SZ therapy, the benefits of a combination of VPA and atypical antipsychotics in SZ treatment prompted us to study whether epigenetic mechanisms are also included in these treatments with antipsychotics. In preliminary studies, we observed that clozapine (2.5–10 mg/kg) and sulpiride (2.5–10 mg/kg), but not haloperidol (0.5–2.0 mg/kg), exhibited a dose-related increase in the cortical and hippocampal content of acetylated histone-3 (Ac-H3), in 2 hours. Clozapine and sulpiride injected into MET-pretreated mice in doses that increase brain Ac-H3 induce a rapid demethylation of hypermethylated reelin and GAD67 promoters. Furthermore, when clozapine or sulpiride at relatively low doses (i.e., 1.25 mg/kg) was given together with threshold HDAC inhibitor doses of VPA (i.e., 0.75 g/kg; Tremolizzo *et al.*, 2002), the two atypical antipsychotics dramatically accelerated the demethylation of the hypermethylated reelin and GAD67 promoters. In contrast, haloperidol, a typical antipsychotic, was ineffective.

We are currently studying other typical and atypical antipsychotics. So far, the results suggest that: (a) atypical antipsychotics may enhance DNA demethylase activity, (b) direct and indirect activators of nuclear DNA-demethylase may have a beneficial action in relieving psychotic symptoms via DNA demethylation of promoters regulating the expression of genes such as reelin and GAD67 downregulated by hypermethylation in SZ, and (c) the coadministration of antipsychotics with HDAC inhibitors may increase their potency and reduce their side effects during SZ treatment.

Hence, a pharmacological strategy with great potential to normalize the reduced amount of reelin, GAD67, or other protein expression in cortical GABAergic neurons of SZ or BP patients is to use drugs that by inhibiting HDACs can reduce the pathology of hypermethylation of reelin and GAD67 promoters via an induction of DNA-demethylases.

The possible success of such a strategy is supported by a report that the short-chain fatty acid VPA (used as an adjunctive with antipsychotics in the medication of SZ morbidity) and the benzamides MS-275 or sulpiride given to animals in doses that increase acetylation of brain chromatin histones (Tremolizzo *et al.*, 2002, 2005; Simonini *et al.*, 2006), induce reelin and GAD67 promoter demethylation and thereby antagonize MET-induced reelin and GAD67 expression downregulation (Tremolizzo *et al.*, 2002, 2005; Weaver *et al.*, 2006; Dong *et al.*, 2007).

9 Conclusions

Evidence is accumulating that the hypermethylation of reelin and GAD67 promoters is part of the etiopathogenetic process that leads to their transcriptional inactivation and to the GABAergic dysfunction and cognitive deficits found in SZ. This evidence encourages the development of new treatment procedures that can reverse the epigenetically induced reelin and GAD67 transcriptional inactivation by acting on the dynamic interplay of chromatin remodeling processes, including DNA promoter methylation, DNA promoter demethylation, and covalent histone modifications.

Recent studies suggest that DNA methylation and demethylation of genes encoding for reelin and GAD67 and perhaps of other genes are dynamic processes ongoing in the adult brain (Costa *et al.*, 2001; Guidotti *et al.*, 2005; Dong *et al.*, 2007; Miller and Sweatt, 2007). In SZ, the downregulation of reelin and GAD67 expression elicited by hypermethylation of reelin and GAD67 promoters points to a role for epigenetic mechanisms in memory formation.

Presently, one may speculate that a future SZ medication to be considered for normalizing reelin and GAD67 downregulation in the treatment protocols of SZ patients might be the use of Dnmt1 activity antagonists (i.e., procainamide) (Brueckner and Lyko, 2004). These antagonists may be administered with VPA or with its more active HDAC inhibitor analogues in an attempt to induce the activity of neuronal DNA-demethylase.

In fact, so far, the more promising drugs to target epigenetic disorders of cortical GABAergic neurons in SZ are the HDAC inhibitors which have the ability to induce DNA demethylation (Dong *et al.*, 2007). Two of these drugs, namely, VPA and sulpiride, are already known to be beneficial in SZ when administered in combination with atypical antipsychotics in patients resistant to antipsychotic monotherapy (Wassef *et al.*, 2003; Munro *et al.*, 2004).

The analogies of sulpiride action with that of VPA on HDAC activity, which modify reelin and GAD67 expression, suggest that in the treatment of SZ symptomatology, their adjuvant action with atypical antipsychotics may be mediated via an epigenetic modification of GABAergic tone.

To use the GABAergic neurotransmitter system as a target for SZ treatment is now becoming almost mandatory (Guidotti *et al.*, 2005); it does not make any sense, and it is not even justified to insist on the continuation of monotherapy with dopamine antagonists, which are not as potent as VPA or HDAC inhibitors combined.

Our data on reelin and GAD67 gene expression regulation point out numerous directions for future research on a therapy targeted to epigenetic mechanisms, including the actions of typical and atypical antipsychotics on the catalytic activities of Dnmts and DNA-demethylases.

References

Abdolmaleky, H. M., Cheng, K. H., Russo, A., Smith, C. L., Faraone, S. V., Wilcox, M., Shafa, R., Glatt, S. J., Nguyen, G., Ponte, J. F., Thiagalingam, S., and Tsuang, M. T. (2005). Hypermethylation of the reelin (RELN) promoter in the brain of schizophrenic patients; a preliminary report. *Am. J. Med. Genet. B Neuropsychiatry Genet.* 134:60–66.

Akbarian, S., Kim, J. J., Potkin, S. G., Hagman, J. O., Tafazzoli, A., Bunney, W. E., Jr., and Jones, E. G. (1995). Gene expression for glutamic acid decarboxylase is reduced without loss of neurons in prefrontal cortex of schizophrenics. *Arch. Gen. Psychiatry* S2:258–266.

Alcantara, S., Ruiz, M., D'Arcangelo, G., Ezan, F., de Lecea, L., Curran, T., Sotelo, C., and Soriano, E. (1998). Regional and cellular patterns of reelin mRNA expression in the forebrain of the developing and adult mouse. *J. Neurosci.* 18:7779–7799.

Andres, C. (2002). Molecular genetics and animal models in autistic disorder. *Brain Res. Bull.* 57:109–119.

Beffert, U., Weeber, E. J., Durudas, A., Qiu, S., Masiulis, I., Sweatt, J. D., Li, W. P., Adelmann, G., Frotscher, M., Hammer, R. E., and Herz, J. (2005). Modulation of synaptic plasticity and memory by reelin involves differential splicing of the lipoprotein receptor Apoer2. *Neuron* 47:567–579.

Benes, F. M., and Beretta, S. (2001).GABAergic interneurons: implications for understanding schizophrenia and bipolar disorder. *Neuropsychopharmacology* 25:1–27.

Black, J. E., Kodish, I. M., Grossman, A. W., Klintsova, A. Y., Orlovskaya, D., Vostrikov, V., Uranova, N., and Greenough, W. T. (2004). Pathology of layer V pyramidal neurons in the prefrontal cortex of patients with schizophrenia. *Am. J. Psychiatry* 161:742–744.

Brueckner, B., and Lyko, F. (2004). DNA methyltransferase inhibitors: old and new drugs for an epigenetic cancer therapy. *Trends Pharmacol. Sci.* 25:551–554.

Burgers, W. A., Fuks, F., and Kouzarides, T. (2002). DNA methyltransferases get connected to chromatin. *Trends Genet.* 18:275–277.

Carboni, G., Tueting, P., Tremolizzo, L., Sugaya, I., Davis, J., Costa, E., and Guidotti, A. (2004). Enhanced dizocilpine efficacy in heterozygous reeler mice relates to GABA turnover downregulation. *Neuropsychopharmacology* 46:1070–1081.

Chen, Y., Sharma, R., Costa, R. H., Costa, E., and Grayson, D. R. (2002). On the epigenetic regulation of the human reelin promoter. *Nucleic Acids Res.* 30:2930–2939.

Chen, Y., Beffert, U., Ertunc, M., Tang, T. S., Kavalali, E. T., Bezprozvanny, I., and Herz, J. (2005). Reelin modulates NMDA receptor activity in cortical neurons. *J. Neurosci.* 25:8209–8216.

Chen, Y., Kundakovic, M., Agis-Balboa, R. C., Pinna, G., and Grayson, D. R. (2007). Induction of the reelin promoter by retinoic acid is mediated by Sp1. *J. Neurochem.* 103:650–665.

Costa, E., Davis, J., Grayson, D. R., Guidotti, A., Pappas, G. D., and Pesold, C. (2001). Dendritic spine hypoplasticity and downregulation of reelin and GABAergic tone in schizophrenia vulnerability. *Neurobiol. Dis.* 8:723–742.

D'Arcangelo, G., Miao, G. C., Chen, S. C., Soares, H. D., Morgan, J. I., and Curran, T. (1995). A protein related to extracellular matrix proteins deleted in the mouse mutant reeler. *Nature* 374:719–723.

DeSilva, U., D'Arcangelo, G., Braden, V. V., Chen, J., Miao, G. G., Curran, T., and Green, E. D. (1997). The human reelin gene; isolation, sequencing, and mapping on chromosome 7. *Genome Res.* 7:157–164.

Dong, E., Caruncho, H., Liu, W. S., Smalheiser, N. R., Grayson, D. R., Costa, E., and Guidotti, A. (2003). The reelin–integrin interaction regulates Arc mRNA translation in synaptoneurosomes. *Proc. Natl. Acad. Sci. USA* 100:5479–5484.

Dong, E., Agis-Balboa, C., Simonini, M. V., Grayson, D. R., Costa, E., and Guidotti, A. (2005). Reelin and glutamic acid decarboxylase67 promoter remodeling in an epigenetic methionine-induced mouse model of schizophrenia. *Proc. Natl. Acad. Sci. USA* 102:12578–12583.

Dong, E., Grayson, D. R., Guidotti, A., and Costa, E. (2007). Histone hyperacetylation induces demethylation of reelin and 67-kDa glutamic acid decarboxylase promoters. *Proc. Natl. Acad. Sci. USA* 104:4676–4681.

Eastwood, S. L., and Harrison, P. J. (2003). Interstitial white matter neurons express less reelin and are abnormally distributed in schizophrenia: towards an integration of molecular and morphologic aspects of the neurodevelopmental hypothesis. *Mol. Psychiatry* 8:821–831.

Fatemi, S. H., Earle, J. A., and McMenomy, T. (2000). Reduction in reelin immunoreactivity in hippocampus of subjects with schizophrenia, bipolar disorder and major depression. *Mol. Psychiatry* 5:654–663.

Glantz, L. A., and Lewis, D. A. (2000). Decreased dendritic spine density on prefrontal cortical pyramidal neurons in schizophrenia. *Arch. Gen. Psychiatry* 57:65–73.

Goll, M. G., and Bestor, T. H. (2005). Eukaryotic cytosine methyltransferases. *Annu. Rev. Biochem.* 74:481–514.

Gray, J. A. (1998). Integrating schizophrenia. *Schizophr. Bull.* 24:249–266.

Grayson, D. R., Jia, X., Chen, Y., Sharma, R. P., Mitchell, C. P., Guidotti, A., and Costa, E. (2005). Reelin promoter hypermethylation in schizophrenia. *Proc. Natl. Acad. Sci. USA* 102:9341–9346.

Guidotti, A., Auta, J., Davis, J. M., DiGiorgi-Cerenini, V., Dwivedi, J., Grayson, D. R., Impagnatiello, F., Pandey, G. N., Pesold, C., Sharma, R. F., Uzunov, D. P., and Costa, E. (2000). Decreased reelin and glutamic acid decarboxylase 67 (GAD_{67}) expression in schizophrenia and bipolar disorders: a postmortem brain study. *Arch. Gen. Psychiatry* 57:1061–1069.

Guidotti, A., Auta, J., Davis, J. M., Dong, E., Grayson, D. R., Veldic, M., Zhang, X., and Costa, E. (2005). GABAergic dysfunction in schizophrenia; new treatment strategies on the horizon. *Psychopharmacology (Berlin)* 180:191–205.

Hall, H., Lawyer, G., Sillen, A., Johnsson, E. G., Agartz, I., Terenius, L., and Arnborg, S. (2007). Potential variants in schizophrenia: a Bayesian analysis. *World J. Biol. Psychiatry* 8:12–22.

Herz, J., and Chen, Y. (2006). Reelin, lipoprotein receptors and synaptic plasticity. *Nature Rev. Neurosci.* 7:850–859.

Huang, C. H., and Chen, C. H. (2006). Absence of association of a polymorphic GGC repeat at 5' untranslated region of the reelin gene with schizophrenia. *Psychiatry Res.* 142:89–92.

Ignatova, N., Sindic, C. J., and Goffinet, A. M. (2004). Characterization of the various forms of the reelin protein in the cerebrospinal fluid of normal subjects and in neurological diseases. *Neurobiol. Dis.* 15:326–330.

Impagnatiello, F., Guidotti, A., Pesold, C., Dwivedi, Y., Caruncho, H., Pisu, M. G., Smalheiser, N. R., Davis, J. M., Pandey, G. N., Pappas, G. D., Tueting, P., and Costa, E. (1998). A decrease of reelin expression as a putative vulnerability factor in schizophrenia. *Proc. Natl. Acad. Sci. USA* 95:15718–15723.

Jenuwein, T. (2002). Molecular biology. An RNA-guided pathway for the epigenome. *Science* 297:2215–2218.

Johnstone, R. W. (2002). Histone-deacetylase inhibitors; novel drugs for the treatment of cancer. *Nature Rev. Drug Discov.* 1:287–299.

Kilts, C. D. (2001). The changing roles and targets for animal models of schizophrenia. *Biol. Psychiatry* 50:845–855.

Krystal, J. H., Belger, A., D'Souza, D. C., Anand, A., Charney, D. S., Aghajanian, G. K., and Moghaddam, B. (1999). Therapeutic implications of the hyperglutamatergic effects of NMDA antagonists. *Neuropsychopharmacology* 21:S143–S157.

Lacor, P. H., Grayson, D. R., Auta, J., Sugaya, I., Costa, E., and Guidotti, A. (2000). Reelin secretion from glutamatergic neurons in culture is independent from neurotransmitter regulation. *Proc. Natl. Acad. Sci. USA* 97:3556–3561.

Larson, J., Hoffman, J. S., Guidotti, A., and Costa, E. (2003). Olfactory discrimination learning deficit in heterozygous reeler mice. *Brain Res.* 971:40–46.

Levenson, J. M., and Sweatt, J. D. (2005). Epigenetic mechanisms in memory formation. *Nature Rev. Neurosci.* 6:108–118.

Lewis, D. A., Hashimoto, T., and Volk, D. W. (2005). Cortical inhibitory neurons and schizophrenia. *Nature Rev. Neurosci.* 6:312–324.

Lipska, B. K., and Weinberger, D. R. (2000). To model a psychiatric disorder in animals: schizophrenia as a reality test. *Neuropsychopharmacology* 23:223–239.

Liu, W. S., Pesold, C., Rodriguez, M. A., Carboni, G., Auta, J., Lacor, P., Larson, J., Condie, B., Guidotti, A., and Costa, E. (2001). Downregulation of dendritic spine and glutamic acid decarboxylase$_{67}$ expressions in reelin haploinsufficient heterozygous reeler mouse. *Proc. Natl. Acad. Sci. USA* 98:3477–3482.

Marcotte, E. R., Pearson, D. M., and Srivastava, L. K. (2001). Animal models of schizophrenia: a critical review. *J. Psychiatry Neurosci.* 26:395–410.

Miller, C. H., and Sweatt, J. D. (2007). Covalent modification of DNA regulates memory formation. *Neuron* 53:857–869.

Moser, P. C., Hitchcock, J. M., Lister, S., and Moran, P. M. (2000). The pharmacology of latent inhibition as an animal model of schizophrenia. *Brain Res. Rev.* 33:275–307.

Munro, J., Matthiasson, P., Osborne, S., Travis, M., Purcell, S., Cobb, A. M., Launer, M., Beer, M. D., and Kerwin, R. (2004). Amisulpride augmentation of clozapine: an open non-randomized study in patients with schizophrenia partially responsive to clozapine. *Acta Psychiatr. Scand.*110:292–298.

Murcia, C. L., Gulden, F., and Herrup, K. (2005). A question of balance: a proposal for new mouse models of autism. *Int. J. Dev. Neurosci.* 23(2-3):265–275.

Niu, S., Renfro, A., Quattrocchi, C. C., Sheldon, M., and D'Arcangelo, G. (2004). Reelin promotes hippocampal dendrite development through the VLDLR/ApoER2-Dab1 pathway. *Neuron* 41:71–84.

Pappas, G. D., Kriho, V., and Pesold, C. (2001). Reelin in the extracellular matrix and dendritic spines of the cortex and hippocampus: a comparison between wild type and heterozygous reeler mice by immunoelectron microscopy. *J. Neurocytol.* 30:413–425.

Persico, A. M., Levitt, P., and Pimenta, A. F. (2006). Polymorphic GGC repeat differentially regulates human reelin gene expression levels. *J. Neural Transm.* 113:1373–1382.

Pesold, C., Impagnatiello, F., Pisu, M. G., Uzunov, D. P., Costa, E., Guidotti, A., and Caruncho, H. J. (1998). Reelin is preferentially expressed in neurons synthesizing γ-aminobutyric acid in cortex and hippocampus of adult rats. *Proc. Natl. Acad. Sci. USA* 95:3221–3226.

Pesold, C., Liu, W. S., Guidotti, A., Costa, E., and Caruncho, H. J. (1999). Cortical bitufted, horizontal, and Martinotti cells preferentially express and secrete reelin into perineuronal nets, nonsynaptically modulating gene expression. *Proc. Natl. Acad. Sci. USA* 96:3217–3222.

Qiu, S., and Weeber, E. F. (2007). Reelin signaling facilitates maturation of CA1 glutamatergic synapses. *J. Neurophysiol.* 97:2312–2321.

Qiu, S., Korwek, D. M., Pratt-Davis, A. R., Peters, M., Bergman, M. Y., and Weeber, E. J. (2006). Cognitive disruption and altered hippocampus synaptic function in reelin haploinsufficient mice. *Neurobiol. Learn. Mem.* 85:228–242.

Richter, J. D., and Lorenz, L. J. (2002). Selective translation of mRNAs at synapses. *Curr. Opin. Neurobiol.* 12:300–304.

Rodriguez, M.A., Pesold, C., Liu, W. S., Kriho, V., Guidotti, A., Pappas, G., and Costa, E. (2000). Colocalization of integrin receptors and reelin in dendritic spine postsynaptic densities of adult non-human primate cortex. *Proc. Natl. Acad. Sci. USA* 97:3550–3555.

Rodriguez, M. A., Caruncho, H. J., Costa, E., Pesold, C., Liu, W. S., and Guidotti, A. (2002). In Patas monkey GAD67 and reelin mRNA coexpression varies in a manner dependent of layer and cortical area. *J. Comp. Neurol.* 451:279–288.

Rosoklija, G., Toomayan, G., Ellis, S. P., Keilp, J., Mann, J. J., Latov, N., Hays, A. P., and Dwork, A. J. (2000). Structural abnormalities of subicular dendrites in subjects with schizophrenia and mood disorders; preliminary findings. *Arch. Gen. Psychiatry* 57:349–356.

Royaux, I., Lambert de Rouvroit, C., D'Arcangelo, G., Demirov, D., and Goffinet, A. M. (1997). Genomic organization of the mouse reelin gene. *Genomics* 46:240–250.

Ruzicka, W., Zhubi, A., Veldic, M., Grayson, D. R., Costa, E., and Guidotti, A. (2007). Selective epigenetic alteration of layer I GABAergic neurons isolated from prefrontal cortex of schizophrenia patients using laser-assisted microdissection. *Mol. Psychiatry* 12:385–397.

Selemon, L. D., and Goldman-Rakic, P. S. (1999). The reduced neuropil hypothesis: a circuit-based model of schizophrenia. *Biol. Psychiatry* 45:17–25.

Simonini, M. V., Carmargo, L. M., Dong, E., Maloku, E., Veldic, M., Costa, E., and Guidotti, A. (2006). The benzamide MS-275 is a potent long-lasting brain region-selective inhibitor of histone deacetylases. *Proc. Natl. Acad. Sci. USA*103:1587–1592.

Spencer, D. M., Nestor, P. G., Perlmutter, R., Niznikiewicz, M. A., Klump, M. C., Frumin, M., Shenton, M. E., and McCarley, R. W. (2004). Neural synchrony indexes disordered perception and cognition in schizophrenia. *Proc. Natl. Acad. Sci. USA* 101:17288–17293.

Steward, O., and Schuman, E. M. (2001). Protein synthesis at synaptic sites on dendrites. *Annu. Rev. Neurosci.* 24:299–325.

Szyf, M. (2005). Therapeutic implications of DNA methylation. *Future Oncol.* 1(1):125–135.

Tremolizzo, L., Carboni, G., Ruzicka, W. B., Mitchell, C. O., Sugaya, I., Tueting, P., Sharma, R., Grayson, D. R., Costa, E., and Guidotti, A. (2002). An epigenetic mouse model for epigenetic molecular and behavioral neuropathologies related to schizophrenia vulnerability. *Proc. Natl. Acad. Sci. USA* 99:17095–17100.

Tremolizzo, L., Doueiri, M. S., Dong, E., Grayson, D. R., Pinna, G., Tueting, P., Rodriguez-Menendez, V., Costa, E., and Guidotti, A. (2005). Valproate corrects the schizophrenia-like epigenetic behavioral modifications induced by methionine in mice. *Biol. Psychiatry* 57:500–509.

Tueting, P., Costa, E., Dwivedi, Y., Guidotti, A., Impagnatiello, F., Manev, R., and Pesold, C. (1999). The phenotypic characteristics of heterozygous reeler mouse. *NeuroReport* 10:1329–1334.

Tueting, P., Doueiri, M. S., Guidotti, A., Davis, J. M., and Costa, E. (2006). Reelin downregulation in mice and psychosis endophenotypes. *Neurosci. Biobehav. Rev.* 30:1065–1077.

Veldic, M., Caruncho, H. M., Liu, W. S., Davis, J., Satta, R., Grayson, D. R., Guidotti, A., and Costa, E. (2004). DNA methyltransferase-1 (DNMT1) is selectively overexpressed in telencephalic GABAergic interneurons of schizophrenia brains. *Proc. Natl. Acad. Sci. USA* 101:348–353.

Veldic, M., Guidotti, A., Maloku, E., Davis, J. M., and Costa, E. (2005). In psychosis, cortical interneurons overexpress DNA-methyltransferase 1. *Proc. Natl. Acad. Sci. USA* 102:2152–2157.

Veldic, M., Kadriu, B., Maloku, E., Agis-Balboa, R. C., Guidotti, A., Davis, J. M., and Costa, E. (2007). Epigenetic mechanisms expressed in basal ganglia GABAergic neurons differentiate schizophrenia from bipolar disorder. *Schizophr. Res.* 91:51–61.

Wassef, A., Baker, J., and Kochan, L. D. (2003). GABA and schizophrenia: a review of basic science and clinical studies. *J. Clin. Psychopharmacol.* 23:601–640.

Weaver, I.C., Meaney, M. J., and Szyf, M. (2006). Maternal care effects on hippocampal transcriptome and anxiety-mediated behaviors in the offspring that are reversible in adulthood. *Proc. Natl. Acad. Sci. USA* 103:3480–3485.

Weeber, E. J., Beffert, U., Jones, C., Christian, J. M., Forster, E., Sweatt, J. D., and Herz, J. (2002). Reelin and ApoE receptors cooperate to enhance hippocampal synaptic plasticity and learning. *J. Biol. Chem.* 277:39944–39952.

Woo, T. U., Walsh, J. P., and Benes, F. M. (2004). Density of glutamic acid decarboxylase 67 messenger RNA-containing neurons that express the N-methyl-D-aspartate receptor subunit NR_{2A} in the anterior cingulate cortex in schizophrenia and bipolar disorder. *Arch. Gen. Psychiatry* 61:649–657.

Wyatt, R. J., Benedict, A., and Davis, J. (1971). Biochemical and sleep studies of schizophrenia: a review of the literature 1960–1970. *Schizophr. Bull.* 4:10–44.

Yin, Y., Edelman, G. M., and Vanderklish, P. W. (2002). The brain-derived neurotrophic factor enhances synthesis of Arc in synaptoneurosomes. *Proc. Natl. Acad. Sci. USA* 99:2368–2374.

Chapter 24
Epigenetic Modulation of Reelin Function in Schizophrenia and Bipolar Disorder

Hamid Mostafavi Abdolmaleky, Cassandra L. Smith, Jin-Rong Zhou, and Sam Thiagalingam

Contents

H. M. Abdolmaleky
Biomedical Engineering Department, Boston University, Boston; Laboratory of Nutrition and Metabolism at BIDMC, Department of Surgery, Harvard Medical School, Boston; Departments of Medicine (Genetics Program), Genetics & Genomics, and Pathology & Laboratory Medicine, Boston University School of Medicine, Boston, MA; Department of Psychiatry, Tehran Psychiatric Institute and Mental Health Research Center, Iran University of Medical Sciences, Tehran, Iran
e-mail: hamostab@bu.edu and hamostafavi@yahoo.com

C. L. Smith
Biomedical Engineering Department, Boston University, Boston, MA 02215

J. -R. Zhou
Laboratory of Nutrition and Metabolism at BIDMC, Department of Surgery, Harvard Medical School, Boston, MA 02115

S. Thiagalingam
Departments of Medicine (Genetics Program), Genetics & Genomics, and Pathology & Laboratory Medicine, Boston University School of Medicine, Boston, MA 02118

S. H. Fatemi (ed.), *Reelin Glycoprotein: Structure, Biology and Roles in Health and Disease.*

1 Introduction

Studies from several laboratories have provided convincing data to support the notion that altered DNA methylation in response to varying physiological and environmental conditions may play a critical role in the fine-tuning of gene expression. However, the establishment of abnormal gene promoter DNA methylation patterns resulting from environmental insults or dysfunctional genes of the DNA methylation machinery may destabilize the normal epigenetic modification of genes. This may affect the equilibrium in the differential gene expression patterns in the normal differentiated cells and tilt the balance toward the disease phenotype. The individuals with genetic susceptibility to specific diseases are likely to be more prone to abnormal DNA methylation. Thus, it is highly likely that the lack of a direct relationship between genotype and phenotype in major psychiatric disorders and the variability in the manifestation of diseases in individuals with identical genetic makeup could be derived from the changes in the DNA methylation patterns.

Recent studies have reported that hypermethylation of DNA at different sites of the promoter region of reelin *(RELN)* with corresponding alterations in gene expression is associated with schizophrenia (SCZ) pathogenesis. It appears that hypermethylation of proximal cytosines of the promoter may recruit inhibitory proteins (i.e., MeCP and MBD2), while the methylation of the distal sites may inhibit binding of the stimulatory factors (SP-1 and CRE), impacting the expression of *RELN*, a hypoexpressed gene in SCZ and mood disorders. In addition to the effects of micro- and macroenvironment during early or later life, dysfunction of dopaminergic and serotonergic systems may also be responsible for the aberrant DNA methylation pattern of *RELN* promoter in SCZ and bipolar disorder (BD).

The findings obtained from the *RELN* studies provide compelling reasons for future validation of these initial observations, as well as for the extension of these studies to other genes with significant roles that are still to be uncovered. In the long term, this line of research will likely play a significant role in helping to develop strategies for early diagnosis, prevention, and therapy of SCZ, BD, and other mental diseases. Importantly, these studies strongly suggest that, in addition to genetic analyses, epigenetic analyses of the candidate genes are necessary before one can formulate a comprehensive picture of the molecular basis of the pathogenesis of complex diseases.

About 10 years ago, after nearly three decades of epigenetic research in medicine, the concept of epigenetics, defined by Waddington in 1940 (reviewed by Morange, 2002), was introduced to the field of psychiatry, and it was followed by extensive discussions and research to establish its potential roles in the pathogenesis of mental disorders (Tasman *et al.*, 1997; Petronis, 2000; Singh *et al.*, 2003; Abdolmaleky *et al.*, 2004a). Based on the current views, epigenetics refers to modifications in gene expression that are controlled by heritable, but potentially reversible, changes in DNA methylation and/or chromatin structure, RNA editing, and RNA interference, which are not accompanied by any change in DNA sequences (Bird, 2002; Jiang *et al.*, 2004; Lavorgna *et al.*, 2004). Although all of the cells of an organism have the same genetic makeup, each tissue and even individual cell may elicit specified functions in multicellular organisms. Throughout cell differentiation, as well as evolution, epigenetic modifications of the DNA structure could provide the cells with unique identity and function in each cellular network and environmental condition, however in a dynamic manner (Russo *et al.,* 1996; Bird, 2002). Although epigenetics marking is established and developed, due to interactions with other cells and the environment, the cell-specific epigenetic profiles, which govern cell differentiation during embryogenesis, are retained in the genomic memory to be transferred to the next generation of cells (Monk, 1995; Russo *et al.,* 1996; Bird, 2002).

DNA methylation is one of the best-known epigenetic mechanisms for encoding the micro-/macroenvironmental exposures, as well as the inherited epigenetic properties to the developmental memory (Monk, 1995; Russo *et al.*, 1996; Bird, 2002). Methylation of DNA is mediated by methyltransferase enzymes (Bestor, 2000; Kim *et al.*, 2002) by catalyzing the addition of a methyl group (CH_3) to the cytosines which are followed by guanine (CpG). While S-adenosyl methionine is the major methyl donor, folic acid and B12 are involved in remethylation/recruitment of the demethylated S-adenosyl methionine (Fenech, 2001). It has been shown that manipulation of any of these contributors/players can change DNA methylation patterns of genes and influence their expression levels (Bertino *et al.*, 1996; Cooney *et al.*, 2002; Dong *et al.*, 2005; Waterland *et al.*, 2006). Furthermore, it is also well documented that the manner of nurturing in early life could modulate DNA methylation pattern and gene expression levels in later life (Weaver *et al.*, 2004). Therefore, environmental insults on DNA methylation pattern could impact the epigenetic memory leading to the disease phenotype, particularly in individuals with genetic susceptibility to specific diseases. However, flexibility and dynamics of DNA methylation could also allow adaptive fine-tuning of the gene expression in variable environmental conditions or in individuals with dysfunctional polymorphisms (Abdolmaleky *et al.*, 2006).

The emergence of this new branch of science, as a new paradigm to the field of molecular genetics, has promoted a worldwide interest in examining the potential roles of epigenetics in mental illnesses. Accordingly, several studies have directly addressed the epigenetic alterations of specific genes in psychiatric disorders as summarized in the next section.

1.1 A Summary of Epigenetic Aberrations in Major Psychiatric Diseases

Epigenetic modification of DNA, resulting in differential disease phenotypes, has been well documented in other complex diseases, such as cancer. The first observations in neuroscience came from studies of fragile X and Rett syndrome (reviewed by Robertson and Wolffe, 2000), and the brain laterality of DRD2 promoter DNA methylation (Popendikyte *et al.*, 1999). In fragile X, DNA hypermethylation of the expanded CGG repeats in the *FMR1* gene, and its association with the severity of the disease was reported (Weinhausel and Haas, 2001). In Rett syndrome, a mutation in *MECP2* influences the interplay of MECP2 protein with methylated DNA (reviewed by Akbarian, 2003; Akbarian *et al.*, 2006) that normally inhibits the binding of transcription factors (Kaludov and Wolffe, 2000). Following these observations, activity-dependent DNA methylation of BDNF promoter at CRE binding sites was reported (Martinowich *et al.*, 2003) as one of the most important observations of the current decade in the field of neuroscience that was followed by elegant studies addressing the effects of early life experiments on steroid receptor promoter DNA methylation status in the hippocampus (Weaver *et al.*, 2004). Furthermore, the established epigenetic modifications of the DNA structure during early life periods were manipulated in adult animals, causing change in the corresponding phenotype (Weaver *et al.*, 2005).

In addition, several investigators reported the following methylation changes in various genes and diseases: (1) DNA hypermethylation of *RELN* promoter (Abdolmaleky *et al.*, 2005; Grayson *et al.*, 2005) and *SOX10* (Iwamoto *et al.*, 2005) in SCZ; (2) hyperexpression of DNMT1 (Veldic *et al.*, 2004, 2007; Ruzicka *et al.*, 2007) and increase in S-adenosyl methionine content (Guidotti *et al.*, 2007) in cortical interneurons of patients with SCZ and psychotic BD; (3) hypermethylation of genomic DNA for alpha synuclein (Bonsch *et al.*, 2005), as well as HERP gene promoters in alcoholism (Bleich *et al.*, 2006); and (4) hypomethylation of *MB-COMT* promoter in SCZ and BD (Abdolmaleky *et al.*, 2006).

These observations have now provided a convincing rationale in accepting the emerging science of epigenomics and methylomics, as complementary to the study of genomics, in uncovering the dilemmas of pathogeneses in major mental diseases. As 30% of human genes expressed in the brain (Davies and Morris, 1997) and ~50% of all genes harbor CG-rich promoters (Costello and Plass, 2001), approximately 5000 genes are legitimate candidates for methylation analyses in the field of neuroscience. Since methylation pattern of each gene could vary in each region of the brain, epigenetic profiling of the gene methylome, particularly targeted to the regulatory elements, could be an expanding field of research in the next decade and may serve a bridge to the next revolution in psychiatry.

1.2 Epigenetic Modulation of RELN Functions

RELN is mainly expressed by GABAergic interneurons of the brain and encodes a large extracellular matrix protein that acts on apolipoprotein E receptor 2 (ApoER2) and very-low-density lipoprotein receptor (VLDLR). *RELN* is involved in neuronal migration, cellular positioning, axonal branching, synaptogenesis, and memory formation throughout development of the brain and in later life (Pesold *et al.*, 1999; Fatemi *et al.*, 2000; Costa *et al.*, 2001, 2006; Beffert *et al.*, 2005). Binding of RELN to lipoprotein receptors (LPR) activates a tyrosine kinase (TK)-dependent cascade leading to Dab1 phosphorylation (Fig. 24.1) and activation of Src-family

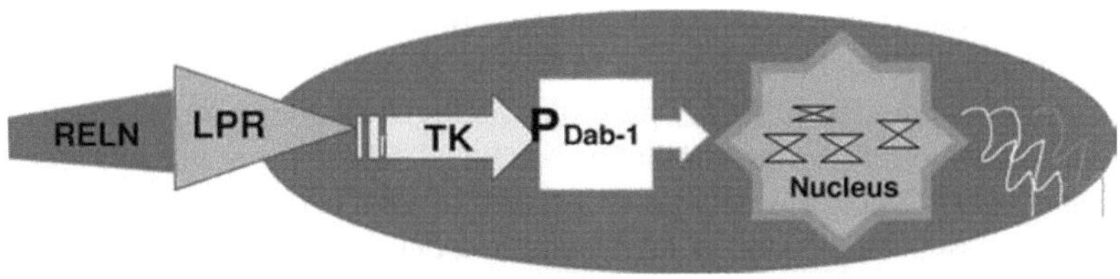

Fig. 24.1 Binding of RELN to lipoprotein receptors (LPR) activates a tyrosine kinase (TK)-dependent cascade leading to Dab1 phosphorylation and expression of several genes that lead to long-lasting structural changes (*See Color Plates*)

kinases and Akt that coordinately orchestrate the expression of several genes that lead to long-lasting structural changes (Pesold *et al.*, 1999; Homayouni *et al.*, 2003; Beffert *et al.*, 2005, 2006). RELN also causes an increase in *DRD2* expression (Ballmaier *et al.*, 2002) and modulates NMDA activity mediated by Dab1 (Chen *et al.*, 2005) and ApoER2 (Beffert *et al.*, 2005, 2006). ApoER2 also forms a functional complex with NMDA receptors and facilitates memory formation in adulthood (Beffert *et al.*, 2005). It has been reported that NMDA receptors influence acetylation of histone H3 in hippocampus (Levenson *et al.*, 2004), which is likely to be tied to DNA methylation changes. NMDA signaling is also known to participate in the regulation of several neurotransmitters, including dopamine and norepinephrine, through intracellular mechanisms involving CREB (Martucci *et al.*, 2003; Marrone *et al.*, 2006). Thus, any malfunction of the *RELN* gene could affect glutamatergic and dopaminergic pathway functions as well. Consistent with this notion, reeler mice, which are haploinsufficient for *RELN* expression, showed a decrease in dopamine transporter (DAT1) and DRD2, an increase in DRD3 levels, and a failure of dopaminergic neuron migration (Ballmaier *et al.*, 2002; Nishikawa *et al.*, 2003).

Currently, there are no polymorphisms for *RELN* that are known to be associated with SCZ or BD. Although the long allele of polymorphic GGC repeat in the 5' untranslated region of *RELN* is correlated with reduced gene expression compared to the common 8- and 10-repeat alleles (Persico *et al.*, 2006), this polymorphism has not yet been positively linked to SCZ (Akahane *et al.*, 2002; Huang and Chen, 2006). However, it is interesting to note that while animal studies showed that chronic use of antipsychotic drugs, such as olanzapine, could increase *RELN* expression and protein level (Fatemi *et al.*, 2006), individuals carrying the 10-repeat allele may have a better response to antipsychotic treatment (Goldberger *et al.*, 2005).

Human postmortem studies showed that *RELN* expression is reduced in the brains of patients diagnosed with SCZ, BD, depression, and autism (Impagnatiello *et al.*, 1998; Guidotti *et al.*, 2000; Fatemi *et al.*, 2000, 2005a, 2005b; Eastwood and Harrison, 2003; Ruzicka *et al.*, 2007; Veldic *et al.*, 2007). These observations, supported by methionine-induced exacerbation of psychotic symptoms in SCZ and BD, led to the conclusion that hypoactivity of *RELN* may be due to hypermethylation of the gene's promoter (Chen *et al.*, 2002). Animal studies also showed that L-methionine, a methyl donor, is linked to a decrease in *RELN* mRNA, which was associated with an increase in the degree of *RELN* promoter methylation (Tremolizzo *et al.*, 2002), while valproate could prevent methionine-induced *RELN* promoter hypermethylation (Chen *et al.*, 2002; Tremolizzo *et al.*, 2002). Other, more recent studies have also found that hypoexpression of *RELN* was associated with the hyperexpression of *DNMT1*, and a corresponding increase in S-adenosyl methionine content in postmortem brains of patients with SCZ and BD (Guidotti *et al.*, 2007).

RELN harbors one of the most CpG rich promoters in the human genome with 72 candidate cytosines for methylation and several regulatory elements (e.g., CRE, SP1, and the consensus GC box) in 450 base pairs upstream of the first exon (Fig. 24. 2). Most of these binding elements harbor **CG** islands in their sequence, such as CRE (TGA**CG**TCA), SP1 (GGG**CG**G), and consensus GC box (GGGG**CG**GG**CG**CC), that affect cAMP-induced activities (Park-Sarge and Sarge, 1995).

Human studies have revealed that various human tissues, such as liver, stomach, and breast, express *RELN* mRNA (Abdolmaleky *et al.*, 2005) indicating that *RELN* has a pleiotropic role in human development. Further studies on some of these tissues show an inverse correlation between *RELN* promoter methylation and expression levels (Fig. 24.3), providing direct support that *RELN* promoter methylation modulates gene expression. Our analyses of several cancer cell lines have also shown that, unlike normal tissues, *RELN* promoter is extensively hypermethylated and underexpressed in some types of cancers. For example, the MCF7 cell line contains a totally methylated *RELN* promoter (Fig. 24.3) and, as a result, the gene expression is almost 15 times less than in normal breast tissue (Abdolmaleky, Zhou, and Thiagalingam, unpublished). These observations indicate that *RELN* has diverse roles in human development and diseases that are yet to be uncovered.

GCCCTCTGCGGGGCTTTGACGTCCCTCGCAGAAGAGTCGCGGGCTCAGCGGTC
CTCGACAGCGTCCCGTCCCGCTCCCCGGCGGGCGCCCCTCCCTGTCCTCCCGG
GTGCGAACCGGGCGCTGGCCGGGGACTCCGGGGACGCGTGCGCCCCTCGCC
GCGCGAGGTGCCGCCGAGCCAGCCCGAGAGGGCGGGGGGCGGGCGGGGCG
GCGCGCGGGGGCGGGGGAGCGGCCGGGACACGTGTGGCGGCGGCGGGGGG
GACGCGGCGCCCGGGGCTTTAAGAAGGTGTGGAGCGGGGCGGGCGCTTTCCC
AGGCCTGGCCGAGGGGCGTCGCGCAGAGGCGGCGGCGGCGCACGGAGGCG
GCAGACGACGCGCTCTCGGCGCCCGCAGCCCCGGTCCCGCGCTCCCGCGGC
CCAAAGTAACTTTGGGAGCCGCCGTCTCCCGCGGAAACTT-exon

Fig. 24.2 A view of *RELN* promoter sequence. *RELN* harbors a CG-rich promoter with 72 candidate cytosine (C) sites for methylation and several regulatory binding sites located in 450 base pairs upstream of the coding region. A CRE binding site is underlined in the first line and several SP1 binding sites (GGGCGG) and a consensus GC box are underlined in other locations. The boldface **C**s that are followed by G are candidates for methylation, while other Cs or unmethylated Cs will be converted to T during bisulfite treatment (*See Color Plates*)

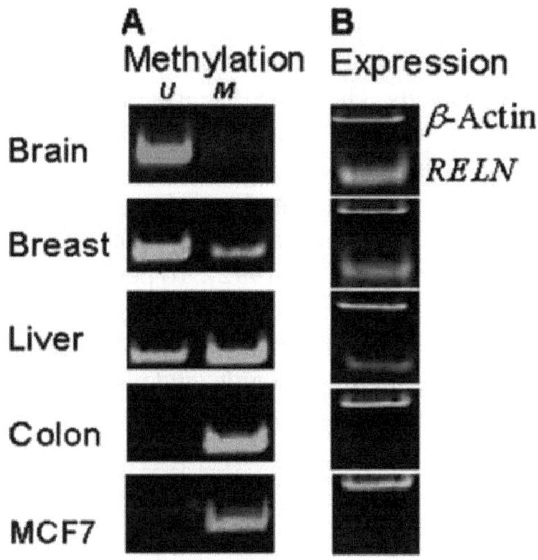

Fig. 24.3 *RELN* expression as a function of promoter methylation. *RELN* methylation and expression analyses in different human tissues showed a direct correlation between methylation of CRE and SP1 binding sites and expression levels and the brain, with the least methylation and greatestt expression levels compared to the other tissues. *U* and *M* in panel A refer to unmethylated and methylated PCR products, respectively. The upper and lower bands in panel B show β-actin and *RELN* expression levels, respectively

Examination of postmortem brains of patients with SCZ versus control subjects, analyzed independently by two groups using completely different sets of samples, although provided from the same brain banks [Harvard Brain Tissue Resource Center (HBTRC) and the Stanley Medical Research Institute (SMRI)], revealed that the *RELN* promoter DNA is hypermethylated in SCZ with a concomitant decrease in the transcript levels (Abdolmaleky *et al.*, 2005; Grayson *et al.*, 2005). Here, details of the existing data and methodology for the analyses of the methylome of *RELN* gene promoter are outlined to promote the worldwide search for this and other epigenetically altered candidate genes in psychiatric diseases.

2 Methods for the Analysis of DNA Methylation Status

2.1 Bisulfite Sequencing and Methylation-Specific PCR (MSP)

Bisulfite-treated genomic DNA sequencing and MSP have been successfully used to map the differentially methylated CpG islands in the promoter regions of the *DRD2, RELN, COMT*, and other genes in psychiatric disorders (e.g., Popendikyte *et al.*, 1999; Abdolmaleky *et al.*, 2005, 2006; Grayson *et al.*, 2005; Murphy *et al.*,

2005). To accomplish this goal, initially, the methylation status of candidate CpG islands could be evaluated using bisulfite sequencing to assess the overall methylation pattern of the target CpG islands. Then, MSP analyses could be used to screen and determine the frequency of DNA methylation (methylated product) of the selected group of specific CpGs. In brief, test genomic DNA is chemically modified by sodium bisulfite to convert the unmethylated cytosines to uracils, while methylated cytosines remain unaffected. Then, primers are synthesized to amplify the target fragment for sequencing (Frommer *et al.*, 1992). To perform MSP, primers are synthesized to selectively amplify methylated and unmethylated DNA in separate PCR reactions (Herman *et al.*, 1996). PCR reactions are resolved on nondenaturing polyacrylamide gels, stained with ethidium bromide, and visualized under UV illumination. The bisulfite-modified placental DNA and *in vitro* methylated placental DNA are used as negative and positive controls, respectively, for methylation.

2.2 SYBR Green-Based Quantitative MSP (qMSP) and Quantitative Multiplex MSP (QM-MSP)

Although the standard MSP is an efficient approach for screening the presence or absence of promoter DNA methylation of genes, an evaluation of the degree of methylation could be quantified by performing the qMSP when both controls and the subjects have exhibited different degrees of methylation with significant physiological endpoints. The detailed methodology for this type of analysis is presented elsewhere (Abdolmaleky *et al.*, 2008). In summary, in order to evaluate the differential DNA methylation levels of the candidate genes, we established a real-time PCR-based qMSP and quantitative multiplex MSP (QM-MSP) using SYBR green in our laboratory.

In general, for qMSP or QM-MSP, bisulfite-treated DNA was used as the template, and unmethylated or methylated DNA-specific primers (50–100 nM) were used for PCR amplification in separate reactions (Table 24.1) (Abdolmaleky *et al.*, 2008). For relative quantification of the methylated product, the method of $\Delta\Delta C_T$ was used and normalized with the C_T (cycle threshold) for β-actin gene (Fackler *et al.*, 2004). An additional step in the PCR cycling (72–77°C) was introduced to eliminate the potential confounding effects of nonspecific product or primer dimer formation (Chan *et al.*, 2004).

For QM-MSP, first, the promoter regions of the several candidate genes were amplified in the same reaction using primes (Table 24.1), which correspond to the CG free regions of the gene promoters (Swift-Scanlan *et al.*, 2006). Then, 1 μL of the 40-fold diluted PCR product was used (as the template) to amplify the methylated and unmethylated templates, using methylation- or nonmethylation-specific primers in separate reactions. By employing this approach, the generation of nonspecific products that may compromise the reliability of SYBR green-based real-time PCR was efficiently eliminated, in addition to the remedy that prevented dimer formation due to the use of the minimal amount of MSP primers (<10 pg) necessary for the second-round PCR (Abdolmaleky *et al.*, 2008).

Table 24.1 Primers for methylation analyses of RELN (M: methylated; U: unmethylated specific primers)

Genes and primer type	Forward (5'-3')	Reverse (5'-3')	Ann. Tem.°C (fragment size, bp)
β -actin Promoter Amplification	TTGGGAGGGTAGTT TAGTTGTGGT	CAAAACAAAACA CCTTTTACCCTAA	60 (197)
β -actin for qMSP	GGTGGGTTTAGATTT AGGTTGTGTA	CTACCTACTTTTA AAAATAACAATCAC	60 (125)
RELN Promoter Amplification	GTATTTTTTTAGGAAA ATAGGGT	CTCCCAAAAT TACTTTAAA	56 (506)
RELN MSP M1	CGGGGTTTTGACGT TTTTC	CGCCCTCACG AACTCGACG	60 (184)
RELN MSP U1	TATTTTGGTTA TTGTTGTGT	CACCCTCACA AACTCAACA	60 (184)
RELN MSP M2	CGGGAGGTGTTTTT TGCGGGGTTTTGAC	CCGAAAAAAC AAAAAAAA ACGCCCG	60 (115)
RELN MSP U2	TGGGAGGTGTTTTTT GTGGGGTTTTGAT	CCCAAAAAAA CAAAAAAAA ACACCCA	60 (115)
RELN MSP M3	GTCGTCGAGTTAG TTCGAGAGGGC	GACCAAACCTAAA AAAACGCCCG	60 (150)
RELN MSP U3	GTTGTTGAGTTAGTT TGAGAGGGT	AACCAAACCTAAA AAAACACCCA	60 (150)
RELN, nested for sequencing	GTTAAAGGGGTTGGTT (or reverse of RELN promoter amplification primer)		57

3 Hypermethylation of *RELN* Promoter Localized to the CRE and SP1 Binding Sites in SCZ and BD

3.1 MSP Analyses for Evaluation of RELN Promoter Methylation Status in SCZ

In a preliminary study, bisulfite sequencing and MSP analysis were employed (Abdolmaleky *et al.*, 2005) to map the promoter methylation status in 10 postmortem brain samples from the frontal lobe Brodmann's area 10 (BA10) of male patients with SCZ and controls with a mean age of 46 (SD = 2.7) donated by HBTRC. These studies using tissue samples from four different locations showed a significantly higher frequency of promoter methylation in SCZ when compared to the control subjects (Abdolmaleky *et al.*, 2005). Subsequent studies using an additional nine brain samples donated by HBTRC (four from SCZ patients and five from the control subjects) with a narrow range of age (mean = 38 and 45, SD = 11 and 5, respectively) showed a similar pattern of promoter methylation and corresponding gene expression changes in BA10.

The same experiments, using a different sample set (105 postmortem brain-derived DNA and RNA samples, including 35 patients with SCZ, 35 BD, and 35 control subjects) from BA46 provided by the SMRI, showed the same trend but at a higher degree and frequency of promoter methylation in both the patients and control subjects (SCZ 83% methylated, versus controls 68%). It is to be noted that the dissected brain tissues from the gray matter (BA10) were used for the isolation of DNA from the HBTRC brain samples, while homogenized brain tissues from the cortical brain regions (BA46), which likely contained more cells from the white matter, were used to extract DNA for the SMRI samples by the provider. As the higher level of methylation in SMRI samples could have arisen from the cells of the white matter (the high sensitivity of MSP enabled it to be detected as a positive methylated signal in both the patients and controls), the level of *RELN* promoter methylation was determined using qMSP, as detailed under methods.

3.2 qMSP Analyses of RELN Promoter Methylation in SCZ

qMSP analysis showed that in SCZ, the level of *RELN* promoter methylation, particularly at the CRE binding site, was significantly increased in comparison to the controls subjects (Fig. 24.4). Furthermore, these analyses revealed that alcohol abuse was associated with a higher degree of *RELN* methylation in SCZ. With the use of qMSP analyses, we also uncovered that the degree of promoter methylation increased with age in both SCZ and the control subjects.

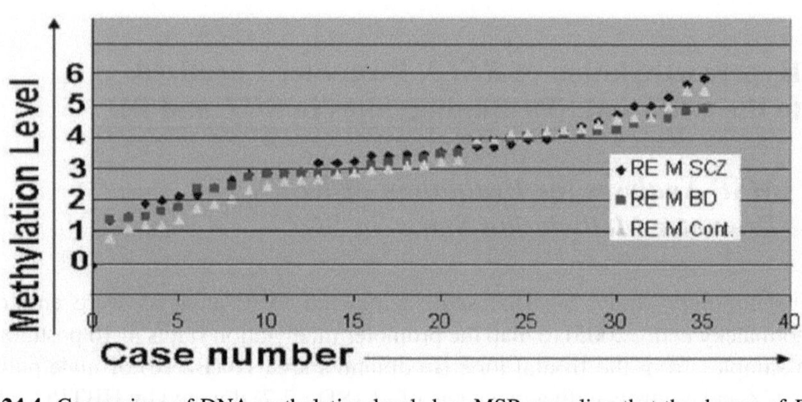

Fig. 24.4 Comparison of DNA methylation levels by qMSP, revealing that the degree of *RELN* methylation in SCZ and BD is almost twice that of the controls. To visualize the differential levels of *RELN* promoter methylation in the patients and controls, the ΔC_T of methylated product for *RELN*, normalized with the C_T of β-actin, was sorted from minimum to maximum. Thus, the increase in the percent of methylation would be exponential. As shown, the base level of *RELN* promoter DNA methylation was greater in SCZ and BD compared to the control subjects (almost twofold). This difference remained nearly the same across the entire samples; however, patients with BD showed a lesser degree of *RELN* methylation in the last part of the curve, where the level of methylation was relatively high (*See Color Plates*)

3.3 MSP and qMSP Analyses of the RELN Promoter Methylation in BD

By performing MSP analyses, we found that while 100% (5/5) of manic patients had a severe degree of *RELN* methlyation, those patients who were classified as depressed had a significantly lower frequency and intensity of *RELN* promoter methylation in CpGs proximal to the first SP1 binding site. Additionally, from the evaluation of the variables that could be associated with a relatively lower frequency of *RELN* promoter methylation in BD patients, we detected a highly significant association between the use of serotonin-specific reuptake inhibitors (SSRIs) and decrease in the frequency of *RELN* promoter methylation of these CpG sites [38% in SSRI users versus 86% in nonusers ($p=0.004$)]. However, this was not associated with a significant change in gene expression level due to small sample size. Furthermore, qMSP analysis that targeted the CRE binding site revealed that patients with BD who were less than 50 years of age showed a drastically high level of DNA hypermethylation of *RELN* promoter compared to the controls and the other BD patients, although the methylation level generally exhibited a tendency to decrease after age 45 in BD. The level of methylation reached a minimum after age 50 compared to the younger ages (less than 40%). Excluding those patients who were older than 50 (73% of them were under valproate and/or SSRI treatment at the time of death), the methylation level of *RELN* promoter in BD was 50% more than the control subjects. In addition, there was a clear correlation between the degree of *RELN* promoter methylation and the younger age of disease onset in BD.

3.4 Influence of RELN Promoter Methylation Localized to the CRE and SP1 Binding Sites on Gene Expression

Examination of several human tissues provided further validation that methylation of the *RELN* promoter (particularly at the CRE and SP1 binding sites) is associated with reduced gene expression (Abdolmaleky *et al.*, 2005) (Fig. 24.3). Similarly, in the normal human brain, the expression of *RELN* in half of the samples who exhibited lower levels of methylation was 1.5 times higher than the other half with higher level of methylation ($p=0.04$), as determined by quantitative real-time PCR (qRT-PCR) and qMSP, respectively. Expression analysis of HBTRC samples using qRT-PCR also showed a significant degree (almost 40%) of hypoexpression of *RELN* in SCZ compared to the control subjects ($p=0.015$). The same analysis on SMRI samples revealed that SCZ and BD patients had a lower expression of *RELN* compared to the controls (almost 20%). However, this difference reached a significant level only in half of the patients with a high level of *RELN* methlyation, compared to the entire group of control subjects ($p=0.04$; two-tailed *t*-test). In the other words, the decrease in *RELN* expression in SCZ patients with low levels of *RELN* promoter methylation was insignificant compared to the control subjects, but the mean level of expression was decreased by at least 15% in this group compared to the same group of the controls.

3.5 Inverse Correlation Between the Expression of *RELN* Versus *DRD1*, *DRD2*, and *MB-COMT*

Since our previous studies (Abdolmaleky *et al.*, 2006) revealed that *MB-COMT* promoter hypomethylation was associated with *DRD2* and *RELN* promoter hypermethylation, we analyzed the expression of *DRD2* and *RELN* genes in the same samples. In order to quantify the correlation between *MB-COMT* versus *DRD2 or RELN* expression, we stratified the total samples, controls, and patients into two subgroups (low and high *MB-COMT* expression) as sorted by *MB-COMT* expression levels (Fig. 24.5). The expression of *MB-COMT* was inversely correlated with the expression of *DRD2* or *RELN* in the entire samples ($p=0.001$ for both *DRD2* and *RELN*, in 52 low- versus 52 high-*MB-COMT*-expressing groups of the SMRI samples; two-tailed *t*-test). We also examined the trend within the patients and the control groups of the SMRI samples. The expression of *DRD2 and RELN* in 17 SCZ patients with high levels of *MB-COMT* showed significant reduction compared to the same group with low levels of *MB-*

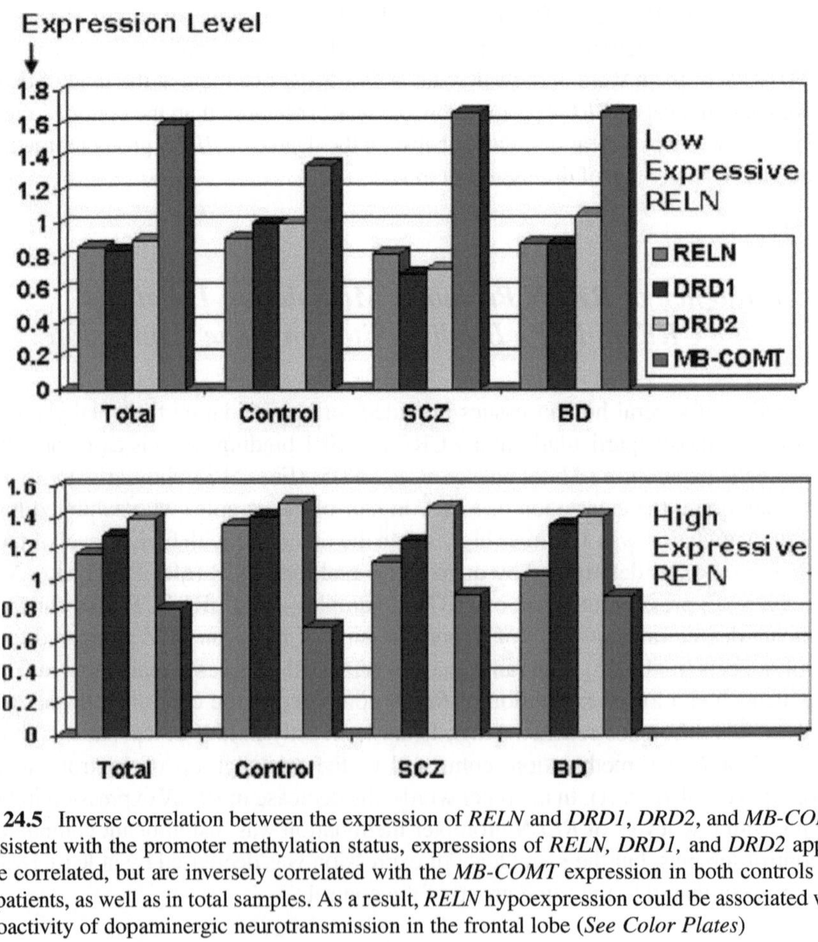

Fig. 24.5 Inverse correlation between the expression of *RELN* and *DRD1*, *DRD2*, and *MB-COMT*. Consistent with the promoter methylation status, expressions of *RELN, DRD1,* and *DRD2* appear to be correlated, but are inversely correlated with the *MB-COMT* expression in both controls and the patients, as well as in total samples. As a result, *RELN* hypoexpression could be associated with hypoactivity of dopaminergic neurotransmission in the frontal lobe (*See Color Plates*)

COMT expression ($p=0.01$ and 0.04, respectively; two-tailed *t*-test). A similar trend for such relationship in the control subjects was also detected ($p=0.05$ and 0.03, respectively). Furthermore, it is noteworthy that the expression of *DRD2* and *RELN*, in high-*MB-COMT*-expressing SCZ and BD patients, was significantly lower than in the control subjects ($p=0.005$ and 0.03, respectively).

Additionally, a direct correlation between the expression of *DRD1* and that of *DRD2* and *RELN* was detected ($p=0.0000005$ and 0.025, respectively, in 50% of the total samples with minimum level of *DRD1* expression versus the other 50% with maximum level of *DRD1* expression). This association remained significant in SCZ and controls, as well as in BD for *DRD2* ($p=0.0005$, 0.003, and 0.05, respectively). Overall, as shown in Fig. 24.5, the expressions of *DRD1*, *DRD2*, and *RELN* were highly correlated, whereas they were inversely correlated with that of *MB-COMT*. These observations suggest that *MB-COMT* hyperactivity causing a reduction in the synaptic dopamine level and *DRD1* or *DRD2* understimulation are likely to influence *RELN* promoter methylation/hypoexpression contributing to the pathogenesis of SCZ and BD. However, the possibility of a converse correlation remains to be excluded, i.e., *RELN* hypermethylation may lead to *DRD1/2* hypoexpression and/or *MB-COMT* hyperexpression. In support of the latter possibility, SCZ subjects with a low level of *RELN* methylation assessed by qMSP showed almost 95% and 60% higher expression of *DRD1* and *DRD2*, respectively, and 50% lower expression of *MB-COMT* compared to those individuals with a maximum level of *RELN* methylation (half of the samples with low versus high *RELN* methylation as determined by sorting the methylation level, $p=0.016$, 0.03, and 0.005, respectively; two-tailed *t*-test). The same trend for *DRD1*, as well as *DRD2* in the control samples, was also detected (60% less, $p=0.06$). Furthermore, the expression of *MB-COMT* was significantly increased in low versus high *RELN* methylation in SCZ, BD, and the entire samples ($p=0.005$, 0.05, and 0.002, respectively; two-tailed *t*-test). Overall, the observations suggest that there could either be a cause–effect relationship between the expressions of these genes or they are simultaneously influenced by the same unknown factor(s).

3.6 The Effect of the Functional Status of the Dopaminergic System on the Modulation of RELN Promoter Methylation at SP1 and CRE Binding Sites

As has been reported elsewhere (Abdolmaleky *et al.*, 2006), our studies showed a significantly high frequency of *MB-COMT* hypomethylation in SCZ and BD compared to control subjects. Additionally, there was a significant correlation between *MB-COMT* hypomethylation and concurrent hypermethylation of *DRD2* and *RELN* promoters in SCZ (11/35) and BD (12/35) versus control subjects (0/35) ($p=0.001$). The qMSP analysis of the same samples revealed that the level of *RELN* promoter methylation in SCZ and BD with a methylated *DRD2* promoter was greater than that in other SCZ or BD patients (30% and 100%, respectively). In contrast, in controls an inverse relationship was detected (>100% less in individuals with methylated *DRD2*, compared to unmethylated *DRD2*). Furthermore, as shown in Fig. 24.5, there was also a direct correlation between *RELN, DRD2,* and *DRD1* expression, and an inverse correlation between the expression of these genes and *MB-COMT*.

Altogether, these data along with the significant correlation observed between the presence of valine allele of *COMT* (the overactive allele) and the frequency of *RELN* promoter methylation (Abdolmaleky *et al.,* 2006) suggest that methylation of the *RELN* promoter could be under the influence of brain dopaminergic systems. However, the likelihood of alternative conclusions, such as *RELN* coordinately orchestrating the expression of dopaminergic genes or the existence of a reciprocal interaction, needs to be investigated.

3.7 The Effects of the Brain Serotonergic System on RELN Promoter Methylation

As the use of SSRIs was correlated with a significant decrease in *RELN* promoter methylation, we examined the association of the T102C polymorphism of *HTR2A* with the *RELN* promoter methylation level, as determined by qMSP. In SCZ and BD, the degree of *RELN* promoter methylation in individuals with the CC genotype of the T102C polymorphism of *HTR2A* was three and four times higher than the control subjects, respectively. This difference was not drastic in the TC heterozygotes or TT homozygotes. The expression of *RELN* in SCZ and BD with the CC genotype was 30% and 48% lower, respectively, compared to the control subjects with the same genotype. However, the differences were only significant in BD, and these observations indicate that the reported association between the C allele of the T102C polymorphism of *HTR2A* and SCZ, as confirmed by two meta-analyses (Williams *et al.*, 1997; Abdolmaleky *et al.,* 2004b), could be related to effects mediated by the CC genotype on *RELN* promoter methylation and expression levels.

4 RELN Promoter Methylation Localized to −139 to −131 Cytosines

Grayson *et al.* (2005), a group with a decade of research interest in the *RELN* signaling pathway analysis, reported that increased methylation at positions −134 and −139 are critically important in determining the promoter activity of *RELN* through recruitment of MeCP2, MBD2, and/or Dnmt1 to the methylated bases. However, their approach to examining *RELN* promoter methylation status was conducted on a completely different set of brain samples, provided by the same brain banks. They showed that the *RELN* promoter DNA is hypermethylated in SCZ with a concomitant decrease in the transcript levels (Grayson *et al.,* 2005). Although this group reported the highest levels of *RELN* promoter methylation of cytosines at positions −139 and −134 compared to the other cytosines in the promoter region, our earlier and recent studies showed low levels of cytosine methylation at these sites. A careful examination of the samples used by Grayson *et al.* (2005) revealed that discrepancies in the results between the two studies could be primarily due to a large difference in demographics (age, in particular) of our HBTRC samples versus their samples, despite both having been provided by the same brain bank (HBTRC). A comparison of demographic data of their (supplementary Table 2 in Grayson *et al.*, 2005) and our samples (Table 1 in Abdolmaleky *et al.*, 2005) clearly indicated that almost all of their HBTRC samples were different from our samples. For example, all of our samples were from males and under age 49 (mean age 45.5 and SD=2.7), while 60% of their samples were female and all were over 49 with a mean age of 65 (60% over 65). Furthermore, while 60% of our samples were from the left brain, 30% of their samples were from the left brain. However, there was one case in their HBTRC samples from a male patient with an age of 49. This case exhibited the heaviest methylation level among all of their samples, including the samples from SMRI (see supplementary Table 2 in Grayson *et al.*, 2005). Considering that the level of DNA methylation of *RELN* changes with age and that the mean age of their HBTRC samples was 20 years older that ours, it is highly likely that the discrepancy could have arisen from the mean age differences between the two sample sets.

Although a new study on the epigenetic aberration of human *RELN* in SCZ reported an age-related increase in *RELN* methylation only in the control subjects (Tamura *et al.*, 2007), our statistical analysis for the effect of age on *RELN* methylation level of the Grayson *et al.* (2005) samples (supplementary Table 2 in Grayson *et al.*, 2005) showed that consistent with our data the degree of DNA methylation of *RELN* promoter is significantly increased in both SCZ and the control subjects (Fig. 24.6) (p=0.025 for either SCZ or normal controls; student's t-test). Thus, it would be interesting to perform follow-up studies to uncover the potential impact

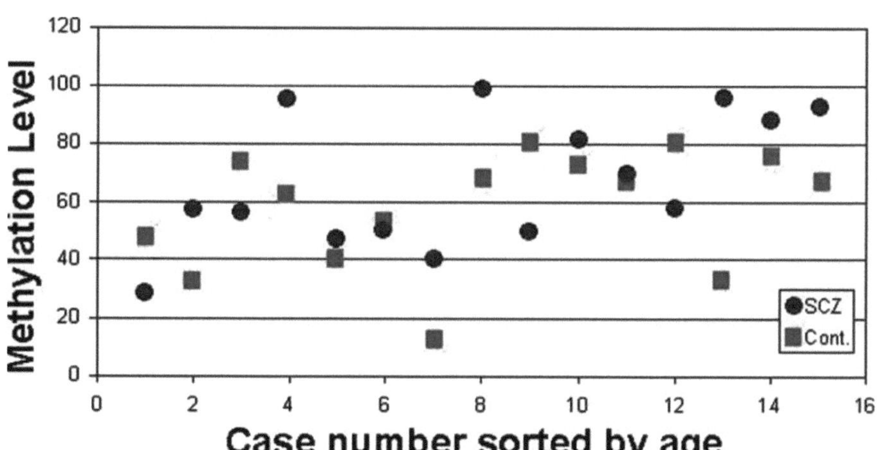

Fig. 24.6 Age-dependent increase in *RELN* promoter methylation. The degree of *RELN* promoter methylation (*Y* axis), extracted from Grayson *et al.* (2005, supplementary Table 2), was sorted by age (*X* axis). As shown, the degree of promoter methylation increased by age in both SCZ and the control subjects (*See Color Plates*)

of *RELN* methylation status on the age of disease onset. Our studies so far have established a clear correlation between the degree of *RELN* promoter methylation and the younger age of disease onset in BD.

5 Summary

Studies from our and other laboratories support that altered DNA methylation in response to environmental insults or dysfunctional genes may destabilize the normal epigenetic fine-tuning of genes in psychiatric disorders (Abdolmaleky *et al.*, 2005, 2006; Bonsch *et al.*, 2005; Grayson *et al.*, 2005; Iwamoto *et al.*, 2005; Bleich *et al.*, 2006). Aberrant DNA methylation patterns of *RELN* promoter in SCZ and BD, its association with the dysfunctions of dopaminergic and serotonergic systems and similarity of the promoter DNA methylation status of the patients at young age with the normal controls at old age, and occurrence of an early onset and age-inappropriate *RELN* promoter DNA methylation in the patients, have provided compelling evidence that calls for more investigations in this arena. Thus, further studies will be needed for validation of these initial observations using other sets of brain samples to provide a comprehensive molecular insight for the genesis of SCZ and BD. These studies will have future implications for early diagnosis, prognosis, and management of these diseases, and will provide clues for preventive measures and identification of nodal points for effective therapeutic interventions.

Acknowledgments Postmortem DNA and RNA samples were donated by The Stanley Brain Collection courtesy of Drs. Michael B. Knable, E. Fuller Torrey, Maree J. Webster, and Robert H. Yolken. Dr. Francine Benes provided the brain tissues from the Harvard Brain Tissue Resource Center, which is supported in part by PHS grant number MH/NS 31862. This work was supported by grants from the NIH (CA101773), IUMS (Tehran Institute of Psychiatry and Mental Health Research Center), and NARSAD. Dr. Sam Thiagalingam is a Dr. Walter F. Nichols Investigator supported by NARSAD Independent Investigator Award.

References

Abdolmaleky, H. M., Smith, C. L., Faraone, S. V., Shafa, R., Stone, W., Glatt, S. J., and Tsuang, M. T. (2004a). Methylomics in psychiatry: modulation of gene-environment interactions may be through DNA methylation. *Am. J. Med. Genet.* 127B(1):51–59.

Abdolmaleky, H. M., Faraone, S. V., Glatt, S. J., and Tsuang, M. T. (2004b). Meta-analysis of association between the T102C polymorphism of the 5HT2a receptor gene and schizophrenia. *Schizophr. Res.* 67(1):53–62.

Abdolmaleky, H. M., Cheng, H. H., Russo, A., Smith, C. L., Faraone, S. V., Shafa, R., Wilcox, M., Glatt, S., Stone, W. S., Nguyen, G., Ponte, J. F., Thiagalingam, S., and Tsuang, M. (2005). Hypermethylation of the reelin (*RELN*) promoter in the brain of schizophrenic patients: a preliminary report. *Am. J. Med. Genet. B Neuropsychiatr. Genet.* 134B:60–66.

Abdolmaleky, H. M., Cheng, K. H., Faraone, S. V., Wilcox, M., Glatt, S. J., Gao, F., Smith, C. L., Shafa, R., Aeali, B., Carnevale, J., Pan, H., Papageorgis, P., Ponte, J. F., Sivaraman, V., Tsuang,

M. T., and Thiagalingam, S. (2006). Hypomethylation of MB-COMT promoter is a major risk factor for schizophrenia and bipolar disorder. *Hum. Mol. Genet.* 15(21):3132–3145.

Abdolmaleky, H. M., Smith, C. L., Zhou, R. J., Thiagalingam, S. (2008). "Epigenetic alterations of the dopaminergic system in major psychiatric disorders" in Pharmacogenomics in drug discovery and development, Yan, Q., Humana Press. London, U.K. 187–212.

Akahane, A., Kunugi, H., Tanaka, H., and Nanko, S. (2002). Association analysis of polymorphic CGG repeat in 5′ UTR of the reelin and VLDLR genes with schizophrenia. *Schizophr. Res.* 58(1):37–41.

Akbarian, S. (2003). The neurobiology of Rett syndrome. *Neuroscientist.* 9(1):57–63.

Akbarian, S., Jiang, Y., and Laforet, G. (2006). The molecular pathology of Rett syndrome: synopsis and update. *Neuromol. Med.* 8(4):485–494.

Ballmaier, M., Zoli, M., Leo, G., Agnati, L. F., and Spano, P. (2002). Preferential alterations in the mesolimbic dopamine pathway of heterozygous reeler mice: an emerging animal-based model of schizophrenia. *Eur. J. Neurosci.* 15(7):1197–1205.

Beffert, U., Weeber, E. J., Durudas, A., Qiu, S., Masiulis, I., Sweatt, J. D., Li, W. P., Adelmann, G., Frotscher, M., Hammer, R. E., and Herz, J. (2005). Modulation of synaptic plasticity and memory by reelin involves differential splicing of the lipoprotein receptor Apoer2. *Neuron* 47(4):567–579.

Beffert, U., Durudas, A., Weeber, E. J., Stolt, P. C., Giehl, K. M., Sweatt, J. D., Hammer, R. E., and Herz, J. (2006). Functional dissection of reelin signaling by site-directed disruption of disabled-1 adaptor binding to apolipoprotein E receptor 2: distinct roles in development and synaptic plasticity. *J. Neurosci.* 26(7):2041–2052.

Bertino, L., Ruffini, M. C., Copani, A., Bruno, V., Raciti, G., Cambria, A., and Nicoletti, F. (1996). Growth conditions influence DNA methylation in cultured cerebellar granule cells. *Dev. Brain Res.* 95:38–43.

Bestor, T. H. (2000). The DNA methyltransferases of mammals. *Hum. Mol. Genet.* 9:2395–2402.

Bird, A. (2002). DNA methylation patterns and epigenetic memory. *Genes Dev.* 16:6–21.

Bleich, S., Lenz, B., Ziegenbein, M., Beutler, S., Frieling, H., Kornhuber, J., and Bonsch, D. (2006). Epigenetic DNA hypermethylation of the HERP gene promoter induces down-regulation of its mRNA expression in patients with alcohol dependence. *Alcohol. Clin. Exp. Res.* 30:587–591.

Bonsch, D., Lenz, B., Kornhuber, J., and Bleich, S. (2005). DNA hypermethylation of the alpha synuclein promoter in patients with alcoholism. *Neuroreport* 16:167–170.

Chan, M. W., Chu, E. S., To, K. F., and Leung, W. K. (2004). Quantitative detection of methylated SOCS-1, a tumor suppressor gene, by a modified protocol of quantitative real time methylation-specific PCR using SYBR green and its use in early gastric cancer detection. *Biotechnol. Lett.* 26:1289–1293.

Chen, Y., Sharma, R. P., Costa, R. H., Costa, E., and Grayson, D. R. (2002). On the epigenetic regulation of the human reelin promoter. *Nucleic Acids Res.* 30:2930–2939.

Chen, Y., Beffert, U., Ertunc, M., Tang, T. S., Kavalali, E. T., Bezprozvanny, I., and Herz, J. (2005). Reelin modulates NMDA receptor activity in cortical neurons. *J. Neurosci.* 25(36):8209–8216.

Cooney, C. A., Dave, A. A., and Wolff, G. L. (2002). Maternal methyl supplements in mice affect epigenetic variation and DNA methylation of offspring. *J. Nutr.* 132:2393–2400S.

Costa, E., Davis, J., Grayson, D. R., Guidotti, A., Pappas, G. D., and Pesold, C. (2001). Dendritic spine hypoplasticity and downregulation of reelin and GABAergic tone in schizophrenia vulnerability. *Neurobiol. Dis.* 8:723–742.

Costa, E., Dong, E., Grayson, D. R., Ruzicka, W. B., Simonini, M. V., Veldic, M., and Guidotti, A. (2006). Epigenetic targets in GABAergic neurons to treat schizophrenia. *Adv. Pharmacol.* 54:95–117.

Costello, J. F., and Plass, C. (2001). Methylation matters. *J. Med. Genet.* 38(5):285–303.

Davies, R., and Morris, B. (1997). *Molecular Biology of Neuron.* Oxford University Press, London.

Dong, E., Agis-Balboa, R. C., Simonini, M. V., Grayson, D. R., Costa, E., and Guidotti, A. (2005). Reelin and glutamic acid decarboxylase67 promoter remodeling in an epigenetic methionine-induced mouse model of schizophrenia. *Proc. Natl. Acad. Sci. USA* 102(35):12578–12583.

Eastwood, S. L., and Harrison, P. J. (2003). Interstitial white matter neurons express less reelin and are abnormally distributed in schizophrenia: towards an integration of molecular and morphologic aspects of the neurodevelopmental hypothesis. *Mol. Psychiatry* 8(9):769, 821–831.

Fackler, M. J., McVeigh, M., Mehrotra, J., Blum, M. A., Lange, J., Lapides, A., Garrett, E., Argani, P., and Sukumar, S. (2004). Quantitative multiplex methylation-specific PCR assay for the detection of promoter hypermethylation in multiple genes in breast cancer. *Cancer Res.* 64:4442–4452.

Fatemi, S. H., Earle, J. A., and McMenomy, T. (2000). Reduction in reelin immunoreactivity in hippocampus of subjects with schizophrenia, bipolar disorder and major depression. *Mol. Psychiatry* 5:654–663.

Fatemi, S. H., Stary, J. M., Araghi-Niknam, M., and Egan, E. (2005a). GABAergic dysfunction in schizophrenia and mood disorders as reflected by decreased levels of reelin and GAD 65 & 67 kDa proteins in cerebellum. *Schizophr. Res.* 72:109–122.

Fatemi, S. H., Snow, A. V., Stary, J. M., Araghi-Niknam, M., Reutiman, T. J., Lee, S., Brooks, A. I., and Pearce, D.A. (2005b). Reelin signaling is impaired in autism. *Biol. Psychiatry* 57:777–787.

Fatemi, S. H., Reutiman, T. J., Folsom, T. D., Bell, C., Nos, L., Fried, P., Pearce, D. A., Singh, S., Siderovski, D. P., Willard, F. S., and Fukuda, M. (2006). Chronic olanzapine treatment causes differential expression of genes in frontal cortex of rats as revealed by DNA microarray technique. *Neuropsychopharmacology* 31(9):1888–1899.

Fenech, M. (2001). The role of folic acid and vitamin B12 in genomic stability of human cells. *Mutat. Res.* 475:57–67.

Frommer, M., McDonald, L. E., Millar, D. S., Collis, C. M., Watt, F., Grigg, G. W., Molloy, P. L., and Paul, C. L. (1992). A genomic sequencing protocol that yields a positive display of 5-methylcytosine residues in individual DNA strands. *Proc. Natl. Acad. Sci. USA* 89(5):1827–1831.

Goldberger, C., Gourion, D., Leroy, S., Schurhoff, F., Bourdel, M. C., Leboyer, M., and Krebs, M. O. (2005). Population-based and family-based association study of 5′UTR polymorphism of the reelin gene and schizophrenia. *Am. J. Med. Genet. B Neuropsychiatr. Genet.* 137(1):51–55.

Grayson, D. R., Jia, X., Chen, Y., Sharma, R. P., Mitchell, C. P., Guidotti, A., and Costa, E. (2005). Reelin promoter hypermethylation in schizophrenia. *Proc. Natl. Acad. Sci. USA* 102(26):9341–9346.

Guidotti, A., Auta, J., Davis, J. M., Di-Giorgi-Gerevini, V., Dwivedi, Y., Grayson, D. R., Impagnatiello, F., Pandey, G., Pesold, C., Sharma, R., Uzunov, D., and Costa, E. (2000). Decrease in reelin and glutamic acid decarboxylase67 (GAD67) expression in schizophrenia and bipolar disorder: a postmortem brain study. *Arch. Gen. Psychiatry* 57(11):1061–1069.

Guidotti, A., Ruzicka, W., Grayson, D. R., Veldic, M., Pinna, G., Davis, J. M., and Costa, E. (2007). S-adenosyl methionine and DNA methyltransferase-1 mRNA overexpression in psychosis. *Neuroreport* 18(1):57–60.

Herman, J. G., Graff, J. R., Myohanen, S., Nelkin, B. D., and Baylin, S. B. (1996). Methylation-specific PCR: a novel PCR assay for methylation status of CpG islands. *Proc. Natl. Acad. Sci. USA* 93(18):9821–9826.

Homayouni, R., Magdaleno, S., Keshvara, L., Rice, D. S., and Curran, T. (2003). Interaction of disabled-1 and the GTPase activating protein Dab2IP in mouse brain. *Brain Res. Mol. Brain Res.* 115(2):121–129.

Huang, C. H., and Chen, C. H. (2006). Absence of association of a polymorphic GGC repeat at the 5' untranslated region of the reelin gene with schizophrenia. *Psychiatry Res.* 142(1):89–92.

Impagnatiello, F., Guidotti, A. R., Pesold, C., Dwivedi, Y., Caruncho, H., Pisu, M. G., Uzunov, D. P., Smalheiser, N. R., Davis, J. M., Pandey, G. N., Pappas, G. D., Tueting, P., Sharma, R. P., and Costa, E. (1998). A decrease of reelin expression as a putative vulnerability factor in schizophrenia. *Proc. Natl. Acad. Sci. USA* 95(26):15718–15723.

Iwamoto, K., Bundo, M., Yamada, K., Takao, H., Iwayama-Shigeno, Y., Yoshikawa, T., and Kato, T. (2005). DNA methylation status of SOX10 correlates with its downregulation and oligodendrocyte dysfunction in schizophrenia. *J. Neurosci.* 25(22):5376–5381.

Jiang, Y. H., Bressler, J., and Beaudet, A. L. (2004). Epigenetics and human disease. *Annu. Rev. Genomics Hum. Genet.* 5:479–510.

Kaludov, N. K., and Wolffe, A. P. (2000). MeCP2 driven transcriptional repression in vitro: selectivity for methylated DNA, action at a distance and contacts with the basal transcription machinery. *Nucleic Acids Res.* 28(9):1921–1928.

Kim, G. D., Ni, J., Kelesoglu, N., Roberts, R. J., and Pradhan, S. (2002). Co-operation and communication between the human maintenance and de novo DNA (cytosine-5) methyltransferases. *EMBO J.* 21(15):4183–4195.

Lavorgna, G., Dahary, D., Lehner, B., Sorek, R., Sanderson, C. M., and Casari, G. (2004). In search of antisense. *Trends Biochem. Sci.* 29(2):88–94.

Levenson, J. M., O'Riordan, K. J., Brown, K. D., Trinh, M. A., Molfese, D. L., and Sweatt, J. D. (2004). Regulation of histone acetylation during memory formation in the hippocampus. *J. Biol. Chem.* 79(39):40545–40559.

Marrone, M. C., Marinelli, S., Biamonte, F., Keller, F., Sgobio, C. A., Ammassari-Teule, M., Bernardi, G., and Mercuri, N. B. (2006). Altered cortico-striatal synaptic plasticity and related behavioural impairments in reeler mice. *Eur. J. Neurosci.* 24(7):2061–2070.

Martinowich, K., Hattori, D., Wu, H., Fouse, S., He, F., Hu, Y., Fan, G., and Sun, Y. E. (2003). DNA methylation-related chromatin remodeling in activity-dependent BDNF gene regulation. *Science* 302(5646):890–893.

Martucci, L., Wong, A. H., Trakalo, J., Cate-Carter, T., Wong, G. W., Macciardi, F. M., and Kennedy, J. L. (2003). N-methyl-D-aspartate receptor NR1 subunit gene (GRIN1) in schizophrenia: TDT and case-control analyses. *Am. J. Med. Genet. B Neuropsychiatr. Genet.* 119(1):24–27.

Monk, M. (1995). Epigenetic programming of differential gene expression in development and evolution. *Dev. Genet.* 17(3):188–197.

Morange, M. (2002). The relations between genetics and epigenetics: a historical point of view. *Ann. N.Y. Acad. Sci.* 981:50–60.

Murphy, B. C., O'Reilly, R. L., and Singh, S. M. (2005). Site-specific cytosine methylation in S-COMT promoter in 31 brain regions with implications for studies involving schizophrenia. *Am. J. Med. Genet. B Neuropsychiatr. Genet.* 133(1):37–42.

Nishikawa, S., Goto, S., Yamada, K., Hamasaki, T., and Ushio, Y. (2003). Lack of reelin causes malpositioning of nigral dopaminergic neurons: evidence from comparison of normal and Reln(rl) mutant mice. *J. Comp. Neurol.* 461:166–173.

Park-Sarge, O. K., and Sarge, K. D. (1995). Cis-regulatory elements conferring cyclic 3′,5′-adenosine monophosphate responsiveness of the progesterone receptor gene in transfected rat granulosa cells. *Endocrinology* 136(12):5430–5437.

Persico, A. M., Levitt, P., and Pimenta, A. F. (2006). Polymorphic GGC repeat differentially regulates human reelin gene expression levels. *J. Neural Transm.* 113(10):1373–1382.

Pesold, C., Liu, W. S., Guidotti, A., Costa, E., and Caruncho, H. J. (1999). Cortical bitufted, horizontal, and Martinotti cells preferentially express and secrete reelin into perineuronal nets, nonsynaptically modulating gene expression. *Proc. Natl. Acad. Sci. USA* 96(6):3217–3222.

Petronis, A. (2000). The genes for major psychosis: aberrant sequence or regulation? *Neuropsychopharmacology* 23:1–12.

Popendikyte, V., Laurinavicius, A., Paterson, A. D., Macciardi, F., Kennedy, J. L., and Petronis, A. (1999). DNA methylation at the putative promoter region of the human dopamine D2 receptor gene. *Neuroreport* 10(6):1249–1255.

Robertson, K. D., and Wolffe, A. P. (2000). DNA methylation in health and disease. *Nature Rev. Genet.* 1(1):11–19.

Russo, V., Martienssen, R., and Riggs, A. (1996). *Epigenetic Mechanisms of Gene Regulation.* Cold Spring Harbor Laboratory Press, Plainview, NY.

Ruzicka, W. B., Zhubi, A., Veldic, M., Grayson, D. R., Costa, E., and Guidotti, A. (2007). Selective epigenetic alteration of layer I GABAergic neurons isolated from prefrontal cortex of schizophrenia patients using laser-assisted microdissection. *Mol. Psychiatry* 12(4):385–397.

Singh, S. M., Murphy, B., and O'Reilly, R. L. (2003). Involvement of gene-diet/drug interaction in DNA methylation and its contribution to complex diseases: from cancer to schizophrenia. *Clin. Genet.* 64(6):451–460.

Swift-Scanlan, T., Blackford, A., Argani, P., Sukumar, S., and Fackler, M. J. (2006). Two-color quantitative multiplex methylation-specific PCR. *Biotechniques* 40:210–219.

Tamura, Y., Kunugi, H., Ohashi, J., and Hohjoh, H. (2007). Epigenetic aberration of the human REELIN gene in psychiatric disorders. *Mol. Psychiatry* 12:519.

Tasman, A., Key, J., and Lieberman, J. A. (1997). *Psychiatry*. W. B. Saunders, Philadelphia.

Tremolizzo, L., Carboni, G., Ruzicka, W. B., Mitchell, C. P., Sugaya, I., Tueting, P., Sharma, R., Grayson, D. R., Costa, E., and Guidotti, A. (2002). An epigenetic mouse model for molecular and behavioral neuropathologies related to schizophrenia vulnerability. *Proc. Natl. Acad. Sci. USA* 99(26):17095–17100.

Veldic, M., Caruncho, H. J., Liu, W. S., Davis, J., Satta, R., Grayson, D. R., Guidotti, A., and Costa, E. (2004). DNA-methyltransferase 1 mRNA is selectively overexpressed in telencephalic GABAergic interneurons of schizophrenia brains. *Proc. Natl. Acad. Sci. USA* 101(1):348–353.

Veldic, M., Kadriu, B., Maloku, E., Agis-Balboa, R. C., Guidotti, A., Davis, J. M., and Costa, E. (2007). Epigenetic mechanisms expressed in basal ganglia GABAergic neurons differentiate schizophrenia from bipolar disorder. *Schizophr. Res.* 91(1–3):51–61.

Waterland, R. A., Lin, J. R., Smith, C. A., and Jirtle, R. L. (2006). Post-weaning diet affects genomic imprinting at the insulin-like growth factor 2 (Igf2) locus. *Hum. Mol. Genet.* 15:705–716.

Weaver, I. C., Cervoni, N., Champagne, F. A., D'Alessio, A. C., Sharma, S., Seckl, J. R., Dymov, S., Szyf, M., and Meaney, M. J. (2004). Epigenetic programming by maternal behavior. *Nature Neurosci.* 8:847–854.

Weaver, I. C., Champagne, F. A., Brown, S. E., Dymov, S., Sharma, S., Meaney, M. J., and Szyf, M. (2005). Reversal of maternal programming of stress responses in adult offspring through methyl supplementation: altering epigenetic marking later in life. *J. Neurosci.* 25(47):11045–11054.

Weinhausel, A., and Haas, O. A. (2001). Evaluation of the fragile X (FRAXA) syndrome with methylation-sensitive PCR. *Hum. Genet.* 108(6):450–458.

Williams, J., McGuffin, P., Nothen, M., and Owen, M. J. (1997). Meta-analysis of association between the 5-HT2a receptor T102C polymorphism and schizophrenia. EMASS Collaborative Group. European Multicentre Association Study of Schizophrenia. *Lancet* 349(9060):1221.

Chapter 25
Reelin Gene Polymorphisms in Autistic Disorder

Carla Lintas and Antonio Maria Persico

Contents

1 Introduction

Migratory streams occur throughout the central nervous system (CNS) during development. Neuronal and glial cell populations migrate out of proliferative zones to reach their final location, where neurons soon establish early intercellular connections. Reelin plays a pivotal role in cell migration processes, acting as a stop signal for migrating neurons in several CNS districts, including the neocortex, the cerebellum, and the hindbrain (Rice and Curran, 2001). At the cellular level, Reelin acts by binding to a variety of receptors, including the VLDL receptors, ApoER2, and $\alpha3\beta1$ integrins, and also by exerting a proteolytic activity on extracellular matrix proteins, which is critical to neuronal migration (D'Arcangelo *et al.*, 1999; Hiesberger *et al.*, 1999; Quattrocchi *et al.*, 2002). Indeed, neuronal migration is profoundly altered in *reeler* mice, lacking Reelin protein due to spontaneous deletions of the reelin

C. Lintas
Laboratory of Molecular Psychiatry and Neurogenetics, University "Campus Bio-Medico," and Department of Experimental Neurosciences, I.R.C.C.S. "Fondazione Santa Lucia," Via Longoni 83, I-00155 Rome, Italy

A. M. Persico
Laboratory of Molecular Psychiatry and Neurogenetics, University "Campus Bio-Medico," and Department of Experimental Neurosciences, I.R.C.C.S. "Fondazione Santa Lucia," Via Longoni 83, I-00155 Rome, Italy
e-mail: a.persico@unicampus.it

S. H. Fatemi (ed.), *Reelin Glycoprotein: Structure, Biology and Roles in Health and Disease.*

(*RELN*) gene (D'Arcangelo *et al.*, 1995). Their brains display major cytoarchitectonic alterations, yielding a behavioral phenotype characterized by action tremor, dystonic posture, and ataxic gait (Goffinet, 1984). Interestingly, despite significant interindividual differences, postmortem studies of brains of autistic patients have consistently found neuropathological evidence of altered neuronal migration, including ectopic neurons, altered cytoarchitectonics, and aberrant fiber tracts, as recently reviewed by Persico and Bourgeron (2006). Furthermore, the *RELN* gene maps to human chromosome 7q22, in a region hosting one or more autism genes, according to converging evidence from multiple genetic linkage studies (Muhle *et al.*, 2004; Persico and Bourgeron, 2006). These findings provided initial suggestions that Reelin may play relevant roles in neurodevelopmental disorders, such as autism. Yet, autistic patients are not "*reeler*" humans: *RELN* gene mutations resulting in the absence of Reelin protein yield a much more severe phenotype, the Norman-Roberts syndrome (Hong *et al.*, 2000). This rare autosomal recessive neurological disease is characterized by lissencephaly and cerebellar hypoplasia, with severe mental retardation, abnormal neuromuscular connectivity, and congenital lymphedema. Therefore, *RELN* gene variants potentially conferring genetic liability to neuropsychiatric disorders, such as autism and schizophrenia, were predicted to more likely modulate gene expression levels and/or protein function, rather than to produce a complete loss of function. And indeed, no mutation resulting in premature stop codons and no triplet repeat expansions halting *RELN* gene expression have been identified to date in autistic or schizophrenic patients. This chapter will thus review current knowledge on *RELN* gene polymorphisms influencing gene expression and summarize the results of studies addressing the possible genetic association between functional *RELN* gene variants and autism. Genetic and epigenetic *RELN* gene variants possibly involved in other neurodevelopmental disorders, such as schizophrenia, will be described elsewhere in this book.

2 *RELN* Gene Polymorphisms and Autism

The *RELN* gene encompasses approximately 450 kb, including 65 exons, with alternative splicing of exon 64 and two different polyadenylation sites (Royaux *et al.*, 1997). Several polymorphisms present in the 5'UTR, in the coding region, and in the splice junctions have been assessed for association with autistic disorder (Table. 25.1); an additional promoter variant (G-888C) was assessed only in schizophrenia (Chen *et al.*, 2002). To date, no *de novo* mutations have been identified in autistic individuals. Missense coding variants inherited by autistic children from a heterozygous parent, co-segregating with autism in the family, and not found in normal controls, include N1159K in exon 25, R1742Q and V1762I in exon 35, R2290H in exon 44, and T2718A in exon 51 (Bonora *et al.*, 2003). These rare variants, though interesting for their disease-specificity, have not yet been investigated from a functional standpoint and cannot explain the linkage peak detected in the same families around chromosomal region 7q22 (Bonora *et al.*, 2003). The P1703R

Table 25.1 Reelin missense, splicing, and 5'UTR variants in autistic and control samples (n.d., not determined)

Localization	Variation in the 5'UTR or amino acid change	Allelic frequencies*	
		Patients	Controls
5'UTR	GGC triplet repeat†	17.9%[a]	9.1%[a]
		9.2%[b]	6.8%[b]
		5.9%[c]	n.d.
		5.6%[d]	n.d.
Intron 5	A84446G‡ (rs607755)	49.5%[a]	54.3%[a]
Exon 10	Gly370Arg	1.8%	0
Exon 10	Val338Gly	<1%	0
Exon 15	Ser630Arg	7.2%	8.3%
Exon 22	Leu997Val (rs362691)	14.5%[e]	21.4%[e]
		11.0%[f]	n.d.
Exon 25	Asn1159Lys	1.8%	0
Exon 27	Gly1280Glu	3.6%	3.1%
Exon 34	Pro1703Arg (rs2229860)	<1%	n.d.
Exon 34	Ser1719Leu	1.8%	<1%
Exon 35	Arg1742Trp	<1%	0
Exon 35	Arg1742Gln	<1%	0
Exon 35	Val1762Ile	3.6%	0
Exon 44	Arg2290His	1.8%	0
Exon 47	Gly2480Ser	1.8%	1.6%
Exon 51	Thr2718Ala	1.8%	0

* Allelic frequencies are from: [a]Persico *et al.* (2001), [b]Zhang *et al.* (2002), [c]Li *et al.* (2004), [d]Krebs *et al.* (2002), [e]the IMGSAC sample in Bonora *et al.* (2003), [f]Serajee *et al.* (2005). Rare missense variants without footnote are from the IMGSAC sample (Bonora *et al.*, 2003).

† Allelic frequencies of "long" alleles (i.e., ≥ 11 GGC repeats) are reported.

‡ bp numbering refers to GenBank acc. n. AC000121; G allele frequencies are reported.

missense variant, present in exon 34 (rs2229860), was found in one family with an autistic proband (Serajee *et al.*, 2005) and was not encountered in other control samples (Bonora *et al.*, 2003). An A/G transversion present in intron 5 (rs607755), 3 base pairs 5' of the exon 6 splice junction, is predicted to affect the probability of splicing and represents a common polymorphism not associated with autism (Bonora *et al.*, 2003; Persico *et al.*, 2001; Serajee *et al.*, 2005). Finally, a polymorphic trinucleotide GGC repeat was identified in the 5'UTR, immediately adjacent to the ATG start site (Persico *et al.*, 2001), and represents to date the only *RELN* gene polymorphism characterized at both the genetic and functional level.

The 5'UTR GGC triplet repeat alleles range between 4 and 23 repeats (Fig. 25.1, panel A). In our initial study, the 8 and 10 GGC repeats represented the most frequent alleles both in autistic patients (8-repeats = 44.2%; 10-repeats = 45.3%) and in normal controls (8-repeats = 44.4%; 10-repeats = 51.1%). Interestingly, longer GGC alleles (i.e., alleles encompassing 11 or more GGC repeats) were found in

A

4 repeats: 5'-agggcgcgc*ggcggcggcggc*c**AUG**gagcgcagu-3'

7 repeats: 5'-agggcgcgc*ggcggcggcggcggcggcggc*c**AUG**gagcgcagu-3'

8 repeats: 5'-agggcgcgc*ggcggcggcggcggcggcggcggc*c**AUG**gagcgcagu-3'

10 repeats: 5'-agggcgcgc*ggcggcggcggcggcggcggcggcggcggc*c**AUG**gagcgcagu-3'

12 repeats: 5'-agggcgcgc*ggcggcggcggcggcggcggcggcggcggcggcggc*c**AUG**gagcgcagu-3'

13 repeats: 5'-agggcgcgc*ggcggcggcggcggcggcggcggcggcggcggcggcggc*c**AUG**gagcgcagu-3'

23 repeats: 5'-agggcgcgc*ggcggcggcggcggcggcggcggcggcggcggcggcggcggcggcggcggcggcggc*
*ggcggcggcggc*c**AUG**gagcgcagu-3'

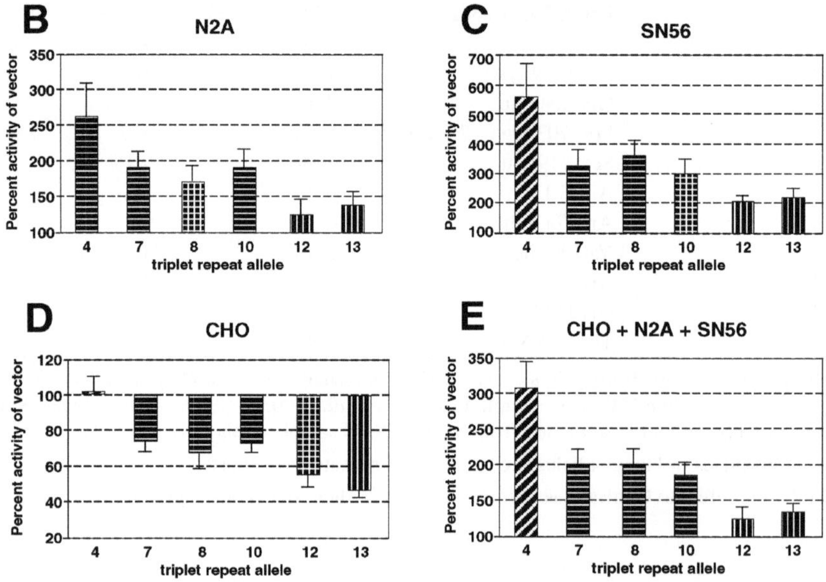

Fig. 25.1 (**A**) Schematic representation of GGC repeat alleles 4-to-23 (repeat underlined and in italics) located in the 5' UTR of the *RELN* mRNA, adjacent to the AUG translation start site, highlighted in bold. Mean luciferase activity of pGL3-Promoter Vector constructs with different 5'UTR GGC alleles in (**B**) N2A, (**C**) SN56, (**D**) CHO, and (**E**) all cell lines together (Persico *et al.*, 2006). Statistically homogeneous sets of alleles differing from one another by at least $p<0.05$ are designated by oblique, horizontal, and vertical lines. Overlapping vertical and horizontal lines designate samples reaching a p value <0.1

17.9% of autistic patients versus 9.1% of normal controls ($p<0.05$). The existence of a significant association between these "long" alleles and autism was confirmed using family-based association tests, showing the preferential transmission of "long" alleles from heterozygous parents to their autistic offspring (31 transmissions versus 11 nontransmission: $p<0.05$ after Bonferroni's correction). The preferential transmission of these putative risk-conferring alleles from heterozygous parents to their autistic offspring was paralleled in these same families by the preferential nontransmission of "long" alleles to the unaffected offspring (6 transmissions versus 13 nontransmission). Consequently, allelic transmission rates differed very significantly between autistic patients and their unaffected siblings ($p<0.001$). Also, the frequency distribution of *RELN* gene alleles marked by haplotypes estimated after genotyping the 5'UTR GGC repeat, the intron 5 SNP (rs607755), and a synonymous coding SNP found in exon 50 (rs2229864), were significantly different between 94 autistic patients and 186 normal controls ($p<0.01$). Several haplotypes encompassing "long" GGC alleles were more frequent among autistic patients than in controls, more frequently transmitted from heterozygous parents to these autistic patients, and less frequently transmitted to their unaffected siblings, than expected by chance (Persico *et al.*, 2001).

3 Functional Studies of *RELN* GGC Alleles

The polymorphic GGC triplet repeat is immediately adjacent to the AUG translation start site and occupies an ideal position to influence gene expression rates. *In vitro* and *in vivo* studies have indeed shown that this GGC variant plays functional roles in modulating *RELN* gene expression. *In vitro* experiments were performed transfecting nonneuronal CHO and neuronal SN56 and N2A cell lines with constructs encompassing: (a) the SV40 promoter, (b) 136 bp of *RELN* 5'UTR sequence, (c) either 4, 7, 8, 10, 12, or 13 GGC repeats, and (d) the luciferase reporter gene, in this order (Persico *et al.*, 2006). Both neuronal and nonneuronal cell lines showed statistically significant reductions in reporter gene expression with "long" allele, compared to the "normal" 8- and 10-repeat alleles (Fig. 25.1, panels B–E). Levels of gene expression separate GGC alleles into three statistically homogeneous subsets, namely, 4 repeats > 7–10 repeats > 12 or longer. Neural cell lines display step-like 50–60% reductions in luciferase activity with "long" 12- and 13-repeat alleles, compared to the more common 8- and 10-repeat alleles (Fig. 25.1, panels B and C); instead, CHO cells show a roughly linear, progressive decrease in luciferase activity with increasing GGC repeat number (Fig. 25.1, panel D). This progressive trend closely resembles the outcome of computer-based simulations, predicting that the number of GGC repeats would positively correlate with increasingly more stable mRNA secondary structures, characterized by progressively lower free energy (Persico *et al.*, 2006). Increasing triplet repeat numbers could presumably decrease the efficiency of *RELN* gene expression in CHO cells, as ribosomes would require more time and energy to scan progressively longer and thermodynamically more stable mRNA secondary structures, before reaching the AUG translational start site. In neuronal cell lines, displaying much higher *RELN* 5'UTR-driven luciferase activity compared to CHO cells (Fig. 25.1, panels B–D), additional complexity could be contributed by the presence of neural-specific factors modulating *RELN* gene expression. Putative factors could include CAGER-1 and CAGER-2, i.e., the $(CAG)_n$ and $(CGG)_n$ repeat-binding proteins 1 and 2, which are selectively expressed in postmitotic neurons and are known to bind single-stranded triplet repeats (Yano *et al.*, 1999). Conceivably, their affinity for *RELN* mRNA could be influenced by the length of the GGC repeat, as suggested by preliminary *in vitro* experiments (Dennis Grayson and colleagues, unpublished data discussed in Zhang *et al.*, 2002). Alternatively, or in synergy with cell-specific profiles of RNA-binding proteins, differential DNA methylation patterns could also contribute to differences in basal *RELN* gene expression between neuronal and fibroblast cell lines.

A significant correlation between *RELN* gene expression and the number of GGC repeats was also found *in vivo* (Lugli *et al.*, 2003) by assessing archival plasma samples from a subset of autistic patients studied in our initial genetic study (Persico *et al.*, 2001). Peripherally, Reelin is expressed by the hepatocytes and secreted into the bloodstream (Smalheiser *et al.*, 2000). Reelin's peripheral functions have not yet been fully elucidated but may relate to the immune system, since patients with Norman-Roberts syndrome display congenital lymphedema (Hong *et al.*, 2000). Reelin blood levels were initially shown to be significantly reduced in autistic individuals compared to normal controls (Fatemi *et al.*, 2002). We then demonstrated a "genotype" effect superimposed on this "disease status" effect by studying 10 pairs of autistic patients matched by sex and age, while differing only at the genotypic level, with one member of each pair carrying the common 8 and/or 10 GGC allele (i.e., genotypes were 8/8, 8/10, or 10/10), while the other member had one "long" GGC allele (for example, 8/12 or 10/13). All 10 pairs consistently showed lower Reelin plasma levels in the patient carrying one "long" GGC allele: the correlation between GGC repeat genotypes and the intensity of the 310-kDa band visualized by Western blotting was highly significant ($p < 0.001$). Overall, patients carrying "long" repeat alleles displayed a mean 24.5 ± 3.8% reduction in Reelin plasma levels compared to the matched counterpart carrying "normal" GGC alleles (Lugli *et al.*, 2003).

4 *RELN* GGC Alleles and Autism: Replication Studies

Following the initial genetic findings by Persico *et al.* (2001), several replication studies were performed, as summarized in Table 25.2. Three studies either replicated the initial association of "long" GGC alleles with autism or provided evidence supporting *RELN* gene contributions to autism liability through gene variants in linkage disequilibrium with these and with other known polymorphisms.

Zhang *et al.* (2002) tested the GGC repeat for association with autism using both a case–control and a family-based association test. Selecting one autistic patient per multiplex family (i.e., including two or more siblings affected with autism) and contrasting allelic and genotypic distributions in 126 patients with those of 347 unassessed controls, the authors found no significant association, but only a minor trend ("long" GGC allele frequencies: 9.2% versus 6.9% in autistic patients and controls, respectively; $p=0.370$). Applying the family-based

Table 25.2 Summary of genetic association studies on *RELN* gene variants and autism. All *RELN* gene polymorphisms assessed in each study are listed, except for Serajee *et al.* (2005), where only the 2 SNPs found associated with autism are reported here, out of 34 SNPs assessed; AGRE, Autism Genetic Resource Exchange

Reference	Polymorphisms	Experimental design	Race and ethnicity	Outcome
Persico *et al.* (2001)	5' UTR: GGC repeat Intron 5: rs607755 Exon 50: rs2229864	Case-control Family-based	Italians; U.S.-Caucasians	Association with GGC repeat and with haplotypes formed by GGC + rs607755 + rs2229864
Zhang *et al.* (2002)	5'UTR: GGC repeat	Case-control Family-based	Not specified (families from Canada and AGRE)	Association with GGC repeat
Krebs *et al.* (2002)	5'UTR: GGC repeat	Family-based	Mostly (94%) EU-Caucasians	No association with GGC repeat
Bonora *et al.* (2003)	5'UTR: GGC repeat Intron 5: rs607755 Exon 22: rs362691 Intron 31: RELNint31 Exon 50: rs2229864	Family-based	EU-Caucasians: Sample I, IMGSAC families; Sample II, German families	No association with any common variant; rare missense variants are present (see Table. 25. 1)
Li *et al.* (2004)	5'UTR: GGC repeat	Family-based	Not specified	No association with GGC repeat
Devlin *et al.* (2004)	5'UTR: GGC repeat	Family-based	Not specified (families from the NIH CPEA network)	No association with GGC repeat
Skaar *et al.* (2004)	5'UTR: GGC repeat Intron 5: rs607755 Exon 44: rs2075043 Exon 45: rs362746 Exon 50: rs2229864 Intron 59: rs736707	Family-based	U.S.-Caucasians from Duke Univ., AGRE, and Tufts Univ.	Association with GGC triplet and with specific haplotypes
Serajee *et al.* (2005)	Exon 22: rs362691 Intron 59: rs736707	Family-based	U.S.-Caucasians from AGRE	Association with rs362691 and rs736707

association test (FBAT), which uses information from both affected siblings in multiplex families, a statistically significant overtransmission of the "long" alleles to the affected offspring was found ($p<0.05$). Autistic children without delayed onset of phrase speech tended to carry "long" alleles more frequently than children with onset of phrase speech later than 36 months (59.1% versus 78.3%, respectively; $p=0.06$).

An association between the GGC variant and autism was also found by Skaar *et al.* (2004), who studied 371 Caucasian-American families ascertained through Duke University (217 families), AGRE (86 families), and Tufts University (68 families). In addition to the GGC variant, these authors genotyped five SNPs in the *RELN* gene, including the SNPs in intron 5 (rs607755) and exon 50 (rs2229864), previously tested by Persico *et al.* (2001), and SNPs found in the *ORC5L* and *PSMC2* genes, flanking the *RELN* gene at the 5′ and 3′ ends, respectively. The strongest single-marker family-based association was detected at the 5'UTR GGC variant ($p=0.002$), followed by the exon 44 SNP ($p=0.028$). Different subsamples displayed different patterns of association, with the AGRE subsample largely driving the association with the GGC variant. An overall haplotypic analysis defined an association between autism and a 6-marker *RELN* gene haplotype, encompassing ≥ 10 repeats at the GGC variant (FBAT $p<0.002$). Unlike the work by Persico *et al.* (2001), this group found an overtransmission of the 10-repeat allele and not of "long" alleles, which were present at low frequency (approximately 5%) in these patients. This discrepancy was thus interpreted as likely stemming from interethnic differences in genetic structure and in allelic frequencies.

Serajee *et al.* (2005) investigated 34 SNPs in 196 Caucasian families from the AGRE collection. Two SNPs located in intron 59 and in exon 22 showed overtransmission to affected individuals (TDT p-values$=0.0005$ and 0.03, respectively). Applying strict diagnostic criteria for autism, only the intron 59 variant remained significant, with a preferential transmission of the common C allele. These two variants had been previously reported by Skaar *et al.* (2004), who also found a significant association with the intron 59 SNP in the AGRE sample, and by Bonora *et al.* (2003), who did not further investigate the exon 22 missense variant due to its low frequency in their sample.

Four studies have failed to replicate the initial association findings. Krebs *et al.* (2002) performed a family-based study with 117 simplex families (i.e., only one affected individual per family) and 50 multiplex families, mainly recruited throughout Europe. These authors genotyped only the GGC repeat and found no significant overtransmission of the "long" alleles to affected children. In a thorough mutational search of the entire *RELN* gene performed on two separate samples encompassing IMGSAC and German families, Bonora *et al.* (2003) identified the missense variants described above, concluding that their low frequencies could not explain the strong linkage results on 7q22 obtained in these same families. Furthermore, these authors found no evidence of association between autism and more common *RELN* gene polymorphisms, including the polymorphic GGC repeat. Nonetheless, all affected individuals carrying the rare missense variants displayed severe language impairment, possibly providing evidence converging on the genotype–phenotype correlation between *RELN* gene variants and language development, initially proposed by Zhang *et al.* (2002). Finally, two U.S.-based studies also described a lack of association between the polymorphic GGC repeat and autism: Li *et al.* (2004) assessed the GGC repeat and two SNPs located in the 3′ UTR of the RELN gene in 107 multiplex families, whereas Devlin *et al.* (2004) assessed the GGC repeat in a larger sample, comprising 202 simplex and 183 multiplex families recruited by the NIH Collaborative Programs of Excellence in Autism (CPEA) Network. The latter study also found the GGC repeat associated neither with age at first word nor with the age at first phrase.

Several methodological issues must be briefly considered in order to evaluate these studies and their outcome. First, a case–control design is more powerful than family-based designs, but it is also less reliable in the presence of population substructure (i.e., interethnic differences in linkage disequilibrium and allelic frequencies). In the latter scenario, false-positive and false-negative differences in allelic or genotypic distributions between cases and controls could reflect differences in the ethnic composition of the case and control samples, rather than the existence of a true genetic association. Therefore, studies using both approaches, and where both approaches display a significant association or at least similar trends, are more reliable than studies using only either approach. Second, sample size and statistical power are a major issue, especially with family-based designs: most replication studies performed to date, with the exception of studies by Bonora *et al.* (2003), Skaar *et al.* (2004), and Devlin *et al.* (2004), lack the power necessary to find modest-to-moderate-size single-gene contributions within the framework of a polygenic disorder, like autism. Third, the genetic underpinnings of multiplex families may partly differ from those of simplex families. Multiplex families likely encompass patients with more genetically driven forms of autism, whereas simplex families may represent a mix of families with more environmentally driven forms of the disease, and families that, despite prominent genetic liability, have not evolved to become multiplex only due to stoppage (i.e., parents choosing to have no more children once their first child is diagnosed with autism). Merging multiplex and simplex families into a single sample may thus not be entirely appropriate. Finally, several groups have assessed subsets of families recruited by the AGRE consortium (Geschwind *et al.*, 2001), without providing a complete list of their AGRE family identification numbers. It is thus impossible to determine to what extent different studies have really assessed "independent" samples and to perform a reliable meta-analysis of all available data.

The most plausible interpretation of genetic, biochemical, and neurodevelopmental studies of Reelin in autism is that *RELN* gene variants may provide contributions to autism pathogenesis, neither necessary nor sufficient to cause the disease (Bartlett *et al.*, 2005). *RELN* gene variants could enhance susceptibility in interaction with gene variants at other loci and/or with environmental factors. In particular, "long" GGC alleles are functionally correlated with decreased *RELN* gene expression, but several studies have pointed toward risk haplotypes encompassing "normal" alleles (Persico *et al.*, 2001; Skaar *et al.*, 2004). Genetic contributions to autism pathogenesis could thus come not only from "long" GGC alleles but also from additional polymorphisms in linkage disequilibrium with SNPs located in intron 1 or more toward the 3′ end of the *RELN* gene, or perhaps even in the nearby *ORC5L* gene, which is in linkage disequilibrium with the 5′ GGC repeat (Skaar *et al.*, 2004). In particular, the large intron 1 present in the *RELN* gene could host functionally relevant sequences, similar to intronic sequences exerting profound influences on gene expression at the acetylcholinesterase (De Jaco *et al.*, 2005) and β-casein loci (Lenasi *et al.*, 2006).

5 Modeling *RELN* Gene Contributions to Autism: The Challenge of Complexity

Genetic and clinical heterogeneity is often seen as the main source of nonreproducibility in psychiatric genetics. Syndromic autism can, indeed, be produced by a variety of genetic and environmental causes (Persico and Bourgeron, 2006). Consistent with the hypothesis of genetic heterogeneity, in our initial study, "long" *RELN* GGC alleles were present only in approximately 20% of autistic patients (Persico *et al.*, 2001). We thus concluded that this variant could play a role in a relatively limited subset of patients. In order to move now from the generic notion of "genetic heterogeneity" to more heuristic and hypothesis-generating pathogenetic models, it may be critical to consider gene × gene and gene × environment interactions, which display some geographical and ethnic specificity, possibly contributing to the discrepancies recorded among association studies. In our original study, the genetic association was almost entirely carried by our Caucasian-American families, with only a modest nonsignificant trend present in our Italian families (odd ratios=19.2 and 1.6 for Caucasian-Americans and Italians, respectively) (Persico *et al.*, 2001). One possible interpretation of this interethnic difference is offered by *in vitro* studies, showing that Reelin exerts a proteolytic activity which is crucial for neuronal migration; this proteolytic activity is inhibited by diisopropylphosphofluoridate (Quattrocchi *et al.*, 2002), one of many toxic organophosphate (OP) compounds routinely used as pesticides in agriculture and as household insecticides. Based on this observation and on the significantly more widespread use of OPs inside American homes compared to Europe, we proposed a gene × gene × environment interaction model (Fig. 25.2), which predicts that individuals carrying genetic or epigenetic variants resulting in reduced *RELN* gene expression, if exposed prenatally to OPs during critical periods in neurodevelopment, will more likely suffer from altered neuronal migration resulting in autistic disorder (Persico and Bourgeron, 2006). Additional evidence in favor of this model comes from the demonstration that autism is associated, again in our Caucasian-American but not in our Italian families, with genetic variants of the *PON1* gene encoding for paraxonase, the organophosphate-detoxifying enzyme present in human serum bound to HDL (D'Amelio *et al.*, 2005). Furthermore, we have recently shown that the *PON1 R192* allele associated with autism yields in autistic patients, but not in normal controls, prominent reductions in serum PON1 arylesterase activity (Gaita and Persico, 2006), as predicted by our model (Fig. 25.2).

The biological roles of the Reelin protein are likely broader than currently appreciated and may eventually justify or even require that other models be generated, in addition to or in substitution of this proposed model (Fig. 25.2). As an example, recent neuroanatomical and brain imaging studies have provided intriguing evidence supporting an abnormal activation of the immune system in autism (Laurence and Fatemi, 2005; Vargas *et al.*, 2005; Petropoulos *et al.*, 2006). These findings are sur-

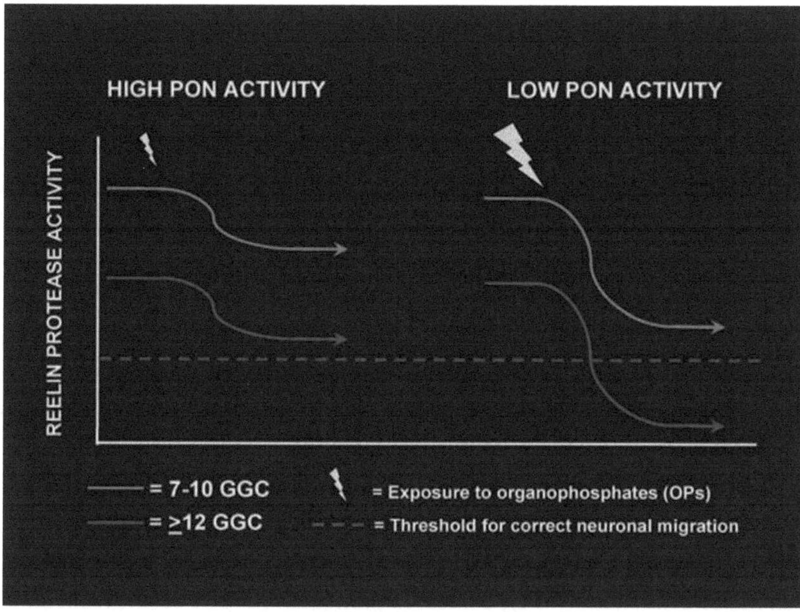

Fig. 25.2 Putative gene–environment interaction model involving the Reelin and PON1 genes, and prenatal exposure to organophosphates (OPs). Reelin gene variants genetically determine normal or reduced levels of Reelin, associated with normal or "long" GGC alleles, respectively. Both conditions are compatible with normal neurodevelopment, but prenatal exposure to OPs can transiently inhibit Reelin's proteolytic activity, which may or may not fall below the threshold critical to neuronal migration, also depending on baseline levels of Reelin. Furthermore, exposure to identical doses of OPs can affect Reelin to a different extent, depending on the amount and affinity spectrum of the OP-inactivating enzyme paraoxonase produced by the *PON1* gene alleles carried by each subject (Gaita and Persico, 2006; Persico and Bourgeron, 2006). (Modified from *Trends Neurosci.*, Vol. 29, Persico, A.M., and Bourgeron, T., Searching for ways out of the autism maze: genetic, epigenetic and environmental clues, pages 349–358, copyright 2006, with permission from Elsevier) (*See Color Plates*)

prisingly convergent with our results indicating that macrocephaly (i.e., a head circumference > 97th percentile) is significantly correlated with a positive history of "allergies" both in the autistic patient and in his/her first-degree relatives (Sacco *et al.*, 2006). Also, the significant decrease in serum PON1 enzymatic activity we find in autistic patients and first-degree relatives (Gaita and Persico, 2006) parallels similar decreases found in the presence of viral hepatitis C (Ferré *et al.*, 2005; Kilic *et al.*, 2005), influenza (Van Lenten *et al.*, 2002), and HIV infections (Parra *et al.*, 2007). These results are thus compatible with the presence of a persistent, virally triggered immune reaction in a subgroup of genetically predisposed autistic children displaying macrosomic features. Within this scenario, the proteolytic activity exerted by Reelin on extracellular matrix proteins, such as fibronectin, could play multiple roles in modulating inflammatory mechanisms at the extracellular level and the recruitment of lymphocytes to the site of inflammation. The latter phenomenon is interestingly mediated by the binding of α4β1 integrin receptors found on lymphocyte membranes, with the CS-1 fragment of fibronectin, present on the membranes of endothelial cells (Munoz *et al.*, 1997).

Finally, geographically diversified environmental factors could also differentially influence epigenetic mechanisms including DNA methylation, histone acetylation, and higher-order chromatin organization. It is well known that DNA methylation is profoundly influenced by factors like nutrition, smoking habits, and aging (Jaenisch and Bird, 2003; Feil, 2006). Epigenetic mechanisms have been demonstrated to play a role in the pathogenesis of several neurological disorders often associated with autism. In fragile-X syndrome, a CGG triplet repeat expansion present in the 5'UTR of the *FMR1* gene is accompanied by an aberrant *de novo* methylation of the CpG islands located in the *FMR1* promoter, yielding transcription silencing (Chelly and Mandel, 2001). Mutations in the methyl CpG binding protein 2 (MeCP2) are associated with Rett syndrome, an X-linked pervasive developmental disorder characterized by autism, loss of speech, seizures, microcephaly, and hand wringing (Amir, 1999). A recent study by Horike *et al.* (2005) identified Dlx5 as a direct target of MeCP2 and showed substantial chromatin differences at the Dlx5-Dlx6 locus between MeCP2 null and wild-type mice. In particular, MeCP2 seems to mediate the formation of a silent-chromatin loop through histone modifications, which was absent in MeCP2 null mice. There is strong *in vitro* evidence that the methylation status of the promoter modulates *RELN* gene transcription rates (Chen *et al.*, 2002). Downregulation of Reelin mRNA and protein levels in schizophrenic patients have been directly correlated to *RELN* promoter hypermethylation (Abdolmaleky *et al.*, 2005). These observations point toward the possibility that genetic polymorphisms associated with autism may mark *RELN* gene alleles carrying abnormal epigenetic variants. Experiments are ongoing in our laboratory to test this hypothesis.

References

Abdolmaleky, H. M., Cheng, K. H., Russo, A., Smith, C. L., Faraone, S. V., Wilcox, M., Shafa, R., Glatt, S. J., Nguyen, G., Ponte, J. F., Thiagalingam, S., and Tsuang, M. T. (2005). Hypermethylation of the reelin (RELN) promoter in the brain of schizophrenic patients: a preliminary report. *Am. J. Med. Genet. B Neuropsychiatr. Genet.* 134:60–66.

Amir, R. E. (1999). Rett syndrome is caused by mutations in X-linked MECP2, encoding methyl-CpG-binding protein 2. *Nature Genet.* 23:185–188.

Bartlett, C. W., Gharani, N., Millonig, J. H., and Brzustowicz, L. M. (2005). Three autism candidate genes: a synthesis of human genetic analysis with other disciplines. *Int. J. Dev. Neurosci.* 23:221–234.

Bonora, E., Beyer, K. S., Lamb, J. A., Parr, J. R., Klauck, S. M., Benner, A., Paolucci, M., Abbott, A., Ragoussis, I., Poustka, A., Bailey, A. J., and Monaco, A. P. (2003). Analysis of reelin as a candidate gene for autism. *Mol. Psychiatry* 8:885–892.

Chelly, J., and Mandel, J. L. (2001). Monogenic causes of X-linked mental retardation. *NatureRev. Genet.* 9:669–680.

Chen, M. L., Chen, S. Y., Huang, C. H., and Chen, C. H. (2002). Identification of a single nucleotide polymorphism at the 5′ promoter region of human reelin gene and association study with schizophrenia. *Mol. Psychiatry* 7:447–448.

D'Amelio, M., Ricci, I., Sacco, R., Liu, X., D'Agruma, L., Muscarella, L. A., Guarnirei, V., Militerni, R., Bravaccio, C., Elia, M., Schneider, C., Melmed, R., Trillo, S., Pascucci, T., Puglisi-Allegra, S., Reichelt, K. L., Macciardi, F., Holden, J. J., and Persico, A. M. (2005). Paraoxonase gene variants are associated with autism in North America, but not in Italy: possible regional specificity in gene–environment interactions. *Mol. Psychiatry* 10:1006–1016.

D'Arcangelo, G., Miao, G. G., Chen, S. C., Soares, H. D., Morgan, J. I., and Curran, T. (1995). A protein related to extracellular matrix proteins deleted in the mouse mutant reeler. *Nature* 374:719–723.

D'Arcangelo, G., Homayouni, R., Keshvara, L., Rice, D. S., Sheldon, M., and Curran, T. (1999). Reelin is a ligand for lipoprotein receptors. *Neuron* 24:471–479.

De Jaco, A., Camp, S., and Taylor, P. (2005). Influence of the 5′ intron in the control of acetylcholinesterase gene expression during myogenesis. *Chem. Biol. Interact.* 157–158:372–373.

Devlin, B., Bennett, P., Dawson, G., Figlewicz, D.A., Grigorenko, E. L., McMahon, W., Minshew, N., Pauls, D., Smith, M., Spence, M. A., Rodier, P. M., Stodgell, C., The CPEA Genetics Network, and Schellenberg, G. D. (2004). Alleles of a reelin CGG repeat do not convey liability to autism in a sample from the CPEA network. *Am. J. Med. Genet. B Neuropsychiatr. Genet.* 126:46–50.

Fatemi, S. H., Stary, J. M., and Egan, E. A. (2002). Reduced blood levels of reelin as a vulnerability factor in pathophysiology of autistic disorder. *Cell. Mol. Neurobiol.* 22:139–152.

Feil, R. (2006). Environmental and nutritional effects on the epigenetic regulation of genes. *Mutat. Res.* 600:46–57.

Ferré, N., Marsillach, J., Camps, J., Rull, A., Coll, B., Tous, M., and Joven, J. (2005). Genetic association of paraoxonase-1 polymorphisms and chronic hepatitis C virus infection. *Clin. Chim. Acta* 361:206–210.

Gaita, L., and Persico, A. M. (2006). The R192 allele of the PON1 gene is associated with reduced serum arylesterase activity in autistic patients and in their first-degree relatives [abstract 2122]. Presented at the 56th Annual Meeting of the American Society of Human Genetics, October 11, 2006, New Orleans, Louisiana. Available from http://www.ashg.org/genetics/ashg05s/

Geschwind, D. H., Sowinski, J., Lord, C., Iversen, P., Shestack, J., Jones, P., Ducat, L., Spence, S. J., and AGRE Steering Committee. (2001). The autism genetic resource exchange: a resource for the study of autism and related neuropsychiatric conditions. *Am. J. Hum. Genet.* 69:463–466.

Goffinet, A. M. (1984). Events governing organization of postmigratory neurons: studies on brain development in normal and reeler mice. *Brain Res. Rev.* 7:261–296.

Hiesberger, T., Trommsdorff, M., Howell, B. W., Goffinet, A., Mumby, M. C., Cooper, J. A., and Herz, J. (1999). Direct binding of reelin to VLDL receptor and ApoE receptor 2 induces tyrosine phosphorylation of disabled-1 and modulated tau phosphorylation. *Neuron* 24:481–489.

Hong, S. E., Shugart, Y. Y., Huang, D. T., Shahwan, S. A., Grant, P. E., Hourihane, J. O., Martin, N. D., and Walsh, C. A. (2000). Autosomal recessive lissencephaly with cerebellar hypoplasia is associated with RELN mutations. *Nature Genet.* 26:93–96.

Horike, S., Cai, S., Miyano, M., Cheng, J. F., and Kohwi-Shigematsu, T. (2005). Loss of silent-chromatin looping and impaired imprinting of DLX5 in Rett syndrome. *Nature Genet.* 37:31–40.

Jaenisch, R., and Bird, A. (2003). Epigenetic regulation of gene expression: how the genome integrates intrinsic and environmental signals. *Nature Genet.* 33:245–254.

Kilic, S. S., Aydin, S., Kilic, N., Erman, F., Aydin, S., and Celik, I. (2005). Serum arylesterase and paraoxonase activity in patients with chronic hepatitis. *World J. Gastroenterol.* 11:7351–7354.

Krebs, M. O., Betancur, C., Leroy, S., Bourdel, M. C., Gillberg, C., Leboyer, M., and the Paris Autism Research International Sibpair (PARIS) Study (2002). Absence of association between a polymorphic GGC repeat in the 5′ untranslated region of the reelin gene and autism. *Mol. Psychiatry* 7:801–804.

Laurence, J. A., and Fatemi, S. H. (2005). Glial fibrillary acidic protein is elevated in superior frontal, parietal and cerebellar cortices of autistic subjects. *Cerebellum* 4:206–210.

Lenasi, T., Peterlin, B. M., and Dovc, P. (2006). Distal regulation of alternative splicing by splicing enhancer in equine beta-casein intron 1. *RNA* 12:498–507.

Li, J., Nguyen, L., Gleason, C., Lotspeich, L., Spiker, D., Risch, N., and Myers, R. M. (2004). Lack of evidence for an association between WNT2 and RELN polymorphisms and autism. *Am. J. Med. Genet. B Neuropsychiatr. Genet.* 126:51–57.

Lugli, G., Krueger, J. M., Davis, J. M., Persico, A. M., Keller, F., and Smalheiser, N. R. (2003). Methodological factors influencing measurement and processing of plasma reelin in humans. *BMC Biochem.* 4:9.

Muhle, R., Trentacoste, V., and Rapin, I. (2004). The genetics of autism. *Pediatrics* 113:e472–486.

Munoz, M., Serrador, J., Nieto, M., Luque, A., Sanchez-Madrid, F., and Teixido, J. (1997). A novel region of the alpha4 integrin subunit with a modulatory role in VLA-4-mediated cell adhesion to fibronectin. *Biochem. J.* 327:727–733.

Parra, S., Alonso-Villaverde, C., Coll, B., Ferre, N., Marsillach, J., Aragones, G., Mackness, M., Mackness, B., Masana, L., Joven, J., and Camps, J. (2007). Serum paraoxonase-1 activity and concentration are influenced by human immunodeficiency virus infection. *Atherosclerosis* 194(1):175–181.

Persico, A. M., and Bourgeron, T. (2006). Searching for ways out of the autism maze: genetic, epigenetic and environmental clues. *Trends Neurosci.* 29:349–358.

Persico, A. M., D'Agruma, L., Maiorano, N., Totaro, A., Militerni, R., Bravaccio, C., Wassink, T. H. for the CLSA, Schneider, C., Melmed, R., Trillo, S., Montecchi, F., Palermo, M., Pascucci, T., Puglisi-Allegra, S., Reichelt, K. L., Conciatori, M., Marino, R., Quattrocchi, C. C., Baldi, A., Zelante, L., Gasparini, P., and Keller, F. (2001). Reelin gene alleles and haplotypes as a factor predisposing to autistic disorder. *Mol. Psychiatry* 6:150–159.

Persico, A. M., Levitt, P., and Pimenta, A. F. (2006). Polymorphic GGC repeat differentially regulates human reelin gene expression levels. *J. Neural Transm.* 113:1373–1382.

Petropoulos, H., Friedman, S. D., Shaw, D. W., Artru, A. A., Dawson, G., and Dager, S. R. (2006). Gray matter abnormalities in autism spectrum disorder revealed by T2 relaxation. *Neurology* 67:632–636.

Quattrocchi, C. C., Wannenes, F., Persico, A. M., Ciafrè, S. A., D'Arcangelo, G., Farace, M. G., and Keller, F. (2002). Reelin is a serine protease of the extracellular matrix. *J. Biol. Chem.* 277:303–309.

Rice, D. S., and Curran, T. (2001). Role of the reelin signaling pathway in central nervous system development. *Annu. Rev. Neurosci.* 24:1005–1039.

Royaux, I., Lambert de Rouvroit, C., D'Arcangelo, G., Demirov, D., and Goffinet, A. M. (1997). Genomic organization of the mouse reelin gene. *Genomics* 46:240–250.

Sacco, R., Militerni, R., Bravaccio, C., Frolli, A., Elia, M., Trillo, S., Curatolo, P., Manzi, B., Lenti, C., Saccani, M., and Persico, A. M. (2006). Clinical and morphological characterization of the macrocephalic endophenotype in autism [abstract 2124]. Presented at the 56th Annual Meeting of the American Society of Human Genetics, October 11, 2006, New Orleans, Louisiana. Available from http://www.ashg.org/genetics/ashg05s/

Serajee, F. J., Zhong, H., and Huq, A. H. M. M. (2005). Association of reelin gene polymorphisms with autism. *Genomics* 87:75–83.

Skaar, D. A., Shao, Y., Haines, J. L., Stenger, J. E., Jaworski, J., Martin, E. R., DeLong, G. R., Moore, J. H., McCauley, J. L., Sutcliffe, J. S., Ashley-Koch, A. E., Cuccaro, M. L., Folstein, S. E., Gilbert, J. R., and Pericak-Vance, M. A. (2004). Analysis of the RELN gene as a genetic risk factor for autism. *Mol. Psychiatry* 10:563–571.

Smalheiser, N. R., Costa, E., Guidotti, A., Impagnatiello, F., Auta, J., Lacor, P., Kriho, V., and Pappas, G. D. (2000). Expression of reelin in adult mammalian blood, liver, pituitary pars intermedia, and adrenal chromaffin cells. *Proc. Natl. Acad. Sci. USA* 97:1281–1286.

Van Lenten, B. J., Wagner, A. C., Anantharamaiah, G. M., Garber, D. W., Fishbein, M. C., Adhikary, L., Nayak, D. P., Hama, S., Navab, M., and Fogelman, A. M. (2002). Influenza infection promotes macrophage traffic into arteries of mice that is prevented by D-4F, an apolipoprotein A-I mimetic peptide. *Circulation* 106:1127–1132.

Vargas, D. L., Nascimbene, C., Krishnan, C., Zimmerman, A. W., and Pardo, C. A. (2005). Neuroglial activation and neuroinflammation in the brain of patients with autism. *Ann. Neurol.* 57:67–81.

Yano, H., Wang, B. E., Ahmad, I., Zhang, J., Abo, T., Nakayama, J., Krempen, K., and Kohwi, Y. (1999). Identification of $(CAG)_n$ and $(CGG)_n$ repeat-binding proteins, CAGERs expressed in mature neurons of the mouse brain. *Exp. Cell Res.* 251:388–400.

Zhang, H., Liu, X., Zhang, C., Mundo, E., Macciardi, F., Grayson, D. R., Guidotti, A. R., and Holden, J. J. A. (2002). Reelin gene alleles and suscetibility to autism spectrum disorders. *Mol. Psychiatry* 7:1012–1017.

Chapter 26
Alzheimer's Disease and Reelin

Arancha Botella-López and Javier Sáez-Valero

Contents

1 Introduction

Alzheimer's disease (AD) is the most common cause of dementia among elderly people and is characterized by loss of memory and cognitive functions. The pathological hallmarks include extensive synaptic and neuronal loss, astrogliosis, and accumulation of fibrillar deposits. The amyloid plaques are extracellular deposits mainly composed of a small insoluble protein called β-amyloid protein or Aβ that is derived from the β-amyloid precursor protein (APP) (Masters *et al.*, 1985). The neurofibrillary tangles are composed of intracellular paired helical filaments containing an abnormally phosphorylated form of the tau protein (Grundke-Iqbal *et al.*, 1986). Specific genetic factors are also linked closely to AD. Thus, despite the occurrence of missense mutations in APP, the most common mutations in AD to date are in presenilin (PS1 and PS2) genes, membrane proteins which play a critical role in the γ-secretase processing of APP (Selkoe, 2001). Whereas these mutations are quite infrequent causes of AD, the major known genetic risk factor for the disorder in the typical late-onset period is the ε4 allele of apolipoprotein E (ApoE) (Strittmatter *et al.*, 1993).

A. Botella-López
Instituto de Neurociencias de Alacant, Universidad Miguel Hernández-CSIC, E-03550 Sant Joan d'Alacant, and Centro de Investigación Biomédica en Red sobre Enfermedades Neurodegenerativas (CIBERNED), Spain

J. Sáez-Valero
Instituto de Neurociencias de Alacant, Universidad Miguel Hernández-CSIC, E-03550 Sant Joan d'Alacant, and Centro de Investigación Biomédica en Red sobre Enfermedades Neurodegenerativas (CIBERNED), Spain
e-mail: j.saez@umh.es

S. H. Fatemi (ed.), *Reelin Glycoprotein: Structure, Biology and Roles in Health and Disease.*

As extensively reviewed in this book, Reelin is a signaling protein that regulates the migration of neurons during encephalic development and is essential for the correct organization, development, and plasticity of the cerebral cortex. The Reelin pathway involves a cascade of intracytoplasmic events that ends with limitations of the extent to which the tau protein is phosphorylated. Reelin binds to the transmembrane liporeceptors apolipoprotein E receptor 2 (ApoER2) and very-low-density lipoprotein receptor (VLDLR) (D'Arcangelo *et al.*, 1999; Hiesberger *et al.*, 1999), which relay the signal into the cell via the adapter Dab1 (disabled-1; Bar and Goffinet 1999; Cooper and Howell, 1999; Trommsdorff *et al.*, 1999). Co-receptors, such as cadherin-related neuronal receptors (Senzaki *et al.*, 1999), $\alpha 3\beta 1$ integrin protein (Dulabon *et al.*, 2000), are also likely involved in this process, as well as the c-Jun N-terminal kinase (JNK)-interacting proteins (JIP)-1 and -2 (Stockinger *et al.*, 2000; Verhey *et al.*, 2001).

Although the role of Reelin pathway in the adult brain is not precisely known, the complex pattern of cellular and regional Reelin expression is consistent with Reelin having multiple roles in adult mammalian brain function (Ikeda and Terashima, 1997; Alcántara *et al.*, 1998; Pesold *et al.*, 1998; Rodriguez *et al.*, 2000; ; Smalheiser *et al.*, 2000; Martínez-Cerdeño *et al.*, 2002; Roberts *et al.*, 2005). In addition, recent studies have suggested a connection between the Reelin/ApoE receptor system and human neuropsychiatric disorders, which will have exhaustive review in other chapters of this book. The possibility of an involvement of the Reelin signaling pathway in neurodegeneration has merited extensive review (D'Arcangelo *et al.*, 1999; Bothwell and Giniger, 2000; Herz and Beffert, 2000; Rice and Curran, 2001; Grilli *et al.*, 2003; Fatemi, 2005). Recent studies have also shown that Reelin itself can modulate synaptic function and that disruption of Reelin receptors results in learning and memory deficits (Weeber *et al.*, 2002; Beffert *et al.*, 2005), suggesting that impairment of Reelin/ApoE receptor-dependent neuromodulation may contribute to cognitive impairment and synaptic loss in AD.

The purposes of this chapter are to review the links between Reelin and elements of its signaling pathway with the main hallmarks of AD pathology and summarize our recent findings, including the first evidence of altered Reelin expression in the AD brain.

2 Altered Reelin Expression in Brains of Subjects with Alzheimer's Disease and Transgenic Mouse Models

The first evidence for the association of Reelin with AD features comes from a transgenic mouse model. In APP/PS1 double transgenic mice, Reelin immunostaining was found together with human APP in the neuritic component of many AD-like plaques (Wirths *et al.*, 2001). However, in a second APP/PS1 transgenic mouse model, despite the occurrence of occasional Reelin-immunoreactive AD-like plaques, the distribution and intensity of Reelin immunoreactivity in the hippocampal formation was similar to that in the wild-type (Miettinen *et al.*, 2005). This puzzling scenario is completed by a couple of studies in the human AD brain, both focused on

Reelin-immunoreactive Cajal-Retzius cells. First, Riedel *et al.* (2003) did not find Reelin immunoreactivity in neuritic plaques and described that Cajal-Retzius cells appear marginally affected by the formation of paired helical filaments in the AD brain, suggesting that these subtle changes are a result rather than a cause of the pathogenetic cascade of AD. On the other hand, Baloyannis (2005) reported a dramatic decline of the number of Cajal-Retzius cells in early cases of AD and suggested that Reelin loss may be implicated in the synaptic pathology and the multifactorial pathogenetic pathways of AD. Despite these controversial reports, the expression of Reelin in cortical interneurons of the AD brain warrants further study.

In this context, using SDS-PAGE and Western blotting, the presence of detectable levels of the three Reelin forms (full-length 420-, and 310- and 180-kDa N-terminal fragments) was reported in cerebrospinal fluid (CSF) (Sáez-Valero *et al.*, 2003). A significant increase of 180-kDa Reelin levels was found in AD patients compared to healthy individuals (Sáez-Valero *et al.*, 2003). Increased levels of this major 180-kDa CSF-Reelin fragment have recently been confirmed in a different cohort of AD samples (Botella-López *et al.*, 2006; see also Fig. 26.1). The concentrations of the full-length 420-kDa Reelin and 310-kDa bands in CSF from AD patients did not differ significantly from those of nondemented subjects, demonstrating that altered Reelin processing is unlikely to account for the increased abundance of this protein in the CSF of patients affected by AD. Another study failed to confirm our previous report of altered levels of the 180-kDa CSF-Reelin fragment in AD samples (Ignatova *et al.*, 2004). However, the smaller sample sizes analyzed and the handling of the samples may contribute to this divergence of results. In fact,

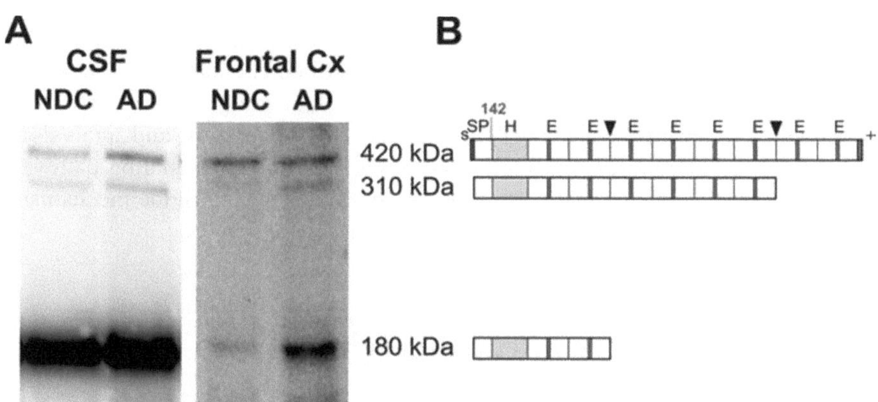

Fig. 26.1 (**A**) Comparison of the banding pattern of CSF and brain (frontal cortex extracts) Reelin identified with antibody 142 (for details see Botella-López *et al.*, 2006), from AD and nondemented control individuals (NDC). The N-terminal 180-kDa Reelin was always the predominant fragment, although faint 420- and 310-kDa bands were also stained in most cases. (**B**) Schematic representation of the Reelin protein, its 310- and 180-kDa fragments generated by the two main processing sites (arrowheads), and recognized by the 142 antibody (the epitope is approximately located as indicated). s, signal peptide; SP, spondin similarity region; H, unique region; E, EGF-like motifs which separated two related subdomains in the eight internal repeats; +, terminal basic region

and in good agreement with the results obtained for human plasma by Lugli *et al.* (2003), methodological factors, such as storage temperature, thawing–freezing cycles, and heating before electrophoresis, influenced Reelin level assessment (Botella-López *et al.*, 2006).

By SDS-PAGE analysis of brain extracts, a marked increase (~40%) in Reelin protein levels was also found in AD frontal cortex (see also Fig. 26.1) compared to nondemented controls (Botella-López *et al.*, 2006). The increase in protein levels was accompanied by a similar increase (~60%) in Reelin transcript levels, supporting the notion that Reelin levels are intrinsically altered in this pathology. In contrast, the protein and mRNA for Reelin appeared unaffected in the cerebellum of AD, an area which normally does not have compact neuritic plaques, demonstrating that Reelin expression is only affected in brain areas targeted by AD. Finally, an abnormality in the glycosylation pattern of Reelin in AD CSF was identified. The pattern of Reelin–lectin binding was altered for two mannose-specific lectins, *Lens culinaris* agglutinin (LCA) and *Canavalia ensiformis* lectin (Con A). Whether the altered glycosylation pattern in AD is a direct consequence of altered metabolism or reflects changes in differentiation state warrants further study.

Taken together, these results suggest an altered Reelin expression in AD. However, these results do not distinguish whether altered Reelin signaling is a mechanism participating in the pathogenesis or whether it is secondary to the degenerative process itself. Interestingly, other types of neurological disorders, such as frontotemporal dementia and progressive supranuclear palsy, which, together with AD, belong to the group of diseases referred to as tauopathies, also show increased Reelin levels in CSF compared to nondemented controls (Sáez-Valero *et al.*, 2003; Botella-López *et al.*, 2006). Whether this reflects the participation of Reelin in the pathogenesis of these diseases, and the potential molecular mechanisms by which Reelin contributes to AD, remains to be elucidated. The next step will be to demonstrate whether Reelin overexpression in adult brains is involved in the sequence of events that occurs during AD. The generation and analysis of transgenic mice overexpressing Reelin may shed light on new signaling pathways associated with Reelin, and may also improve our understanding of the mechanisms related to AD and other neurodegenerative diseases.

3 Reelin, Amyloid, and Neurodegeneration

The relation between Reelin and AD has been highlighted by recent findings suggesting that Reelin signaling affects APP processing or Aβ deposition. Interestingly, it has been reported that Reelin can function as a serine protease of the extracellular matrix (Quattrocchi *et al.*, 2002). Based on this proteolytic activity of Reelin, it has been suggested that Reelin may be involved in the proteolytic processing of APP or in the clearance of amyloid aggregates generated by APP processing (Grilli *et al.*, 2003). However, this serine protease activity of Reelin is currently doubted and questioned (Quattrocchi *et al.*, 2004). Alternatively, sprouting of Reelin-expressing interneurons might be induced by Aβ.

Dab1, which is considered an essential component of the Reelin signaling pathway, also binds to APP (Trommsdorff *et al.*, 1998; Howell *et al.*, 1999). Recent studies examined the effect of Dab1 on APP processing. Whereas Parisiadou and Efthimiopoulos (2006) found that Dab1 increases surface APP and its processing by secretases, Hoe *et al.* (2006) reported that APP and secreted Aβ decrease in Dab1 transfected cells. More interestingly, Reelin treatment increases cleavage of APP and ApoER2 and decreases production of Aβ, suggesting that the effect of Dab1 on APP and ApoER2 trafficking and processing is influenced by Reelin (Hoe *et al.*, 2006). Finally, in the context of AD, it has been recently proposed that increased formation of APP cytoplasmic domain in the cytosol released after cleavage of Aβ could inhibit the Reelin signaling pathway and influence synaptic plasticity (Hoareau *et al.*, 2008).

Perhaps the most intriguing indication thus far that Reelin is involved in AD is the potential relationship between disturbed synaptic plasticity and loss of differentiation control in neurodegenerative processes. Considerable effort is concentrated on gaining insights into the basic mechanisms of the Reelin/ApoE pathway in adult synaptic transmission and plasticity. Conversely, AD appears as a disorder of brain self-organization associated with morphodysregulation at the synaptic level and is characterized by a complex reactivation of developmental molecular mechanisms and synaptic dysregulation (Bothwell and Giniger, 2000; Arendt, 2003; Grilli *et al.*, 2003). Reelin and ApoE receptors fulfill critical functions during brain development and may influence the pathogenesis of AD as part of this common mechanism of aberrant neuronal plasticity which changes neuronal morphology and synaptic contacts.

Moreover, presenilins are involved in the Notch and Wnt/beta-catenin signaling pathways linking many of the players involved in neuronal maturation and neurodegeneration. Interestingly, in agreement with models in which neuronal migration disorders have been linked to a defect in Reelin-expressing Cajal-Retzius cells, the altered properties and loss of most of these cells in PS1-deficient mice lead to cortical dysplasia (Hartmann *et al.*, 1999; Kilb *et al.*, 2004). Thus, the link between presenilins and Reelin appears to be of interest.

4 Reelin Signaling Pathway, ApoE Receptors, and the Regulation of Tau Phosphorylation

The most robust circumstantial evidence linking Reelin with neurodegeneration is the binding of Reelin to ApoE receptors and the identity of the downstream target of the Reelin pathway as mediators of tau hyperphosphorylation. Indeed, Reelin binding to the ApoE receptors leading to a cascade of phosphorylations ultimately inhibits glycogen synthase kinase-3β (GSK3β) (Beffert *et al.*, 2002), an enzyme that regulates tau phosphorylation (Mandelkow *et al.*, 1992; Ishiguro *et al.*, 1993). The cyclin-dependent kinase 5 (CDK5), another kinase that can phosphorylate tau (Baumann *et al.*, 1993), is speculatively considered to be a downstream partner for the Reelin signaling pathway (Rice and Curran, 2001). Thus, lack of Reelin is associated with increased tau phosphorylation (D'Arcangelo *et al.*, 1999; Hiesberger *et al.*,

1999; Tueting *et al.,* 1999), and this hyperphosphorylation apparently leads to cytoskeletal disruption and neuronal degeneration. Mutations that prevent the Reelin-dependent induction of Dab1 tyrosine phosphorylation also cause tau hyperphosphorylation (Brich *et al.*, 2003).

On the other hand, the *in vitro* interaction of Reelin with lipoprotein receptors is inhibited in the presence of ApoE3 and ApoE4 alleles (D'Arcangelo *et al.*, 1999), and the presence of ApoE also limits phosphorylation of tau protein and protein kinase activity in Reelin-deficient mice (Ohkubo *et al.*, 2003). It has further been shown that a secreted soluble isoform of ApoER2 can also inhibit Reelin signaling (Koch *et al.*, 2002). A direct role for ApoER2 in amyloid deposition and neurodegeneration in AD has also been suggested (Motoi *et al.*, 2004).

All of the findings together raise the intriguing possibility that Reelin, ApoE, and its receptors contribute to some of the complex processes involved in neurodegeneration.

Acknowledgments The authors acknowledge the support received from la Caixa Foundation and FIS (Grants 03/0038 & 06/0181, and CIBERNED) from Spain.

References

Alcántara, S., Ruiz, M., D'Arcangelo, G., Ezan, F., de Lecea, L., Curran, T., Sotelo, C., and Soriano, E. (1998). Regional and cellular patterns of reelin mRNA expression in the forebrain of the developing and adult mouse. *J. Neurosci.* 18:7779–7799.

Arendt, T. (2003). Synaptic plasticity and cell cycle activation in neurons are alternative effector pathways: the 'Dr. Jekyll and Mr. Hyde concept' of Alzheimer's disease or the yin and yang of neuroplasticity. *Prog. Neurobiol.* 71:83–248.

Baloyannis, S. J. (2005). Morphological and morphometric alterations of Cajal-Retzius cells in early cases of Alzheimer's disease: a Golgi and electron microscope study. *Int. J. Neurosci.* 115:965–980.

Bar, I., and Goffinet, A. M. (1999). Developmental neurobiology. Decoding the reelin signal. *Nature* 399:645–646.

Baumann, K., Mandelkow, E. M., Biernat, J., Piwnica-Worms, H., and Mandelkow, E. (1993). Abnormal Alzheimer-like phosphorylation of tau-protein by cyclin-dependent kinases cdk2 and cdk5. *FEBS Lett.* 336:417–424.

Beffert, U., Morfini, G., Bock, H. H., Reyna, H., Brady, S. T., and Herz, J. (2002). Reelin-mediated signaling locally regulates protein kinase B/Akt and glycogen synthase kinase 3beta. *J. Biol. Chem.* 277:49958–49964.

Beffert, U., Weeber, E. J., Durudas, A., Qiu, S., Masiulis, I., Sweatt, J. D., Li, W. P., Adelmann, G., Frotscher, M., Hammer, R. E., and Herz, J. (2005). Modulation of synaptic plasticity and memory by reelin involves differential splicing of the lipoprotein receptor Apoer2. *Neuron* 47:567–579.

Botella-López, A., Burgaya, F., Gavín, R., García-Ayllón, M. S., Gómez-Tortosa, E., Peña-Casanova, J., Ureña, J. M., Del Río, J. A., Blesa, R., Soriano, E., and Sáez-Valero, J. (2006). Reelin expression and glycosylation patterns are altered in Alzheimer's disease. *Proc. Natl. Acad. Sci. USA* 103:5573–5578.

Bothwell, M., and Giniger, E. (2000). Alzheimer's disease: neurodevelopment converges with neurodegeneration. *Cell* 102:271–273.

Brich, J., Shie, F. S., Howell, B. W., Li, R., Tus, K., Wakeland, E. K., Jin, L. W., Mumby, M., Churchill, G., Herz, J., and Cooper, J. A. (2003). Genetic modulation of tau phosphorylation in the mouse. *J. Neurosci.* 23:187–192.

Cooper, J. A., and Howell, B. W. (1999). Lipoprotein receptors: signaling functions in the brain? *Cell* 97:671–674.

D'Arcangelo, G., Homayouni, R., Keshvara, L., Rice, D. S., Sheldon M., and Curran, T. (1999). Reelin is a ligand for lipoprotein receptors. *Neuron* 24:471–479.

Dulabon, L., Olson, E. C., Taglienti, M. G., Eisenhuth, S., McGrath, B., Walsh, C. A., Kreidberg, J. A., and Anton, E. S. (2000). Reelin binds alpha3beta1 integrin and inhibits neuronal migration. *Neuron* 27:33–44.

Fatemi, S. H. (2005). Reelin glycoprotein: structure, biology and roles in health and disease. *Mol. Psychiatry* 10:251–257.

Grilli, M., Ferrari Toninelli, G., Uberti, D., Spano, P., and Memo, M. (2003). Alzheimer's disease linking neurodegeneration with neurodevelopment. *Funct. Neurol.* 18:145–148.

Grundke-Iqbal, I., Iqbal, K., Tung, Y. C., Quinlan, M., Wisniewski, H. M., and Binder, L. I. (1986). Abnormal phosphorylation of the microtubule-associated protein tau in Alzheimer cytoskeletal pathology. *Proc. Natl. Acad. Sci. USA* 83:4913–4917.

Hartmann, D., De Strooper, B., and Saftig, P. (1999). Presenilin-1 deficiency leads to loss of Cajal-Retzius neurons and cortical dysplasia similar to human type 2 lissencephaly. *Curr. Biol.* 9:719–727.

Herz, J., and Beffert, U. (2000). Apolipoprotein E receptors: linking brain development and Alzheimer's disease. *Nature Rev. Neurosci.* 1:51–58.

Hiesberger, T., Trommsdorff, M., Howell, B. W., Goffinet, A., Mumby, M. C., Cooper, J. A., and Herz, J. (1999). Direct binding of reelin to VLDL receptor and ApoE receptor 2 induces tyrosine phosphorylation of disabled-1 and modulates tau phosphorylation. *Neuron* 24:481–489.

Hoareau, C., Borrell, V., Soriano, E., Krebs, M. O., Prochiantz, A., and Allinquant, B. (2008). APP cytoplasmic domain antagonizes reelin neurite outgrowth inhibition of hippocampal neurons. *Neurobiol. Aging* 29:542–553.

Hoe, H. S., Tran, T. S., Matsuoka, Y., Howell, B. W., and Rebeck, G. W. (2006). Dab1 and reelin effects on APP and ApoEr2 trafficking and processing. *J. Biol. Chem.* 281:35176–35185.

Howell, B. W., Lanier, L. M., Frank, R., Gertler, F. B., and Cooper, J. A. (1999). The disabled 1 phosphotyrosine-binding domain binds to the internalization signals of transmembrane glycoproteins and to phospholipids. *Mol. Cell. Biol.* 19:5179–5188.

Ignatova, N., Sindic, C. J., and Goffinet, A. M. (2004). Characterization of the various forms of the reelin protein in the cerebrospinal fluid of normal subjects and in neurological diseases. *Neurobiol. Dis.* 15:326–330.

Ikeda, Y., and Terashima, T. (1997). Expression of reelin, the gene responsible for the reeler mutation, in embryonic development and adulthood in the mouse. *Dev. Dyn.* 210:157–172.

Ishiguro, K., Shiratsuchi, A., Sato, S., Omori, A., Arioka, M., Kobayashi, S., Uchida, T., and Imahori, K. (1993). Glycogen synthase kinase 3 beta is identical to tau protein kinase I generating several epitopes of paired helical filaments. *FEBS Lett.* 325:167–172.

Kilb, W., Hartmann, D., Saftig, P., and Luhmann, H. J. (2004). Altered morphological and electrophysiological properties of Cajal-Retzius cells in cerebral cortex of embryonic presenilin-1 knockout mice. *Eur. J. Neurosci.* 20:2749–2756.

Koch, S., Strasser, V., Hauser, C., Fasching, D., Brandes, C., Bajari, T. M., Schneider, W. J., and Nimpf, J. (2002). A secreted soluble form of ApoE receptor 2 acts as a dominant-negative receptor and inhibits reelin signaling. *EMBO J.* 21:5996–6004.

Lugli, G., Krueger, J. M., Davis, J. M., Persico, A. M., Keller, F., and Smalheiser, N. R. (2003). Methodological factors influencing measurement and processing of plasma reelin in humans. *BMC Biochem.* 4:9.

Mandelkow, E. M., Drewes, G., Biernat, J., Gustke, N., Van Lint, J., Vandenheede, J. R., and Mandelkow, E. (1992). Glycogen synthase kinase-3 and the Alzheimer-like state of microtubule-associated protein tau. *FEBS Lett.* 314:315–321.

Martínez-Cerdeño, V., Galazo, M. J., Cavada, C., and Clasca, F. (2002). Reelin immunoreactivity in the adult primate brain: intracellular localization in projecting and local circuit neurons of the cerebral cortex, hippocampus and subcortical regions. *Cereb. Cortex* 12:1298–1311.

Masters, C. L., Simms, G., Weinman, N. A., Multhaup, G., McDonald, B. L., and Beyreuther, K. (1985). Amyloid plaque core protein in Alzheimer disease and Down syndrome. *Proc. Natl. Acad. Sci. USA* 82:4245–4249.

Miettinen, R., Riedel, A., Kalesnykas, G., Kettunen, H. P., Puolivali, J., Soininen, H., and Arendt, T. (2005). Reelin-immunoreactivity in the hippocampal formation of 9-month-old wildtype mouse: effects of APP/PS1 genotype and ovariectomy. *J. Chem. Neuroanat.* 30:105–118.

Motoi, Y., Itaya, M., Mori, H., Mizuno, Y., Iwasaki, T., Hattori, H., Haga, S., and Ikeda, K. (2004). Apolipoprotein E receptor 2 is involved in neuritic plaque formation in APP sw mice. *Neurosci. Lett.* 368:144–147.

Ohkubo, N., Lee, Y. D., Morishima, A., Terashima, T., Kikkawa, S., Tohyama, M., Sakanaka, M., Tanaka, J., Maeda, N., Vitek, M. P., and Mitsuda, N. (2003). Apolipoprotein E and reelin ligands modulate tau phosphorylation through an apolipoprotein E receptor/disabled-1/glycogen synthase kinase-3beta cascade. *FASEB J.* 17:295–297.

Parisiadou, L., and Efthimiopoulos, S. (2007). Expression of mDab1 promotes the stability and processing of amyloid precursor protein and this effect is counteracted by X11alpha. *Neurobiol. Aging* 28:377–388.

Pesold, C., Impagnatiello, F., Pisu, M. G., Uzunov, D. P., Costa, E., Guidotti, A., and Caruncho, H. J. (1998). Reelin is preferentially expressed in neurons synthesizing gamma-aminobutyric acid in cortex and hippocampus of adult rats. *Proc. Natl. Acad. Sci. USA* 95:3221–3226.

Quattrocchi, C. C., Wannenes, F., Persico, A. M., Ciafre, S. A., D'Arcangelo, G., Farace, M. G., and Keller, F. (2002). Reelin is a serine protease of the extracellular matrix. *J. Biol. Chem.* 277:303–309.

Quattrocchi, C. C., Huang, C., Niu, S., Sheldon, M., Benhayon, D., Cartwright, J., Jr, Mosier, D. R., Keller, F., and D'Arcangelo, G. (2004). Retraction of 'Quattrocchi et al. *Science* 2003; 301:649–653'. *Science* 303:1974.

Rice, D. S., and Curran, T. (2001). Role of the reelin signaling pathway in central nervous system development. *Annu. Rev. Neurosci.* 24:1005–1039.

Riedel, A., Miettinen, R., Stieler, J., Mikkonen, M., Alafuzoff, I., Soininen, H., and Arendt T. (2003). Reelin-immunoreactive Cajal-Retzius cells: the entorhinal cortex in normal aging and Alzheimer's disease. *Acta Neuropathol.* 106:291–302.

Roberts, R. C., Xu, L., Roche, J. K., and Kirkpatrick, B. (2005). Ultrastructural localization of reelin in the cortex in post-mortem human brain. *J. Comp. Neurol.* 482:294–308.

Rodriguez, M. A., Pesold, C., Liu, W. S., Kriho, V., Guidotti, A., Pappas, G. D., and Costa, E. (2000). Colocalization of integrin receptors and reelin in dendritic spine postsynaptic densities of adult nonhuman primate cortex. *Proc. Natl. Acad. Sci. USA* 97:3550–3555.

Sáez-Valero, J., Costell, M., Sjögren, M., Andreasen, N., Blennnow, K., and Luque, J. M. (2003). Altered levels of cerebrospinal fluid reelin in frontotemporal dementia and Alzheimer's disease. *J. Neurosci. Res.* 72:132–136.

Selkoe, D. J. (2001). Alzheimer's disease: genes, proteins, and therapy. *Physiol. Rev.* 81:741–766.

Senzaki, K., Ogawa, M., and Yagi, T. (1999). Proteins of the CNR family are multiple receptors for reelin. *Cell* 99:635–647.

Smalheiser, N. R., Costa, E., Guidotti, A., Impagnatiello, F., Auta, J., Lacor, P., Kriho, V., and Pappas, G. D. (2000). Expression of reelin in adult mammalian blood, liver, pituitary pars intermedia, and adrenal chromaffin cells. *Proc. Natl. Acad. Sci. USA* 97:1281–1286.

Stockinger, W., Brandes, C., Fasching, D., Hermann, M., Gotthardt, M., Herz, J., Schneider, W. J., and Nimpf, J. (2000). The reelin receptor ApoER2 recruits JNK-interacting proteins-1 and -2. *J. Biol. Chem.* 275:25625–25632.

Strittmatter, W. J., Saunders, A. M., Schmechel, D., Pericak-Vance, M., Enghild, J., Salvesen, G. S., and Roses, A. D. (1993). Apolipoprotein E: high-avidity binding to beta-amyloid and increased frequency of type 4 allele in late-onset familial Alzheimer disease. *Proc. Natl. Acad. Sci. USA* 90:19771981.

Trommsdorff, M., Borg, J. P., Margolis, B., and Herz, J. (1998). Interaction of cytosolic adaptor proteins with neuronal apolipoprotein E receptors and the amyloid precursor protein. *J. Biol. Chem.* 273:33556–33560.

Trommsdorff, M., Gotthardt, M., Hiesberger, T., Shelton, J., Stockinger, W., Nimpf, J., Hammer, R. E., Richardson, J. A., and Herz, J. (1999). Reeler/disabled-like disruption of neuronal migration in knockout mice lacking the VLDL receptor and ApoE receptor 2. *Cell* 97:689–701.

Tueting, P., Costa, E., Dwivedi, Y., Guidotti, A., Impagnatiello, F., Manev, R., and Pesold, C. (1999). The phenotypic characteristics of heterozygous reeler mouse. *Neuroreport* 10:1329–1334.

Verhey, K. J., Meyer, D., Deehan, R., Bleni, J., Schnapp, B. J., Rapoport, T. A., and Margolis, B. (2001). Cargo of kinesin identified as JIP scaffolding proteins and associated signaling molecules. *J. Cell Biol.* 152:959–970.

Weeber, E. J., Beffert, U., Jones, C., Christian, J. M., Forster, E., Sweatt, J. D., and Herz, J. (2002). Reelin and ApoE receptors cooperate to enhance hippocampal synaptic plasticity and learning. *J. Biol. Chem.* 277:39944–39952.

Wirths, O., Multhaup, G., Czech, C., Blanchard, V., Tremp, G., Pradier, L., Beyreuther, K., and Bayer, T. A. (2001). Reelin in plaques of beta-amyloid precursor protein and presenilin-1 double-transgenic mice. *Neurosci. Lett.* 316:145–148.

Chapter 27
Reelin and Stroke

Kunlin Jin

Contents

1 Introduction

In 1951, Falconer reported the spontaneous occurrence of a disorder that produced ataxia, incoordination, and tremor in mice, and which came to be designated the *reeler* phenotype (Falconer, 1951). These mice showed neuropathological changes consisting of malpositioned neurons in a variety of brain regions, including cerebral neocortex, hippocampus, and cerebellum (D'Arcangelo *et al.*, 1995). The gene defect was discovered subsequently to affect reelin (Reln), a serine protease produced by developing neurons and found in the extracellular matrix (ECM). Reln expression regulates the migration and settling of central neurons in the developing brain (Tissir and Goffinet, 2003) and spinal cord (Yip *et al.*, 2000). Mutations in the *RELN* gene on chromosome 7q22 in patients account for rare cases of autosomal recessive, Norman-Roberts type lissencephaly with developmental delay, epilepsy, and nystagmus (Hong *et al.*, 2000). At least two allelic variants have been reported: a splice acceptor site mutation (IVS37AS, G-A, -1) and an exon deletion (EX43 DEL).

In addition to its developmental role, Reln appears to function in the adult brain, and decreased *RELN* expression has been implicated in some cases of temporal

K. Jin
Buck Institute for Age Research, 8001 Redwood Boulevard, Novato, CA 94945
e-mail: kjin@buckinstitute.org

S. H. Fatemi (ed.), *Reelin Glycoprotein: Structure, Biology and Roles in Health and Disease.*

lobe epilepsy (Haas *et al.*, 2002). In this chapter, we review effects of Reln on proliferation and migration of neural stem/progenitor cells (NPCs) in normal and ischemic rodent brain and on outcome from experimental stroke.

2 Stroke

Clinical stroke usually results from cerebral ischemia due to occlusion of a cerebral blood vessel, most often an artery. Less common causes include venous occlusion and intracerebral hemorrhage. The three main classes of *in vivo* rodent models of cerebral ischemia are global ischemia, focal ischemia, and combined hypoxia/ ischemia. In the latter, which is typically used to study neonatal brain ischemia, vascular occlusion is combined with hypoxia (Levine, 1960). Focal ischemia, a model of stroke, is usually modeled by middle cerebral artery occlusion (MCAO), and gives rise to localized brain infarction (Ginsberg and Busto, 1989). This model incorporates pathophysiological and histopathological features of clinical stroke, although drugs that protect mice or rats in this model have often failed in clinical trials of stroke therapy. Global ischemia, which produces pancerebral hypoperfusion leading to death of selectively vulnerable neurons, such as those in the CA1 region of the hippocampus, recapitulates the neuropathology that may follow cardiac arrest.

In focal ischemia, all cells (neurons, glia, and endothelium) within the affected vascular territory are deprived of oxygen and glucose, leading to energy failure, glutamate release, and loss of transmembrane ion gradients. These cells may survive or die, depending on the severity and duration of the insult and the efficacy of endogenous neuroprotective programs (Lipton, 1999). Brain injury from experimental focal ischemia is characteristically quantified in terms of infarct volume (Osborne *et al.*, 1987), although this does not necessarily correlate with the degree of functional neurological impairment, as measured by neurobehavioral tests of cognitive or sensorimotor performance.

3 Reln and Neurogenesis in Normal Adult Brain

Altman first observed the proliferative potential of adult rodent brain in the 1960s (Altman, 1962; Altman and Das, 1965). It is now generally accepted that the subventricular zone (SVZ) surrounding the lateral ventricles and the subgranular zone (SGZ) of the hippocampal dentate gyrus (DG) are active proliferative regions that generate neurons (Chiasson *et al.*, 1999), astrocytes (Doetsch *et al.*, 1999), and oligodendrocytes (Chiasson *et al.*, 1999; Johansson *et al.*, 1999) throughout life in mice (Yoshimura *et al.*, 2001), rats (Jin *et al.*, 2001), nonhuman primates (McDermott and Lantos, 1991), and humans (Eriksson *et al.*, 1998). Neural precursor cells (NPCs) in these regions of adult brain can be identified by administering

[^{3}H]thymidine or the thymidine analog 5-bromo-2′-deoxyuridine-5′-monophosphate (BrdU), which are incorporated into DNA during S-phase of the cell cycle (Altman and Das, 1965; del Rio and Soriano, 1989). Because these labels are nonspecific with regard to cell type, identification of newborn neurons requires the use of cell type-specific markers as well.

Cells derived from the adult SVZ or SGZ appear to be capable of developing into mature, functional neurons, although this normally occurs in small numbers. Evidence for neuronal differentiation of these cells includes the observations that they: (1) express neuronal markers such as neuron-specific enolase (NSE) (Cameron *et al.*, 1993; Jin *et al.*, 2001), (2) receive synaptic inputs (Bayer, 1985), develop neuronal electrical properties and synaptic transmission (Song *et al.*, 2002), (3) have axons that can be backfilled (Cameron *et al.*, 1993), (4) exhibit rapid, reversible increases in intracellular Ca^{2+} in response to depolarization with K^+, typical of neuronal voltage-gated Ca^{2+} channels (Kirschenbaum *et al.*, 1994), and (5) develop neurotransmitter-specific phenotypes *in vitro* and *in vivo* (Sawamoto *et al.*, 2001).

The production of new neurons in adult brain is regulated by physiological factors, such as exercise (van Praag *et al.*, 1999) and stress (Gould *et al.*, 1998), as well as by pathological conditions, such as stroke (Jin *et al.*, 2001) and epilepsy (Yoshimura *et al.*, 2001). We found that Reln may also be involved in neurogenesis in the hippocampal DG of normal adult brain (Won *et al.*, 2006). Proliferation of NPCs in DG of *reeler* mice (B6C3Fe-a/a-Reln^{rl}) was identified by BrdU incorporation and immunostaining for doublecortin (Dcx), a microtubule-stabilizing factor found in newborn and migrating neurons, and a generally reliable marker of new neurons in the adult rodent brain (Nacher *et al.*, 2001). Knockdown of Dcx expression inhibits the transit of newborn neurons from the SVZ along the rostral migratory stream (RMS) en route to the olfactory bulb (Jin *et al.*, 2004). We found that the number of BrdU-labeled cells that also expressed Dcx was reduced in the SGZ, but not in the SVZ, of *reeler* mice. In contrast to wild-type mice, *reeler* mice showed an aberrant, disorganized distribution of BrdU-labeled cells in the hippocampus, consistent with the absence of an anatomically identifiable SGZ. The mechanism for impaired hippocampal neurogenesis in adult *reeler* mice is unclear, but could relate to the absence of normal SGZ-derived signaling.

4 Reln and Migration of Neural Stem/Progenitor Cells in Normal Adult Brain

NPCs arising in SGZ and SVZ of adult brain must migrate to the regions in which they will become functional neurons. The SGZ, located between the hilus and the granule cell layer (GCL) of the hippocampal DG, retains the potential to form new neurons into adulthood (Gage *et al.*, 1998; Cameron and McKay, 1999). These cells migrate into the GCL, where they become granule neurons. In the SVZ, immature neurons aggregate in a network of neuroblast chains that line the lateral wall of the lateral ventricles (Doetsch and Alvarez-Buylla, 1996), and form a restricted migratory route, the RMS, from the anterior SVZ into the olfactory bulb (OB). Unlike the radial glia-guided migration of young neurons during early brain development (Rakic, 1990), chain migration in the adult SVZ/RMS involves interactions between migrating cells and tubelike structures formed by specialized astrocytes (Lois *et al.*, 1996). When neuroblasts enter the OB, they differentiate into interneurons (Luskin, 1993; Kornack and Rakic, 2001). However, the OB is not essential for proliferation or directed migration of these cells, since the number of cells in the RMS is not significantly affected after olfactory bulbectomy (Kirschenbaum *et al.*, 1999). In adult *reeler* mice, the number of BrdU- and Dcx-positive cells is reduced, and the normal chainlike structure of the RMS is lost, suggesting that migration of neural stem/progenitor cells in the SVZ-to-OB pathway via the RMS is impaired (Won *et al.*, 2006).

Reeler mice (Rakic and Caviness, 1995), mouse mutants deficient in VLDLR and ApoER2, and mice deficient in Disabled-1 (Dab1) (Howell *et al.*, 1997; Borrell *et al.*, 1999; Trommsdorff *et al.*, 1999) all show dentate granule cell migration defects, consistent with an important role for the Reln signaling pathway in the neuronal migration during cerebral cortical development (Forster *et al.*, 1998). Reln binding to VLDLR and ApoER2 on migrating neurons induces tyrosine phosphorylation of Dab1, activating PI3K/Akt and inhibiting glycogen synthase kinase-3β (GSK3β) (Tissir and Goffinet, 2003). PI3K/Akt is implicated in neuronal migration during both development (Bock *et al.*, 2003) and adulthood (Katakowski *et al.*, 2003), and inhibiting PI3K impairs migration of new neurons from adult rat SVZ explants *in vitro*. GSK3β has not been clearly implicated in SVZ neurogenesis, but GSK3β activation reduces granule neuron migration *in vitro* (Tong *et al.*, 2001). Consequently, the effects of Reln deficiency on PI3K and GSK3β signaling could account for impaired migration of newborn neurons in the RMS of adult brain.

5 Reln and Neurogenesis After Stroke

There is substantial evidence for increased proliferation of NPCs in the adult brain after brain injuries. In global cerebral ischemia in the gerbil, neurogenesis was increased in the SGZ (Liu *et al.*, 1998), with enhanced BrdU labeling of cells coexpressing the neuronal markers NeuN, MAP-2, and calbindin. These cells migrated into the GCL, where they took on phenotypic attributes of mature neurons. Global ischemia also enhanced proliferation of BrdU-labeled NPCs in mouse DG (Takagi *et al.*, 1999). MCAO selectively affects the ipsilateral hemisphere, leaving the contralateral hemisphere for a yoked control. MCAO increased the proliferation of NPCs, identified by BrdU labeling and the expression of Dcx and cell proliferation (PCNA) markers, in both DG and SVZ (Jin *et al.*, 2001). Notably, neurogenesis occurred in areas that were not themselves affected by the injury, requiring the existence of a mechanism linking injury to neurogenesis and operating at a distance. In addition, unilateral injury increased neurogenesis bilaterally. Other studies have demonstrated increased numbers of BrdU/Musahi1 (an RNA-binding protein that is highly expressed in NPCs), immunopositive cells in DG (Takasawa *et al.*, 2002), or of BrdU- or PSA-NCAM-immunoreactive cells in SVZ and cerebral cortex ipsilateral to MCAO (Zhang *et al.*, 2001). At postischemic intervals of 1–2 months, BrdU-labeled cells that coexpressed neuronal marker proteins were also increased in cerebral cortex (Jiang *et al.*, 2001; Takasawa *et al.*, 2002). Although SGZ neurogenesis was reduced in *reeler* mice, ischemia-induced SVZ neurogenesis was preserved (Won *et al.*, 2006).

6 Effects of Reln on Migration of Neural Stem/Progenitor Cells After Stroke

Following ischemia, newborn neurons migrate from SVZ into affected brain areas. In transient global forebrain ischemia in rats, BrdU/NeuN-immunopositive cells appeared to migrate to the hippocampus and replace CA1 pyramidal neurons damaged by ischemia (Nakatomi *et al.*, 2002). Transient MCAO in the rat was associated with migration of BrdU-labeled cells that coexpressed Dcx, and later NeuN, from SVZ into the ischemic striatum (Arvidsson *et al.*, 2002). In another rat MCAO study, SVZ neurogenesis, identified by BrdU labeling and immunostaining for neuronal markers, was markedly increased 10–21 days postischemia, and newborn neurons appeared to migrate in chains from the SVZ to the ischemic striatum. Within the striatum, some of these cells expressed markers of medium spiny neurons, which are preferentially affected in ischemia, suggesting differentiation toward the phenotype of dead or damaged cells (Sato *et al.*, 2001; Parent *et al.*, 2002). We examined the ipsilateral hemisphere of rat brain after MCAO using dual-label immunohistochemistry and found new neurons migrated from SVZ, either directly or via RMS, to the striatum. A time-course mapping study showed a progressive increase in the number of newborn neurons and the extent of their penetration into the striatum over 72 hours (Jin *et al.*, 2003). In the contralateral hemisphere, this was not the case (Arvidsson *et al.*, 2002; Jin *et al.*, 2003). In other studies, SVZ-derived cells have been shown to migrate in a "ventral migratory mass" via the nucleus accumbens into the basal forebrain (De Marchis *et al.*, 2004) and, following MCAO, into the cortex (Jin *et al.*, 2003). These observations suggest the existence of endogenous guidance mechanisms that are mobilized in the ischemic brain. However, in *reeler* mice, Dcx-positive migrating cells were not observed in the cortical penumbra after MCAO, confirming that postischemic neuromigration is impaired.

7 Effects of Reln on Functional Outcome After Stroke

In *reeler* mice subjected to MCAO, neurobehavioral deficits were more severe, and cerebral infarcts were larger than in wild-type mice (Fig. 27.1) (Won *et al.*, 2006). The explanation for this finding is unclear, but several possibilities merit consideration. Impaired neuroproliferation and migration might contribute to worsened outcome, because ischemia appears to stimulate the generation of functional neurons (Nakatomi *et al.*, 2002), and because ablation of SGZ neurogenesis in guinea pigs by whole-brain ionizing radiation (two 5-Gy doses separated by 7 days) produced a less favorable outcome after ischemia (Raber *et al.*, 2004). This schedule of radiation seems to inhibit neurogenesis without affecting microvascular morphology or dendritic profiles (Monje *et al.*, 2002; Mizumatsu *et al.*, 2003). Another possibility is that postischemic neurogenesis does not yield new functional neurons but does lead to the release of neuroprotective growth factors.

Finally, Reln deficiency could worsen outcome after stroke by reducing inhibitory control of excitatory transmission, thereby exacerbating excitotoxic damage

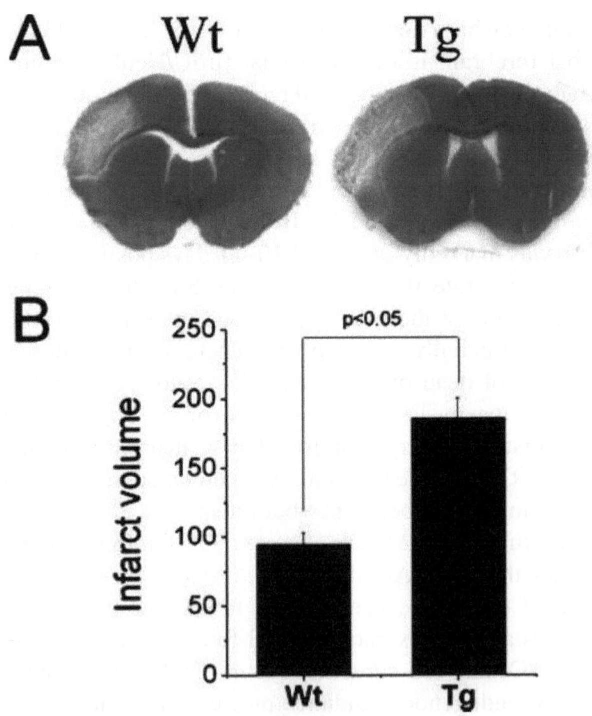

Fig. 27.1 Ischemic brain injury in wild-type (Wt) and transgenic mice with Reln deficiency (Tg) following focal cerebral ischemia. (**A**) HE staining shows an increased area of ischemic injury in *reeler* mice compared to WT mice. (**B**) Quantification of infarct volume in WT and *reeler* mice. $p<0.05$ compared to WT (Student's *t*-test) (*See Color Plates*)

(Costa *et al.*, 2004). A mechanism for such an effect is suggested by the observation that GABA turnover is decreased in cerebral cortex, hippocampus, and striatum of *reeler* mice (Carboni *et al.*, 2004).

8 Summary

Reln is an extracellular matrix-associated serine protease, which helps to regulate the migration of newborn neurons in development and adulthood. Neurogenesis is decreased in the dentate gyrus but not the subventricular zone of *reeler* mice, and neuromigration in the rostral migratory stream is also reduced. Similar findings pertain to the ischemic *reeler* brain, which fails to mount a normal neuroproliferative and neuromigratory response to injury. Unexpectedly, infarct volume after MCAO was also increased in *reeler* compared to wild-type mice, and neurobehavioral deficits were increased. The latter finding suggests that Reln exerts a neuroprotective effect in the adult brain, the basis for which remains to be determined.

References

Altman, J. (1962). Are new neurons formed in the brains of adult mammals? *Science* 135:1127–1128.

Altman, J., and DasG, D. (1965). Autoradiographic and histological evidence of postnatal hippocampal neurogenesis in rat. *J. Comp. Neurol.* 124:319–336.

Arvidsson, A., Collin, T., Kirik, D., Kokaia, Z., and Lindvall, O. (2002). Neuronal replacement from endogenous precursors in the adult brain after stroke. *Nature Med.* 8:963–970.

Bayer, S. A. (1985). Neuron production in the hippocampus and olfactory bulb of the adult rat brain: addition or replacement? *Ann. N.Y. Acad. Sci.* 457:163–172.

Bock, H. H., Jossin, Y., Liu, P., Forster, E., May, P., Goffinet, A. M., and Herz, J. (2003). Phosphatidylinositol 3-kinase interacts with the adaptor protein dab1 in response to reelin signaling and is required for normal cortical lamination. *J. Biol. Chem.* 278:38772–38779.

Borrell, V., Del Rio, J. A., Alcantara, S., Derer, M., Martinez, A., D'Arcangelo, G., Nakajima, K., Mikoshiba, K., Derer, P., Curran, T., and Soriano, E. (1999). Reelin regulates the development and synaptogenesis of the layer-specific entorhino-hippocampal connections. *J. Neurosci.* 19:1345–1358.

Cameron, H. A., and McKay, R. D. (1999). Restoring production of hippocampal neurons in old age. *Nature Neurosci.* 2:894–897.

Cameron, H. A., Woolley, C. S., McEwen, B. S., and Gould, E. (1993). Differentiation of newly born neurons and glia in the dentate gyrus of the adult rat. *Neuroscience* 56:337–344.

Carboni, G., Tueting, P., Tremolizzo, L., Sugaya, I., Davis, J., Costa, E., and Guidotti, A. (2004). Enhanced dizocilpine efficacy in heterozygous reeler mice relates to GABA turnover downregulation. *Neuropharmacology* 46:1070–1081.

Chiasson, B. J., Tropepe, V., Morshead, C. M., and van der Kooy, D. (1999). Adult mammalian forebrain ependymal and subependymal cells demonstrate proliferative potential, but only subependymal cells have neural stem cell characteristics. *J. Neurosci.* 19:4462–4471.

Costa, E., Davis, J. M., Dong, E., Grayson, D. R., Guidotti, A., Tremolizzo, L., and Veldic, M. (2004). A GABAergic cortical deficit dominates schizophrenia pathophysiology. *Crit. Rev. Neurobiol.* 16:1–23.

D'Arcangelo, G., Miao, G. G., Chen, S. C., Soares, H. D., Morgan, J. I., and Curran, T. (1995). A protein related to extracellular matrix proteins deleted in the mouse mutant reeler. *Nature* 374:719–723.

del Rio, J. A., and Soriano, E. (1989). Immunocytochemical detection of 5′-bromodeoxyuridine incorporation in the central nervous system of the mouse. *Brain Res. Dev. Brain Res.* 49:311–317.

De Marchis, S., Fasolo, A., and Puche, A. C. (2004). Subventricular zone-derived neuronal progenitors migrate into the subcortical forebrain of postnatal mice. *J. Comp. Neurol.* 476:290–300.

Doetsch, F., and Alvarez-Buylla, A. (1996). Network of tangential pathways for neuronal migration in adult mammalian brain. *Proc. Natl. Acad. Sci. USA* 93:14895–14900.

Doetsch, F., Caille, I., Lim, D. A., Garcia-Verdugo, J. M., and Alvarez-Buylla, A. (1999). Subventricular zone astrocytes are neural stem cells in the adult mammalian brain. *Cell* 97:703–716.

Eriksson, P. S., Perfilieva, E., Bjork-Eriksson, T., Alborn, A. M., Nordborg, C., Peterson, D. A., and Gage, F. H. (1998). Neurogenesis in the adult human hippocampus. *Nature Med.* 4:1313–1317.

Falconer, D. S. (1951). Two congenital defects of the limb bones in the mouse. *Acta Anat. (Basel)* 13:371–378.

Forster, E., Kaltschmidt, C., Deng, J., Cremer, H., Deller, T., and Frotscher, M. (1998). Lamina-specific cell adhesion on living slices of hippocampus. *Development* 125:3399–3410.

Gage, F. H., Kempermann, G., Palmer, T. D., Peterson, D. A., and Ray, J. (1998). Multipotent progenitor cells in the adult dentate gyrus. *J. Neurobiol.* 36:249–266.

Ginsberg, M. D., and Busto, R. (1989). Rodent models of cerebral ischemia. *Stroke* 20:1627–1642.

Gould, E., Tanapat, P., McEwen, B. S., Flugge, G., and Fuchs, E. (1998). Proliferation of granule cell precursors in the dentate gyrus of adult monkeys is diminished by stress. *Proc. Natl. Acad. Sci. USA* 95:3168–3171.

Haas, C. A., Dudeck, O., Kirsch, M., Huszka, C., Kann, G., Pollak, S., Zentner, J., and Frotscher, M. (2002). Role for reelin in the development of granule cell dispersion in temporal lobe epilepsy. *J. Neurosci.* 22:5797–5802.

Hong, S. E., Shugart, Y. Y., Huang, D. T., Shahwan, S. A., Grant, P. E., Hourihane, J. O., Martin, N. D., and Walsh, C. A. (2000). Autosomal recessive lissencephaly with cerebellar hypoplasia is associated with human reln mutations. *Nature Genet.* 26:93–96.

Howell, B. W., Hawkes, R., Soriano, P., and Cooper, J. A. (1997). Neuronal position in the developing brain is regulated by mouse disabled-1. *Nature* 389:733–737.

Jiang, W., Gu, W., Brannstrom, T., Rosqvist, R., and Wester, P. (2001). Cortical neurogenesis in adult rats after transient middle cerebral artery occlusion. *Stroke* 32:1201–1207.

Jin, K., Minami, M., Lan, J. Q., Mao, X. O., Batteur, S., Simon, R. P., and Greenberg, D. A. (2001). Neurogenesis in dentate subgranular zone and rostral subventricular zone after focal cerebral ischemia in the rat. *Proc. Natl. Acad. Sci. USA* 98:4710–4715.

Jin, K., Sun, Y., Xie, L., Peel, A., Mao, X. O., Batteur, S., and Greenberg, D. A. (2003). Directed migration of neuronal precursors into the ischemic cerebral cortex and striatum. *Mol. Cell. Neurosci.* 24:171–189.

Jin, K., Mao, X. O., Cottrell, B., Schilling, B., Xie, L., Row, R. H., Sun, Y., Peel, A., Childs, J., Gendeh, G., Gibson, B. W., and Greenberg, D. A. (2004). Proteomic and immunochemical characterization of a role for stathmin in adult neurogenesis. *FASEB J.* 18:287–299.

Johansson, C. B., Momma, S., Clarke, D. L., Risling, M., Lendahl, U., and Frisen, J. (1999). Identification of a neural stem cell in the adult mammalian central nervous system. *Cell* 96:25–34.

Katakowski, M., Zhang, Z. G., Chen, J., Zhang, R., Wang, Y., Jiang, H., Zhang, L., Robin, A., Li, Y., and Chopp, M. (2003). Phosphoinositide 3-kinase promotes adult subventricular neuroblast migration after stroke. *J. Neurosci. Res.* 74:494–501.

Kirschenbaum, B., Nedergaard, M., Preuss, A., Barami, K., Fraser, R. A., and Goldman, S. A. (1994). In vitro neuronal production and differentiation by precursor cells derived from the adult human forebrain. *Cereb. Cortex* 4:576–589.

Kirschenbaum, B., Doetsch, F., Lois, C., and Alvarez-Buylla, A. (1999). Adult subventricular zone neuronal precursors continue to proliferate and migrate in the absence of the olfactory bulb. *J. Neurosci.* 19:2171–2180.

Koliatsos, V. E., Dawson, T. M., Kecojevic, A., Zhou, Y., Wang, Y. F., and Huang, K. X. (2004). Cortical interneurons become activated by deafferentation and instruct the apoptosis of pyramidal neurons. *Proc. Natl. Acad. Sci. US.A*101:14264–14269.

Kornack, D. R., and Rakic, P. (2001). The generation, migration, and differentiation of olfactory neurons in the adult primate brain. *Proc. Natl. Acad. Sci. USA* 98:4752–4757.

Levine, S. (1960). Anoxic-ischemic encephalopathy in rats. *Am. J. Pathol.* 36:1–17.

Lipton, P. (1999). Ischemic cell death in brain neurons. *Physiol. Rev.* 79:1431–1568.

Liu, J., Solway, K., Messing, R. O., and Sharp, F. R. (1998). Increased neurogenesis in the dentate gyrus after transient global ischemia in gerbils. *J. Neurosci.* 18:7768–7778.

Lois, C., Garcia-Verdugo, J. M., and Alvarez-Buylla, A. (1996). Chain migration of neuronal precursors. *Science* 271:978–981.

Luskin, M. B. (1993). Restricted proliferation and migration of postnatally generated neurons derived from the forebrain subventricular zone. *Neuron* 11:173–189.

McDermott, K. W., and Lantos, P. L. (1991). Distribution and fine structural analysis of undifferentiated cells in the primate subependymal layer. *J. Anat.* 178:45–63.

Mizumatsu, S., Monje, M. L., Morhardt, D. R.; Rola, R., Palmer, T. D., and Fike, J. R. (2003). Extreme sensitivity of adult neurogenesis to low doses of x-irradiation. *Cancer Res.* 63:4021–4027.

Monje, M. L., Mizumatsu, S., Fike, J. R., and Palmer, T. D. (2002). Irradiation induces neural precursor-cell dysfunction. *Nature Med.* 8:955–962.

Nacher, J., Crespo, C., and McEwen, B. S. (2001). Doublecortin expression in the adult rat telencephalon. *Eur. J. Neurosci.* 14:629–644.

Nakatomi, H., Kuriu, T., Okabe, S., Yamamoto, S., Hatano, O., Kawahara, N., Tamura, A., Kirino, T., and Nakafuku, M. (2002). Regeneration of hippocampal pyramidal neurons after ischemic brain injury. *Cell* 110:429–441.

Osborne, K. A., Shigeno, T., Balarsky, A. M., Ford, I., McCulloch, J., Teasdale, G. M., and Graham, D. I. (1987). Quantitative assessment of early brain damage in a rat model of focal cerebral ischaemia. *J. Neurol. Neurosurg. Psychiatry* 50:402–410.

Parent, J. M., Vexler, Z. S., Gong, C., Derugin, N., and Ferriero, D. M. (2002). Rat forebrain neurogenesis and striatal neuron replacement after focal stroke. *Ann. Neurol.* 52:802–813.

Raber, J., Fan, Y., Matsumori, Y., Liu, Z., Weinstein, P. R., Fike, J. R., and Liu, J. (2004). Irradiation attenuates neurogenesis and exacerbates ischemia-induced deficits. *Ann. Neurol.* 55:381–389.

Rakic, P. (1990). Principles of neural cell migration. *Experientia* 46:882–891.

Rakic, P., and Caviness, V. S., Jr. (1995). Cortical development: view from neurological mutants two decades later. *Neuron* 14:1101–1104.

Sato, K., Hayashi, T., Sasaki, C., Iwai, M., Li, F., Manabe, Y., Seki, T., and Abe, K. (2001). Temporal and spatial differences of PSA-NCAM expression between young-adult and aged rats in normal and ischemic brains. *Brain Res.* 922:135–139.

Sawamoto, K., Nakao, N., Kakishita, K., Ogawa, Y., Toyama, Y., Yamamoto, A., Yamaguchi, M., Mori, K., Goldman, S. A., Itakura, T., and Okano, H. (2001). Generation of dopaminergic neurons in the adult brain from mesencephalic precursor cells labeled with a nestin-GFP transgene. *J. Neurosci.* 21:3895–3903.

Song, H. J., Stevens, C. F., and Gage, F. H. (2002). Neural stem cells from adult hippocampus develop essential properties of functional CNS neurons. *Nature Neurosci.* 5:438–445.

Takagi, Y., Nozaki, K., Takahashi, J., Yodoi, J., Ishikawa, M., and Hashimoto, N. (1999). Proliferation of neuronal precursor cells in the dentate gyrus is accelerated after transient forebrain ischemia in mice. *Brain Res.* 831:283–287.

Takasawa, K., Kitagawa, K., Yagita, Y., Sasaki, T., Tanaka, S., Matsushita, K., Ohstuki, T., Miyata, T., Okano, H., Hori, M., and Matsumoto, M. (2002). Increased proliferation of neural progenitor cells but reduced survival of newborn cells in the contralateral hippocampus after focal cerebral ischemia in rats. *J. Cereb. Blood Flow Metab.* 22:299–307.

Tissir, F., and Goffinet, A. M. (2003). Reelin and brain development. *Nature Rev. Neurosci.* 4:496–505.

Tong, N., Sanchez, J. F., Maggirwar, S. B., Ramirez, S. H., Guo, H., Dewhurst, S., and Gelbard, H. A. (2001). Activation of glycogen synthase kinase 3 beta (gsk-3beta) by platelet activating factor mediates migration and cell death in cerebellar granule neurons. *Eur. J. Neurosci.* 13:1913–1922.

Trommsdorff, M., Gotthardt, M., Hiesberger, T., Shelton, J., Stockinger, W., Nimpf, J., Hammer, R. E., Richardson, J. A., and Herz, J. (1999). Reeler/disabled-like disruption of neuronal migration in knockout mice lacking the VLDL receptor and ApoE receptor 2. *Cell* 97:689–701.

van Praag, H., Christie, B. R., Sejnowski, T. J., and Gage, F. H. (1999). Running enhances neurogenesis, learning, and long-term potentiation in mice. *Proc. Natl. Acad. Sci. USA* 96:13427–13431.

Won, S. J., Kim, S. H., Xie, L., Wang, Y., Mao, X. O., Jin, K., and Greenberg, D. A. (2006). Reelin-deficient mice show impaired neurogenesis and increased stroke size. *Exp. Neurol.* 198:250–259.

Yip, J. W., Yip, Y. P. L., Nakajima, K., and Capriotti, C. (2000). Reelin controls position of autonomic neurons in the spinal cord. *Proc. Natl. Acad. Sci. USA* 97:8612–8616.

Yoshimura, S., Takagi, Y., Harada, J., Teramoto, T., Thomas, S. S., Waeber, C., Bakowska, J. C., Breakefield, X. O., and Moskowitz, M. A. (2001). Fgf-2 regulation of neurogenesis in adult hippocampus after brain injury. *Proc. Natl. Acad. Sci. USA* 98:5874–5879.

Zhang, R. L., Zhang, Z. G., Zhang, L., and Chopp, M. (2001). Proliferation and differentiation of progenitor cells in the cortex and the subventricular zone in the adult rat after focal cerebral ischemia. *Neuroscience* 105:33–41.

Chapter 28
Reelin and Pancreatic Cancer

Kimberly Walter[1] and Michael Goggins

Contents

1 Overview: Reelin and Pancreatic Cancer

Classically, the *RELN* gene has been known for its role in neuronal migration and positioning during central nervous system development. Absence of *RELN* expression results in the characteristic reeler phenotype in rodents, marked by severe defects in cortical layer formation and an uncoordinated, unsteady gait. In humans, loss of reelin expression causes a type of lissencephaly with severe cortical and cerebellar malformation. *RELN* is also expressed in peripheral tissues, including the liver, kidney, adrenal glands, and pancreas, suggesting an additional role for reelin in development and possibly in structural maintenance of these organs

K. Walter
Department of Pathology, The Sol Goldman Pancreatic Research Center, The Johns Hopkins Medical Institutions, 1550 Orleans Street, CRB II, Room 342, Baltimore, MD 21231

M. Goggins
Departments of Pathology, Medicine, and Oncology, The Sol Goldman Pancreatic Research Center, The Johns Hopkins Medical Institutions, 1550 Orleans Street, CRB II, Room 342, Baltimore, MD 21231
e-mail: mgoggins@jhmi.edu

[1]Kimberly Walter is a graduate student in the Pathobiology and Disease Mechanisms Program at Johns Hopkins University.

(Smalheiser *et al.*, 2000). Recent findings indicate that *RELN* is expressed in the normal duct cells of the adult pancreas, and that *RELN* expression is frequently lost in pancreatic ductal adenocarcinomas and in precursor neoplasms in association with epigenetic silencing (Sato *et al.*, 2006). *In vitro* studies suggest that loss of *RELN* contributes to the ability of pancreatic cancer cells to migrate and invade surrounding tissues. These findings support the notion that the effect of reelin pathway status on cell migration may depend on the cell type affected, perhaps depending on the downstream effects of reelin-mediated signaling on the cell's cytoskeleton. For example, reelin loss stimulates migration in some cell types (Gong *et al.*, 2007), even though the phenotype of RELN gene inactivation in the brain is a failure of migration (Kim *et al.*, 2002; Trommsdorff *et al.*, 1999). Although epigenetic mechanisms appear to be responsible for *RELN* silencing in pancreatic neoplasms, the mechanism directing this epigenetic silencing of *RELN* expression is uncertain (Sato *et al.,* 2006).

2 Epigenetic Alterations in Cancer

Cancer is initiated and driven by genetic changes that include mutations in tumor-suppressor genes and oncogenes, and chromosomal abnormalities, such as deletions, amplifications, and rearrangements. A growing body of evidence indicates that in addition to these genetic modifications, cancer is also driven by epigenetic alterations—heritable modifications in DNA-associated information that do not involve the primary DNA sequence itself (Baylin and Ohm, 2006). One of the best-characterized epigenetic alterations is DNA methylation, which occurs particularly at gene promoter sequences and may result in changes in gene expression that can drive tumorigenesis. Methylation occurs at CpG islands, regions of DNA that contain a high frequency of CG dinucleotides, and leads to a closed chromatin state which results in repression of gene transcription. The mechanisms responsible for the aberrant DNA methylation in pancreatic and other cancers are still not well understood, but multiple mechanisms are likely to be responsible for the full spectrum of methylation alterations associated with cancer.

Overall, it is likely that alterations in methylation in cancers likely reflect the combined effects of aging, nutritional and environmental influences, imprinting effects, genetic alterations associated with tumor development, and cell environmental changes due to tumor stromal interactions or chronic inflammation (Anway *et al.*, 2005; Bachman *et al.*, 2003; Blewitt *et al.*, 2006; Dolinoy *et al.*, 2006; Feinberg *et al.*, 2006; Fraga *et al.*, 2005; Huusko *et al.*, 2004; Ishihara *et al.*, 2006; Morgan *et al.*, 1999; Pruitt *et al.*, 2006; Waterland and Jirtle, 2003). Some of these influences act by regulating normal DNA methylation and chromatin modifications; others influence changes at the local gene level in the context of normal epigenetic machinery. DNA methylation likely initially arises in discrete CpG sites independent of gene expression but then spreads into promoter CpG islands, presumably through a loss of balance between factors that promote and those that protect against methylation spreading (Song *et al.*, 2002). There is also a close interplay between DNA methylation and histone modifications. In some instances, cancer-associated DNA methylation changes may be a secondary event that occurs as a consequence of genetic or other events, such as loss of transcription factor(s) that alter the transcriptional activity of an affected promoter (Di Croce *et al.*, 2002; Huusko *et al.*, 2004). In addition, certain sequences in the human genome are probably more prone to DNA methylation (Bock *et al.*, 2006). For example, transposons are prone to methylation, as exemplified by the transposon that undergoes variable methylation in the agouti mouse (Blewitt *et al.*, 2006; Dolinoy *et al.*, 2006; Morgan *et al.*, 1999; Waterland and Jirtle, 2003). The variable methylation at this locus influences expression of the agouti gene and results in the variable coat color phenotype in these mice (Morgan *et al.*, 1999). Importantly, nutritional or environmental influences on DNA methylation can influence agouti phenotype (Morgan *et al.*, 1999). In addition, certain chromosomal regions are targeted for DNA methylation in cancer cells (Frigola *et al.*, 2006). In contrast, in certain model systems histone modifications (methylation of histone H3 lysine-9) are the primary events that occur prior to DNA methylation (Bachman *et al.*, 2003). In addition, a certain overall histone code (the presence of bivalent and trivalent lysine 9 methylation in H3 histones) appears to predispose to the development of DNA methylation in cancers (Ohm *et al.*, 2007). These findings have suggested that, at least at some genetic loci, initial silencing events lead to chromatin modifications that may predispose promoter CpG islands to hypermethylation. In certain experimental settings, alterations in chromatin insulator function can also give rise to methylation of imprinted or other genes (Ishihara *et al.*, 2006). That chromatin alterations can lead to changes in DNA methylation is interesting not least because it is well known that signaling pathways can alter chromatin modifications (Cha *et al.*, 2005). Therefore, in theory, environmental influences that mediate changes in signaling pathways and thus chromatin modifications can leave permanent epigenetic marks (West and van Attikum, 2006).

3 Reelin Silencing and Inflammation

One important change in cell environment that could lead to epigenetic alterations is chronic inflammation. Previous studies have found that maternal inflammation can suppress Reelin expression in the postnatal brain (Meyer *et al.*, 2006). Reduced Reelin expression in the cortex and hippocampus has been reported in neonatal offspring from dams having been infected with influenza virus at midpregnancy (Fatemi *et al.*, 1999, 2002). The cellular mechanism by which *RELN* expression is reduced by inflammation is not known. Since several inflammatory conditions are associated with altered HDAC expression (Ito *et al.*, 2005), it is possible that inflammatory stimuli alter histone acetylation marks on the *RELN* promoter causing gene silencing that may be associated with promoter methylation. Indeed, *RELN* promoter methylation has previously been shown to be reduced in cells treated with

HDACs (Mitchell *et al.*, 2005). The role of inflammation in *RELN* silencing may be relevant to cancer development because the risk of pancreatic and other cancers increases in the setting of chronic inflammation, such as chronic pancreatitis (Lowenfels *et al.*, 1997).

4 Epigenetic Silencing of *RELN* in Human Cancer

Several genomewide strategies are available to identifying genes which are targets of epigenetic silencing in cancer. One approach involves treating cancer cell lines with epigenetic modifying drugs, followed by microarray expression analysis. Genes which are silenced by epigenetic mechanisms in cancers are reactivated on treatment with modifying drugs. By this strategy, *RELN* was identified as a gene that is silenced by aberrant methylation in the majority of pancreatic cancers (Sato *et al.*, 2006). Sato *et al.* compared the gene expression patterns between 17 pancreatic neoplasms (including 5 pancreatic cancer cell lines and 12 primary pancreatic neoplasms) and 5 normal pancreatic ductal epithelial samples (Sato *et al.*, 2006). *RELN* was underexpressed in pancreatic neoplasms and its expression, methylation status, and functional significance were further examined in additional pancreatic neoplasms and pancreatic cancer cell lines. These studies revealed that *RELN* silencing in pancreatic cancers is associated with promoter methylation, as demonstrated by methylation-specific PCR, combined bisulfite restriction analysis (COBRA), and bisulfite genomic sequencing. Hypermethylation of *RELN* was detected in 14 (61%) of 23 pancreatic cancer cell lines, 17 (85%) of 20 high-grade (carcinoma *in situ*) IPMNs (intraductal papillary mucinous neoplasm), and 9 (47%) of 19 pancreatic cancer xenografts (Sato *et al.,* 2006). Thus, aberrant methylation of the *RELN* promoter is a frequent epigenetic alteration in pancreatic cancer and its precursor lesions, and probably mediates gene silencing.

The role of *RELN* alterations in the pathogenesis of other cancers is largely unknown. Increased *RELN* transcripts have been demonstrated in esophageal cancer cells and tissues (Wang *et al.*, 2002). Recently, reelin expression was shown to be present in about one-half of prostate cancers of advanced grade but not in normal prostate or in low-grade prostate cancers (Perrone *et al.*, 2007). Although these findings raise the possibility that reelin pathway alterations contribute to prostate and esophageal cancer progression, it is not yet known if the reelin pathway is active in these cancers or if the aberrant expression of reelin is simply a manifestation of nonspecific alterations in gene expression associated with cancer.

5 Silencing of Downstream Reelin Pathway Genes

Mutations in *DAB1*, an intracellular adapter protein which mediates the *RELN* signaling pathway, result in a phenotype similar to the reeler mouse phenotype. Therefore, we investigated the possibility that *DAB1* is also a target of silencing in pancreatic cancer. Interestingly, the *DAB1* promoter CpG island is unmethylated in normal pancreas and hypermethylated in 14 (64%) of 22 pancreatic cancer cell lines, 10 (59%) of 17 pancreatic xenografts, and 15 (71%) of 21 primary pancreatic adenocarcinomas (Sato *et al.*, 2006). Promoter hypermethylation correlated with loss of expression of *DAB1*. This observation represents a unique phenomenon in cancer epigenetics, where two members of the same signaling pathway are both targets of epigenetic inactivation. Typically, cancers only harbor gene mutations or inactivation of one member of a particular signaling pathway. Since the loss of expression of one gene is normally sufficient to inactivate the pathway of that gene, inactivation of other genes in the same pathway is not likely to confer any additional growth advantage to the tumor. The reason for the *RELN/DAB1* silencing pattern is not clear; however, it suggests that either reelin signals through alternative downstream targets other than DAB1, or that the *RELN* pathway has additional functions which can be inactivated only through silencing of multiple pathway members.

6 Reelin Parallel Pathways in Human Cancer

In addition to control by the reelin pathway, neuronal migration is also regulated by cyclin-dependent kinase 5 (cdk5) and its coactivators p35 and p39. While cdk5 is ubiquitously expressed, p35 and p39 are neuronal proteins. cdk5 is a serine-threonine kinase that phosphorylates Dab1 independently of reelin signaling (Keshvara *et al.*, 2002). It also phosphorylates other cytoskeletal proteins to contribute to the regulation of cell motility. p35 and p39 are neuronal proteins, whereas cdk5 is ubiquitously expressed. Therefore, this pathway was previously thought to be active only in neuronal tissue. However, we recently observed overexpression of p35 by immunohistochemistry and RT-PCR analysis in pancreatic cancer cell lines and primary pancreatic cancers. We observed an absence of p35 expression in normal pancreas and in a normal pancreatic duct (HPDE) immortalized cell line (unpublished data). Furthermore, specific inhibition of cdk5 in MiaPaCa2, a pancreatic cancer cell line lacking reelin expression, resulted in a marked decrease in cell migration. Although the mechanism of p35 overexpression in pancreatic cancers has not yet been determined, these data suggest that genes in both the RELN and cdk5 pathways are targets for aberrant expression and may contribute to the migratory ability of pancreatic cancer cells.

7 Effects of *RELN* Silencing in the Pancreas

RELN is expressed in normal duct cells and in normal islet cells of the pancreas (Sato *et al.*, 2006). However, the functional role of RELN in normal pancreas has not been elucidated. The receptor VLDLR and downstream effectors of RELN, including DAB1 and LIS1, are expressed in normal pancreatic ductal cells, suggesting that the RELN signaling pathway is active in pancreas and may play a regulatory role in cell positioning, as it does in neuronal tissues (Sato *et al.*, 2006; unpublished data). In fact, targeted knockdown of RELN expression in pancreatic cancer cells which retain RELN expression results in increased migration and invasion of these cells *in vitro* (Sato *et al.*, 2006). By measuring migration of a pancreatic cancer cell line (Su8686) in a Transwell system, we observed a 35-fold increase in the migratory capacity of RELN-siRNA-transfected cells compared to cells transfected with control nontargeting siRNA (Sato *et al.,* 2006). The ability of these cells to invade through a reconstituted basement membrane (Matrigel) in a Boyden chamber assay was increased by ~15-fold when transfected with RELN-siRNA. These data suggest a functional role for reelin in controlling cell migration and inhibiting pancreatic tumorigenesis. Further studies will help to better define the role of reelin expression in normal pancreas and the functional effects of *RELN* silencing on carcinogenesis.

8 Reelin and Pancreatic Cancer Precursor Lesions

The finding of epigenetic silencing of RELN in the majority of primary pancreatic cancers raises the question of whether RELN silencing occurs in early or late stages of pancreatic ductal carcinogenesis. IPMN and PanIN (pancreatic intraepithelial neoplasm) are two distinct precursor lesions of pancreatic ductal adenocarcinoma. Immunohistochemical analysis of RELN in IPMN lesions revealed that 17 (85%) of 20 high-grade (carcinoma in situ) IPMNs had lost RELN expression (Sato *et al.*, 2006).

Preliminary analysis of PanIN lesions including PanIN1a, PanIN1b, PanIN2, and PanIN3 suggest that RELN expression is also lost in a percentage of these precancerous lesions. Examples of reelin expression in pancreatic precursor lesions are shown in Fig. 28.1. Furthermore, *RELN* knockdown increases the ability of pancreatic cancer cells to form colonies *in vitro* (Sato *et al.*, 2006). Thus, epigenetic inactivation of RELN occurs in relatively early stages of pancreatic ductal carcinogenesis.

The effect of RELN silencing on the ability of pancreatic cancer cells to migrate and invade *in vitro* suggests that RELN loss could also facilitate metastasis of primary pancreatic cancers. Future studies including immunohistochemical analysis of tissue microarrays, including matching primary and metastatic pancreatic cancer cases, as well as functional analyses of the effects of *RELN* silencing *in vivo*, are needed to address this question.

9 Epigenetic Modifying Drugs as Cancer Therapeutics

Epigenetic silencing of gene transcription occurs through a sequence of events which convert the DNA from an open, accessible conformation to a compact, heterochromatic state. These events are initiated by methylation of CpG dinucleotides by

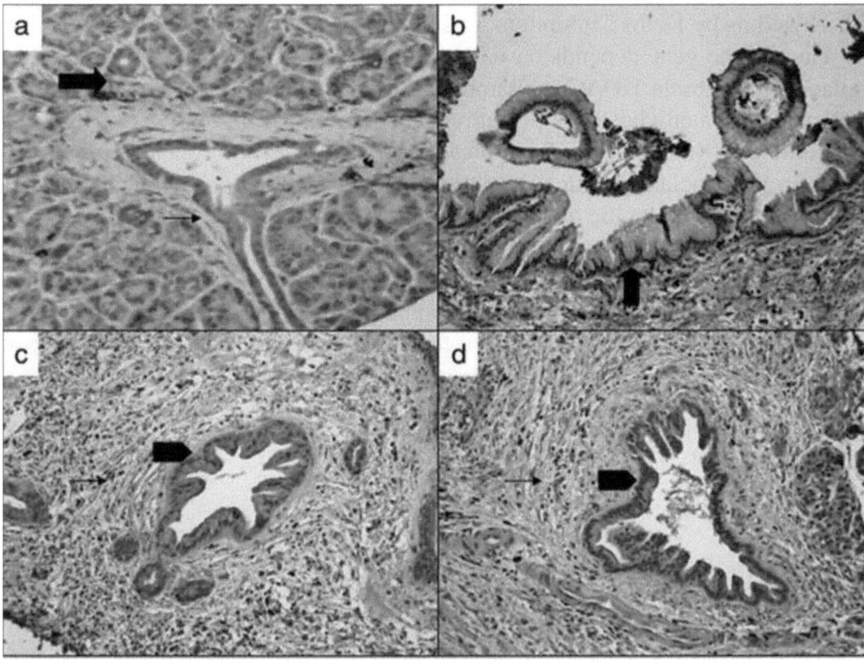

Fig. 28.1 Immunohistochemical analysis of RELN in normal pancreas (**a**) and IPMN (**b**) and PanIN lesions (**c**,**d**). The thin arrow in **a** is pointing to pancreatic ductal epithelium, while the thick arrow is pointing to pancreatic acinar cells. In **b**, the arrow is pointing to the abnormal ductal epithelium of an IPMN; in **c** and **d**, the thick arrowhead is pointing to the abnormal ductal epithelium of a PanIN. The thin arrow in **c** and **d** is pointing to the surrounding fibrosis (*See Color Plates*)

DNA methyltransferases (DNMT). Specific proteins, such as MeCP2, can then bind the CpG dinucleotides and recruit transcriptional corepressors including histone deactylases (HDAC). Cancer-associated epigenetic alterations are attractive therapeutic targets because such epigenetic alterations, unlike genetic changes, are potentially reversible. Inhibitors of DNA methylation and HDAC have been shown to suppress tumor growth *in vitro* and *in vivo*, and some of the inhibitors are being tested in clinical trials for patients with different types of solid and hematological cancers (Egger *et al.*, 2004). For example, the DNMT inhibitor decitabine is an FDA-approved drug useful for the treatment of myeloid neoplasms (Oki *et al.*, 2007). HDAC inhibitors are also undergoing extensive clinical trials for a variety of cancers (Marks and Jiang, 2005). For example, SAHA (suberoylanilide hydroxamic acid) has been FDA approved for the treatment of cutaneous lymphoid neoplasms (http://www.fda.gov/ohrms/dockets/98fr/84n-0102-lst0101-01.pdf). Most aberrantly hypermethylated genes require DNMT inhibition in order to reverse their epigenetic silencing; inhibition of HDACs alone is usually insufficient. Interestingly, epigenetic silencing of reelin can be reversed by HDAC inhibitors, such as SAHA and valproic acid, as well as by DNMT inhibitors, raising the possibility that the DNA methylation status of the gene depends on its chromatin modifications (Sato *et al.*, 2006). Similarly, in the brain HDAC inhibitors can limit the effects of methyl donors on *RELN* promoter methylation by demethylating the promoter (Dong *et al.*, 2007).

In summary, the recent findings that pancreatic cancers frequently lose reelin expression in association with epigenetic silencing and that pancreatic cancer cell motility is influenced by the reelin pathway highlight an important overlap between the biology of neurodevelopmental disorders and cancer development. Understanding the mechanisms of *RELN* silencing may provide clues as to the environmental influences that predispose to cancer development. The ability of epigenetic modifying drugs to modulate the expression of reelin pathway components highlights the potential of these drugs to treat patients with pancreatic and other cancers.

Acknowledgments This work was supported by the National Cancer Institute (grants CA90709, CA120432, CA62924) and the Michael Rolfe Foundation.

References

Anway, M. D., Cupp, A. S., Uzumcu, M., and Skinner, M. K. (2005). Epigenetic transgenerational actions of endocrine disruptors and male fertility. *Science* 308:1466–1469.

Bachman, K. E. P. B., Rhee, I., Rajagopalan, H., Herman, J. G., Baylin, S. B., Kinzler, K. W., and Vogelstein, B. (2003). Histone modifications and silencing prior to DNA methylation of a tumor suppressor gene *Cancer Cell* 3:89–95.

Baylin, S. B., and Ohm, J. E. (2006). Epigenetic gene silencing in cancer—a mechanism for early oncogenic pathway addiction? *Nature Rev. Cancer* 6:107–116.

Blewitt, M. E., Vickaryous, N. K., Paldi, A., Koseki, H., and Whitelaw, E. (2006). Dynamic reprogramming of DNA methylation at an epigenetically sensitive allele in mice. *PLoS Genet.* 2:e49.

Bock, C., Paulsen, M., Tierling, S., Mikeska, T., Lengauer, T., and Walter, J. (2006). CpG island methylation in human lymphocytes is highly correlated with DNA sequence, repeats, and predicted DNA structure. *PLoS Genet.* 2:e26.

Cha, T. L., Zhou, B. P., Xia, W., Wu, Y., Yang, C. C., Chen, C. T., Ping, B., Otte, A. P., and Hung, M. C. (2005). Akt-mediated phosphorylation of EZH2 suppresses methylation of lysine 27 in histone H3. *Science* 310:306–310.

Di Croce, L., Raker, V. A., Corsaro, M., Fazi, F., Fanelli, M., Faretta, M., Fuks, F., Coco, F. L., Kouzarides, T., Nervi, C., Minucci, S., and Pelicci, P. G. (2002). Methyltransferase recruitment and DNA hypermethylation of target promoters by an oncogenic transcription factor. *Science* 295:1079–1082.

Dolinoy, D. C., Weidman, J. R., Waterland, R. A., and Jirtle, R. L. (2006). Maternal genistein alters coat color and protects Avy mouse offspring from obesity by modifying the fetal epigenome. *Environ. Health Perspect.* 114:567–572.

Dong, E., Guidotti, A., Grayson, D. R., and Costa, E. (2007). Histone hyperacetylation induces demethylation of reelin and 67-kDa glutamic acid decarboxylase promoters. *Proc. Natl. Acad. Sci. USA* 104:4676–4681.

Egger, G., Liang, G., Aparicio, A., and Jones, P. A. (2004). Epigenetics in human disease and prospects for epigenetic therapy. *Nature* 429:457–463.

Fatemi, S. H., Emamian, E. S., Kist, D., Sidwell, R. W., Nakajima, K., Akhter, P., Shier, A., Sheikh, S., and Bailey, K. (1999). Defective corticogenesis and reduction in reelin immunoreactivity in cortex and hippocampus of prenatally infected neonatal mice. *Mol. Psychiatry* 4:145–154.

Fatemi, S. H., Earle, J., Kanodia, R., Kist, D., Emamian, E. S., Patterson, P. H., Shi, L., and Sidwell, R. (2002). Prenatal viral infection leads to pyramidal cell atrophy and macrocephaly in adulthood: implications for genesis of autism and schizophrenia. *Cell. Mol. Neurobiol.* 22:25–33.

Feinberg, A. P., Ohlsson, R., and Henikoff, S. (2006). The epigenetic progenitor origin of human cancer. *Nature Rev. Genet.* 7:21–33.

Fraga, M. F., Ballestar, E., Paz, M. F., Ropero, S., Setien, F., Ballestar, M. L., Heine-Suner, D., Cigudosa, J. C., Urioste, M., Benitez, J., Boix-Chornet, M., Sanchez-Aguilera, A., Ling, C., Carlsson, E., Poulsen, P., Vaag, A., Stephan, Z., Spector, T. D., Wu, Y. Z., Plass, C., and Esteller, M. (2005). Epigenetic differences arise during the lifetime of monozygotic twins. *Proc. Natl. Acad. Sci. USA* 102:10604–10609.

Frigola, J., Song, J., Stirzaker, C., Hinshelwood, R. A., Peinado, M. A., and Clark, S. J. (2006). Epigenetic remodeling in colorectal cancer results in coordinate gene suppression across an entire chromosome band. *Nature Genet.* 38:540–549.

Gong, C., Wang, T. W., Huang, H. S., and Parent, J. M. (2007). Reelin regulates neuronal progenitor migration in intact and epileptic hippocampus. *J. Neurosci.* 27:1803–1811.

http://www.fda.gov/ohrms/dockets/98fr/84n-0102-lst0101-01.pdf.

Huusko, P., Ponciano-Jackson, D., Wolf, M., Kiefer, J. A., Azorsa, D. O., Tuzmen, S., Weaver, D., Robbins, C., Moses, T., Allinen, M., Hautaniemi, S., Chen, Y., Elkahloun, A., Basik, M., Bova, G. S., Bubendorf, L., Lugli, A., Sauter, G., Schleutker, J., Ozcelik, H., Elowe, S., Pawson, T., Trent, J. M., Carpten, J. D., Kallioniemi, O. P., and Mousses, S. (2004). Nonsense-mediated decay microarray analysis identifies mutations of EPHB2 in human prostate cancer. *Nature Genet.* 36:979–983.

Ishihara, K., Oshimura, M., and Nakao, M. (2006). CTCF-dependent chromatin insulator is linked to epigenetic remodeling. *Mol. Cell* 23:733–742.

Ito, K., Ito, M., Elliott, W. M., Cosio, B., Caramori, G., Kon, O. M., Barczyk, A., Hayashi, S., Adcock, I. M., Hogg, J. C., and Barnes, P. J. (2005). Decreased histone deacetylase activity in chronic obstructive pulmonary disease. *N. Engl. J. Med.* 352:1967–1976.

Keshvara, L., Magdaleno, S., Benhayon, D., and Curran, T. (2002). Cyclin-dependent kinase 5 phosphorylates disabled 1 independently of reelin signaling. *J. Neurosci.* 22:4869–4877.

Kim, H. M., Qu, T., Kriho, V., Lacor, P., Smalheiser, N., Pappas, G. D., Guidotti, A., Costa, E., and Sugaya, K. (2002). Reelin function in neural stem cell biology. *Proc. Natl. Acad. Sci. USA* 99:4020–4025.

Lowenfels, A. B., Maisonneuve, P., DiMagno, E. P., Elitsur, Y., Gates, L. K., Jr., Perrault, J., and Whitcomb, D. C. (1997). Hereditary pancreatitis and the risk of pancreatic cancer. International Hereditary Pancreatitis Study Group. *J.Natl. Cancer Inst.* 89:442–446.

Marks, P. A., and Jiang, X. (2005). Histone deacetylase inhibitors in programmed cell death and cancer therapy. *Cell Cycle* 4:549–551.

Meyer, U., Nyffeler, M., Engler, A., Urwyler, A., Schedlowski, M., Knuesel, I., Yee, B. K., and Feldon, J. (2006). The time of prenatal immune challenge determines the specificity of inflammation-mediated brain and behavioral pathology. *J. Neurosci.* 26:4752–4762.

Mitchell, C. P., Chen, Y., Kundakovic, M., Costa, E., and Grayson, D. R. (2005). Histone deacetylase inhibitors decrease reelin promoter methylation in vitro. *J. Neurochem.* 93:483–492.

Morgan, H. D., Sutherland, H. G., Martin, D. I., and Whitelaw, E. (1999). Epigenetic inheritance at the agouti locus in the mouse. *Nature Genet.* 23:314–318.

Ohm, J., McGarvey, K., Yu, X., Cheng, L., Schuebel, K., Cope, L., Mohammad, H., Chen, W., Daniel, V., Yu, W., Berman, D., Jenuwein, T., Pruitt, K., Sharkis, S., Watkins, D. N., Herman, J., and Baylin, S. (2007). A stem cell-like chromatin pattern may predispose tumor suppressor genes to DNA hypermethylation and heritable silencing. *Nature Genet.* 39:237–242.

Oki, Y., Aoki, E., and Issa, J.P. (2007). Decitabine—Bedside to bench. *Crit. Rev. Oncol. Hematol.* 61:140–152.

Perrone, G., Vincenzi, B., Zagami, M., Santini, D., Panteri, R., Flammia, G., Verzi, A., Lepanto, D., Morini, S., Russo, A., Bazan, V., Tomasino, R. M., Morello, V., Tonini, G., and Rabitti, C. (2007). Reelin expression in human prostate cancer: a marker of tumor aggressiveness based on correlation with grade. *Mod. Pathol.* 20:344–351.

Pruitt, K., Zinn, R. L., Ohm, J. E., McGarvey, K. M., Kang, S. H., Watkins, D. N., Herman, J. G., and Baylin, S. B. (2006). Inhibition of SIRT1 reactivates silenced cancer genes without loss of promoter DNA hypermethylation. *PLoS Genet*. 2:e40.

Sato, N., Fukushima, N., Chang, R., Matsubayashi, H., and Goggins, M. (2006). Differential and epigenetic gene expression profiling identifies frequent disruption of the RELN pathway in pancreatic cancers. *Gastroenterology* 130:548–565.

Smalheiser, N. R., Costa, E., Guidotti, A., Impagnatiello, F., Auta, J., Lacor, P., Kriho, V., and Pappas, G. D. (2000). Expression of reelin in adult mammalian blood, liver, pituitary pars intermedia, and adrenal chromaffin cells. *Proc. Natl. Acad. Sci. USA* 97:1281–1286.

Song, J. Z. S. C., Harrison, J., Melki, J. R., and Clark, S. J. (2002). Hypermethylation trigger of the glutathione-S-transferase gene (GSTP1) in prostate cancer cells. *Oncogene* 21:1048–1061.

Trommsdorff, M., Gotthardt, M., Hiesberger, T., Shelton, J., Stockinger, W., Nimpf, J., Hammer, R. E., Richardson, J. A., and Herz, J. (1999). Reeler/disabled-like disruption of neuronal migration in knockout mice lacking the VLDL receptor and ApoE receptor 2. *Cell* 97:689–701.

Wang, Q., Lu, J., Yang, C., Wang, X., Cheng, L., Hu, G., Sun, Y., Zhang, X., Wu, M., and Liu, Z. (2002). CASK and its target gene Reelin were co-upregulated in human esophageal carcinoma. *Cancer Lett*. 179:71–77.

Waterland, R. A., and Jirtle, R. L. (2003). Transposable elements: targets for early nutritional effects on epigenetic gene regulation. *Mol. Cell Biol*. 23:5293–5300.

West, A. G., and van Attikum, H. (2006). Chromatin at the crossroads. Meeting on signalling to chromatin epigenetics. *EMBO Rep*. 7:1206–1210.

Index

D

E

S

Color Plates

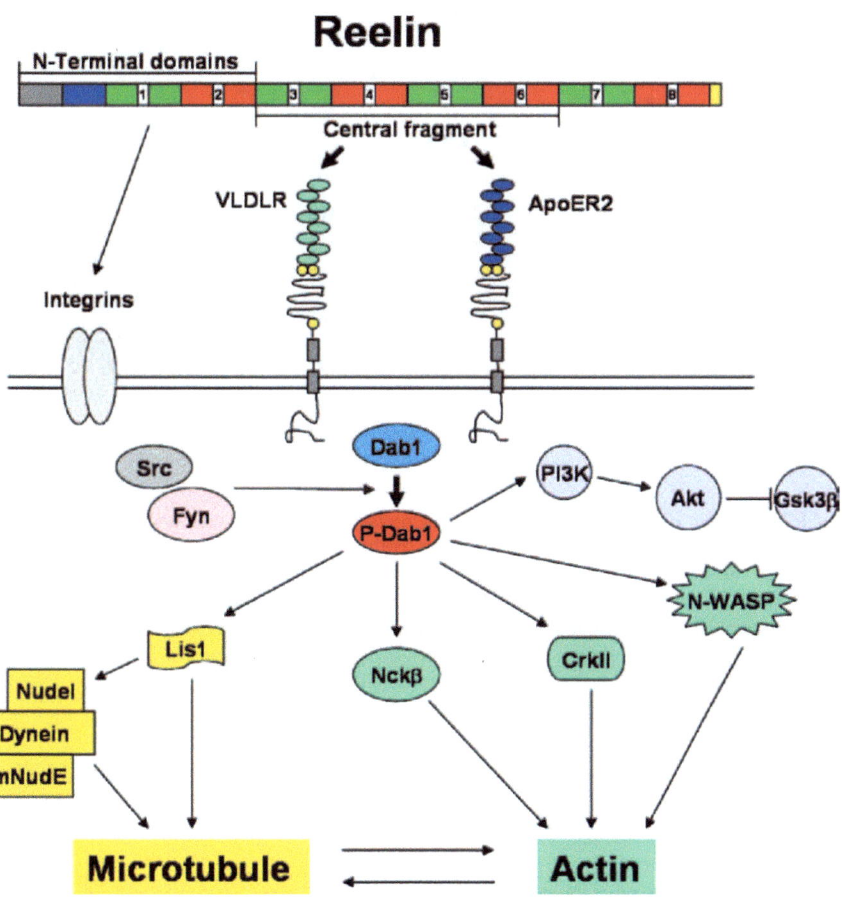

Fig. 1.1 The Reelin signaling pathway

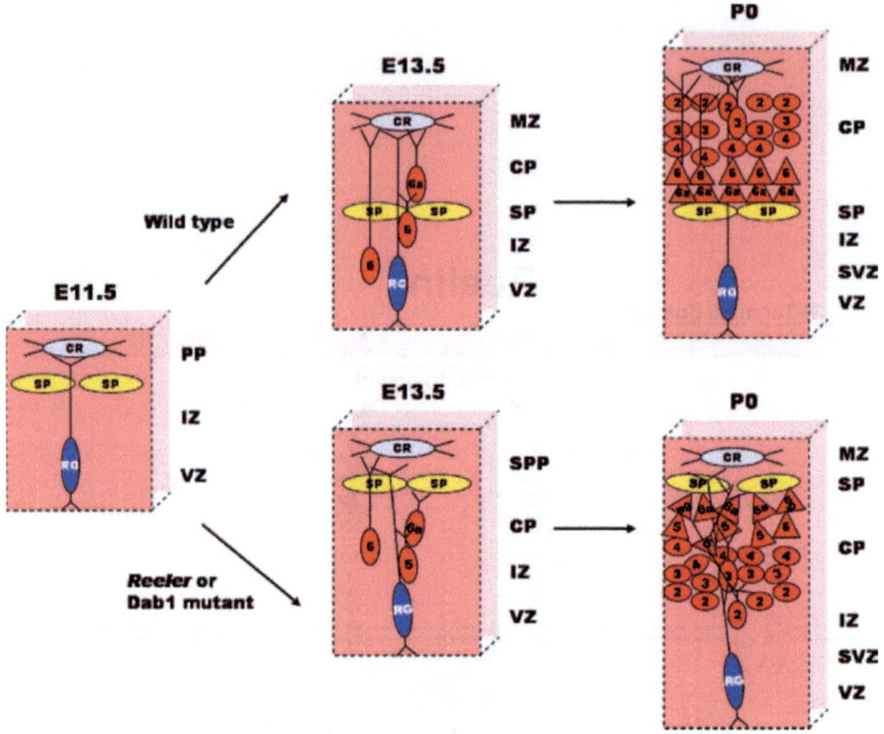

Fig. 1.2. Cortical development in normal and *reeler* and Dab1 mutant mice. In the embryonic cortex of normal mice, the preplate (PP) is split by the arrival of early radially migrating neurons, whereas in the *reeler* cortex, this does not happen and cells form a superplate structure (SPP). Cellular layers in the cortical plate (CP) are also disrupted in *reeler*. Other abbreviations: MZ, marginal zone; IZ, intermediate zone; VZ, ventricular zone; SVZ, subventricular zone; RG, radial glia; CR, Cajal-Retzius cells

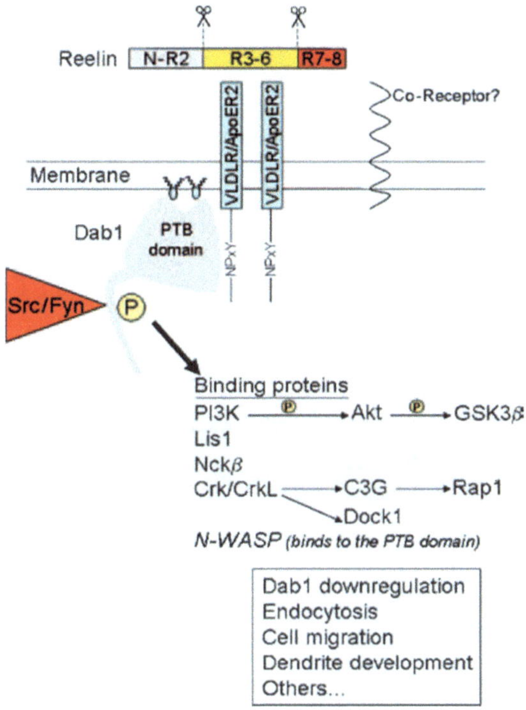

Fig. 3.1 Summary of the Reelin signaling pathway. See text for details

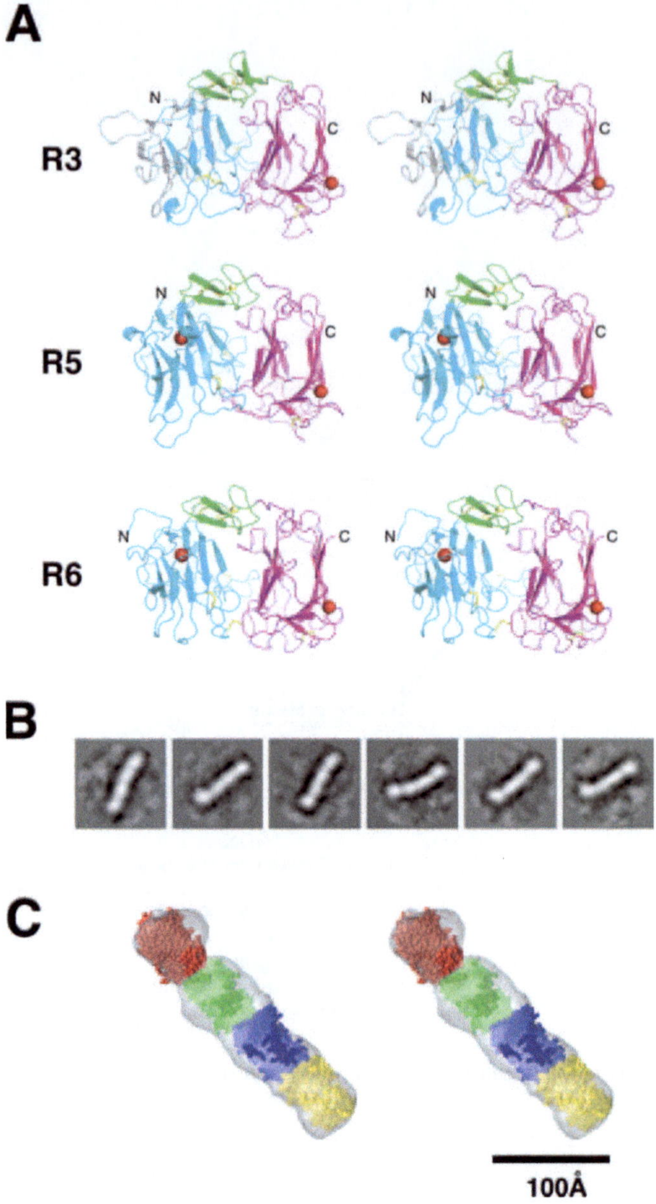

Fig. 5.2 Reelin repeat structure. (**A**) Crystal structures of single reelin repeat domains. Each panel shows a stereo presentation of R3 (top), R5 (middle), and R6 (bottom) structures. Subdomains are differently colored; subrepeat A (cyan), EGF (green), subrepeat B (magenta), and N- and C-termini are labeled. Bound calcium ions and disulfide bridges are shown as red spheres and yellow stick models, respectively. In R3, segments missing in the crystal structure are modeled and shown in gray. (**B**) Two-dimensional averages from representative particle classes obtained from the untilted electron micrographs of the R3–6 fragment. The width of each panel corresponds to 376 Å. (**C**) Three-dimensional volume map of an R3–6 fragment derived from single-particle tomography (gray) in a stereo representation. Four complete space-filling models for reelin repeats (R3, red; R4, green; R5, blue; and R6, yellow) are fitted into the envelope

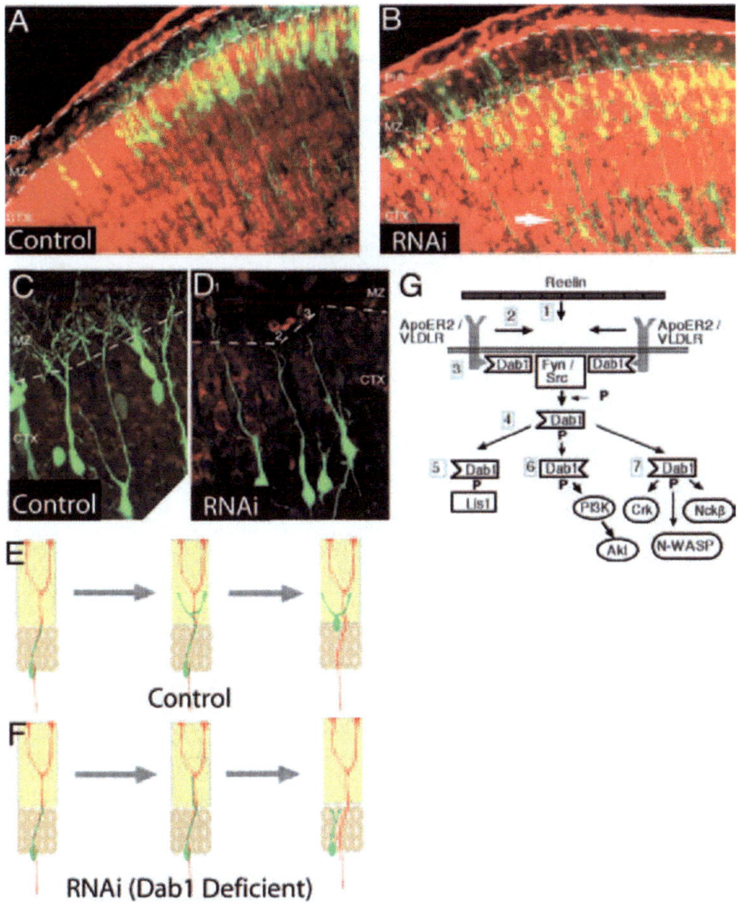

Fig. 7.1 Reelin Dab1 signaling in upper layer cortical neurons. (**A**, **B**) Low-magnification images of layer 2/3 cortical neurons on postnatal day 2 (P2), 7 days after *in utero* electroporation on E16 with either RNAi that suppresses Dab1 (RNAi) or control RNAi vector (Control). (**A**) Control electroporated neurons show precise lamination and exuberant dendritic growth in the MZ (dashed lines) on P2, whereas (**B**) Dab1-suppressed cells (RNAi) show disrupted lamination with occasional ectopic deep cells (arrow) and sparse dendrites in the MZ. (**C**, **D**) Higher-magnification images revealing extensive dendrites in (**C**) control cells and stunted dendrites in (**D**) RNAi-treated cells that either do not penetrate the MZ (cells 2 and 3) or stunted dendrites that do not show extensive secondary and tertiary branching in the MZ (cell 1). Scale bars: 50 μm (**A**, **B**); 20 μm (**D**). (**E**, **F**) Model of cell positioning and dendritogenesis in the developing cortex. (**E**) A control neuron (dark green) migrating on a radial glial process (red) extends a branched leading process into the MZ and then translocates through the upper ~50 μm of the CP, arresting migration at the first branch point of the leading process. (**F**) Dab1-deficient cells extend a leading process into the MZ but it remains simplified and the neuron does not translocate efficiently. (**G**) Dab1 interactions (after D'Arcangelo, 2006). Reelin secreted by CR cells (1) binds Reelin receptors (ApoER2 and VLDLR) in the migrating neuron causing (2) the clustering of Reelin receptors and Dab1. (3) The cytoplasmic clustering of Dab1 activates two SFKs (Fyn and Src) leading to (4) tyrosine phosphorylation of Dab1. (5) Phospho-Dab1 binds Lis1, a cytoplasmic dynein interacting protein encoded by *Lis1*, the gene underlying Miller-Dieker lissencephaly. (6) Phospho-Dab1 also activates PI3 kinase and Akt kinase and (7) binds adapter proteins Crk, Nckβ as well as N-WASP. Reelin signaling may regulate multiple cellular events including glial adhesion, somal positioning, and dendritogenesis. Panels **A–F** modified from Olson *et al.* (2006), copyright 2006 by the Society for Neuroscience

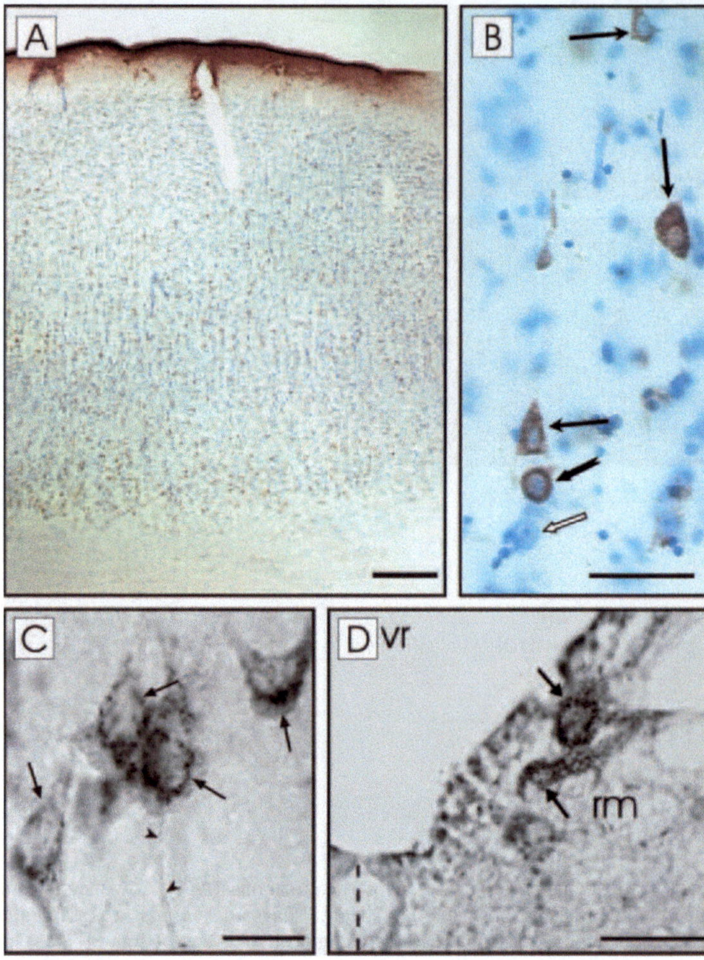

Fig. 8.1 Reelin-labeled neurons in the vertebrate brain. (**A**) Low magnification image of reelin labeling in the adult human cortex (BA39) demonstrating the abundant presence of reelin-labeled cells in all layers of the cortex (brown-stained cells). The section is counterstained with cresyl violet. (**B**) High magnification of the same cortical area as in **A** showing reelin-labeled pyramidal (plain black arrows) and nonpyramidal (notched arrow) cells. An unlabeled pyramidal cell is indicated with a white arrow. (**C**) Reelin-labeled cells of the adult rat entorhinal cortex. Arrows indicate the particle reelin labeling present in the cytoplasm, while arrowheads indicate reelin-labeled processes. (**D**) Reelin-labeled cells of the reticular rhombencephalic nucleus of the lamprey. Note the high similarity of the intracytoplasmic staining of these cells with the staining shown in **C**. vr, rhombencephalic ventricle; rm, nucleus reticularis medius. Scale bars: 500 μm (**A**); 50 μm (**B**); 15 μm (**C**); 150 μm (**D**). [**A, B** extracted from Roberts *et al.* (2005) *J. Comp. Neurol.* 482:294–308; **C** extracted from Perez-Costas (2002) Doctoral Thesis, p. 143; **D** extracted from Perez-Costas *et al.* (2004) *J. Chem. Neuroanat.* 27:7–21]

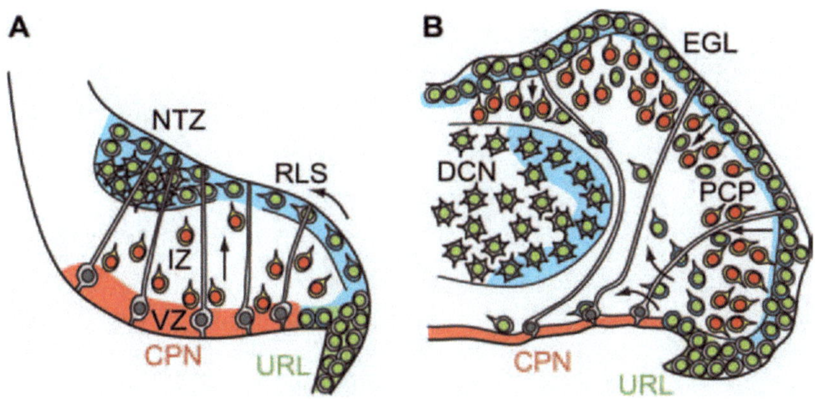

Fig. 10.1 Reelin signaling and cell migrations in cerebellar development. The diagrams show schematic views of the developing cerebellum in sagittal sections through the vermis, oriented with rostral to the left and dorsal to the top. (**A**) Early stage of cerebellar development (mouse E13.5). Cells derived from the upper rhombic lip (URL) (green nuclei) migrate nonradially (curved arrow) through the rostral rhombic lip migratory stream (RLS) to the nuclear transitory zone (NTZ). Reelin (blue) is expressed by many cells in the RLS and NTZ. At the same time, Purkinje cells (red nuclei) migrate radially (straight arrow) from the ventricular zone (VZ) of the cerebellar plate neuroepithelium (CPN) along radial glial cells (gray) through the intermediate zone (IZ), toward the RLS and NTZ. The Purkinje cells express cytoplasmic Dab1 (yellow). (**B**) Later stage of cerebellar development (mouse E17.5). The Purkinje cell plate (PCP) has formed, and the external granular layer (EGL) has replaced the RLS. Cells from the EGL migrate radially inward through the PCP (straight arrows), while unipolar brush cells migrate directly from the URL into the IZ (curved arrows). The deep cerebellar nuclei (DCN) contain neurons derived from the NTZ that have migrated radially inward toward the VZ

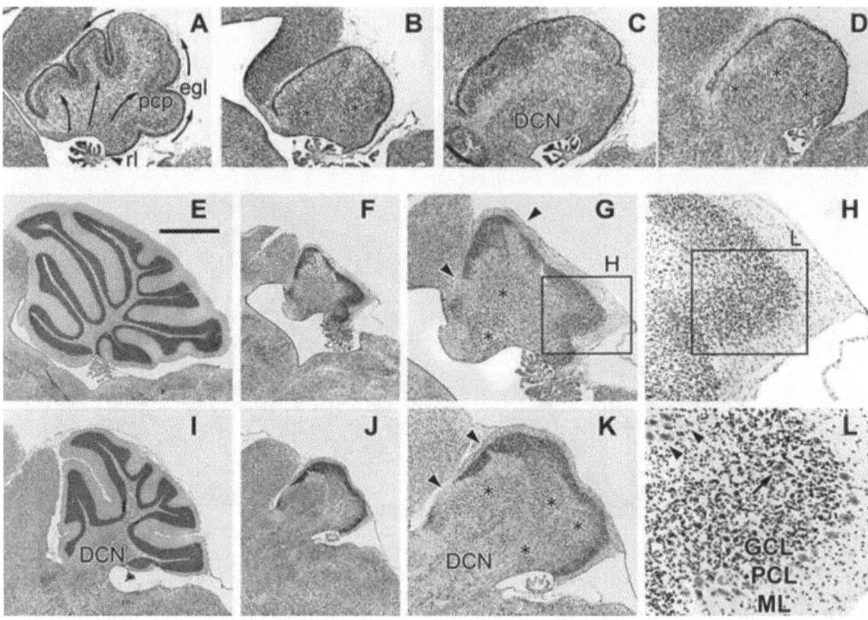

Fig. 10.2 Cerebellar histology in control and *reeler* mice. Sagittal sections through the cerebellar vermis (**A**, **B**, **E–H**, **L**) or hemisphere (**C**, **D**, **I–K**) of control and *reeler* (**B**, **D**, **F–H**, **J–L**) mice were stained with cresyl violet on P0.5 (**A–D**) or P22 (**E–L**). The boxed area in **G** is enlarged in **H**, and the boxed area in **H** is enlarged in **L**. In P0.5 controls, Purkinje cells had migrated to the Purkinje cell plate (pcp), and folia were developing by migration and proliferation of cells in the external granular layer (egl). In P0.5 *reeler* mice, the cerebellum was hypoplastic, no folia were developing, and Purkinje cells formed large, centrally located ectopic clusters (asterisks). The hypoplasia and defective foliation of the *reeler* cerebellum became even more obvious by P22. Most Purkinje cells in the P22 *reeler* cerebellum are located in the large central clusters, although some are isolated ectopically in the granule cell layer (GCL), and others form a nearly normal Purkinje cell layer (PCL) below the molecular layer (ML). In **L**, arrowheads indicate Purkinje cells in deep ectopia, and the arrow indicates a Purkinje cell in the GCL. The GCL in *reeler* consistently shows gaps (arrowheads in **G**, **K**), which may be related to the presumptive locations of fissures (Goldowitz *et al.*, 1997). The deep cerebellar nuclei (DCN) in *reeler* are located near the normal location, but somewhat distorted by the Purkinje cell ectopia (Goffinet, 1983; Goffinet *et al.*, 1984). Sections oriented as described for Figure 1. Scale bar (in **E**): **A–D**, 400 μm; **E**, **F**, **I**, **J**, 1000 μm; **G**, **K**, 500 μm; **H**, 200 μm; **L**, 100 μm

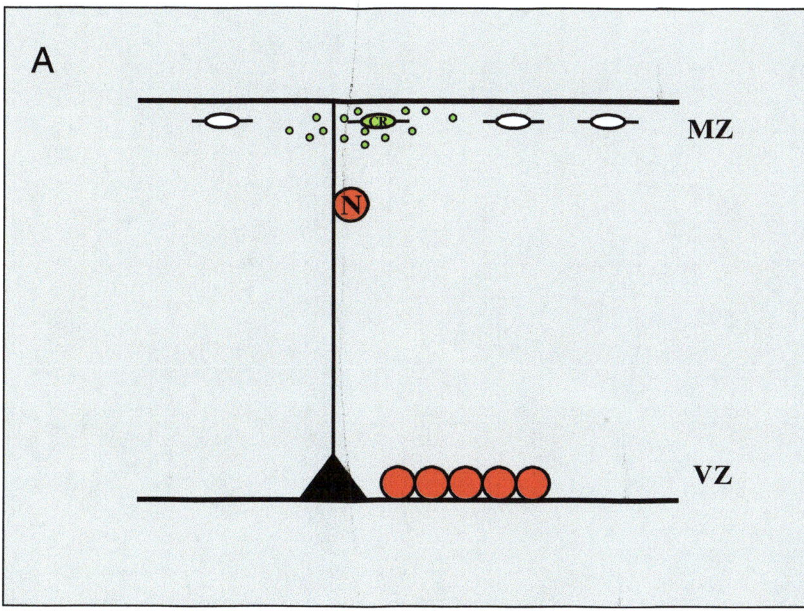

Fig. 11.1 (**A**) Schematic view of the developing cortex. A radial glial cell (black) is shown, extending a radial process from its perikaryon in the ventricular zone (VZ) toward the marginal zone (MZ). Neurons (red) in the ventricular zone are generated by asymmetric division of radial glial cells. A newly generated neuron (N) migrates along the radial glial process toward the marginal zone. Cajal-Retzius cells (CR; green) located in the marginal zone, secrete the glycoprotein Reelin (green dots) into the extracellular matrix. Reelin controls the positioning of radially migrating neurons by acting on both radial glial cells and migrating neurons

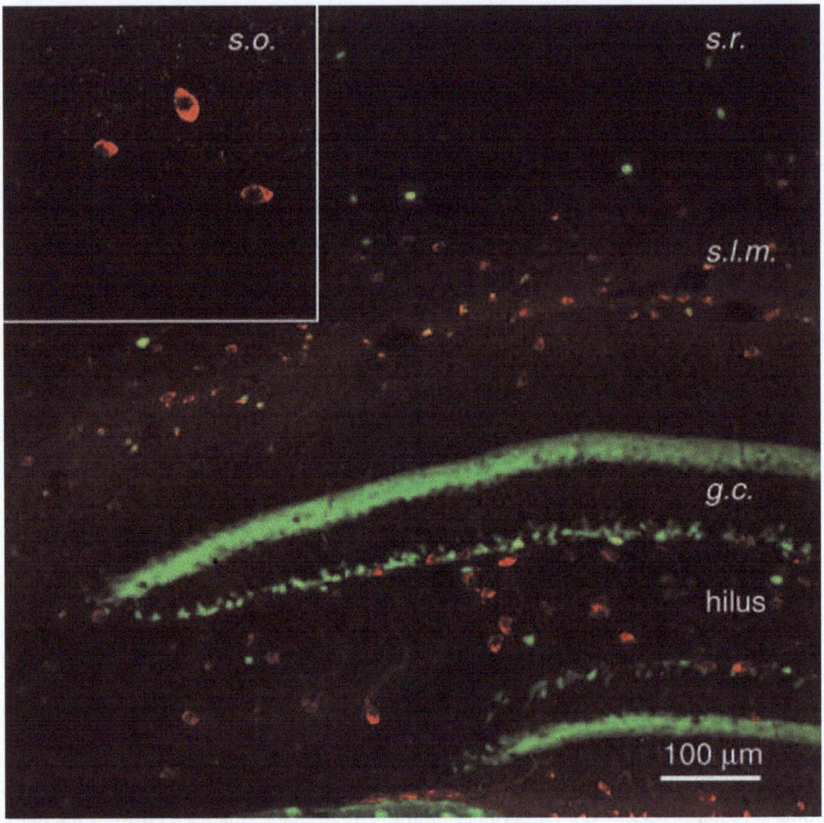

Fig. 12.1 Reelin-expressing cells in adult mouse hippocampus. Double immunofluorescent staining of a hippocampus cryosection obtained from a 6-week-old wild-type mouse. Note that Reelin-containing cells (red) were primarily distributed in the dentate hilar region (hilus) and stratum lacunosum-moleculare (*s.l.m.*) but also can be found in stratum oriens (*s.o.*) and stratum radiatum (*s.r.*) of CA1 region. Immunostaining of the calcium-binding protein calretinin (green) was used to visualize the dentate gyrus layers

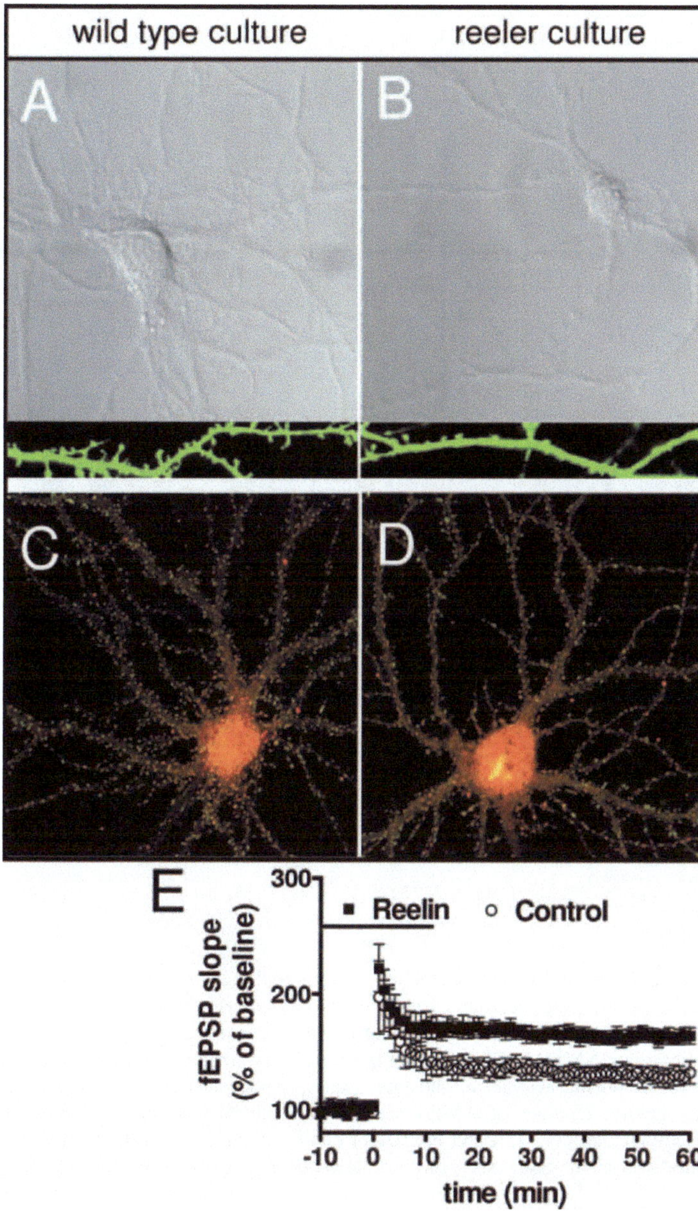

Fig. 12.2 Reelin signaling enhances glutamatergic function in the hippocampus. (**A, B**) In cultured embryonic mouse hippocampal neurons derived from homozygous Reeler embryos, stunted neurite growth and fewer neurite ramifications are seen; in addition, when neurons were filled with fluorophores to reveal dendritic spines, it was observed that neurons from wild-type cultures show significantly more spines in their primary dendrites. (**C, D**) Neurons from both wild-type and Reeler embryos are cultured for 2 weeks and then immunostained with NMDA receptor subunit NR1 and AMPA receptor subunit (GluR1) antibodies. A larger number of puncta that are positive for both NR1 and GluR1 were observed in wild-type cultures compared with Reeler cultures. (**E**) Long-term potentiation experiments using acute hippocampal slices prepared from 6-week-old mice. A 20-min perfusion of Reelin dramatically elevated the magnitude of tetanus-induced LTP

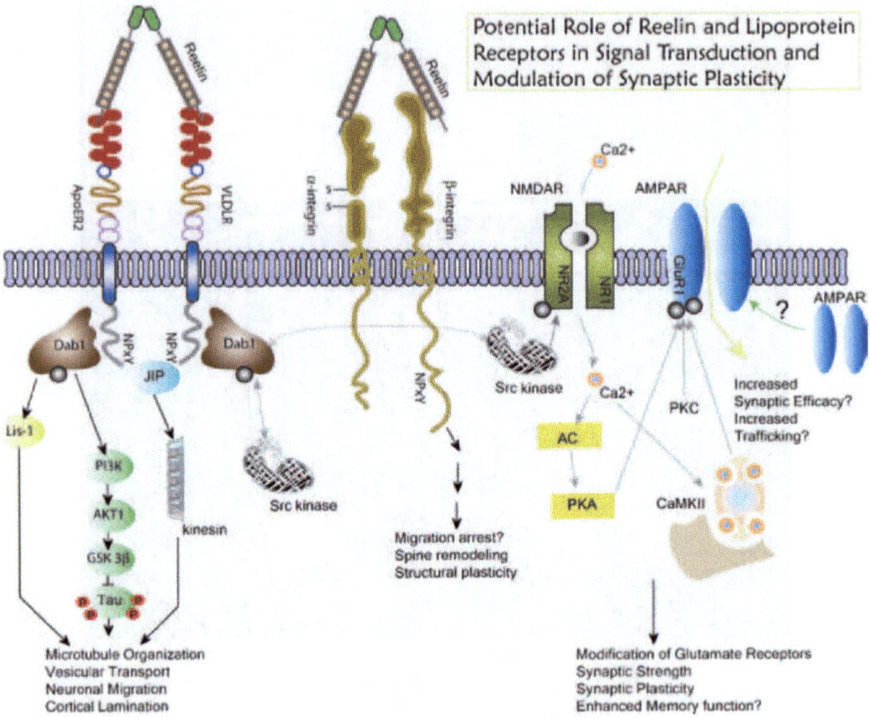

Fig. 12.3 Schematic representation of Reelin signaling and the subsequent enhancement of synaptic function in the adult hippocampus. Reelin binds and activates ApoER2/VLDLR and leads to tyrosine phosphorylation and activation of Dab1 and Src family protein tyrosine kinases. Src kinases phosphorylate NMDA receptor subunits and lead to enhanced channel conductance, augmented Ca^{2+} influx during activation, and increased synaptic plasticity. This increased synaptic plasticity may involve changes of AMPA receptor phosphorylation and trafficking as well. In response to Reelin signaling, PI3K and PKB/AKT can be activated as well, resulting in inhibition of tau phosphorylation. In addition to ApoER2/VLDLR, Reelin also activates integrins

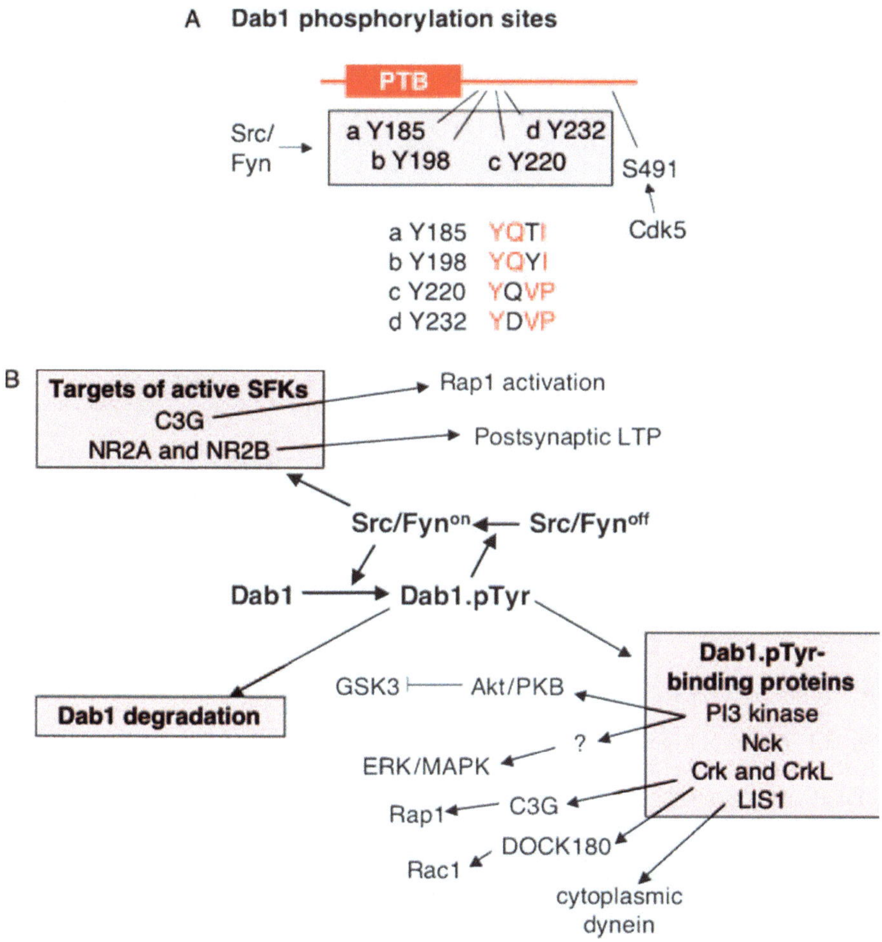

Fig. 13.2 Events downstream of Dab1 phosphorylation and SFK activation. (**A**) Phosphorylation sites in Dab1 that are phosphorylated by SFKs and Cdk5. (**B**) Events that may be important in Reelin signaling are shown separated into two categories: those triggered by active SFKs and those dependent directly on Dab1 phosphorylation

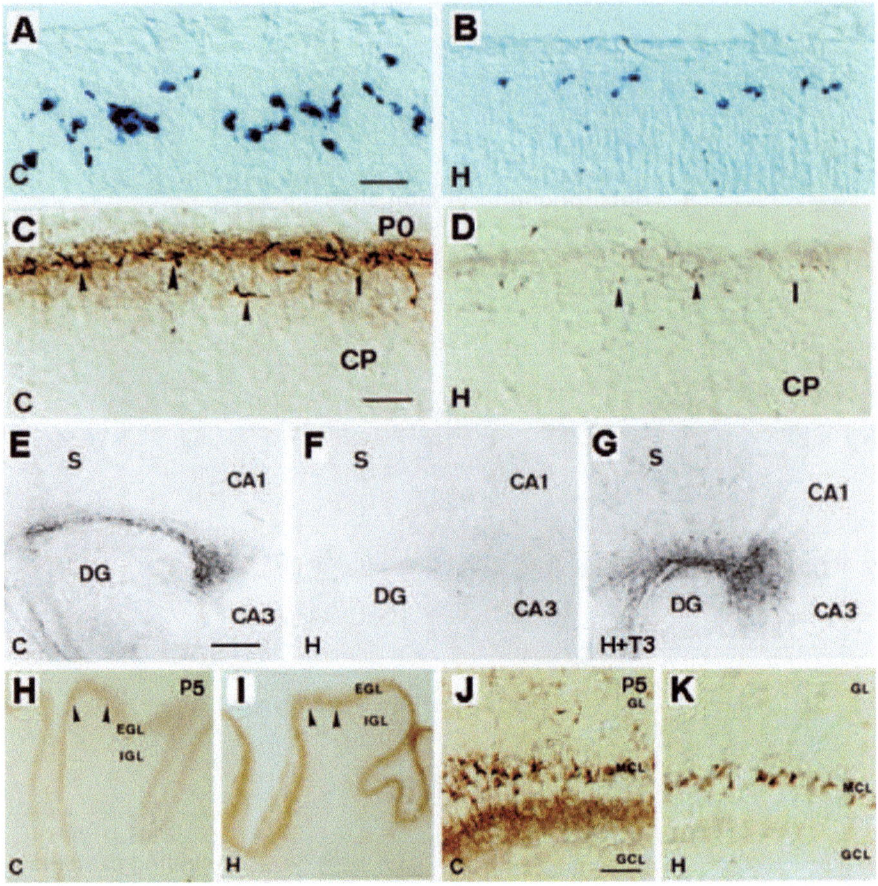

Fig. 15.1 Effects of hypothyroidism on *reelin* RNA and protein expression in the neonatal brain. (**A, B**) Pattern of *reelin* RNA expression in the neocortex of control (**A**) and hypothyroid (**B**) rats at P0. (**C, D**) Photomicrographs showing the distribution of CR50 antibody immunostaining in layer I of control (**C**) and hypothyroid rats (**D**) at P0. Some CR50-positive Cajal-Retzius cells are indicated by arrowheads. Note the decreased staining in hypothyroid animals. Cortical layers are indicated to the right. (**E–G**) Reelin expression detected by CR50 immunostaining in hippocampal organotypic slice cultures. (**E**) Slice from euthyroid rats incubated for 6 days in standard serum. (**F**) Slice from hypothyroid rats incubated for 6 days in thyroid-depleted serum. (**G**) Slices from hypothyroid rats incubated for 6 days in T3/T4-depleted serum supplemented with 500 nM T3. Note that the reduced expression levels in hypothyroid slices are rescued by T3 treatment. (**H–K**) Patterns of Reelin distribution in the cerebellum (**H, I**) and olfactory bulb (**J, K**) of control (**H, J**) and hypothyroid (**I, K**) rats at P5. Note the increased Reelin levels in the hypothyroid cerebellum and the opposite in the olfactory bulb. Abbreviations: C, control; CA3, CA1, hippocampal subdivisions CA3 and CA1; CP, cortical plate; DG, dentate gyrus; EGL, external granule cell layer; GCL, granule cell layer; GL, glomerular cell layer; H, hypothyroid; I, cortical layer I; IGL, internal granule cell layer; MCL, mitral cell layer; ML, molecular layer; S, stratum lacunosum-moleculare. Scale bars: **A**, 40 μm (applies to **A–D**); **E**, 200 μm (applies to **E–I**); **J**, 50 μm (applies to **J** and **K**). (Figure modified from Álvarez-Dolado *et al.*, 1999. © *The Journal of Neuroscience*)

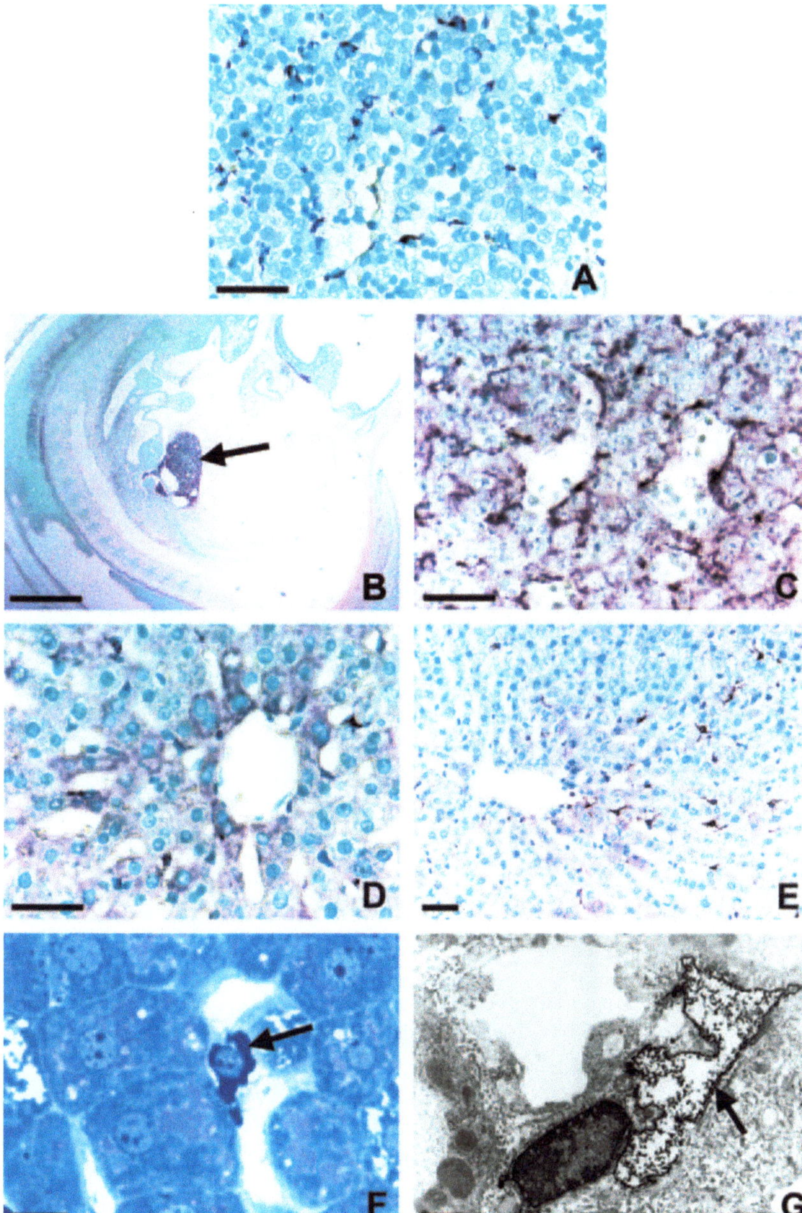

Figure. 17.1. Reelin (**A–D**, **F**, **G**) and GFAP (**E**) expression in human (**A**) and rat (**B–G**) liver. (A) Reelin immunostaining in stellate cells of human fetus at GW7. (B) Reelin immunostaining in liver of rat fetus at E13 (arrow). (C) Reelin immunostaining in stellate cells of rat fetus at E13; **C** is a high magnification of **B**. (D) Reelin immunostaining in adult rat stellate cells. (E) GFAP immunostaining in adult rat stellate cells. (F) Reelin immunostaining in a stellate cell of adult rat observed on a semithin section stained with toluidine blue (arrow). (G) Reelin immunostaining in a stellate cell of adult rat: electron microscopic examination; staining is observed in rough endoplasmic reticulum (arrow). Scale bars = 40 μm (**A**, **C–E**), 800 μm (**B**), 10 μm (**F**), and 2 μm (**G**)

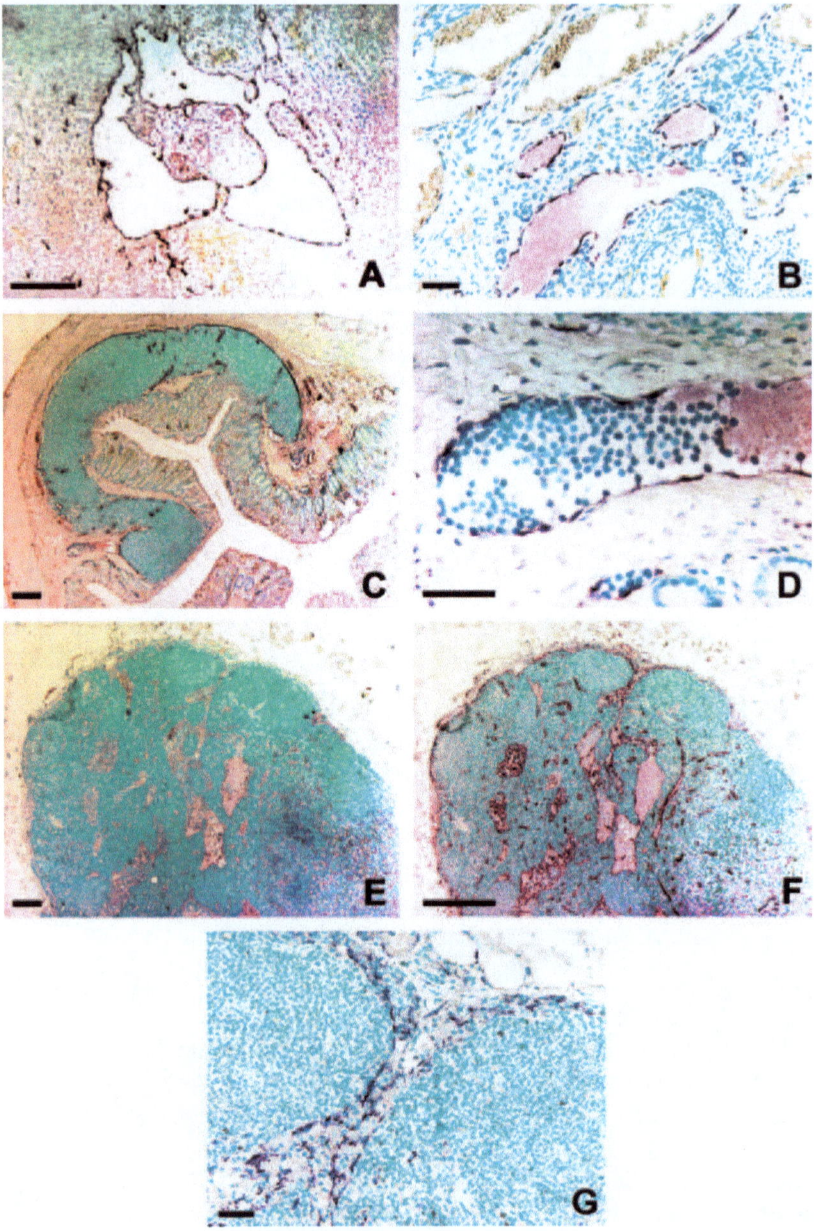

Figure 17.2 Reelin (**A–E**) and CD31 (**F, G**) expression in rat fetus (**A**), adult rat (**B–D**), and adult human (**E–G**). (A) Reelin immunostaining in the jugular lymphatic sac of rat fetus at E13. (B) Reelin immunostaining in lymphatics of adult rat ovarian medulla. **(C, D)** Reelin immunostaining of lymphatics around Peyer's patches in adult rat gut; **D** is a high magnification of **C**. **(E)**Absence of reelin immunostaining in adult human lymph node. **(F, G)** CD31 immunostaining in adult human lymph node; **G** is a high magnification of **F**. Scale bars = 150 μm (**A, C, E**) and 40 μm (**B, D, F, G**)

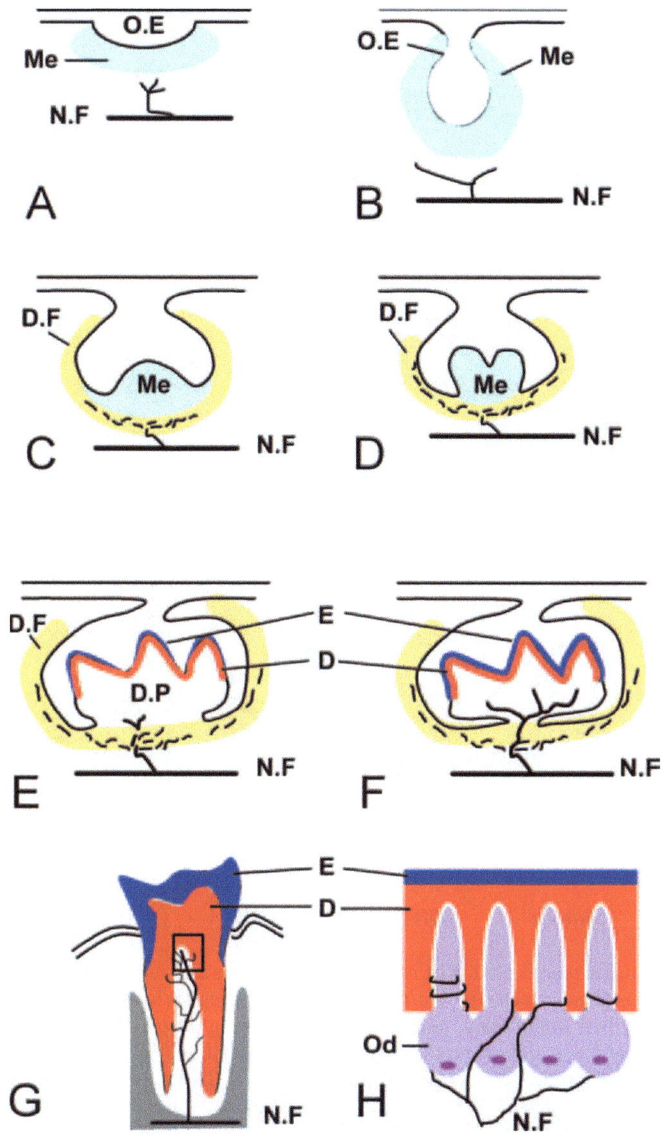

Fig. 19.1 Schematic representation of dental innervation during tooth development from embryonic stages **(A—D)** to postnatal stages **(E—H)**. (**A**) Epithelial thickening stage. A plexus of nerve fibers is observed in the mesenchyme beneath the thickened oral epithelium. (**B**) Bud stage. The oral epithelium thickens and the mesenchyme undergoes a condensation. Axon sprouts grow toward the mesenchyme and continue to the epithelium as lingual and buccal branches. (**C**) Cap stage. Local axons form a plexus at the base of the primitive dental papilla and come into contact with the dental follicle. (**D**) Early bell stage. The number of axons increases in the dental follicle. (**E**) Late bell stage. At the onset of amelogenesis and dentinogenesis, the first sensory axons enter the dental papilla. (**F**) During early root formation, the number of pulpal axons increases. (**G**) During tooth eruption and with the advancing root formation, a rapid development of sensory pulpal axons leads to the formation of the subodontoblastic plexus of Raschkow. (**H**) Enlarged schematic representation of the dentin pulp complex innervation. The sensory nerve endings originating from the plexus of Raschkow coil around the cell bodies and processes of odontoblasts in the dentinal tubules. D, dentin; D.F, dental follicle; D.P, dental papilla; E, enamel; Me, mesenchyme; N.F, nerve fiber; Od, odontoblasts; O.E, oral epithelium

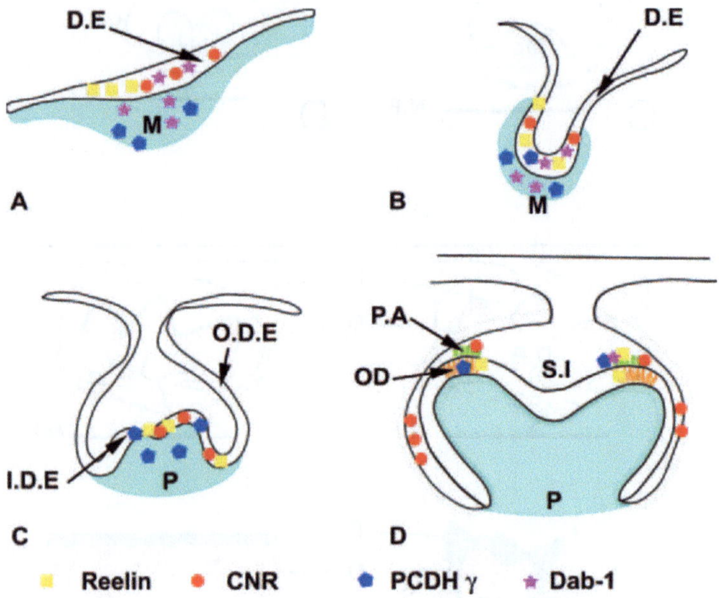

Fig. 19.2 Schematic representation of reelin gene expression and its receptors during successive stages of odontogenesis. Reelin is first detected in the oral epithelium from the initiation stage through the early bell stage. Then, reelin expression shifts in differentiating odontoblasts at the late bell stage. Dab1 is mainly expressed in both oral epithelium and dental mesenchyme during the initiation stages (epithelial thickening and bud stages). CNRs are present in the epithelium through the tooth development whereas PCDH-γ is expressed in both epithelial and mesenchymal compartments. D.E, dental epithelium; I.D.E, inner dental epithelium; M, mesenchyme, OD, odontoblasts; O.D.E, outer dental epithelium; P, dental papilla; P.A, preameloblasts; S.I, stratum intermedium

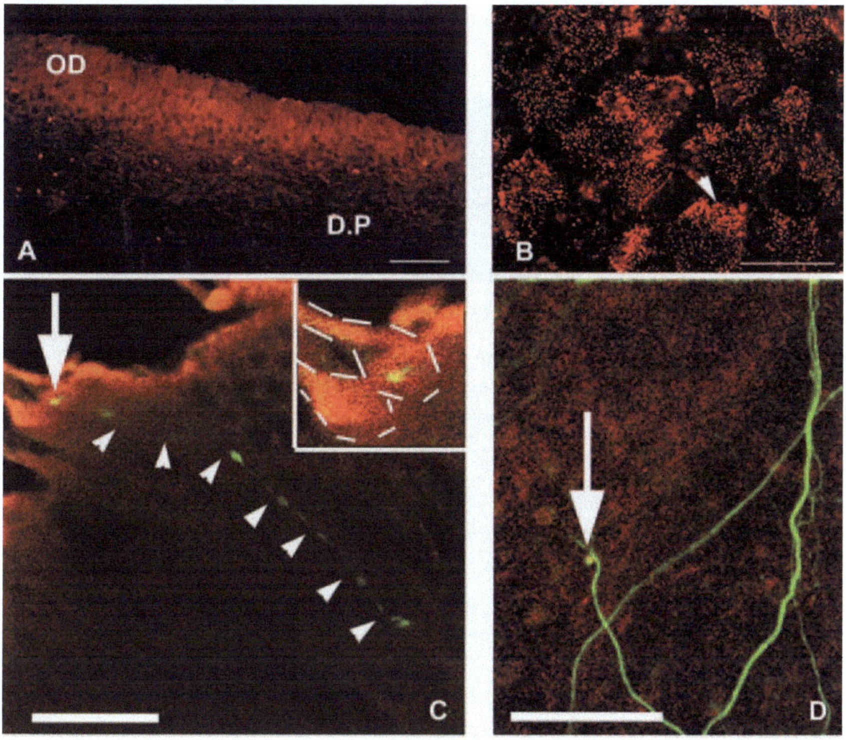

Fig. 19.3 Expression of reelin in human odontoblasts. (**A**) An immunolabeling of reelin performed with anti-reelin antibody 142 shows a signal in the odontoblast layer (OD). No staining is observed in dental pulp cells (D.P) (bar is 100 μm). (**B**) Immunofluorescence labeling with the same antibody, and without permeabilization of the cells, appears as reelin-positive patches localized around the cultured odontoblast cell membrane (arrowhead) (bar is 100 μm). (**C**) A double immunostaining with the monoclonal anti-reelin antibody and a polyclonal anti-neurofilament H on a human dental pulp section was analyzed by confocal microscopy. The nerve fiber course in the pulp can be followed (arrowheads). A yellow patch observed in a nerve varicosity, indicates a colocalization between nerve fiber and reelin close to the odontoblast membrane (arrow and insert) (bar is 20 μm). (**D**) Coculture of human odontoblasts and rat trigeminal ganglion shows the same colocalization (yellow) of reelin (red) and the varicosity (green) in the odontoblast cell layer (bar is 20 μm). [Modified from Maurin *et al.* (2004). Expression and localization of reelin in human odontoblasts. *Matrix Biol.* 23:277–285, with permission from Elsevier]

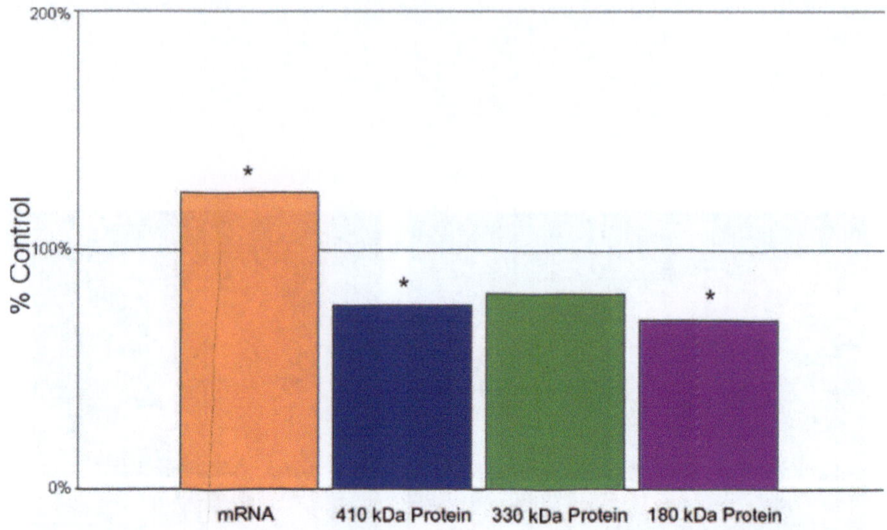

Fig. 22.2 The impact of clozapine on rat brain levels of Reelin. In clozapine-treated rat FC, Reelin protein showed significant downregulation of the 410- and 180-kDa isoforms while Reln mRNA was significantly upregulated versus controls

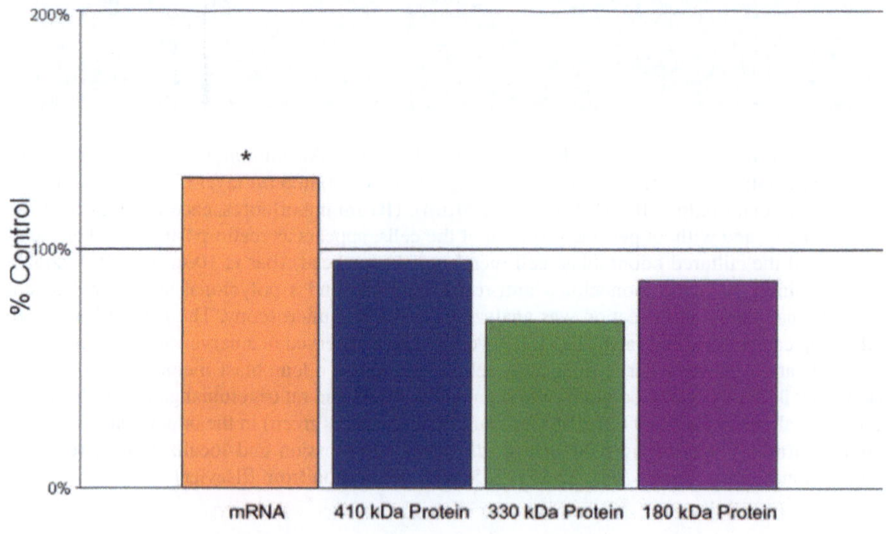

Fig. 22.3 The impact of fluoxetine on rat brain levels of Reelin. Reln mRNA was significantly upregulated in fluoxetine-treated rat FC versus controls

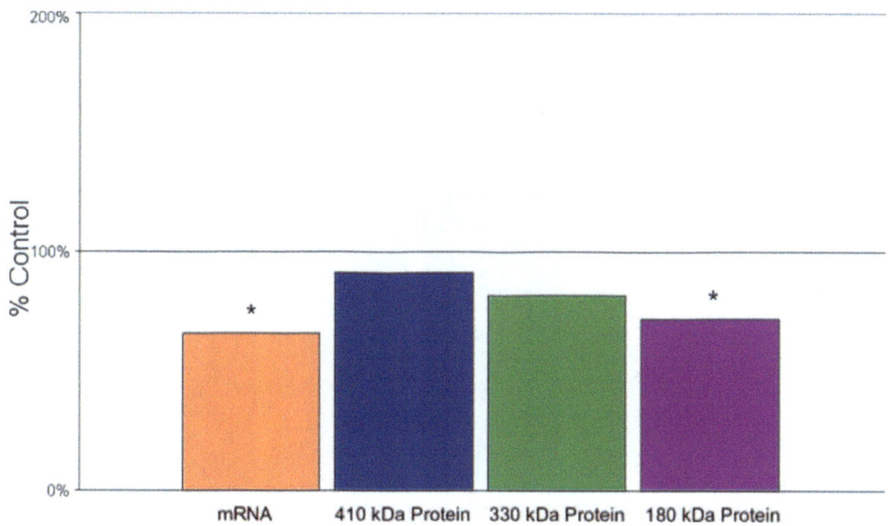

Fig. 22.4 The impact of haloperidol on rat brain levels of Reelin. Reelin protein showed the 180-kDa isoform was significantly downregulated as was Reln mRNA level in haloperidol versus control rat FC

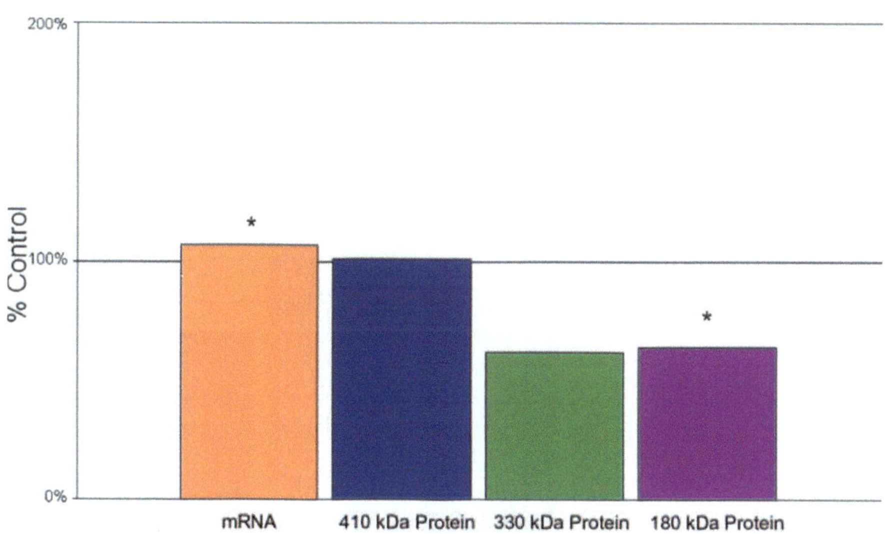

Fig. 22.5 The impact of lithium on rat brain levels of Reelin. The 180-kDa isoform of Reelin was significantly downregulated following chronic treatment with lithium. In contrast, Reln mRNA was significantly upregulated in lithium versus control rat FC

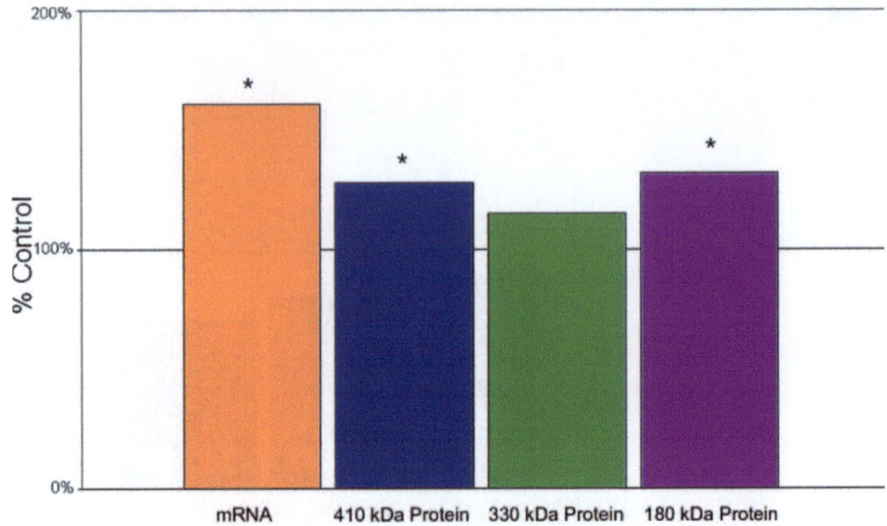

Fig. 22.6 The impact of olanzapine on rat brain levels of Reelin. Olanzapine-treated rat FC showed significant upregulation of the 410- and 180-kDa isoforms of Reelin. Reln mRNA was also significantly upregulated

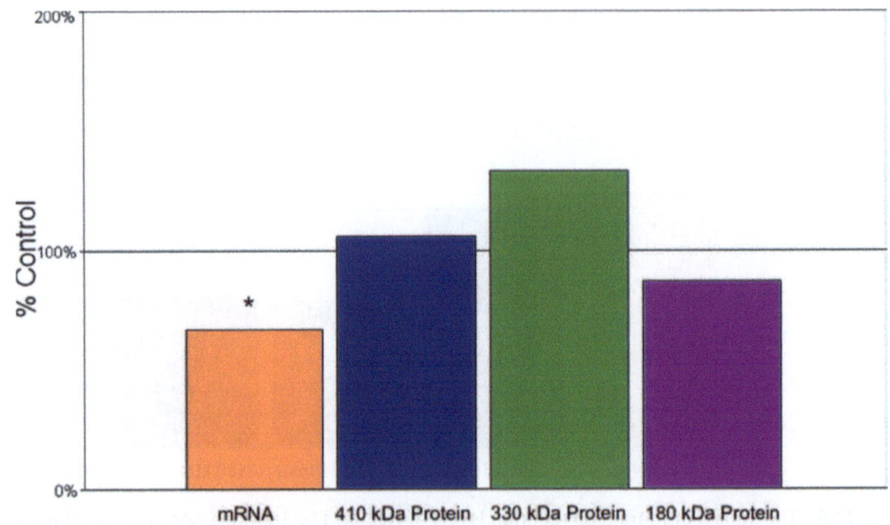

Fig. 22.7 The impact of valproic acid on rat brain levels of Reelin. Reln mRNA was significantly downregulated in rat FC as a result of treatment with VPA

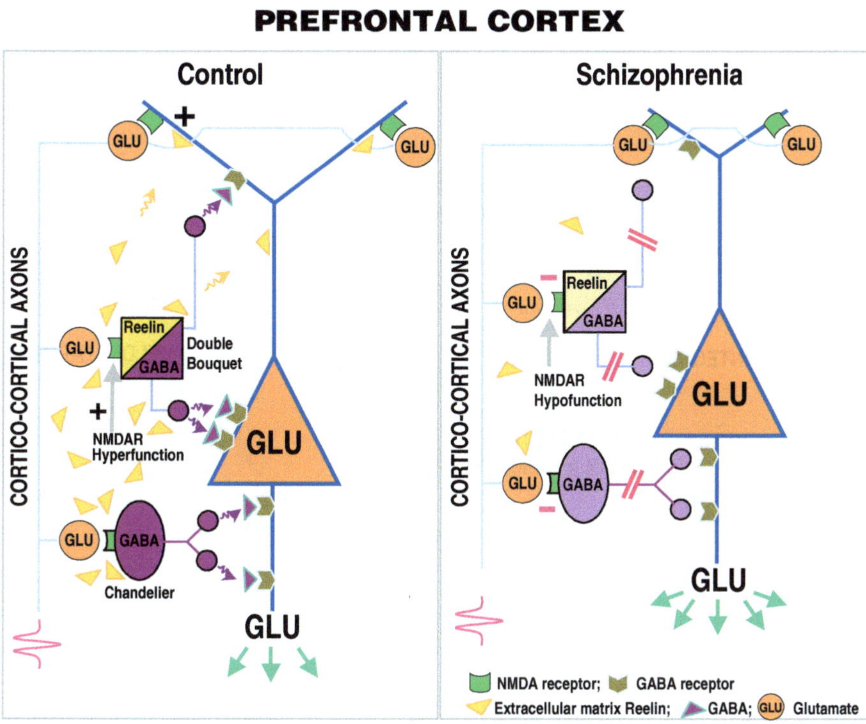

Fig. 23.1 (Control) Reelin expressed in double bouquet or horizontal cells in the upper prefrontal cortex layers is secreted by a constitutive mechanism in the extracellular matrix space and: (a) binds to the apical dendritic branches of pyramidal neurons inducing spine formation by facilitating dendritic resident mRNA translation or (b) binds to dendrites or cell bodies of GABAergic interneurons (double bouquet or chandelier cells), facilitating the action of glutamate at NMDA receptors located on GABAergic interneurons and thereby increasing the release of GABA on apical dendrites, cell bodies, and axon initial segments of pyramidal neurons.

(Schizophrenia) Reelin and GAD67 expression and reelin and GABA release are downregulated. The reelin deficit causes: (a) decreased dendritic spine density on the apical dendrites of pyramidal neurons and (b) hypofunction of NMDA receptors located on double bouquet or chandelier cells, eliciting a further decrease of GABA released on the apical dendrites, cell bodies, or axon initial segments of pyramidal neurons. The deficit of GABAergic neurotransmission results in an increased output of glutamate from the axon terminal of pyramidal neurons

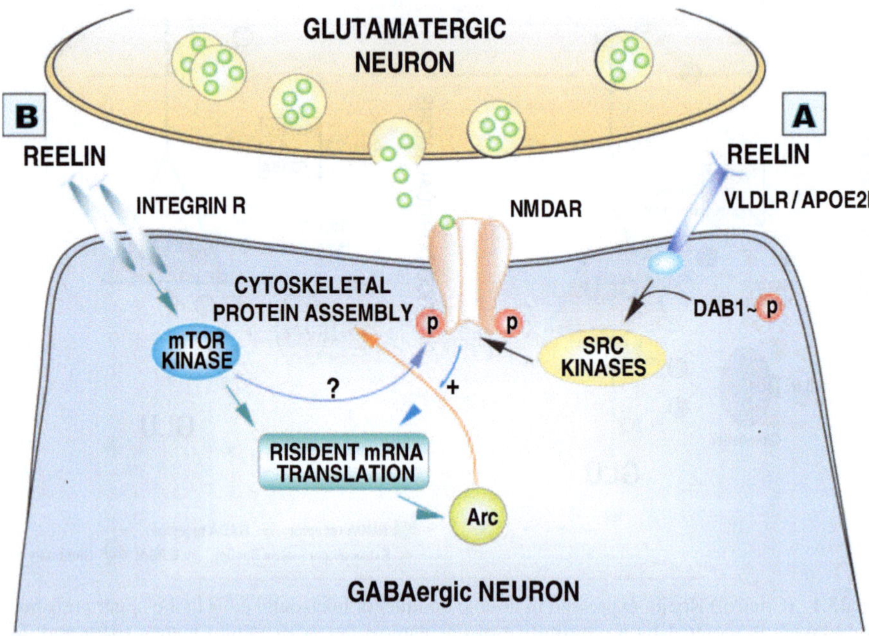

Fig. 23.2 Putative role of reelin in synaptic plasticity. Reelin is depicted binding to a dendritic postsynaptic density of a cortical GABAergic interneuron. Either (A) to VLDL or ApoE2 receptors (VLDLR or APOE2R) or (B) to integrin receptors (INTEGRINR). (A) Reelin modulates NMDA receptor (NMDAR) activity through SRC kinase-mediated tyrosine phosphorylation of the NMDAR intracellular sites (Weeber *et al.*, 2002; Herz and Chen, 2006). (B) Reelin modulates Arc expression and cytoskeletal protein assembly through activation of mTOR kinase (Dong *et al.*, 2003)

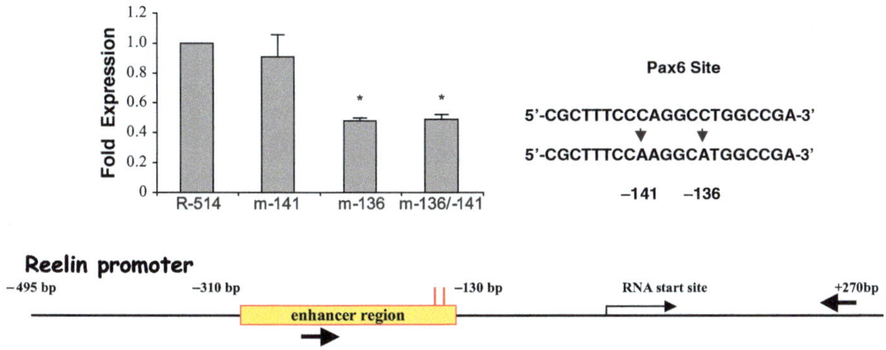

Fig. 23.6 Reelin promoter point mutations. We designed site-directed mutants within the Pax6 binding site that had previously been shown to be more heavily methylated in patients with SZ (Grayson *et al.*, 2005). These corresponded to the double (–141/–136), and single promoter mutants (m –141) and (m –136). These minimal mutants were introduced into NT2 cells using transient transfection assays and reporter activity was measured 36 hr later. NT2 cells transfected with the single mutant (m –136) and double mutant construct (m –136/–141) exhibited 50% of the activity of the –514 promoter. *p, 0.05 expressed as a percent of the SV40 promoter and compared with the reelin –514 promoter for statistical purposes (one-way ANOVA followed by Fisher LSD Method)

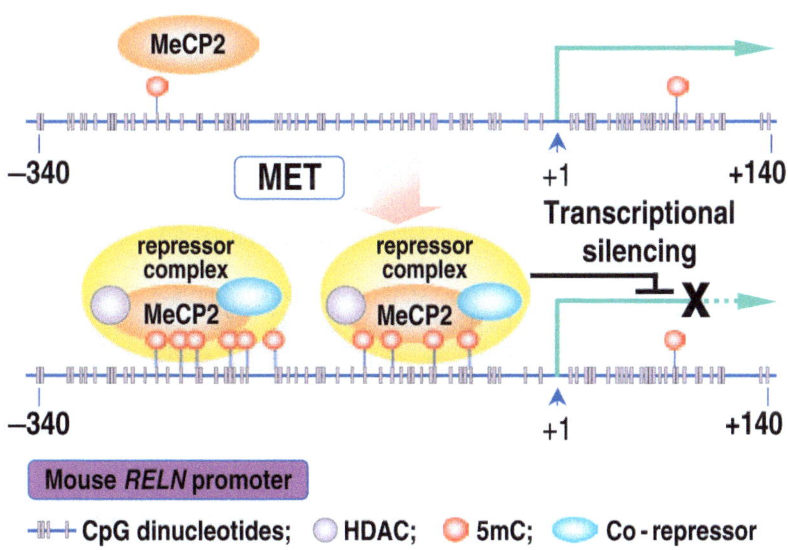

Fig. 23.7 Proposed mechanisms by which mouse *RELN* promoter hypermethylation and recruitment of chromatin remodeling complexes (MeCP2, HDACs, and co-repressors) regulate reelin gene expression. The mouse reelin (*RELN*) promoter region depicted here follows that reported by Tremolizzo *et al.* (2002) and includes the repressor protein complex. Vertical bars represent CpG dinucleotides present in this region. Pink dots denote 5mC present in the sequence. Note the increase of 5mC in MET (methionine)-treated mice. MeCP2 recruits co-repressor complexes including HDACs and induces a state of gene repression

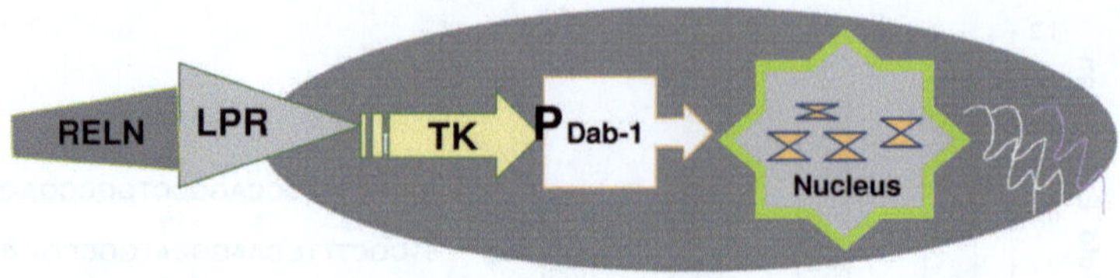

Fig. 24.1 Binding of RELN to lipoprotein receptors (LPR) activates a tyrosine kinase (TK)-dependent cascade leading to Dab1 phosphorylation and expression of several genes that lead to long-lasting structural changes

GCCCTCTGCGGGGCTTTGACGTCCCTCGCAGAAGAGTCGCGGGCTCAGCGGTC
CTCGACAGCGTCCCGTCCCGCTCCCCGGCGGGCGCCCCTCCCTGTCCTCCCGG
GTGCGAACCGGGCGCTGGCCGGGGACTCCGGGGACGCGTGCGCCCCTCGCC
GCGCGAGGTGCCGCCGAGCCAGCCCGAGAGGGCGGGGGCGGGCGGGCG
GCGCGCGGGGGCGGGGGAGCGGCCGGGACACGTGTGGCGGCGGCGGGGGG
GACGCGGCGCCCGGGGCTTTAAGAAGGTGTGGAGCGGGGCGGGCGCTTTCCC
AGGCCTGGCCGAGGGGCGTCGCGCAGAGGCGGCGGCGGCGCACGGAGGCG
GCAGACGACGCGCTCTCGGCGCCCGCAGCCCCGGTCCCGCGCTCCCGCGGC
CCAAAGTAACTTTGGGAGCCGCCGTCTCCCGCGGAAACTT-exon

Fig. 24.2 A view of *RELN* promoter sequence. *RELN* harbors a CG-rich promoter with 72 candidate cytosine (C) sites for methylation and several regulatory binding sites located in 450 base pairs upstream of the coding region. A CRE binding site is underlined in the first line and several SP1 binding sites (GGGCGG) and a consensus GC box are underlined in other locations. The boldface **C**s that are followed by G are candidates for methylation, while other Cs or unmethylated Cs will be converted to T during bisulfite treatment

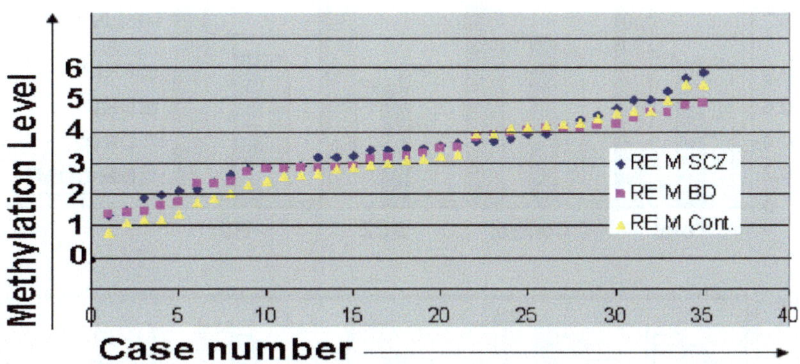

Fig. 24.4 Comparison of DNA methylation levels by qMSP, revealing that the degree of *RELN* methylation in SCZ and BD is almost twice that of the controls. To visualize the differential levels of *RELN* promoter methylation in the patients and controls, the ΔC_T of methylated product for *RELN*, normalized with the C_T of β-actin, was sorted from minimum to maximum. Thus, the increase in the percent of methylation would be exponential. As shown, the base level of *RELN* promoter DNA methylation was greater in SCZ and BD compared to the control subjects (almost twofold). This difference remained nearly the same across the entire samples; however, patients with BD showed a lesser degree of *RELN* methylation in the last part of the curve, where the level of methylation was relatively high

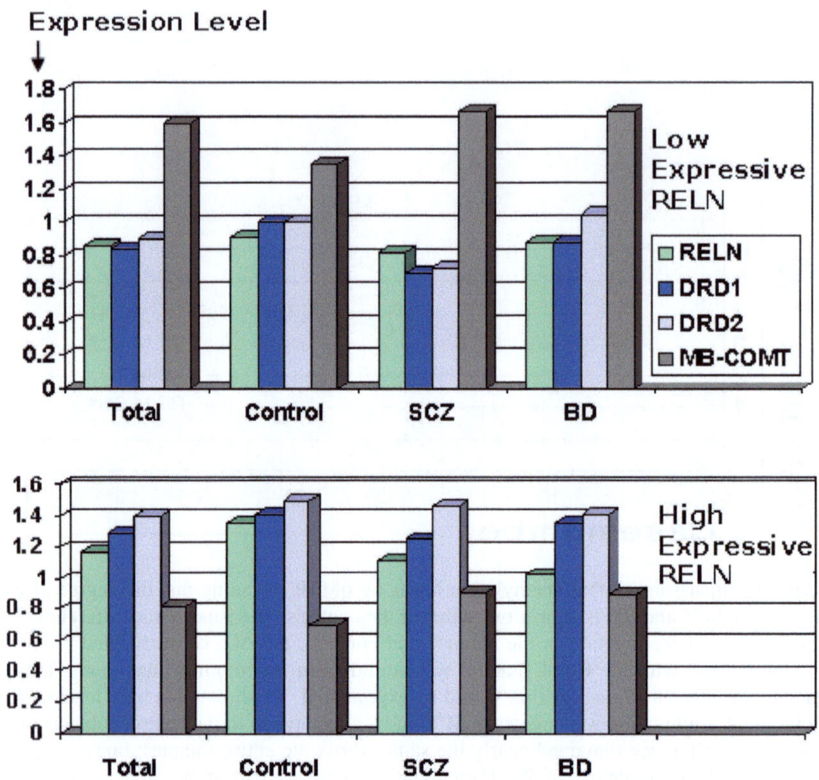

Fig. 24.5 Inverse correlation between the expression of *RELN* and *DRD1*, *DRD2*, and *MB-COMT*. Consistent with the promoter methylation status, expressions of *RELN, DRD1,* and *DRD2* appear to be correlated, but are inversely correlated with the *MB-COMT* expression in both controls and the patients, as well as in total samples. As a result, *RELN* hypoexpression could be associated with hypoactivity of dopaminergic neurotransmission in the frontal lobe

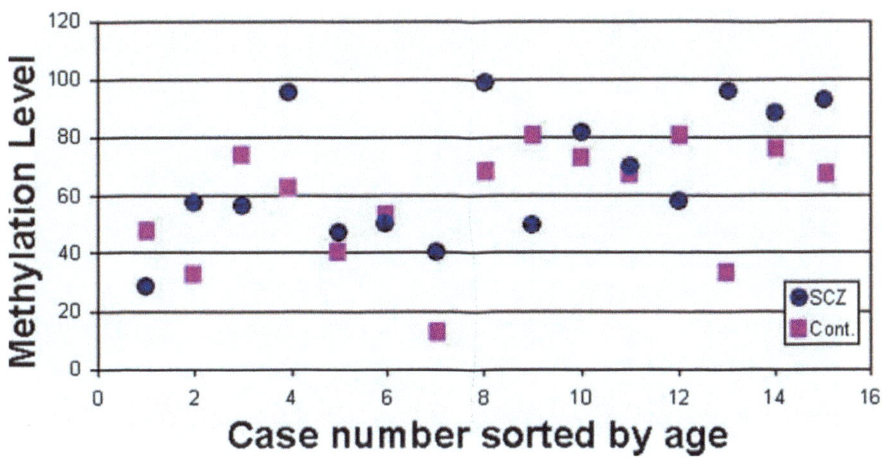

Fig. 24.6 Age-dependent increase in *RELN* promoter methylation. The degree of *RELN* promoter methylation (*Y* axis), extracted from Grayson *et al.* (2005, supplementary Table 2), was sorted by age (*X* axis). As shown, the degree of promoter methylation increased by age in both SCZ and the control subjects

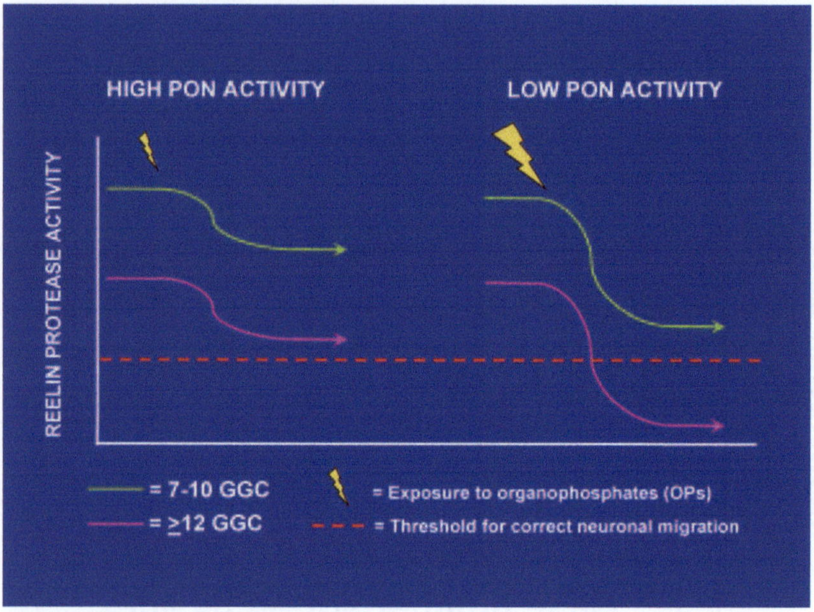

Fig. 25.2 Putative gene–environment interaction model involving the Reelin and PON1 genes, and prenatal exposure to organophosphates (OPs). Reelin gene variants genetically determine normal or reduced levels of Reelin, associated with normal or "long" GGC alleles, respectively. Both conditions are compatible with normal neurodevelopment, but prenatal exposure to OPs can transiently inhibit Reelin's proteolytic activity, which may or may not fall below the threshold critical to neuronal migration, also depending on baseline levels of Reelin. Furthermore, exposure to identical doses of OPs can affect Reelin to a different extent, depending on the amount and affinity spectrum of the OP-inactivating enzyme paraoxonase produced by the *PON1* gene alleles carried by each subject (Gaita and Persico, 2006; Persico and Bourgeron, 2006). (Modified from *Trends Neurosci.*, Vol. 29, Persico, A.M., and Bourgeron, T., Searching for ways out of the autism maze: genetic, epigenetic and environmental clues, pages 349–358, copyright 2006, with permission from Elsevier)

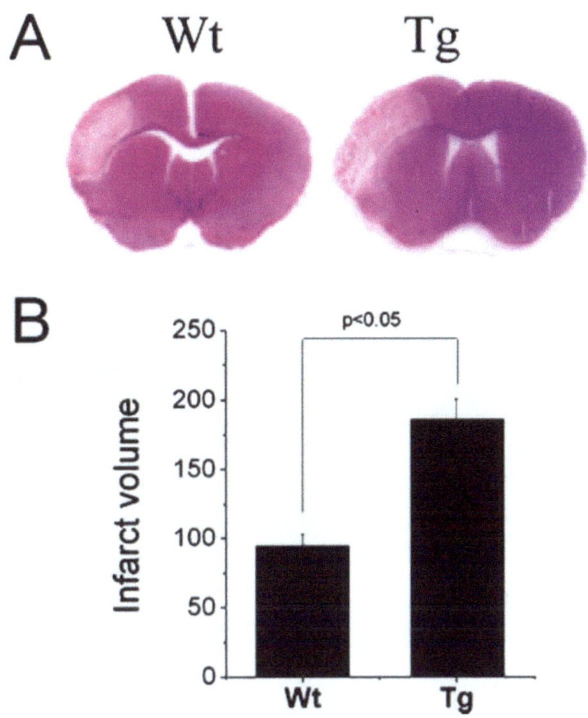

Fig. 27.1 Ischemic brain injury in wild-type (Wt) and transgenic mice with Reln deficiency (Tg) following focal cerebral ischemia. (**A**) HE staining shows an increased area of ischemic injury in *reeler* mice compared to WT mice. (**B**) Quantification of infarct volume in WT and *reeler* mice. $p<0.05$ compared to WT (Student's t-test)

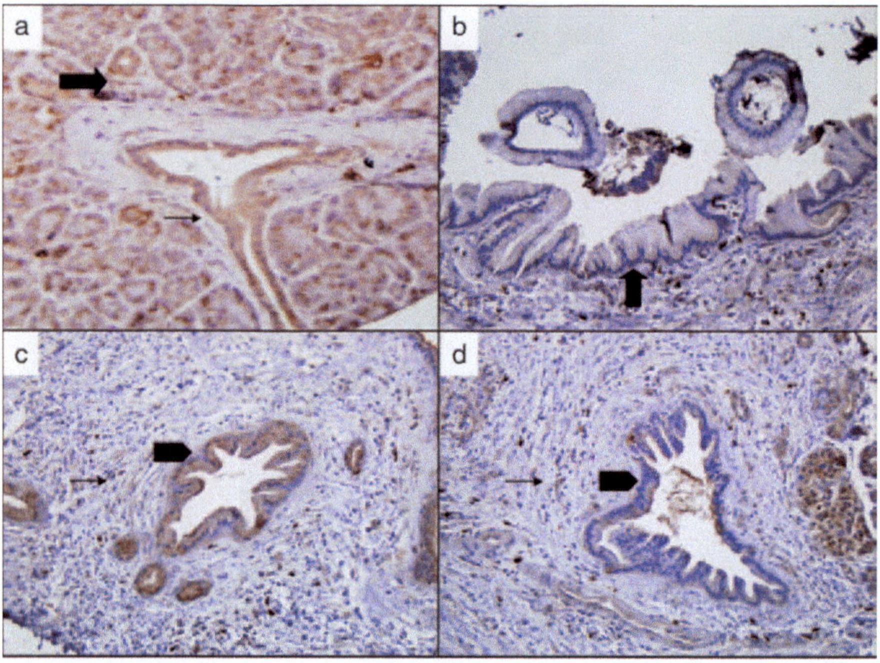

Fig. 28.1 Immunohistochemical analysis of RELN in normal pancreas (**a**) and IPMN (**b**) and PanIN lesions (**c**,**d**). The thin arrow in **a** is pointing to pancreatic ductal epithelium, while the thick arrow is pointing to pancreatic acinar cells. In **b**, the arrow is pointing to the abnormal ductal epithelium of an IPMN; in **c** and **d**, the thick arrowhead is pointing to the abnormal ductal epithelium of a PanIN. The thin arrow in **c** and **d** is pointing to the surrounding fibrosis

MIX
Papier aus verantwortungsvollen Quellen
Paper from responsible sources
FSC® C105338

If you have any concerns about our products,
you can contact us on
ProductSafety@springernature.com

In case Publisher is established outside the EU,
the EU authorized representative is:
Springer Nature Customer Service Center GmbH
Europaplatz 3, 69115 Heidelberg, Germany

Printed by Libri Plureos GmbH
in Hamburg, Germany